MIND CODE

How the Language We Use Influences the Way We Think

Mind Code

How the Language We Use Influences the Way We Think

Charles E. Bailey, M.D.

GIST Publishing • Lake Mary, Florida

Mind Code: How the Language We Use Influences the Way We Think

 Profits from this book to be donated to the Global Institute for Scientific Thinking, a tax-exempt nonprofit organization.

Published in the United States of America by
GIST Publishing Company, Inc.
Lake Mary, Florida
www.gistinc.org/gistpub.html

ISBN: 978-1-936264-25-4 softcover
978-1-936264-26-1 ebook

LCCN: 2013939244

Copyediting: Katharine O'Moore-Klopf, ELS, of KOK Edit; www.kokedit.com
Cover illustration: Phillip Dizick; phillip.dizick@gmail.com
Design and composition: Dick Margulis Creative Services; www.dmargulis.com
Proofreading: Mary L. Tod, PhD, ELS, of Dr. Tod Editing Services; www.drtodediting.com
Index: Lori Holtzinger, of Zinger Indexing; www.zingerindexing.com

MANUFACTURED IN THE UNITED STATES OF AMERICA

Table of Contents

Preface

How and Why This Book Came to Be

This book started as a short article for submission to a conference in Europe. The submission never happened, but the paper uncovered many questions, which initiated a chain reaction for a cascade of thought. The questions slowly morphed into a kernel of an idea that has taken me on quite a search, a journey around the world of scientific thought. The questions and the kernel can be summarized fairly simply: What is happening behind the curtain of life, why do we care, and why is it so perplexing? We can sense the physical world surrounding us with our touch, sight, hearing, smell, and taste. But how does our everyday sense of the world around us correspond to how we respond to the physical reality we encounter? The idea that surfaced from these questions is that we are searching for something invisible, a process rather than a thing. But if processes are invisible, what exactly are we searching for? We can sense the effects but not the process itself. We use our brain and language to try to make sense of it all, but understanding the process somehow eludes us. Does the effectiveness of the language we use play a role in our frustration?

What does it mean for human thinking behavior that the brain we use to decide how and what to think is the same brain we use to evaluate the effectiveness of our thinking? It seems to me that this predicament might lead to problems with circular reasoning, since we base our decisions on the information, experiences, and advice we acquire, and we evaluate that information, those experiences, and that advice with the very same brain. I think that in large part this phenomenon has to do with language: how the brain produces it, how we learn it, and how we use it. As I sought references to provide insight into this question, I came to realize that many disciplines have raised some part of this question concerning their particular subjects, but few if any have taken the issue head on. Do our language habits bias our ability to think and act practically, logically, and effectively, regardless of our discipline, area of study, or our everyday endeavors? Finally, would changing how we think about and use language increase our ability to solve problems, including searching for invisible processes in the world around us?

Furthermore, at the root of this problem, we claim not to know how our language or brains work, alone or together, and we express confusion about how our world works. We do not know how to break the language code to help us understand how language influences the way we think, feel, and behave. We use language and the brain throughout our lives but take both for granted—as if they are invisible.

Mind Code is the ongoing product of several years of contemplation about how to develop a practical understanding of the relationship among language, the human brain, behavior, and nature. I hypothesize that open-minded approaches to language and cognition can produce more effective results by allowing the identification, development, and application of novel, varied alternative solutions; that the world humans live in day to day qualifies as the archetypal complex system; and that we use language to correspond with and make sense of that world. A correspondent relationship between language, brains, thought, and the world around us suggests that by deliberately choosing the most effective language code to represent and understand our relationship with that world, we can learn to think more effectively and improve our ability to predict outcomes. We can then find innovative alternatives and more optimal solutions for knowing, harmoniously interacting with, and adjusting to that world. In my attempt at a scholarly review and integration of the material I have collected over this time, I have relied on my education in and the knowledge I've gained from biology, psychiatry, clinical pharmacology research, observations, readings, seminars, conferences, and occasional brief conversations and correspondence with others.

I wrote this book in the solitude of a clean, well-lighted place. Isolation has its benefits, allowing me to take personal responsibility for introducing any biases, for which I alone am accountable. This book traverses many fields, some of which have fairly difficult and contentious subject matter, and where those fields employ their own particular jargon, I have attempted to define terms and concepts as I applied them in this context and to provide examples.

A lot of the material comes from the myriad astute scientists who have provided unique sections of scaffolding over the past centuries in their efforts to understand nature, ourselves, and our human condition. I extend my gratitude to all of them for their dedication, research, knowledge, and perceptiveness. Without their works, this book would not have been possible. I have attempted, as comprehensively as possible, to reference their works, which make up the foundation of a large part of the book, and any exclusions are unintentional.

While some readers may consider this book ambitious for having crossed the borders of so many disciplines, I hope the trip is nevertheless provocative, intriguing, and maybe even revelatory. In my view, language, at least in some part, underlies or influences nearly all other human behaviors. While we may use different words—or even different language habits—in different disciplines, we use basically the same human brain throughout. It seems logical and reasonable that we can derive practical benefits from understanding that brain and understanding how the language that it makes possible enables us to accomplish what we do regardless of our field of endeavor.

I welcome constructive comments and thoughtful criticisms. Constructive criticism is the backbone of reliability and persistent improvement, especially in learning systems, which benefit from effective feedback. Science realizes the importance and strives to leverage the fundamental role of effective critique in the advance of reliable information and knowledge about nature.

How to Get the Most Out of This Book

Each of us likely has our own opinion about processes in relation to how the world works, the benefits of considering an information-processing model, whether understanding the effects of language concerns us, and whether these topics have any relevance at all. We might ask ourselves, "Why would anyone care?" We likely have already made up our mind to some extent and with the best of intentions. We may look at processes, nature, ideas, and experiments in general terms of different possibilities. But in our everyday lives we often simplify our expressions and beliefs for how we think about the world into shorthand true-or-false categories. Making up our minds beforehand on a "true-or-false" basis can close our minds to new ideas, so that we reject the unfamiliar. If so, and if we believe it is in our best interest, how can we increase the possibility of learning something new?

Perhaps there is a resolution. The trick may be finding a way to temporarily suspend our belief system. A simple method is to strip away preconceived notions of truthfulness a priori, which, in turn allows us to imagine different possibilities in an unfettered manner. I wrote this book with benevolent intentions to open-mindedly search for novel possibilities and informed explanations rather than prove a theory. Consider that what I am going to tell you is neither true nor false but an opinion. Rather than wasting valuable time and effort trying to verify or falsify while reading, consider the option of viewing the material in this book simply as information. But what I'd like to ask of you is this: Even though we are discussing information and opinions rather than true versus false, it might be interesting to consider them and see what effect it might have to think about things in a different light.

This technique encourages a more neutral approach. Then, after the fact, we can do what we will with what, if anything, we have found interesting and worth pondering. In that light, *Mind Code* expresses my pondering of, evidence for, and thoughts about how I think things work in this world, including natural processes and relationships among language, thought, emotions, behavior, and humanity.

Acknowledgments

I extend my gratitude to my daughters for their support. I am also grateful to Dr. Ken Stanley for his valuable time, informative conversations, and thoughtful editorial comments and for allowing me the opportunity to interact with his research group, the EPlex Research Group at the University of Central Florida, http://eplex.cs.ucf.edu/, including Amy Hoover, whose patient tutoring is much appreciated; to Nora Miller, whose initial editorial input helped to make this book more than just a possibility; to Rebecca Lane and Kathleen Bell for their editorial input; to Phillip Dizick for cover illustration and design; and last but not least to Katharine O'Moore-Klopf for final editing.

Some of the information about chaos and dynamic instabilities was drawn from the excellent and concise tutorial on the website of the Ilya Prigogine Center for Studies in Statistical Mechanics and Complex Systems at the University of Texas at Austin: http://order.ph.utexas.edu/chaos/index.html. Some of the information on an engineering approach to problem solving, analog circuits, and oscillator dynamics was drawn from *Basic Electronics*, a Bureau of Naval Personnel Training Manual originally published in 1973 and reprinted in 2004 (New York, NY: Fall River Press).

SECTION I

Behind the Scenes of Language

CHAPTER I

Relations and Processes

HAVE YOU EVER WONDERED now and then about how language weighs on how we think, feel, and behave? In this book, you will encounter questions and answers about how language empowers our brains to perceive and process information about relations with others and the world around us. Keep in mind that throughout the book, language and cognition are typically expressed as co-occurring in parallel, such that the notion of thinking, information, and language is considered a natural process. In other words, the book is primarily about language that includes the relation with cognition as information processing. It is about understanding ourselves—how people think and feel and interact and process the world around them. As language users we apply language daily to communicate about relationships, to process information from messages we take in, and to understand what is going on around us. Does our language impact how we think about relationships and interpret messages, how we make predictions and choices, and what outcomes we get?

In this case, understanding the language we apply to evaluate the relationships helps not only in general but also in particular for evaluating processing faults that can lead to distorted reception of messages and corrupted speech output. That is, how do we get an understandable message, and how can we optimize the information in that message? And can optimizing the message help to optimize our relationships? Finding an effective linguistic method to process and understand our beliefs, values, and relationships might offer us a logical advantage for optimizing communication and collaboration. If we can improve the logical thought processing behind how we value things in the here and now, we can potentially enhance our ability to get along more cooperatively in this world. Life is about relationships. Relationships correspond to processes. We can sense relationships, but the processes behind them are invisible to us.

Process is defined here as a systematic series of actions that lead to some end, such as a continuous action, operation, or a series of changes taking place methodologically, similar to a procedure or method that produces some effect. For example, consider a robotic controller that regulates behavioral effects to approach food and avoid predators. We know that actions and effects require some form of energy or force to produce them. Processes generally range from constructive to destructive and can oscillate between both, depending on the structure of the system. We see a common example in nature where a process defines the ordering of events directed forward in time, such as evolutionary processes that derive energy from the environment and evolve more complex solutions. To understand the processes, we try to make sense of the physical effects they produce, effects we can detect with our senses. Understanding processes depends on having reliable information about the physical structure of the environment and language that can effectively represent and process this information. This connection between reliability, processes, effects, and relationships applies to almost everything, including molecules, the mind, physics, and information processing.

This book covers both relationships and processes, keeping in mind that although we can sense and observe the effects of interactions of objects, we infer the relationships among them, which are a function of an unseen process and forces of nature. It seems logical that to understand relationships and processes, we study the language that we use to define and describe them. How we represent, process, and apply information to thinking about relationships matters. This book takes an information-based approach that examines the functional relationship between language structure, internal brain regulatory processes, and external processes in nature. We will consider how we might best approach the problem of understanding processes, and we will look at some questions it raises, including the cost of accumulating information. Finding information would be a much simpler task if it freely fell out of the sky and onto our lap, but that is unlikely. Instead, we sense that searching for information is anything but free and requires an investment incurred by the cost of spent time, energy, and effort. Nature encompasses a wealth of information and hidden secrets that we can pursue, but we struggle to search them out and make sense of them.

What do we understand about nature, the brain, and language? We know quite a bit, but the bits and pieces of information are scattered to the ends of the Earth, so to speak. The question of where and how we even begin to search quickly becomes overwhelming. The biggest mystery might be "What are we looking for?" This becomes a huge puzzle in its own right. If we knew what we were looking for, there would be nothing to discover. In effect, the more certain we are about what we know, the less we can learn, which equates to closed-mindedness. Why would we consider searching for something we already know? On the other hand, if we do not know what we are looking for, how would we know when we have found it?

Uncertainty means we have room to learn new information and corresponds to open-mindedness. We see an inverse relationship between certainty and learning. If we are 100% certain and 100% closed to learning, then the search becomes a nonissue. If we are uncertain and open to learning, then search enters the problem. The problem gets even more difficult when we consider that what we are looking for might be invisible: We think we can see myriad different disorganized and scattered pieces of the puzzle, but the big picture remains invisible. We sense these pieces but cannot make sense of how to articulate and organize them coherently into a whole. We expect that language can help us to find the relations among the puzzle pieces and that the same language can help our brain to embrace a sense of understanding about our world. This book is about how brains with language can bring our everyday experiences into a clearer focus in a meaningful way—and especially for clarifying and connecting our thinking, feeling, and behaving with the relations and invisible forces and processes in an uncertain world.

Methods and Search Space

Imagine the most obvious things we cannot see are behind a curtain. And once we see them we say, "Of course, I know that." They are things that we take for granted. *We* as used here means us human beings as a group, the general class called *Homo sapiens*. The point is important because our brains work roughly the same way to help us survive in a difficult-to-predict world. Perhaps in the service of parsimony, our brains neglect to inform us about the specific processes going on behind the curtain of life; instead, our automatic thought tends to imagine, assume, and dabble in curious speculation. We tend to almost effortlessly fill in the blanks with our own preformed opinions according to what we have learned about the world.

We can see that the brain is indeed a curious organ. Consider a play or musical where the scenery, props, and costumes change throughout. We are engrossed in enjoying our experience, and focus on the activity happening onstage. The unseen beehive of activity going on behind the scenes escapes us. We can speculate about the unseen happenings and make assumptions from what we see onstage, and often we draw some fairly reasonable conclusions. We might even obtain a free backstage pass and go behind the curtains to see for ourselves what takes place, but typically we don't concern ourselves with those details.

In nature, however, we see what is happening and wonder how things happen, or what the cause is. We wonder why things happen, or what the reasoning behind the cause is. We see and hear the leaves rustling and feel nature's fury when a storm blows through. We feel the warmth of the sun. We stop to peer at the roses, savor their sweet fragrance, and wonder how they came to be. No matter how hard we look, how and why things happen remain invisible, hidden beyond the curtain. It seems that our efforts to peer behind the curtain are futile. It remains an unknown territory where danger could lurk behind every turn; we could fall into a hidden abyss or lose ourselves and get trapped in blind alleys. We can easily become frustrated when it appears that we are no closer to finding the answer to how and why. We know that searching for information has a cost and realize that there are no free backstage passes to go behind the curtain of life; we are left to earn this knowledge.

We know how to search for things and objects we can touch, see, and hear with our senses, which we describe and express with language. How do we search for invisible, silent processes, as opposed to visible objects, when we lack any sense organs with which to detect or capture their presence? We are not talking nonchalantly about life or philosophy in general; instead, we are asking how we can logically understand and relate to our world. How do we find the most effective language to help us to make sense of it all? We have noticed that invisibility is a fundamental characteristic of processes. "Of course, we know that." Even though we think we know that a process resides there, it remains invisible. But even when we lack any evidence whatsoever, our brain fills in the gaps by guessing to explain what is going on, and we assume these guesses about the invisible processes to be factual.

Processes are the invisible strings behind the wind and the rain, the dew on the ground in the morning, the snowflakes we see floating gracefully past, and of life itself. We can stop to ponder a rose's beauty, but when we look for the process hidden behind the rose, we come up empty-handed. Essentially processes are about relationships. We can expand this analogy and say relationships are the essence of life. We might think we have hit the jackpot because we see relationships all around us every day. However, what we see are the effects of relationships rather than the process behind them, rather than the how and why. Of course, seeing only the effects seems to encourage rather than discourage our speculation.

Have we run into a stumbling block and perhaps a disappointment? Our quest for these invisible and illusive processes seems like an impossible one, but how things happen and why they happen still evade us. As with a wisp in the night, when we clutch after it and we think we have it captured, we come up empty-handed again and again. Discovering the processes that go on behind the curtain is a daunting challenge, even when we see the relationships before our very eyes. We speak about what is going on and know there is something there and that it is real, but it hardly seems tangible. It appears that in the end we can only sense the effects indirectly.

Is there something about our language that foils a thorough evaluation of what we sense? Notice that even when language is not mentioned, we are applying language to read and think about the content of this book and the world around us in almost every instance. It is easy for us to forget the power of language, even as we are thinking with it. We tend to take language for granted, but we are unlikely to separate language from a book, problems, or solutions, and especially from the how and why. Language is our food for thought, and the language connection is worth remembering throughout this book. If in doubt, contemplate what thinking would be like without language, books, libraries, speech, and so on.

Identifying the role of language might seem a small consolation for our efforts, but on further inspection it is arguably progress. We have gone from mysterious and haunting to real and daunting. We now know that what we are looking for is invisible and have confirming evidence of our suspicion that language may be involved. This knowledge is anything but trivial. Knowing what we are not searching for becomes relevant. Recall that knowing leaves little reason to search for new knowledge. We now know we are searching for the invisible nature of processes. Knowing that we cannot directly sense processes perhaps leaves us other options. We also see a connection between language, information, and knowledge.

Is there an approach that allows us to search for something novel that we cannot sense? The answer is yes. Searching for the unknown is the forte of science and the accompanying language and routine search methods that make up scientific methodology. Science seems a bit peculiar to someone unfamiliar with its methods, and perhaps for good reason. The language of science seems unusually picky, and science works in a manner that looks like the reverse of how we normally learn: In science, we analyze the whole and learn from the relations among the pieces. But when we learn from our experiences, we do the opposite and tend to synthesize pieces from which we infer relations about the whole. Science admits that it does not know how and why, searches for regularly occurring effects in nature, and then tries to discover what kind of processes could have produced those effects. In the sense that science does not know what exactly it is looking for, we conclude that it is searching in the dark as if blindfolded.

However, science has discovered the human bias toward finding and confirming that for which we are looking. In effect, when we know what we are looking for, we are more likely to find it. "Of course we know that." The discovery of confirmation bias led science to understand that being blinded to what we are searching for can help to subdue the tendency to find evidence that confirms our own beliefs. In other words, when blinded, we can openly search for information as it exists in nature. But even in science, bias does not readily disappear. Science advocates logically searching for information with consistent, orderly methods and reliable language. In effect, science strives for a reliability bias that yields predictable and reproducible solutions. *Search* is a general term often used in the context of finding answers to questions, problem solving, and describing goal-related behavior; a *goal* represents the problem or question we are asking; and solutions resolve the problem or answer the question. In other words, finding solutions is the objective of the search.

Envision this scene: We peer out into the woods and see these scientists hunting while blindfolded and speaking to each other with what seems like a peculiar jargon. We ask what they are searching for, and we find out that they do not know. They are blindly hunting for invisible prey. While this may appear peculiar, it makes sense to science that the solution is invisible, that not knowing the object of the search increases the chances of discovering novel solutions. Blind, open-ended search is a proven method for gaining insight into difficult problems, developed by exploring for and discovering unknown processes hiding in nature. We might ask, "How can we consider blind search insightful?" It is blind in the sense that we are searching not for an object that we can directly sense but instead for a hidden process that we can only sense indirectly by its effects.

This search method from science may be our ticket for finding novel solutions, understanding relationships, and sensing the effects of invisible processes. Recall that we tend to take language for granted, but it is language that provides reliability and defines and propels the fundamental search methods and processes of science. A scientific search method may turn out to be very helpful in our quest for the ever-elusive how and why. We now know that we have another option beyond searching for the familiar known. Instead, now we know we are searching for the unfamiliar unknown, and we have thus admitted our ignorance. We know we are searching for not an object but a process, we know that language is involved, and we have a better idea of where to search. We know where to search because the processes relate to nature itself.

Progress indeed. We have uncovered a search method based on reliable language and revered for its natural ability to sleuth the unknown and discover novel solutions, and with which we can search for tangible effects that may lead to discovering the nature of the hidden processes. Since we know where we are searching, we have narrowed the search to a finite space. Finding the nature of the processes leads us to the mother of all processes, Mother Nature. We know that nature never stands still and changes continually, which presents us with a moving target. If we can observe and capture regularly occurring effects of nature over time, perhaps we can capture the essence. *Essence* means that we are attempting to capture the indirect effects of nature's essential processes.

Since nature is in a constant state of change, we can conclude that instabilities and time are involved. We also know that time invariantly flows to the future. Now we have more clues to help us on our quest. So far, we have acknowledged that we are not searching for a thing but for an invisible process defined and described by language; it is involved in relationships that produce visible effects, change and move continuously, and involve time and instabilities. These clues bring us closer to the how and why hidden behind nature's invisible curtain.

Perhaps that we are searching for an invisible moving target fraught with instabilities doesn't sound like the greatest news, but it seems that we are making some headway. At the very least we have discovered some practical ideas about where and how to search. There are a few other resources we can tap into. We know evolution has been dealing with Mother Nature since life has existed on Earth. Perhaps we can learn from evolution. First of all, evolution is a process rather than a thing. Now we can search for the effects that evolution produces. We may run across a treasure chest of information about processes in nature. We have fossil records, and we have living creatures all around. Better yet, we have brains and language that evolved in nature. Is this really good news, or are we facing an even more difficult challenge?

Perhaps both, but let's consider the different possibilities first—that is, we expect to find equivalent processes corresponding across the lot from nature to evolution to brains and to language. Of course the bad news is that we may have opened up a

Pandora's box of information that will overwhelm us. Perhaps by borrowing a few tricks from computer science we can find a clever way to deal with the large amounts of information we will likely encounter.

Computer science is a field of engineering that routinely deals with large, often massive amounts of information. Computation is related to science, relies on reliable language, and overlaps in general methodology. *Methodology* means a logical and consistent systematic approach to problem solving. An important part of a computational approach is to start by separating the problems from solutions, which creates a problem space and a solution space. Then information processing is separated into three basic components: information input, computation, and output. The logic involved in the computation maps input to output. But we still have not addressed the problem of massive amounts of information.

The field of computer science called machine learning applies a technique called pattern recognition to find patterns in the information and reduce it into a smaller number of chunks. Obviously, we now have fewer packages of information to compute, which can lead to more efficient processing. Does anything come to mind when the word *processing* pops up? We have already proposed that how and why relate to processes. We know that brains and computers both process information. The relationship with information suggests that perhaps we can apply the input-computation-output model to information processing in the brain. This does not mean that we think the brain *is* a computer; instead, it means that the metaphor of a computational brain may lead to some insights about processes in general and in particular about information processing in the brain and the influence of language.

We are finding quite a few practical ways to deal with the problems we have identified so far, including how our language and brain can search the unknown and deal with the information overload. Perhaps we are closer to finding the elusive how and why. While we have answered a few questions, we have also added some, such as how to start and how to go about finding the most effective logic and recognizable patterns to help with the search for the hidden processes. Are things becoming simpler or more complex?

Simple and Easy

Einstein advised scientists to "keep explanations as simple as possible, but not simpler." This notion hearkens back to Occam's razor, the law of parsimony, usually stated as "Do not multiply entities beyond necessity." Simplicity can be a useful concept, especially when trying to explain a complex system. There is a loose connection between simplicity and plausibility, which allows us to mistakenly assume one for the other. Simplicity alone can give us a tacit illusion of plausibility where we conclude that "simple models are more plausible." However, simplicity alone lacks sufficient criteria for evaluating a model or hypothesis. When we turn the previous statement around, we get a different perspective stating that, generally, "more plausible models tend to be simpler." We can then focus on asking about the model's plausibility and evaluate simplicity by whether the model has fewer equally plausible alternatives (Jaynes, 2007, p. 606). Yes, when we consider evaluations of plausibility, the role of language's influence on our thinking comes into play.

Plausible refers to explanations that are reasonable and useful for their intended purpose. A plausible model includes limits that help keep us from violating known physical laws. Occam's razor applies to the simplicity, plausibility, and relative parsimony of the tenets supporting a model or hypothesis.

We now come to the difference between simple and easy. *Simple* means that we can state the concept in a few clearly defined terms, not involving many complicated issues or concepts, while *easy* means that we can employ or implement the concept without concerted effort or study. For instance, a cowboy describes bull riding as a simple sport: He says, "You just keep the bull between you and the ground. You're going to find out the difference between simple and easy."

We see a similar problem when addressing the influence of language on the thinking brain, simple to state but not necessarily easy to solve: When we evaluate the outcomes of our day-to-day decision making, we use the same brain that we used to make the decisions in the first place. This amounts to guessing how long to cut one board by using another board that we measured by eye. The simplicity of the statement may induce us to consider this problem trivial. In the case of the board, the solution is easy: We just use a ruler or yardstick. But how do we separate the contribution of the brain and the language and credit errors that occur, and how do we make corrections? We face the difficulty of evaluating a relatively healthy brain with impaired language.

In terms of our thinking behavior, this raises a series of questions about effectively measuring and considering the problem individually and as a whole:

- Where do we find the ruler for measuring the processing performance of our brains?
- What problems might we encounter in applying our ruler as an analytical tool for evaluating processes that are invisible?
- Where do we measure from?
- How do we make sense of it all as a whole?

Finding the ruler may be fairly simple, but the application may not necessarily be so easy. Furthermore, the way we use language to help with processing our thoughts represents a significant obstacle to applying that same language as an efficient tool for measuring the effectiveness of our familiar, habitual ways of thinking. Our brain tends to act automatically, relegating our familiar linguistic modes of thinking to a less visible level of implicit processing that can cleverly cover up for errors in the measuring process by defending, excusing, or explaining them away. The habitual tendencies lay hidden beneath an implicit blanket of automatic behavior that resists introspection and direct analysis. Dampened introspection can adversely impact our ability to measure and correct for errors, which in turn influence predictions and the effectiveness of our actions.

We expect an effective ruler to produce explicit, tangible, and interpretable measurements. Explicit measurements give us an observable reference for making predictions and detecting and correcting errors. Implicit predictions tend to produce inconsistent results as problems become increasingly complex. We prefer regularity, reliability, and reproducible results. But the ruler we are searching for, with which to measure invisible processes, may in and of itself be invisible. This invisible ruler may create yet another quandary. While we may feel uncomfortable with an unfamiliar ruler, we will likely feel even more uncomfortable with an invisible one.

When we encounter unfamiliar ideas about improving our language or thought processes that do not line up with our familiar beliefs, we find it especially awkward, challenging, and anything but easy to perceive them as worth considering, much less worth embracing their implications. In the sense of familiarity and change, this book is more than an investigation of language; it is about demonstrating that improving our language can improve our thought processing, how we relate and adapt to change, and help us overcome the constraints of familiarity; it is about the benefits of thinking. Learning an unfamiliar perspective and applying it consistently can be hard work. Why would anyone want to expend all that brain power? Of course, the effort will exercise our brain, possibly a benefit in itself. But perhaps less obviously, by thinking and measuring results explicitly, we may learn how to better understand ourselves, others, and the world around us, and as a result learn to have better relationships overall. We may find some fairly simple, if not easy, methods that can become the yardstick by which we measure a more harmonious way of life.

The explanations set forth in this book attempt to follow the rule of keeping it simple. A collection of fairly simple processing rules can extend the idea of keeping it simple in composing general rules. *General* defines consistent, approximate relationships throughout the entire system that supersede subordinate laws. In a large global system such as the Earth and its atmosphere, these general rules represent *systematic* relationships often called the physical laws of nature. Of course, nature is a process rather than a person, so we expect these general laws to define and describe relationships that are, at least in some way, generally amenable to hypothesis, measurement, functional interpretation, and logical explanation. *Function* can be defined operationally as simply describing how something works, and also as defining a relation between structure and action in a system, where the functional relation is equivalent to correspondence, transformation, a map, or mapping. Our language is a powerful modality for understanding our interactions with the physical systems of nature. Understanding language becomes important in its own right because understanding physical systems requires semantic correspondence with meaningful representations that generally map to relations in the external world. In effect, our words and rules for thinking can match with how our world works. General rules extend our insight as to how the world works as a whole. We can consider how things work logically in relation to each other, across the entire system.

General rules work well in an uncertain world that seldom favors us with all of the information we would prefer (Jaynes, 2007, p. 595). Even though in combination the rules may not appear quite as simple, the ability to generalize them to nature can simplify how we process information from the experiences we encounter there. But that does not mean it is an easy task to learn and apply these general processing rules. When we have been thinking one way for a long time, changing our familiar thinking habits presents quite a challenge. Is it worth it? If we think making changes in how we think about ourselves, others, and nature has value, it is our choice to make. It is a simple choice. Perhaps thinking differently can offer new insights, and applying general rules can broaden our worldview. At least, we have an opportunity to give it a try.

Fortunately, a lot of the theoretical groundwork has already been laid from many different scientific perspectives. This accumulation of views gives us an advantage and an opportunity to peruse the collection of evidence, theories, assumptions, hypotheses, and conclusions. We can sort through the accumulated information and knowledge of science, mix and match our findings, and search for general processing features that reliably model what we see in physical relationships found in nature. We can accumulate knowledge gained from observing the effects associated with those physical processes. Importantly, we depend on effective language for describing, defining, and connecting this knowledge to the outside world. In other words, we cannot avoid the key role of effective language for developing a cohesive understanding of our environment.

Language enables knowledge about physical relations in nature and can help us to evaluate the sensibility of our perception, thought processing, and the results of our actions. Perhaps it is an ambitious task, but in this book we will follow the principle of setting high goals along with keeping low expectations. *High goals* means that we would like to find as many novel and exciting ideas as possible. *Low expectations* simply means we expect to get "whatever we get," and is not intended to imply a lack of enthusiasm. On the contrary, we will approach the quest with much gusto. After this search, we can then comb through what we have found and tease out any bounty we find interesting.

High goals and low expectations allow us to explore for the joy of exploring, to enjoy the fruits of our labor either on discovery or at our leisure. While some may disagree with this strategy, it not only works but simply keeps stress to a minimum. High expectations can easily lead to painful disappointment and discouragement, and dampen further exploration. High goals and low expectations help to minimize the *confirmation bias* alluded to earlier, the human tendency to find only that for which we are searching. Low expectations facilitate a more open approach that helps to mitigate our tendency to narrowly search for a known, predetermined objective and feign surprise when we find it.

We can avoid the conspicuous task of prescribing or imperatively mandating an absolute objective to prove a theory or prove we are right. We can instead simply embrace the desire to search for novelty while we enjoy the experience of searching and learning—our eyes wide open to new ideas and solutions that might possibly morph into a reasonable hypothesis. We may come across a hypothesis that corresponds with nature and helps us better understand fine-tuning our language, information processing, and our ability to predict the outcomes of our actions.

The theme of an open-ended search for information and knowledge makes up a large part of the overlying theme of this book; it is open-ended in the sense that the goal is the search process itself, which is in a sense neutral to predetermined notions as to the "goodness" or "badness" of the information or solutions we might find. *Open-ended* as used here represents an open-minded approach to cognitive processing, which informally we call thinking. To those without an engineering or computational background, this open approach may sound peculiar. However, the practical appeal of an open-ended search for information makes logical sense. For example, similar to evolutionary processes, open-ended search algorithms in evolutionary computation can model the search for possibilities as novel information, alternative behavior, and innovative solutions.

Open-ended search allows an open-minded approach through which we can stumble onto solutions "since we are searching anyway" and then test any solutions we find to see how effectively they fit a particular problem, such as finding a process that offers solutions that help in understanding nature. Search has relevant corollaries with information-processing systems that include linguistics, computers, and human brains. "Understanding the brain's search strategies may allow us individuals to have better access to cumulative knowledge of humankind" (Buzsáki, 2006, p. xii).

The concept of search has served engineering well by providing a practical method for exploiting the problem-and-solution search space. Understanding the problem is nontrivial and is requisite to searching for solutions. There is an old saying, "If you cannot find the problem, you will never find the solution." If we misrepresent or remain ignorant about the problem, we can wander aimlessly and default to trivial or imaginary solutions. An engineering approach to problem solving emphasizes first elaborating and understanding the problem as best possible in relation to systematic processes that influence a particular subsystem and the domain in which the problem resides. A broad understanding also can be accomplished simultaneously with open-ended search by accumulating novel alternative solutions as we go and matching them to problems by effectiveness and goodness of fit. Having a consistent method to go about our task seems to make sense, especially having a keen sense of the problem.

We can opt for a structured method that leads to dealing with information in a consistent manner. Recall that science has a consistent methodological approach for identifying problems, searching for hidden processes, and finding answers to scientific questions. We see that methodological processes apply to many tasks involving the search for solutions in a variety of everyday problems we encounter. We have at our disposal a scientific engineering approach to dealing with difficult problems that highlights the benefits of applying logical, orderly methods, especially when searching for invisible processes. Of course, our initial step is to explore and elaborate the problem space to find the dimensions of the problem and more efficiently search for further information. These methods can be applied to most systems involved in processing information in some form or another, including systems as diverse as electronics, mechanics, physics, biological brains, and natural language. Logical processing methods begin with

- Recognizing the problem
- Elaborating the problem space and searching for more information
- Listing possible and probable influences related to the problem
- Localizing likely influences related to the problem
- Localizing specific interactions and functional correlations corresponding to the problem
- Analyzing and testing the result for effectiveness

To maximize our efforts, we benefit from language that enables understanding the larger system and the physical laws involved, knowledge of the subsystems, and knowledge of normal, abnormal, and optimal functioning. For instance, if we are dealing with the brain, we benefit from language that can define and describe what we know about the brain's specifications and the associated operational structure of the applied language. Understanding how the brain evolved, the conditions in which it evolved, and the operations it performs gives us helpful insights for exploring and elaborating the problem space. To understand the brain's minimal and maximal functioning, we likely want to know how it processes information from input to output, what parameters are involved, and the effect that tuning them has on the output. We also want to know the relationship among brains, language, and information and the influence of language on optimizing the effectiveness of input-to-output processing.

In this case, understanding the language we use to evaluate the system helps not only in general but also in particular for evaluating processing faults that can lead to distorted reception and corrupted speech output. That is, how do we get an understandable message, and how can we optimize the information in that message? Recall that with the possibility of impaired language knowledge, we face the difficulty of evaluating a relatively normal brain with degraded language. It may seem like a chicken-and-egg dilemma, but obviously knowing generally works more effectively than guessing and shooting in the dark. We prefer language that enables clarity of thought and reliably informed knowledge. Knowledge helps to minimize running in circles and down dead ends, and in structuring a methodological approach to help dampen our human tendency to jump to unwarranted conclusions.

When we begin by defining the problem, we quickly recognize that efficiently maximizing our efforts benefits from searching for and, when possible, acquiring at least a sufficient amount of plausible and accurate information. Information fidelity helps with reliably identifying, processing, and effectively resolving problems. Generally, maximizing information fidelity enhances efficient and effective elaboration of the problem-and-solution search space in complex domains with influences from many interacting variables. We also would like to know

- The adjustable parameters and what happens when we tune them
- What measuring devices to choose
- What, where, when, and how to measure
- The reasoning process leading to why we are measuring
- The reasoning process leading to the selection of the reference from which we chose to measure

The heavy information load and intensive methodological processing that sometimes may seem tedious and laborious might lead us to consider making a case for the ease of guessing. We would be hard pressed not to acknowledge that guessing is fast and easy, is sometimes the only option, and often works sufficiently. We see here another practical example of the difference between *simple* and *easy*: simple to guess but not so easy to analyze. We notice a trade-off relationship between effortless, fast, and frugal and effortful, slow, and precise. Obviously, the expense and effort of acquiring and analyzing information amounts to anything but frugal. Yet we are hard pressed to find plausible arguments against the benefits of assessing information accuracy and optimizing the fidelity of our knowledge about the system.

When we understand how general processes in the larger system work, solutions can be discovered and matched to specific problems we are trying to solve according to fidelity. Our search for information acts as a primary step toward understanding processes in the domain, and the relationship between the problems and solutions they may address. Then, we can apply an information-processing model to compute solutions, predict results, and measure the effects of output errors. A computational processing model enables

- Comparing different methods, parameters, and variables
- Evaluating measurable differences in processing performance by the effectiveness of the output product
- Accumulating knowledge about the overall system

Processing Information

Language that supports a systematic understanding combined with a methodological approach facilitates consistent information processing and increases the probability of finding more effective solutions. We can define and describe how language, information, information processing, and relationships can be modeled with a computational framework. In this framework, computational modeling encompasses cognitive processing, how we apply language to sense, think, predict, and learn to interact adaptively within the larger system of the world around us. An information-processing perspective provides an operational model that illuminates the neural basis of language and thought and can be applied to most any kind of system, including physical, social, and economic (Feldman, 2006, p. 21).

A computational information-processing model may sound like a somewhat dry, analytical, or even simplistic description of relationships and the life we live. Studying such a model in relation to language and brains can possibly help us to understand how to interact more sensibly with nature. We are a small part of the big picture, and learning about how our language and brains create that big picture can help us to understand how we fit in. A computational model allows us to create a systems analogy that can be applied to the brain, cognition, language, and natural relations in general. A systems model enables a modular approach to information processing across the system that includes the brains of the organisms that live there. We can evaluate relationships, processes, and the efficient flow of information across the system, including the regulatory influence of language on the brain's input-to-output production back to the overall system.

We may to some extent agree or disagree about computational models and their limitations; however, a computational metaphor offers a different but rich and potentially beneficial perspective on understanding language, cognition, and behavior that can possibly enlighten us on how we can come to better know our world. Is it possible that a computational model might yield some surprising but fairly simple solutions to some of the problems we consider to be anything but simple or easy? If so, and even if we are not typically inclined to do so, would we reconsider the potential benefits of computational modeling?

Computation provides us with a processing model for applying language to thinking about how we inform, accumulate knowledge, and understand statistical relationships in large, complex systems such as nature. Thinking and searching comes naturally in computational science, along with data, hardware, software, algorithms, logic, memory, operating systems, and programming languages, which are all involved in computing solutions to problems. In this regard, the brain will be considered the hardware that we inherit. We typically think of computers as hardwired. But our brains are probably best described in general as being softwired. Our brain hardware is *softwired* in the sense that it demonstrates plasticity, which means it can change in relation to our experiences. Plasticity allows our brains to work flexibly and learn as we interact with the world around us.

When we think of software, we tend to take language for granted and think of the software as applied to a particular task we want to deal with, such as word processing, balancing our checkbook and keeping track of our finances, designing autos or buildings, and protecting our computer against viruses. We can think of natural language as a way of configuring our software as programs and algorithms that contributes to how we represent and communicate information, and how we apply that information to computing orderly solutions to problems. In a broad sense, natural language forms a part of our abstract linguistic operating system and software that influences how effectively we can represent, classify, identify, and process relational information about our environment.

Software involves the implementation of one or more algorithms that can be compiled to machine language that the computer can understand. We can think of algorithms as recipes at the root of processing methods for mapping input to output. Algorithms work for solving certain problems. That means that we can have many different algorithms and chose the one best suited for particular tasks. Algorithms are essentially procedures or recipes that tell computers or humans how to solve a problem in an orderly manner, step by step. Having multiple algorithms we can flexibly choose from makes it easier for us to deal with the different experiences and uncertainties we face in a changing world. The rule of thumb in computer science is that no one algorithm is sure to succeed for every problem, so having multiple recipes allows for a broader range of alternatives for problem solving.

Data is defined as a form of information. Computers store information on a hard drive that is somewhat analogous to our brain's memory system. The brain's outermost layer—which in general terms we think of as lying in close proximity to our skull—is called the neocortex, and it accounts for most of the memory storage that we think of as broadly distributed across the brain. As most of us have experienced, we can have lots of data and still not make sense of it. Recall that *pattern recognition* is a computational term from machine learning that helps to organize and partition data into more manageable bundles that enables processing large amounts of information more easily. Recognizing usable patterns in data turns it from meaningless random noise to nonrandom information. Nonrandom patterns relate to information that is meaningful to us in the sense that the patterns help us to compute predictable results. *Random* as used here is defined as disorganized and unpredictable, and *nonrandom* is defined as organized and predictable.

A piece missing from the discussion so far is choosing specific logic for our model. In a computational model, *logic* provides abstract relational rules for managing the flow and processing of information among structural architecture, an operating system, and the work it performs to output solutions. Logical rules establish the operational constraints that bias how internal input information can be weighted and processed to produce output as information, solutions, or behavior. In effect, logic provides the functional processing rules and constraints for how algorithms in a system influence input-to-output mapping. Logic influences how effectively computers and humans solve problems. The types of problems we want to solve will inform us as to what level of logic will work most effectively.

When trying to solve general problems, such as understanding relationships in nature, we can improve our solutions if we choose logic that corresponds with the logical ordering of the entire system. In this case, we are talking about nature as a large, continuous system in which many events and interactions typically occur concurrently and persistently influence each other at many different levels. Large, continuous systems lend themselves to systematic logic that computes probabilities. In other words, either–or two-valued logic that may be suitable for solving idealized or simple true-or-false problems is less than optimal for effectively solving complex general problems related to the entire system.

Whether we realize it or not, language and semantics are a part of our software, algorithms, and logical processing that we use every day for thinking about and dealing with life. For example, we use discrete logic forms—*or*, *and*, *not*—in many of our sentences. We use language to stand in for and logically process information and knowledge we learn from our experiences. Similar to fine-tuning computer languages and programs, understanding how variations in our language influence how logically we process information from our experiences as we think, feel, and behave can give us insights into fine-tuning our language and semantics. Tuning plays a role in how well we take in information, process, understand feedback, make error corrections, and solve problems of adapting to change in a difficult-to-predict world. Language that supports adaptive tuning plays a key role in enhancing flexibility, decreasing errors, and reliably informing our thinking, learning, and behaving as we experience life.

The next issue addressed is defining knowledge. Arguably, information is a form of knowledge. For our purposes, the term *knowledge* will be reserved for understanding regularities related to how the world works in general across the whole of the system, or systematically. Poincaré (1902) argues in favor of scientists focusing on the regularities that tie facts together rather than focusing on the specific facts (Striedter, 2005, p. 14). Knowledge corresponds to the accumulation of external information learned from experience by observing regularities occurring in nature that reliably correlate to our understanding of complex interactions across the entire system. We call this type of systematic understanding global knowledge.

Global and *local* are terms frequently used in complexity theory and nonequilibrium physics and are not usually found in everyday casual conversation. Even in other areas of science, *local* and *global* are fairly uncommon terms, but they are terms that are worth familiarizing ourselves with; we will see that understanding the difference between *local* and *global* is pivotal for making sense of language, brains, and nature. *Global knowledge* represents general knowledge that takes a macroscopic perspective of complex interactions across the entire system and distinguishes it from a localized, discrete, microscopic perspective.

> One of the greatest challenges left for systems neuroscience is to understand the normal and dysfunctional operations of the cerebral cortex by relating local and global patterns of activity at timescales relevant for behavior. This will require monitoring methods that can survey a sufficiently large neuronal space at the resolution of single neurons, and computational solutions that can make sense of complex interactions. (Buzsáki, 2007, p. 267)

Systematic or global knowledge comes about by processing that statistically correlates accumulated information about recurring regularities and event-related environmental interactions across the system. This broad accumulation means that global knowledge has statistical significance about general relations in the entire system over time. The corollary to global knowledge describes isolated parts of the system referred to as local. Language helps us to distinguish between the two perspectives; the term *belief* will be applied here to understanding at a local level, as opposed to *knowledge* at a systematic level. Local and global are two different ways of looking at the world—from a narrow microscopic perspective to a broad macroscopic perspective—two views that yield dramatic and surprisingly different conclusions. *Local language*, also referred to as localized language, defines a specific style of information representation and processing with localized linguistic constraints and does not intend to imply a particular regional, geographical, or cultural dialect. Similarly, *global language*, also referred to as systematic language, defines a generalized style of information representation and processing with systematic linguistic constraints and does not intend to imply a particular dialect per se.

Yet localized language commonly shares closed-ended features found in many dialects. Local dialects can also have open-ended linguistic features that may support systematic descriptions, but in large part these features are ignored rather than applied. The key here is the application of the language, its usage rather than whether it is available. The linguistic distinctions correspond to the structural, representational, and functional processing characteristics rather than the particular regional or geographical distribution. In effect, localized language supports local beliefs and values that typically exhibit high variability for defining relations and impose closed-ended processing constraints that exclude generalization and the accumulation of systematic knowledge.

For example (Figure 1), draw a large circle, which we will define as the open system, systematic or global work space. Draw a line bisecting that space or a smaller circle inside the larger one. We now have two discrete, closed subsections. No matter how

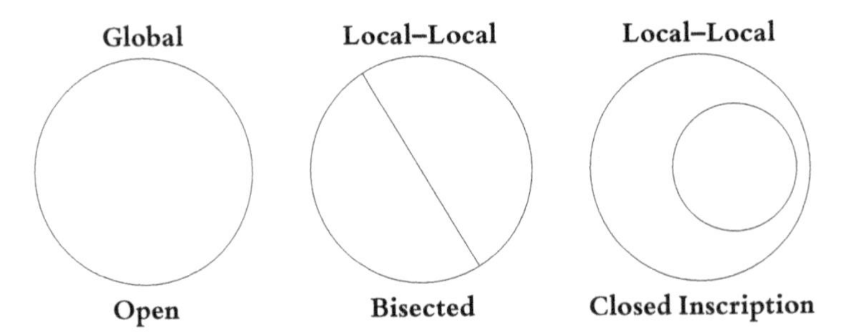

Figure 1. Global and Local Space: The first circle is open and labeled global. In the second and third circles, respectively, bisecting or inscribing a closed space inside a larger circle creates two isolated subspaces labeled local. We can also think of two local spaces as local–local where we designate one local space as true, which automatically relegates the other local space to false.

small or large any one of the closed subsections, each is local by definition. We have two discrete pieces separated and closed off from and obliterating the integrity of the open global space. We know that the global system is dynamic and open, with information constantly moving and changing across permeable boundaries. Arbitrarily partitioning the space closes the system, inhibits the continuous flow of information, and collapses the space to local. Global systems are defined by the ability to transfer information and energy throughout the open system.

Now we apply the same open concepts to words with which we discuss processes, problems, solutions, and information flow across the global space. Problems resolved as possibilities and probabilities are examples of open global language solutions that identify and describe information distributions in changing domains. In changing domains, the information is constantly moving and requires statistical information descriptions as probabilities. Global solutions remain open to other possibilities across the extent of the entire space no matter where they are applied. Possibilities are open-ended and thus do not close off or bisect the global space, and they consistently leave room for alternatives and equivalent solutions. The open-ended nature of possibilities and probabilities is in stark contrast to local discrete solutions and certitudes that close the solution space.

Take localized language, such as *true* or *false* and *right* or *wrong*. You will notice that absolute discrete terms exclude the possibility of alternatives. They close off the solution space and essentially freeze solutions by isolating them from the global space, which is defined as an open system. Discrete terms bisect the global space and stop the flow of information, thus collapsing the global space to a local one. But many localized words can be easily extended to the global space by converting them to possibilities, such as *possibly true* or *probably true*, which leaves the solution space open.

When we isolate a part of the system, we create a subsystem. A subsystem isolates the particular problems and solutions and constrains processing to that particular part of the overall system. In effect, by focusing on a discrete area, we confine discoveries to that particular localized portion of the system and miss seeing the big picture. The harder we try and the harder we focus, the more locally constrained our perception becomes and the more information we miss outside those constraints. In some scientific areas of study, *global* may refer to smaller independent systems such as brains. The systematic definition used here applies only to the global dimensions of the entire system and not to isolated parts of that system, which are defined as local. We will see that this distinction is nontrivial, since the brain can support both localized and global language. With language we also see that any discrete sectioning of the global language system produces localized linguistic subsystems by default. However, globally tuned language generalizes and thus can translate across the domain for resolving localized problems.

Global tuning of language allows general descriptions and translations of relations across the system. Even though we are only a small part of that system, global language provides us with extended analytical methods that can enhance our systematic understanding of nature. Global tuning enables knowledge about how the system works within natural constraints, which in turn facilitates reliable information processing, prediction, action selection, and outcome error assessment. A systematic understanding allows us to expand our worldview by extending our language from black-and-white certitudes to a rainbow of possibilities and probabilities, and thus extend continuity and expand the breadth of our thinking abilities. A linguistic extension allows a progression of information and communication from a circumscribed, narrow view to a broad systems view. A systematic extension changes how we perceive and understand the world around us and, perhaps more importantly, how we relate to that world and other humans.

The computational information-processing model presented here is based on the systematic evaluation of language fitness. The model proposes that a language's effective information-processing accuracy represents the difference in the rate and range of systematic output errors produced or *realized* in relation to *predicted* outcomes as adaptive responses to external change. Language fitness describes how the language, as an explicit information source, fits adaptation to external change; put another way, it describes how well the language processing systematically informs adaptability.

We have identified and discussed some methods and the systematic dimensions of the problem-and-solution search space, including structural and functional relationships among information, language, the brain, and nature. A systematic working model is proposed here that is based on extended linguistic structure and function for effectively representing, supporting, and processing information referenced to nature that includes

- Information accuracy
- Information-processing accuracy
- Event-related information accuracy

- Dynamic physical correspondence
- Logical correspondence
- Classification and parsing correspondence
- Value correspondence
- Error-correction correspondence

Next we will look at how language influences cognitive fitness and problem solving.

CHAPTER 2

Language and Cognitive Fitness

When we think of fitness, we typically think of physical rather than linguistic or cognitive fitness. What do language and semantics have to do with fitness? *Fitness* measures our ability to perform and adapt as we encounter problems and disturbances, formally called perturbations. Adaptation entails some notion of improvement, as in enhanced survival (Feldman, 2006, p. 72). For adaptive fitness, we are talking about systematic processes. In evolutionary biology, fitness describes the ability of organisms to survive and reproduce. The human brain has evolved as a complex structure for cognition, and our cognitive abilities appear to have enhanced our survival, or what we call fitness. Our brain has the computational power to exploit language as an information source for thinking and learning. Learning can enhance our ability to understand our environment, learn from and decrease errors, and adapt our behavior accordingly. We see these concepts of fitness as adaptive biological interrelations of a process we call life. Language plays a role in how we understand these adaptive processes and life itself.

The Language Habit

It appears that language and other cultural aspects work in a concerted way to establish and maintain our habits of thought (Fausey et al, 2010, p. 7). Our language and culture encapsulate the human experiences of thousands of generations. But much of what we learn comes to us already predetermined by and prepackaged in the culture we inherit as members of a linguistic social species. Even though our ancestors did not have access to computers or computational knowledge, we inherited precomputed or preprocessed solutions from them, and some seem quite ancient and well preserved. Much of our cultural history occurred to humans who lacked the scientific resources to accurately explain the natural processes happening around them. Their prescientific explanations, at least those that did not lead to their immediate demise, became encoded in the lexicon of our cultural inheritance that forms the linguistic scaffolding underlying our thought processing, beliefs, and behaviors—errors and all. Over the generations, some more obvious errors were detected and rejected, but many others persist.

Such errors can fall into the category of harmless for many uses. For example, being earthbound, our ancient ancestors could see only as far as their own horizons. Since they could see the appearance of the sun moving across the sky, the concept of the sun as rising and setting was incorporated into our language; to this day, we still use that simple analogy for thinking and speaking. The analogy persists despite our having ample scientific evidence that the turning of the Earth produces this periodic effect because of natural physical processes and relations. According to Jackendoff (2002), the scientific view of sunsets does not say that "in all reality" the sun does not "actually go down" and that therefore, "there is no such thing as a sunset." Rather, science contends that the observed rotation of the Earth yields consistent explanations in terms of physics as a whole. In a case like this, changing our language to match our more recent knowledge might not significantly change most people's experience of daily events. But when NASA scientists describe the trajectory of a rocket destined to leave our atmosphere, they typically use language that accounts more accurately for the scientifically understood processes of logical relationships between planets and their suns.

On the other hand, the unscientific mind can easily view the sun and stars rotating around the Earth and assume that the Earth and its inhabitants are at the center of the universe. It seems that we have nontrivial linguistic choices among many different colloquial and scientific expressions for thinking about the world. We can take a landlocked view from our local position and ponder the mysterious processes of the universe above from the inside out with our colloquial language and thought. With scientific language and thought, we can take a global view from the universe, looking downward from the outside in.

The availability of options for how we speak about processes and relationships suggests that we can choose the type of language we prefer. In this sense, a scientific language type allows for exploration and discovery of new information to update our current knowledge. The idea of updating language for accuracy encapsulates the notion of fitness used in this book: Language fitness increases when we adjust the language we incorporate into processing our thoughts in a way that helps us to increase insight about interactions with nature and decrease prediction errors. Thus we can think as accurately as possible, to accommodate up-to-date, independently validated information about how the world works, which can generally enhance survival. We expect increased fitness to predictably correlate with adaptive outcomes we prefer for ourselves and others.

Improving our language helps us to understand how processes and relationships work in the world and to adjust our language behavior accordingly to accommodate those relationships. We can learn to improve how we interact in a more harmonious way. If we view the world as a place of mysterious processes with random happenings and caprice, as our ancient ancestors tended to do, or even if we operate with the unexamined prescientific explanations of our more recent forebears, we tend to perceive ourselves as subject to nature's unpredictable fury with little recourse except to soothe ourselves by wishful thinking. Alternatively, if we clarify our worldview by updating and incorporating what we have learned about natural processes and relations into our language for daily discourse, we may find that informed decisions produce more effective results in interactions that matter to us.

> We talk about the events of our days, often discussing who did what to whom. We talk to our children, we update our bosses, we negotiate with friends and colleagues. We learn from other people's stories. A huge proportion of what we know about the world comes to us through the medium of language. Human language use is exquisitely structured and systematic, and as a result our conversations carry many regularities in how we communicate "what happened." (Fausey et al, 2010, p. 2)

A consistent and applicable functional model for understanding how language structure influences our cognitive processing can potentially lead to a more effective method for deliberate and reasoned evaluation of the fitness of the language we use. In living organisms, function connects structure to the environment (Prigogine, 1997, p. 62). The word *function* typically refers to operational relations among variables as a process that describes the manner of their interactions. We use functions to map information to the environment and express correspondence with that environment. We use language and semantics to abstractly define and describe these relations logically as mutual information among us, others, and nature. When we include nature as a whole, maximizing mutual information means systematic information.

By taking a functional approach, we can explore how language operates as an information medium that abstractly extends to computational processing. Language serves as an operational platform for understanding and expressing human thought that logically corresponds with processes in nature. Thus we have the option of matching our internal logic with the external logic of natural processes. Describing regularities of processes in nature and coherently correlating our language's logic parameters with nature's logic can enhance logical consistency (Jaynes, 2007, p. 423), optimal information processing (Zellner, 1988), and a practical top-down model for effectively representing and understanding the abstract connection we call language that underpins much of our knowledge.

> Are our top-down concepts, such as thinking, consciousness, motivation, emotions, and similar terms, "real," and therefore can be mapped onto corresponding brain mechanisms with similar boundaries as in our language? Alternatively, do brain mechanisms generate relationships and qualities different from these terms, which could be described properly only with new words whose meanings have yet to be determined? (Buzsáki, 2006, p. 19)

On the other hand, have we overlooked sorting and matching currently available language according to local and global functional characteristics? We can seek to understand language and the brain by defining and exploring the problems, solutions, and constraints encountered in computational brain models as a metaphor for thinking, and then review and integrate information and knowledge from different scientific areas. In turn, we can consider the available evidence for understanding how language, logic, and the human brain work together at a global level to influence human performance as a process of linguistic effectiveness and cognitive fitness. The information is generally available but scattered across many disparate fields. Collating and integrating this information on the basis of how effectively language represents, models, and correlates our knowledge to our interactions with nature may help further thought and research. Collectively, this integration raises questions that may possibly yield answers and help clarify our understanding of the processes and relationships among language, thought, emotion, and behavior. We may find practical applications that help to improve our thinking and behaving in a more harmonious manner.

Borrowing from computer science, we begin with exploration for information and by defining the problem search space, asking questions such as these:

- Do we think about the brain and cognitive processes in the most effective, scientific way?
- How does our use of language inform our knowledge and influence the fitness of our thinking?
- Is it possible to understand the brain and cognition without first establishing the effectiveness of the language used to define, describe, and inform our search for this brain knowledge?
- What role does our understanding or misunderstanding of language play in understanding the overall problem of defining effective language?
- Can these questions and their answers be generally applied across science and society as a whole?

> We know very well what is good for our car, our garden, our stomach, and our heart. But what is good for our brain? Even the question sounds a bit strange if we are thinking not about vitamins, low cholesterol, and moderate alcohol consumption but about the right *food for thought*. Which thoughts, perceptions, sensations, feelings, and so forth are good for our brain? The question makes no sense to us. Why? (Spitzer, 1999, p. 10 [emphasis in original])

Cognitive fitness refers to the effectiveness of the brain's capabilities for effectively processing information related to learning, understanding, predicting, and adapting to a changing environment through plasticity. Language influences the brain's behavioral plasticity for thinking, learning, and communicating—all of which contribute to how effectively we understand, relate, and adapt to environmental change. Language use works through plasticity to help shape the brain's cognitive ability to perceive, represent, learn about, and adapt to our experiences in a persistently fluctuating environment. In this sense, *fitness* can be thought of as the ability to more effectively adapt to and survive the difficult-to-predict changes and disturbances around us. Adaptation facilitated by effective language and behavior can improve outcomes as we deal with the invisible processes we encounter in relations with our uncertain world.

An adaptive model for computational information processing allows us to investigate how different forms of language bias our cognitive fitness, thought processes, and behavior. Then we can take advantage of and augment brain plasticity by deliberately modifying how we choose and use language according to how it influences our thinking and predicting abilities. This fitness-based information-processing model identifies *deterministic language* as a closed-ended, discrete method of processing information and predicting future events and solutions in an "exact" manner, which implicitly assumes that the past determines the future. Applying absolute predictions and problem resolution in a continuously fluctuating world constrains plasticity through rigid language behavior that tends to resist change, including potentially beneficial change. Deterministic language supports the viewpoint that evolution is governed by a set of rules that, input from any particular initial state, can generate one and only one sequence of future states (Prigogine, 1997, p. 201). Deterministic language supports rigid, localized thinking.

In contrast, *nondeterministic language* enables an open-ended, continuous method of processing information and predicting future events and solutions statistically in an "inexact" manner, which explicitly distinguishes the difference between the static past and a dynamic future of possibilities and probabilities. Applying nonabsolute statistical predictions and problem resolution in a continuously fluctuating world augments plasticity through flexible language behavior that tends to facilitate change, including adaptive change. Nondeterministic language takes into account the current system state, current input, and current external changes. Logically extending language behavior to include possibilities and probabilities accounts for fluctuations of future states, and facilitates continuous output adjustments for effectively modulating adaptive actions in response to change. Our brains are adaptive systems in the sense that language, learning, and plasticity facilitate adaptive input-to-output modulation of information as alternative choices, decisions, and actions. Language with dynamic fidelity supports effective information processing and flexible adaptation that generally correspond to the level of cognitive fitness. Nondeterministic language supports flexible, systematic thinking.

Language offers broad descriptive abilities with constructions that may be more or less amenable to describing a variety of cognitive tasks and physical processes (Prigogine & Stengers, 1985, p. 9). *Language fitness* can be defined as a measure of how effectively we observe and understand the structural and functional correspondence of regularities, processes, and relationships in nature that contribute to long-term survival and reproduction; how we select and fine-tune language structure to reliably represent the information and knowledge we acquire about those relationships; and how effectively our language constructions facilitate predictable adaptation to change. The model includes the following premises: We generally enhance our linguistic and cognitive fitness with language constructions that facilitate the production of plausible alternative solutions to flexibly address a broader range of problems. Where language fitness influences how we think, feel, and behave, we expect language that corresponds with our knowledge of nature to enhance informed cognitive analysis and a predictable level of control for effectively modulating adaptive output in response to instabilities. The key idea here is that control in living systems begins with the output (Buzsáki, 2006, p. ix). And in humans, language plays a key role in this regulation.

Boroditsky offers evidence that languages influence many aspects of human cognition about understanding concepts such as space, time, objects, substances, numbers, colors, shapes, events, and the minds of others. She contends that some of the many ways in which languages differ produce analogous differences in the cognition of people who speak those different languages. Beyond showing that the speakers of different languages think differently, Boroditsky suggests that linguistic processes pervade most fundamental domains of thought. It appears that what we normally call "thinking" depends on a complex set of collaborations between linguistic and nonlinguistic representations and processes (Boroditsky, 2003, p. 920).

> Language is a uniquely human gift. When we study language, we are uncovering in part what makes us human, getting a peek at the very nature of human nature. As we uncover how languages and their speakers differ from one another, we discover that human natures too can differ dramatically, depending on the languages we speak. The next steps are to understand the mechanisms through which languages help us construct the incredibly complex knowledge systems we have. Understanding how knowledge is built will allow us to create ideas that go beyond the currently thinkable. This research cuts right to the fundamental questions we all ask about ourselves. How do we come to be the way we are? Why do we think the way we do? An important part of the answer, it turns out, is in the languages we speak. (Boroditsky, 2010, p. 1)

Furthermore, Boroditsky concludes that taken together, experimental results show that linguistic processes shape our thinking from the nuts and bolts of cognition and perception to our loftiest abstract notions and major life decisions. Language is central to our experience of being human, and the way we speak profoundly shapes the way we think, the way we see the world, the way we live our lives (Boroditsky, 2009).

Language provides a tool for deliberately coding information to logically correspond with the environment. In evolutionary computation and machine learning, the encoding influences the representation of the problem, the solution search space, and the logic and algorithms related to adaptive search. In nature, we expect a language code to represent external information compatible with the brain's sensory inputs, logical processing abilities, and predictable adaptive motor output to the environment. Logic that coherently correlates patterns, structure, and behavior with the environment theoretically facilitates adaptation, where fitness and logic converge. Internal processing that corresponds to the logic of the domain enhances consistent learning from experience.

As organisms subject to evolutionary pressure, humans persist via reproductive success. We can expect individuals who more accurately perceive relations with the environment to increase the probability of surviving and transmitting their genes. However, we have learned to some extent to sidestep the natural selective pressures by technologically augmenting survivability and reproductive success, which can diminish the adaptive influence of evolution on fitness. Can the language we use for thinking restrict the reliability and usefulness of our knowledge and impair adaptation? *Reliability* used here and throughout the book formally represents a statistical measure of general effective output performance repeated over time; this is not be confused with a more informal term, *consistency*, which is defined as a measure of agreement, concordance, correspondence, or uniformity among the parts of a complex structure.

Ever since the first shaman said, "I have a message from the spirits," people have leveraged language usage to enhance their well-being and personal comfort and to influence their reproductive success. Over generations, we have developed a variety of deliberate uses of language for thinking about the world, thinking that regularly leads to the promotion of rigidly held beliefs and dogmas and the persuasion of convictions and, in turn, leaves us struggling with local goals and affective utility, often at the expense of global neglect or abuse. Does our inherited language, prone to preprocessed solutions, offer our brain the most effective way to deal with the world adaptively?

We still use many of the genetically inherited habitual behaviors seen in less-complex organisms and other mammals to solve complex interpersonal and societal problems, as if our individual survival depends solely on emotional responses and competitive reactions. While scientific advances have given us health and longevity far beyond those of our early ancestors, we still use language that frames world events in nonscientific terms they would understand. Simply put, our general use of language has not kept up with other technological advances. Furthermore, we tend to overlook the lingering maladaptive effects of linguistic influences on our cognitive processing and cooperative relationships, and the counterproductive effects on our overall fitness.

> Because we use language to talk about so much of our experience, cultural influences through language could end up having pervasive effects throughout the cognitive system. (Fausey et al, 2010, p. 9)

The fitness of our language influences how effectively we process thoughts, learn, and make decisions. Language influences our cognitive fitness by biasing our values, which play a modulatory role in the processing of our preferences, choices, behaviors, and outcomes. As a product of evolutionary processes, value and credit assignment underpin cognitive regulation and survival traits intertwined with natural selection and reproduction. Language and the values it conveys provide a regulatory processing tool that both enables and biases how we understand and adjust to a changing world.

As linguistic beings, we use language to layer our social and cultural values of desires and preferences on top of the physiological mechanisms that distinguish our essential physical requirements for survival. We inherit traditional beliefs and predetermined values from generations past, with little evidence of their adaptive effectiveness or predictive utility. Automatically using these arbitrary and often conflicting traditions as preprocessed solutions to current problems and to evaluate our thought and behavior in the present may not contribute to and could even compromise long-term fitness. Our habitual use of these traditions and values can trap our thought processes in the past and blind us to the here and now in which we live.

Elaborating the Problem

Applying an effective linguistic method to process, understand, and associate value with our beliefs and relationships might offer us an advantage. If we can improve the logical thought processing behind how we assign value in the here and now, we can potentially enhance our ability to get along more cooperatively in this world. A study of language can provide us with a unique linguistic lens for viewing how the input of cultural values biases our patterns of thought and behavior.

> One attractive aspect of studying culture through the lens of language is that because language use is richly structured, it is possible to make precise measurements of patterns in language use, and make specific falsifiable predictions for potential cross-cultural differences. (Fausey et al, 2010, p. 9)

A computational model can possibly help us to understand our elusive natural language code and to learn how language influences information and knowledge processing, establishes reliable referencing, and decreases systematic errors and faulty predictions. This book seeks to expand our understanding of the language with which we communicate, enhance our effective information processing, and help us develop and maintain more harmonious relationships. Natural language may provide the computational key that unlocks hidden value behind the coding process and leads to discovering a language code compatible with more effective thought processing and enhanced cooperative behavior.

Identifying the specifications for the means used to encode and decode language provides us with the key to interpreting and understanding language as an information and communication form. When we consider a message encoded by unknown means, we are ignorant of two things: "the message itself and a specification of the means used to encipher [encode] it, which we may call the key" (Pierce, 1980, p. 271). Cognitive referencing that corresponds to nature can provide the key for unlocking, untangling, and understanding the language code—and more specifically, the key to understanding the processes and relations in the world that language represents.

> It was Leibniz who likened the unraveling of nature by science to the reading of a cryptogram. I am not sure that he himself saw how powerful and exact this analogy is.... We regard nature as composed of processes (and not of single objects or events)... [and] we add the requirement that the code is to make nature as meaningful as possible. That is, science is formally the search for that code which shall maximize the information content of the messages which record the processes of nature.... The code groups have a function space of their own. What the message represents is always a process. (Bronowski, 1977, pp. 49–53)

The properties of our bodies and the structure of our physical and social environment contribute to shaping our language, thought processes, and behavior and establish metaphorical mappings of our embodied experiences represented as concepts coded by vocabulary. Reddy and others have expanded on the concept of metaphorical mapping: Ideas are objects, phrases are surrounding containers (for idea-objects), and communicating is sending (idea-objects in phrase containers) (Feldman, 2006, pp. 7, 195; Johnson & Lakoff, 2003; Reddy, 1979). We can go a step further and state that our brain's cognitive processing reflects our mapping of relational experiences with environmental events that we abstractly construct through language, ideas, and concepts. It follows that we facilitate consistent reflections by acquiring reliable information and processing it into systematic knowledge as effectively as possible. Reliability stems from using the most consistent language available, one that demonstrates domain validity (Prigogine, 1997, p. 29).

How do we make sense of all of this? Can we possibly find a practical method for uncovering the key that unlocks the language code and facilitates communicating more effectively? We can refer back to some of the scientific tools and methods we uncovered earlier:

- Recognizing the problem
- Elaborating the problem space and searching for more information
- Listing possible and probable influences related to the problem
- Localizing most likely influences related to the problem
- Localizing more specific interactions and functional correlations to the problem
- Analyzing and testing results for effectiveness

We can take a methodological approach, start by recognizing and elaborating the problem space, search for more information, and build a hypothesis. We would like to understand the processes and relationships among language, the brain, and nature. Since we are going to use language media as a tool for evaluating our results, we start by developing an understanding of the encodings we will choose for this project. In light of attempting to understand the encoding, let's go back to the concept of metaphorical mapping and take it a step further; we prefer encodings that translate mappings systematically according to domain validity. We will require a language that can break the systematic code across the entirety of these mappings. We are trying to break the global code, and if we are successful, we can compare the effectiveness of globally encoded versus locally encoded language. We can search out a semantic encoding for a vocabulary that coherently maps meaning to the domain. We then have a systematic language with domain validity that represents and equates the structural and functional semantic encodings with the structural and functional features of that domain. Furthermore, we expect the language to demonstrate structural and functional compatibility with the brain.

Even though the brain can act spontaneously without external stimuli, the brain's usefulness depends on how it adapts to the outside world. The brain benefits from calibration to the metrics of the environment it lives in, and its internal connections and processes work most reliably when modified accordingly. If the statistical features abstracted from the environment reflect a particular configuration, the evolving brain can adapt its internal structure to predict most effectively the consequences of the external perturbation forces. The functional connectivity of the brain algorithms effectively generated by continuous modifications derives from interactions with the body, the physical environment, and to a great extent, other beings (Buzsáki, 2006, p. 15).

Algorithms can be thought of as recipes or models that give us information and directions about how to process specific tasks and get the outcomes we prefer. When we use a recipe to produce something, we are concerned about the results, the effectiveness of the outcome. How does the outcome correspond to the ingredients and the methods for processing or mixing those ingredients over time? In the brain, cognitive processing produces behavior and outcomes of variable effectiveness. We can apply our understanding of recipes to help us model and understand how cognitive processes in our brain modulate how we identify and select from available options. A recipe provides information about which ingredients to choose and how to process those ingredients, and informs us of the recommended time to perform each step to produce the preferred results.

We use algorithms in a wide variety of activities, even though we might not think of them as such. For example, recipes for cooking are essentially algorithms. Generally, the closer we follow a recipe, the more predictable the results. We typically prefer the intended outcome to unintended ones related to faulty processing. But sometimes novel changes in the recipe, whether deliberate or inadvertent, have unexpected favorable outcomes. On the other hand, sometimes we acquire recipes that produce variable results: Sometimes they work well, and sometimes not. With recipes like this, even when we rigidly follow each step, we seem to get unwelcomed surprises and confusing random outcomes that defy prediction. Sometimes the recipe leaves out a critical ingredient, step, or time component and thus is unlikely to succeed as written. Of these, timing often plays the crucial role in our results. A basic requirement for describing recipes, experiments, results, or relations in nature is the dimension imposed by time (Bronowski, 1977, p. 53).

> How we approach the problem of time largely determines our view of the outside world around us. (Buzsáki, 2006, p. 6)

Communication experiments with robots have been shown to evolve a locally linked symbol system to internal representations of the physical world (Steels et al, 2007). Functional modeling allows for creating a system to represent physical processes and relationships among entities in a selected domain. We can linguistically represent an information-processing model for cognition and communication as input-computation-output according to effectiveness. We can measure linguistic effectiveness according to predictions of the adaptive output produced under uncertainty. Effective communication is expected to provide a reliable link of internal representations that corresponds with the physical dynamics we observe in nature.

A model with clearly articulated systematic referencing gains the ability to process internal information in correspondence with actions and external events. In turn, a systematically referenced information-processing bias facilitates the correlation and accumulation of systematic knowledge. Then with systematic knowledge we can make informed predictions and decisions for responding adaptively to change. In contrast, in normal discourse, we tend to depend on an implicit internal referencing and affective information we learned as subjects of our experiences. Simply put, we tend to incidentally adopt a localized culturally referenced affective bias into our thinking.

The bias of a system leads it to perceive, process, and act in a particular manner (Siegel, 1999, p. 305). An idiosyncratic internal referent can negatively bias and fundamentally constrain the ability to effectively represent, perceive, and account for externally ordered processes and relations. Consistently applying natural physical laws derived by scientific inquiry provides externally biased referencing that enables orderly parsing and processing of information to systematic knowledge. When possible, scientific language facilitates informed thought and the sensible development of effective communication.

Referencing defines the upper and lower bounds of information distributions and the bias for monitoring systematic error deviations. The systematic error deviation between knowledge and natural physical laws corresponds to the fitness of the language used to represent, process, and output that knowledge. Language processing varies across different frames of reference, including absolute to relative, affective to effective, and local to global. An external systematic referent extends the plausible problem and solution space and addresses the constraints imposed by self-referenced language in terms of language extended and referenced in relation to the entire system (Bronowski, 1977, pp. 67–68). Effective language with systematic correspondence tends to counteract the affective liabilities of localized self-referencing.

Self-referenced language tends to be rigidly restricted by definition. It forms a closed processing system isolated from the larger open-ended system, which recurrently loops back to itself. In a sense, it lacks logical processing methods that enable escaping its own pull of gravity. Typically based on discrete descriptions and certitudes, self-referencing may suffice in particular localities. However, the idiosyncratic nature of localized referencing tends to render it less fit to explain other localized interactions and to account for higher-dimensional processes and relations encountered across the extent of the larger system. Externally referenced language allows an extension to open-ended general descriptions expressed according to possibilities with open-ended processing across the entire domain.

Open-endedness enables generalized tuning of cognitive processes within physical domain limits, such that linguistic representations can effectively extend and fit information to external physical relationships systematically. The physical domain limits prevent open-endedness from extending to infinity. Extended language also facilitates the processing of localized relations as special cases. Put another way, language that supports generalized processing works more effectively across the whole of the system including locally. *Extended language* as used here is equivalent to *effective language* defined as language constructions that correspond to the structural and functional ordering and metrics of the external world.

Fitter language can improve how we adjust our behavioral performance. Effective language facilitates reliably representing, processing, and understanding human experiences. When language is restricted to localized beliefs and usage—whether deliberately or tacitly via inheritance, familiarity, and habit—it excludes the possibility of systematic processing by default. Systematically oriented processing can be deliberately remediated by explicitly extending to, choosing, and using globally fit language that facilitates deliberately thinking about the world as one large, inclusive system.

We can examine and learn from the evolutionary relationship between language, human cognitive processing, and our ever-changing world. Our species uses language to process information from, understand, and interact with our environment. Language with natural correspondence translates to generally reliable information processing, prediction, and decision making. The proposed information-based brain model takes a systematic approach as a logical interaction of computational processing. The model predicts that the fitness of our language influences how effectively we

- Represent and process value
- Correlate value associations with information
- Modulate input and process output solutions
- Regulate adaptation in response to external change

One early observation from this exploration is that how we represent, process, and allocate value matters.

> Viewing our brains as information-processing devices is not demeaning and does not negate human values. If anything, it tends to support them and may in the end help us to understand what, from an information-processing view, human values actually are, why they have select value, and how they are knitted into the capacity for social mores and organization with which our genes have endowed us. (Marr, 2010, p. 361)

In light of the importance of value, this book examines how language fitness biases our understanding of information processing and the accumulation of knowledge. Can we enhance fitness by studying, learning from, and modifying our language usage to illuminate how we make sense of and know our world? Sensible language and semantics fit to effectively process information, and knowledge about relations in the ever-changing world we live in may turn out to be one of our most valuable natural resources.

We will next go over a brief history of language to help orient us to how we got where we are today.

CHAPTER 3

Informing Thinking

WE CAN EASILY FORGET or take for granted that thinking requires information. Of course, searching for, acquiring, and storing information entails expense that we can weigh against the effort required. Since its invention, language has played an influential role in how we inform our thinking. And we can compare language to DNA as modalities for passing along information that contributes to adaptability. Thus we can consider how language came about and how it relates to adaptation.

A Brief History of Language

Understanding the history behind our language, where it comes from, how it got here, and its variable processing characteristics can help us gain a language perspective for today. We can make comparisons between language, information, and the historical evolution of DNA. Current views enable exploiting language as an information and knowledge resource, as a linguistic tool for searching for novel information and solutions, and for understanding the evolutionary underpinnings of cultural beliefs and values. Even though we have accumulated a vast amount of knowledge, our overall perspectives on natural processes and relations have been slow to change over thousands of years.

We have long attempted to explain the world around us. But progress in our knowledge has not come easily. Knowledge requires scaffolding that comes together piece by piece, where discoveries ebb and flow within different cultures over the best and the worst of times. We can consider that the earliest foundation for today's modern scientific thought began with contributions from the early Greeks and Romans, who were prolific thinkers. Because of the state of knowledge available to early thinkers such as Aristotle, Socrates, and Plato, their comprehension was hampered by crude instruments, limited information, and sparse resources that literally confined them to describe the world from a deterministic view—albeit at the forefront of the reasoning curve for that period—constrained by the available knowledge sans scientific literacy (Mazur, 2007, p. 32).

The order of the universe was understood by the ancients as either immutable and fixed for all eternity or as an absolute ideal that the physical world could only seek to approach (Poincaré, 2007, p. 87). The absolute logic and beliefs prevalent at that time were based on constrained language habits that tended to favor concrete certainty and static, true-or-false ideas to explain phenomena observed in the natural world. In the view of the ancients, we live in an absolute world with an idealized referent where the past determines the future. Deterministic language and thought most likely has its roots in our earliest utterances and is marked by wording that implies static, absolute descriptions and inevitable cause and effect.

While these ancient thinkers wondered deeply about how the world and the human mind worked, they simply took their language and thought processes for granted (Korzybski, 1958, p. 371). They sought the absolute "truth" about questions regarding life, nature, and human interactions but failed to recognize the way their own language biased their reasoning and conceptualizations about those questions. Most of their particular prescientific absolute beliefs have since fallen into disfavor (Mitchell, 2009, p. 16), but absolute language and thought pervasively persists as the general rule rather than the exception.

> We must explain many things of which the Greeks were unaware. And we require that our theories harmonize in detail with the wide range of phenomena which they seek to explain. We insist that they provide us with useful guidance rather than rationalizations. (Pierce, 1980, p. 125)

Since the generally favorable intellectual atmosphere enjoyed by the Greeks and Romans, many scholars have often struggled with hostile cultures replete with superstitious beliefs and minimal literacy. One lengthy period of such struggle, from 1,000 to 1,500 years ago, is often called the Dark Ages by historians, in which intellectual pursuits had to endure cultural constraints on thinking and a brutal distaste for novel ideas. The suppression of any ideas other than dogma and the status quo was prevalent; it was the absolute rule without exceptions. The subsequent Renaissance began an era that over the last several hundred years opened the way to renewed interest in thought and discovery, developing into the science of today.

But despite many brilliant discoveries during these times, a deterministic worldview continued to appear in scientific thought, starting with Galileo and continuing on through Kepler, Copernicus, Newton, and Einstein, and even today in many fields (Miresco & Kirmayer, 2006; Mitchell, 2009, pp. 16–20). Even though they did not escape the deterministic philosophies of old, their discoveries built a platform to scrutinize and from which to spring new ideas that made way for nondeterministic thought we have available to us today. Deterministic constraints on thought habits have persisted through centuries of philosophical debate and scientific progress, and continue as the prevalent linguistic mode biasing our thought and communication. Have we overlooked something that spawns and reinforces our proclivity for deterministic thinking? Is it possible that the language we cherish—language that extols our humanity—in and of itself represents a harbinger for habits of thought that constrain us to a narrow-minded worldview?

We have the unique advantage of looking at language, thought processes, and reality through the precisely ground lens of modern science. A lens that enables a systems perspective on language may prove beneficial for constructing an open-minded worldview for today. Today's fine-grained scientific methods allow for more precise balancing between information exploration and exploitation in natural systems (Mitchell, 2009, pp. 183–184, 195). Evolutionary processes explore for and discover novel genetic encodings of solutions by exploiting open-ended search for variability. Similar to open-ended evolution, open-minded exploration is critical to discovery (Lehman & Stanley, 2008; Risi et al, 2009), since "it is difficult to know which possibilities are worth exploring without first exploring them" (Mitchell, 2009, p. 207).

Open-minded reasoning about the world with a global referent and logic based on probability provides a fine-grained method for mitigating the age-old problems and constraints attributable to traditional coarse-grained reasoning (Prigogine, 1997, p. 102). But how does this open-minded approach correlate language to information and computational processing in a perpetually changing environment, and how does it fit into evolutionary theories in a way that can help us understand our world?

In a perpetually changing environment, we develop effective knowledge of relations and predictions from a perspective of possibilities and probabilities, rather than from absolute true-or-false certainties. Since Heisenberg shocked the scientific world with his theory of inexactness embodied in the uncertainty principle, scientific thought has slowly shifted toward a probabilistic worldview (Lindley, 2007) and away from a deterministic Newtonian view. Heisenberg's revelation provided a step toward establishing probability theory as a vital component for advancing from primitive either–or reasoning, but science and society as complex systems continue to this day to struggle with the implications related to uncertainty. However, it seems reasonable to suspect that probability theory represents the more appropriate tool for understanding complex systems (Eliasmith & Anderson, 2003, p. 275).

> Probability plays an essential role in most sciences, from economics to genetics. Still, the idea that probability is merely a state of mind has survived. We now have to go a step farther and show how probability enters the fundamental laws of physics, whether classical or quantum. A new formulation of the laws of nature is now possible. In this way we obtain a more acceptable description in which there is room for both the laws of nature and novelty and creativity. (Prigogine, 1997, p. 17)

Most scientists today believe that evolutionary processes produced language and that the development of language and communication has influenced human thought and behavior. Understanding the evolution of language and how we employ it may lead to improvements in how we perceive, process, and associate value with information, how we make decisions, and how we behave. If language does influence thought and behavior,

- Could a deliberate change in our language use have a meaningful effect on our behavior?
- What kind of changes might we chose?
- How might we measure effectiveness?
- What is the best search method for finding the answers to these questions?

Searching for Novelty

History suggests that experimental efforts to evolve language, words, syntax, or grammar on demand can produce more nonsense than sense. While forcing language to evolve seems problematic, taking a novel open-ended approach to search for and evolve new behaviors, ideas, and solutions based on information, meaning, and usefulness may prove beneficial. For example, open-ended *novelty search* represents an innovative approach to search that is based on machine learning for discovering and accumulating information and behavioral solutions. Novelty search is innovative in the sense that it enables exploration in a complex problem space while avoiding the typical constraints of searching for a predetermined objective. In other words, relaxing the constraints imposed by a predetermined objective facilitates creative search (Lehman & Stanley, 2008, 2010; Risi et al, 2010).

> ...[T]hrough several experiments in interactive evolution and with an algorithm called "novelty search," a picture of innovation is emerging in which objectives...ultimately become handcuffs that...blind us to essential orthogonal discoveries on the road to long-term innovation.... [T]he silver lining is that much can be gained by liberating ourselves from the temptation to frame all our projects in terms of what they ultimately aim to achieve. Instead, with evidence in hand, we can exploit the structure of the unknown by orienting ourselves towards discovery and away from the shackles of mandated outcomes. (Stanley, 2010)

Applying such an open-ended approach to language would mean focusing on words as an information source and using their meanings to creatively search for diversity in new ideas, innovative alternative solutions, and useful applications, rather than attempting to mutate the language itself. Diversity, a fundamental component of evolution and computational neuroscience, supplies variability that tends to selectively enhance adaptability in living creatures (Jaynes, 2007, pp. 230–233; Darwin, 1859; Eliasmith & Anderson, 2003, pp. 210–217). Diversity broadens our repertoire for navigating complex environments. The diversity and complexity of our language can affect the diversification and complexification of concepts about the world through sense-making (Hashimoto et al, 2010, pp. 184–191).

In some machine learning experiments, novelty search outperforms objective search. It seems that not having a prescribed goal or predetermined objective for fitness increases the range of solutions. This means that rather than getting stuck at an unexpected obstacle encountered on a mandated route to a goal, novelty search simply tries something new, avoids obstacles, and runs into new ideas. Thus unrestricted by so-called objective fitness functions, novelty search more often avoids deceptive dead ends and seems to serendipitously find a greater number of approximate solutions. It achieves these results by exploring the search space in an open-ended manner, unlike closed-ended objective searches that often end up trapped at a local optimum (Lehman & Stanley, 2008, 2010; Risi et al, 2009). In machine learning, objective fitness functions are typically included in the algorithm that processes and evolves new behaviors and measures fitness by how well the new behaviors satisfy constraints on the problem resolution according to the prescribed objective.

Objective search, which starts with a predetermined closed-ended goal, is susceptible to deception and prone to becoming trapped at local culs-de-sac. Deception may actually prevent the objective from being reached, suggesting that the objective function does not necessarily reward the stepping-stones that lead to the objective. In contrast, novelty search, which relies on open-ended discovery processes, permits choices that diverge from prior behaviors, tends to avoid deception, and can succeed where objective-based search fails, by rewarding stepping-stones that are genuinely different as alternative jumping-off points for further evolution (Lehman & Stanley, 2008, 2010; Risi et al, 2009).

> Objective fitness, by necessity, instantiates an imposing landscape of peaks and valleys. For complex problems it may be impossible to define an objective function where these peaks and valleys create a direct route through the search space. Yet in novelty search, the rugged landscape evaporates into an intricate web of paths leading from one idea to another; the concepts of higher and lower ground are replaced by an agnostic landscape that points only along the gradient of novelty. (Lehman & Stanley, 2008)

The difference in performance between open- and closed-ended search suggests that objective search, with its imperative fitness function, exhibits a predetermined path-dependent process and, even when prescribed with the best intentions, tends to inherently bias the search space and constrain resultant solutions (Gruau et al, 1996). In a changing environment, prescribing a fixed fitness function usually fails to produce optimal results, since fitness is context specific and variable, not derivable from some predetermined function (Holland, 1995, p. 97). We can relate open- and closed-ended language to flexible and rigid search processes. We expect open-ended language to confer flexible processing fit for searching a changing environment. We expect rigid, closed-ended language to confer rigid processing constraints fit for stationary conditions. Open-ended language provides us with a unique tool, a method for searching, possibly finding, and exploiting new information and solutions.

Conceptually, evolutionary processes develop novel solutions using a flexible open-ended search method where variable individuals in a population are naturally selected as survivors. In the evolution of life on Earth, more adaptable organisms live long enough to reproduce, improving the probability that their genes will persist in the population. Evolutionary processes can also be described as reproduction with mix-and-match diversity and context-dependent selection of alternative solutions. Similar to selecting new language solutions that fit today's knowledge of the modern world, the fitness of the evolved solutions is a product of open-ended trial-and-error search processes with solutions selected in accordance to how they fit with nature.

Selection can modulate genetic diversity by penalizing variations that compromise the organism's fitness for the current environment, while variations that enhance fitness can rapidly become the norm in the gene pool. In this sense, diversity describes the variability threshold in a system (Nowak, 2006, p. 9). An organism's reproductive success and ongoing adaptation in nature under real-world operating conditions demonstrates the current sufficiency of its fitness. Likewise, we would expect fit language to correspond to processes and relationships encountered in the world in which we use it.

Learning by trial and searching commonly occurs throughout organisms from simple to complex (Jennings, 1906, p. 220). Trial-and-error learning in evolution involves elaborating on regularities that have been preserved as previously satisfactory solutions accumulated over time (Jennings, 1906, p. 322).

> Evolutionary searches, which grope blindly in a possible space of solutions to a problem, will most likely find frequent solutions, solutions realized by many variants of the same system. These are solutions with a large neutral space, solutions that can be realized by robust biological systems. (Wagner, 2005, p. 215)

Like the process of open-minded brainstorming with language, novelty search conceptually formalizes the benefit of discovering novel information, innovative ideas, and alternative solutions by generating multiple perspectives on problems through open-ended search. Open-ended language processes can enhance information search and innovate solutions, such that the results can be correlated and selected for effectiveness by comparison with an external referent that is based on general systematic principles and probability (Bailey, 2009). Alternatively, locally referenced closed-ended language can be applied but will likely limit results to absolute problem resolution and circumvent global alternatives. A language with domain-dependent restrictions that match systematic constraints can improve reliable generalization, flexibility, and coherent learning. Generalization requires a restricted (finite) search space (Nowak, 2006, pp. 284–285).

Natural language that extends referencing to the same type of external natural referent that evolution and science rely on provides a sensible and effective open-ended search process. Thus open-ended search enhances the possibility of discovering equivalent alternative solutions, novel information, and innovative behaviors. Global language supports open-ended systematic processes that facilitate exploration in a changing environment and represent a source for diverse novel information and knowledge. Open-endedness enables the exploitation of solutions based on utility as an effective value of their usefulness for adaptation.

Diversity is often equated to entropy by ecologists (Solé & Goodwin, 2000, p. 31), and diversity appears to influence increasing complexity. Diversity and open-ended search processes are natural components of evolutionary complexity. Novelty search climbs the ladder of complexity, and once simple behaviors are exhausted, novelty leads to more diverse behaviors that reveal more complex solutions. Open-ended search has been associated with the arrow of complexity in nature (Lehman & Stanley, 2008, 2010) and offers an analogy to information search. Open-ended search processes that explore for diverse information potentially increase search and information entropy by maximizing useful information (Jaynes, 2007, p. xxiv; Jaynes, 1981; Shannon, 1948).

Preferentially searching for and maximizing available information produces novel discoveries by searching beyond familiar policies and ideas (Holland, 1995, pp. 57, 163). Open-ended search processes encourage the possible discovery of global solutions and more sensible domain knowledge by promoting search beyond the peaks at local optima.

> It is better to search far and wide and eventually reach a summit than to search narrowly and single-mindedly yet never come close. (Lehman & Stanley, 2008)

Somewhat like an open-ended fitness function, as an open-ended product of reproduction, DNA is subject to heterogeneous genetic variations that can serve as alternate solutions. Via natural selection, the environment bypasses variant phenotypes that inhibit adaptation, along with the genes that produced them. On a kind of pass-or-fail basis, the organisms that pass get an opportunity to hang around for another reproductive cycle, suggesting that the fitness function itself has evolved as an adaptive product of natural selective processes in relation to the domain.

Researchers in the field of evolutionary computation have experimented with artificially evolved languages and found that evolving the ability to communicate can indeed enhance fitness (Cangelosi, 2001), and researchers have been working to implement open-ended language and communication in experiments with autonomous robots (Steels & Spranger, 2009; Spranger et al, 2010, p. 297). Language influences fitness by making possible the open-ended evolution of cognitive processing that leads to heterogeneous solutions, which can represent a creative resource for novel ideas and innovation. If searched for, these heterogeneous solutions can provide information and alternative choices for finding, exploring, and exploiting natural regularities in the problem space, potentially yielding more adaptive learning rules and global solutions (Becker, 2007, pp. 1–7).

Novel ideas and concepts, generated by the words that construct their meaning, form an abstract bridge of current information that appears to flow by inference in both directions between the future and the past. Ideas can promote novel open-minded thinking in the present; contribute to simulating, planning, and predicting solutions for the future; and refine and process past information into domain knowledge. The information and knowledge can then be passed on to future generations, somewhat analogous to the transfer of genetic information by evolutionary processes (Ayala, 2009, pp. 132–133).

Genes, Language, and Information

The analysis of DNA has revealed major evolutionary innovations in the past three billion years (Kirschner & Gerhart, 2005). The universality of the genetic code argues for a single origin (Ellner & Guckenheimer, 2006, p. 253). DNA establishes the relatedness among all living things, comprising a discrete and virtually irrefutable record. Drawing from basically the same handful of nucleotides, each species' unique DNA sequence represents an inventory of all the genes involved in building and operating that organism, and with few minor exceptions, the building blocks of similar features appear relatively similar across species (Carroll, 2006, pp. 33, 75).

Complex systems routinely use building blocks to generate internal models (Holland, 1995, pp. 36–37) as demonstrated by DNA. In an abstract sense, these building blocks exploit the natural regularities discovered within the constraints of a complex environment. In evolutionary terms and perhaps as a form of functional modularity, building blocks represent pieces of knowledge "that works" (Murre, 1992, p. 104). They have been found by selective processes according to how well they work or perform, thus contributing to their fitness value. For example, in evolving organisms, the conservation of direct neuronal pathways provides evolutionary scaffolding "that works" from which to further elaborate indirect pathways for processing more complex behavior. New inventions in evolution are built on the back of previously useful functions (Buzsáki, 2006, p. 115).

Genes and proteins contain macromolecular information in much the same way that language and words contain abstract information (Ayala, 2009, pp. 132–133, 136). Even though many scientists consider language an emergent form of nonbiological cultural evolution, language represents a truly human invention, possibly "the biggest invention of the last 600 million years" (Nowak, 2006, p. 249). While we share the majority of our genetic makeup with chimpanzees (Carroll, 2005, pp. 268–270), by providing species-unique linguistic symbols for cognitive representation and processing, language capacity sets human primates apart from all our primate cousins (LeDoux, 2002, p. 198; Broca, 1877).

> Natural selection is one form of interaction between brains that gives rise to the general capacities and predispositions of the brain of a given species. Additionally, given the predispositions an individual has at birth, interactions between conspecific brains as well as between contraspecific brains have much to do with the particular skills, knowledge, and representations an individual brain acquires. For example, the particular language a human learns may empower or limit him, both in how he interacts with other humans, how he solves certain kinds of problems, and how he thinks about things even when not explicitly conversing. (Churchland & Sejnowski, 1994, p. 445)

Similar to DNA's role as a genotypic substrate for phenotypic fitness, the symbols of language provide a scaffold for representing, processing, and building higher-order cognitive abstractions for thought and communication. Structurally, scaffolding is a supportive framework that can support materials, tools, agents, language, and social and cultural processes (Wimsatt & Griesemer, 2007, pp. 227–229). Genetic information is encoded in the form of specific linear sequences of four different nucleotide building blocks in the linear DNA molecule that develops living cells with three-dimensional structure, parts, and components. In effect,

> DNA is translated to proteins that form the three-dimensional structure determining the qualities that create life. (Nüsslein-Volhard, 2006, pp. 35–38)

The simple linear chemistry of DNA's one-dimensional information is processed and translated into three-dimensional protein structure that interacts with the environment. The proteins can recognize each other and group themselves spontaneously into precisely reproducible assemblies because they fit each other in only one specific way (Lehninger, 1975, p. 13). The proteins in living cells can be thought of as the semantic structure that gives logical relational meaning to the process of life as we know it.

> At no point in our examination of the molecular logic of living cells have we encountered any violation of known physical laws, nor has it been necessary to define new ones. The machinery of living cells functions within the same set of laws that governs the operation of man-made machines, but the chemical reactions and processes of cells have been refined far beyond the present capabilities of chemical engineering. (Lehninger, 1975, p. 13)

But how does cell biology help us to understand language? We see that the analogy is anything but tangential because the language of cells is the language of nature. If we look at strings of linear information that make up the syntax of language, we see a similar phenomenon. A one-dimensional language string can easily be translated to structures that fit a three-dimensional world. With appropriate semantics the same syntax can be translated to a three-dimensional world with one-way time dependence and dynamic nonlinear logic. Language describes not only processes found in the molecular logic of the living cell but also the processes found in the semantic logic of living human beings. We would expect processes related to the molecular logic of cells to generally hold for language and also for neural syntax, the language of the brain. Indeed, we see that DNA and RNA are read (translated) in the direction from past to future, analogous to neural syntax.

Our language's construction bestows functional fitness properties that bias the logical effectiveness of our thought processing and problem-solving abilities. Perhaps we can take a hint from the molecular logic in living cells by logically adapting our language constructions to fit the structure of nature such that we can redirect and enhance our thought processing, our cognitive fitness. In turn, enhanced cognitive fitness allows us to

- Improve systematic learning and error-correction skills
- Increase our understanding of cooperative relationships
- Increase the reliability of our predictions
- Modulate our adaptive behavior more effectively

If we make use of language that logically corresponds with the molecular logic and systematic regularities of relationships and processes in the natural world, we can generally expect more effective adaptation to that world. Adaptation, at least in part, can be thought of as a product of how effectively we choose and use the language with which we process information.

> Still another goal [of biochemistry] is the biochemical analysis of neurofunction from the level of simple intercellular communication upward to integration, to memory, to behavior, and ultimately to thought—indeed, all the profound questions posed by man's quest for understanding of his nature. . . . [T]he organizing principles that constitute the molecular logic of cells should serve as a frame of reference. (Lehninger, 1975, p. 14)

Over the last half billion years, natural selection has shaped the vertebrate brain into a remarkably complex organ for acquiring, analyzing, and responding to information from the physical and social environment (Coolidge & Wynn, 2009, p. 57). It appears that fine-tuning adaptive brain regulation increases survival odds in specific environments. Higher functions like language and reasoning evolved as the neural mechanisms that provide that regulation (Moisl, 2001, p. 455).

> Higher-level cognitive processes, also referred to as symbolic processes, typically include memory, attention, imagery, ideation, concept formation, generalization, abstraction, problem solving, thinking, reasoning, and planning. (Logothetis, 2004, p. 849)

In humans, language supports, reinforces, and extends these processes. Once we could speak, we no longer benefited from perpetually larger brains, instead improving the adaptation of the species through functional improvements in our language faculty (Striedter, 2005, p. 322). Human primate communication differs from other animals' by its use of symbolic references involving complex language, and symbolic representation makes it possible to name and describe the environment in terms of categories and objects, their relationships, and their relatively invariant features (Cangelosi, 2001). "Good" symbols, those most fit to the problem, are distinguishable, constructible, compact, and efficacious (Gallistel & King, 2009, p. 72). As a system of codes that designate external objects and their relations, language helps us arrange our experience of these objects into particular systems of categories, leading to the formation of abstract thinking (Luria, 1981, p. 30). In effect, language gives rise to a new evolutionary modality, as we spread our ideas and concepts rather than our genes (Nowak, 2006, pp. 249–251).

Evolution depends on DNA to transfer information to following generations in the form of genes (Nüsslein-Volhard, 2006, p. 7). Through natural selection, DNA accumulates prior information as knowledge about "what works" in the world. Surviving organisms embody a genetic catalog representing world knowledge, which they pass along to their offspring in the form of traits coded in DNA (Floreano & Mattiussi, 2008, p. 5). Evolutionary processes work through information transfer from one individual to another and from one generation to the next (Nowak, 2006, p. 249). Similarly, language and culture make it possible for humans to acquire information and knowledge, passing along the "inherited experience" and beliefs of previous generations (Korzybski, 1958, p. 231; Luria, 1981, p. 35), which is accomplished by replicating, translating, and preserving information in memory as cultural "knowledge."

Science, which is among the greatest enterprises of humankind, also catalogs information about the natural world to share across human populations and pass along to subsequent generations or, put more formally, operates at the population level as a coevolutionary cultural process. More importantly, science works by observing and trying to find out how things operate in nature, "how things really are, not just how they appear to be" (Bloom, 2002, p. 169). Accepting that we interact and relate with the world by accumulating knowledge about it provides reason to develop a plausible basis for this "knowing."

The following chapter addresses the motivation for pursuing this knowledge and further evolutionary correlations with language from a macroscopic perspective, or the big picture.

CHAPTER 4

The Big Picture

WE USE LANGUAGE TO represent and process acquired information into knowledge. The usefulness of the resulting knowledge directly relates to the fitness of the language we use to process and develop that knowledge. Unfortunately, like the proverbial elephant in the room, this fundamental processing relationship remains largely unstudied and overlooked despite its relevance to fitness. To see the elephant we can deliberately apply systematic principles and processes to reason more effectively and insightfully about our own use of language.

The Elephant in the Room

The elephant in the room corresponds to the implicit way we use language without seeing, examining, or understanding how the structure of our language biases what we see and value, what we attend to, and what we ignore. We look outward from our own locally biased internal viewpoint from habit but overlook the role of language in shaping what we see and value. Like a nearsighted person viewing nature without corrective lenses, we run the perceptual risk of myopically interpreting our observations as unassailable, absolute facts about the world without recognizing their shortcomings. The process of thinking is sensitive to the fidelity or accuracies of the language. Language inaccuracies result in processing distortions from perception to behavior in general and in speech behavior in particular.

Language-related processing described in terms of fitness biases how we know and relate to our world. Understanding and applying language to our thinking in an orderly manner corresponding to nature can provide more optimal insight, as if by donning special glasses we gain clarity with which to represent, filter, and interpret what is seen and view the world more broadly, from an open-minded, macroscopic perspective. Effective thinking correlates with language that explicitly fits our world.

> Each of us filters our interactions with others through the lenses of mental models created from patterns of experiences in the past.... Knowing about implicit memory allows us the opportunity to free ourselves from the prison of the past. (Siegel, 1999, p. 34)

Practically speaking, language represents the perceptual lens through which we implicitly and explicitly filter and process information about nature. A local lens or filter gives us a locally biased view. A global lens explicitly extends our view to global dimensions. Like prism glasses, inaccurate language use and the worldview that it creates tend to misinform and distort our perception of that world. The use of scientifically based language allows for corrections or reductions in potential distortions. With closed-ended processing features and absolute semantics that resist fine-tuning, localized language promotes rigidity and a static, narrow-minded worldview and arbitrarily boxes us into a narrow microscopic aspect of our world. Closed-ended thinking amounts to putting on blinders that constrain what we attend to and gives us tunnel vision (Macknik & Martinez-Conde, 2010, p. 60).

Perspective describes the breadth of the aspect window from which we view the world and filter information by both implicit and explicit processing. From this view, we associate value to relevance with stimuli, which in turn biases attention, the assumptions and inferences we make, and what we learn. Unattended information usually fails to be fully processed and is therefore "unlikely to enter awareness" (Vuilleumier, 2009, p. 221). At the bottom line, the breadth of our linguistic perspective matters because it places constraints on informing our attention, thinking, and learning.

Open-ended language enables and encourages cognitive processes conducive to flexibility and innovation. We can think creatively by taking a broad-minded worldview with open language that recognizes and facilitates adaptive change. We can replace rigid, closed-ended thinking and empty command shells, such as *should* and *must*, with informatives and interrogatives that express preferences, opinions, and questions. We can consider the language conundrum by asking ourselves,

- Can we find effective alternatives with an efficient amount of cognitive and physical effort?
- Do we have other options for elaborating and solving this particular problem?
- Does a different representation of the situation reveal an enlightened perspective?
- Does an open-ended approach offer advantages over the way we have handled problems in the past?

We can open our mind to new ideas by restricting our use of the narrow categories *true* or *false*. In effect, trading true-or-false lenses for open lenses admonishes the self-imposed linguistic constraints that tend to limit how we look at and perceive and learn about our continuously changing world. If we could know the "absolute truth," there wouldn't be any more to learn, but in an ever-changing domain, we will likely get along more effectively with flexible ideas and open learning behaviors. Establishments, cultures, groups, and individuals seldom support pursuing novel ideas and knowledge when composed of those who sincerely believe that they are already in possession of the truth (Jaynes, 2007, p. 613). "It is impossible for a man to begin to learn that which he thinks that he knows" (Epictetus). However, an open-minded culture and language spurs thought processes conducive to innovation and learning.

As previously noted, by adopting some processing methods from science and staying current with the latest available knowledge, we can deliberately reformulate the language we use to represent and think about the world in a way that accounts for change and probability. In a changing environment with an evolving future, we can make more effective predictions by contemplating their potential outcomes and phrasing them in terms of possibilities and probabilities rather than certainties. In a dynamic world where certainty falls short, we benefit from probabilities that can change as we change our state of knowledge.

> ... [S]cience has reformulated the laws of physics in accordance with current models of the open, evolving dynamic universe. (Prigogine, 1997, p. 108)

Deliberate reformulation of language to account for probability challenges the absolute view that the world is simply black and white. Common language typically supports dynamically unsupportable absolute true-or-false dichotomies:

- "Either you are with us or against us."
- "We are right and you are wrong."
- "There is only one right way to do this."
- "We must do it this way."
- "We have no other choice."

We tend to tacitly assume there are only two possible options and only one correct choice. These are examples of one-dimensional thinking. The benefits are that one-dimensional language seems almost effortless, fast, and frugal. Reformulation leads to possibilities and probabilities that extend the plausible dimensions for thinking and evaluating problems and solutions. Of course the downside is that possibilities and probabilities require information and effortful cognition, which is slower and more expensive. We live in a high-dimensional world with continuous change, which results in information distributed in an array of changing gradients and many shades of black, white, gray, and continuous spectrums of alternative colors. In a colorful changing world, black-and-white processing constrains the range of accessible information.

There are many possible ways to do things and many equivalent solutions. We are all differentially informed, and the more we know about alternatives, the greater the probability that we can choose more effective solutions for a given problem. Instead of simply telling others that their solutions are wrong, we can take their statements as their opinion. Searching for and accumulating plausible information enables evaluating and choosing solutions that are based on the weight of available evidence for probable effectiveness; criteria for evaluating potential solutions are listed in Table 1.

Table 1. Criteria for Assessing Possible Solutions

From what evidence was the solution derived?
Are the underlying assumptions explicit?
Is the solution supported by the assumptions?
Does the solution address the input of current information and knowledge?
Was the solution derived from reliable information?
Was the information derived from a reliable source?
Does the solution explicitly address context?
Does the solution make sense in the context of the problem?
Is the solution's domain made explicit (open versus closed)?
Are there competing conventions and equivalent alternative solutions in the same domain?
Does the processing of the solution take time into account (one-way time versus two-way time)?
Is the logic of the processing made explicit?
Is the logic effective in the domain that the problem addresses?
Does the logic have adequate dynamic range to address the problem domain (dynamic to static)?
Does the solution address the statistical nature of information as possibilities and probabilities?

We prefer that the conclusion [or solution] represent a preponderance of the available plausible evidence and meet criteria as probably relevant (Jaynes, 2007, p. 507).

> The actual science of logic is conversant at present only with things either certain, impossible or entirely doubtful, none of which (fortunately) we have to reason on. Therefore the true logic for this world is the calculus of Probabilities, which takes account of the magnitude of the probability which is, or ought to be, in a reasonable man's mind. (Maxwell, 1850)

It is easy to assume that being certain about our beliefs is better than being uncertain by expressing them as possibilities. Research shows that people who have not explored and acknowledged the limits of their understanding tend to express undue certainty (Chabris & Simons, 2010, p. 148). Probability addresses the statistical constraints of our knowledge, giving us the option of substituting possibilities and probabilities for certainty in our communications. For example, meteorologists, who realize the probable limits of their knowledge, tend to issue predictions to account for those limits, such as "There is a 75% chance of rain tomorrow."

We make better sense when we talk about our dynamic world in terms of informed possibilities rather than imaginary idealizations of absolute certitudes. For example, at some time in the distant past, some humans believed that demons caused all illness and imagined the only remedy was exorcism. Language that posits rigid belief in a single, unqualified cause for all illnesses narrows the potential for alternative explanations, and an absolute statement of a single solution closes the door on thinking about new alternatives. Today we have the option of updating our language to inform and include a scientific understanding of nature and illnesses we incur.

Thanks to pioneering scientists such as Louis Pasteur and Robert Koch, we currently understand that many infectious diseases result from the transmission of pathogens—such as bacteria, parasites, or viruses—knowledge with which we can make predictions about and generally prevent or effectively treat many of the illnesses. Scientific predictions describe expectations for outcomes as uncertainties derived from knowledge and logical reasoning about various possibilities and probabilities. Yet, outside of science the understanding of illnesses to this day often represents opinions based on superstitions, guessing, or speculation but erroneously expressed as certainties—even though the outcomes have a range of uncertainty. We see this phenomenon persisting in cultural certitudes that preempt an understanding about the possibilities and probabilities found in nature, even though they better fit the changing world in which we live.

Things change over time, and by including a probability measure for predictions we acknowledge the limits of our current state of information and knowledge and its application to future states. Since things change, nonabsolute words such as *many* and *some* help to keep the door open to novel explanations, more precise models of infection, and more effective treatment alternatives. Using relative terms helps to limit the statement to predicting probabilities based on what we believe that we know about how things work, rather than appealing to imaginary certainties. In turn, we then leave room for flexibly updating and revising the future state of treatment as we acquire more current information. But our language can constrain how flexibly we think, how well we predict and choose, and how effectively we adapt to change. The next question to address is how these constraints come about.

We learn our native language as children by absorbing the ambient cultural beliefs, values, and logic systems from the surrounding culture. This cultural infusion carries predetermined values and hand-me-down beliefs encoded as "knowledge" and transmitted as informational building blocks somewhat analogous to physical genes. We inherit not only cultural beliefs but also the language, semantics, and grammar used to formulate those beliefs (Nowak, 2006, p. 263). Cultural learning inevitably influences the way children perceive the world and organize knowledge and behavior—their conceptual system (Feldman, 2006, p. 187).

> Language is good for preserving cultural values in the long term.... Maintaining a culturally-important feature in language use is a good way to ensure its longevity. Further, language ensures universal distribution because learning to speak one's language is a non-optional part of growing up to be a full-fledged participant in a culture. This ensures that any cultural value manifest in language will receive universal distribution within the culture. (Fausey et al, 2010, p. 9)

Our linguistic inheritance directly influences our logic and reasoning abilities (Donald, 2008, pp. 191, 196) in that language defines cultural belief systems (Luria, 1981, pp. 205–209) and social structures (Mitchell et al, 2006). As prior information, these cultural beliefs [implicitly and explicitly] influence the thinking, reasoning, and behavior of the cultural group (Adolphs, 2006, pp. 269–274; Boyd & Richerson, 2005, p. 206; Browne & Keeley, 2007, pp. 53–69; Phelps, 2004, p. 1008). Early and typically uncritical absorption of information influences the scaffolding that structures our brain's linguistic operating system and, whether we realize it or not, sets our initial conditions for how we think about and interact with the complex world around us.

Small errors at early stages can insidiously compromise long-term outcomes and adaptation by dampening learning and robustness to perturbations. It is difficult to add information if our scaffold is rigid and bare of flexible connectors with which to associate new information. We are then susceptible to early constraints on our thinking and learning that leave us clinging to what we believe that we know. But old information may not keep us up to date on what is going on in the world around us today, which can adversely influence whether and how we adapt to change.

As adults, we tend to tenaciously cling to our acquired beliefs as if they were possessions (Jones, 1998, p. 45). Cultural beliefs and values provide an information-priming effect by replicating, translating and interpreting, and representing our prepackaged, preprocessed "knowledge." These predetermined building blocks include tacit assumptions with hidden variables that bias our intrinsic hypotheses, our logic, and the recipes we use to think about how the world works. Perhaps incidentally, we tend to automatically incorporate information, beliefs, and affective values gleaned from the prior cultural experiences of our ancestors into our current worldview.

Cultural inheritance furnishes a significant amount of resilient affective information that imparts a long-lasting cognitive bias to our thought processing, to what we value, to our inferences, decisions, and choices; to our detection and correction of systematic errors; and to how we learn, behave, and adapt to change. These inherited beliefs demonstrate an inherently subjective bias derived from isolated subpopulations with variable histories. The histories have been subjected to inheritance and modification by the collective and subjective inferences of imperfect individuals with a wide variety of world experiences and varying descriptions of their past and the future (Prigogine, 1997, p. 2). These beliefs and affective values are formed from segregated world experiences, typically without benefit of systematically evaluated efficacy. Faulty beliefs and the decisions based on them affect long-term outcomes, not only for our everyday personal lives but also for societal decisions that can impact all humans (Kida, 2006, p. 15).

We tend to think and communicate by habit using preprocessed affective values and inherited beliefs as prior "knowledge." Since we lack an understanding of the linguistic constraints imposed by the implicit encoding, we accept the entire language program by default without questioning the source or inspecting the plausibility and fidelity of the underlying information. Unlike most cultural knowledge systems, science prefers deliberate, explicit processing, repeated systematic inspection of the reliability and validity of conclusions, and measuring results analytically according to effectiveness. Visible and plausible information and explicit supporting evidence for ideas and premises help make reliable introspection possible.

> Science is experiment; science is trying things. It is trying each possible alternative in turn, intelligently and systematically, and throwing away what won't work, and accepting what will, no matter how it goes against our prejudices. And what works adds one more piece to the slow, laborious but triumphant understanding of our world. (Bronowski, 1977, p. 5)

Solutions to scientific problems qualify as such if they satisfy exacting criteria. These constraints, however, do not eliminate creativity—they provoke it (Prigogine, 1997, p. 188). Science puts a high cultural value on effective language, creativity, and innovation by applying logical methods to explicitly expose and limit beliefs deceptively posing as systematic knowledge. Science is open to new ideas and relies on transparency and recurrent inspection of natural ecological relationships; it strives for effective information processing to develop orderly knowledge by asking why and how the world works as a whole.

Science is not about increasing disbelief but instead about decreasing our ignorance about the world in which we live. The practice of science offers reliable processing methods that encourage innovative learning and the searching for knowledge. We can assimilate these innovations into our informal language to enhance our skills for thinking, inquiring, learning, understanding—and for better knowing our world.

> The true opposites of belief, psychologically considered, are doubt and inquiry, not disbelief. (James, 2007, v. 2, p. 284)

Ignorance about how nature works creates systematic blindness by default and leads to deceptive traps at isolated regions of the landscape. *Ignorance* can be loosely defined as a lack of knowledge or learning, the tendency to refrain from noticing or recognizing information, or specifically lacking knowledge, information, or learning—that is, unlearned, uninformed, or unnoticing. These definitions do not separately describe whether ignorance is deliberate or automatic but rather assign a correlation with learning, recognition, knowledge, and an informed state, which can be deliberately extended and applied systematically.

The caveat is that what we ignore inversely correlates with what we pay attention to and thus influences learning. Then we can say that knowledge is indirectly rather than directly associated with our ignorance via attentional processes: What we refrain from noticing or recognizing contributes to what we believe we know and what we believe we don't know. In this sense, ignorance is intimately associated with our attentional processes. Explicit introspection of discrepancies between our knowledge and our ignorance encourages deliberately adjusting our attention to increase our rate of learning.

Lacking insight of language-related problems blinds us to the source and accuracy of our beliefs (Benjafield, 2007, pp. 267, 400; Shilpa et al, 2007, p. 53). Typically, language involves inherited assumptions, potentially faulty premises, and hidden intuitions (Benjafield, 2007, p. 27) exacerbated by the illusion of introspection. We think we have insight and use language adeptly (Polya, 1954, p. 7), but we usually overlook the assumptions, hidden variables, and affective value biases quietly nestled in our language and inherited beliefs. Just as the uncritical use of sight can blind us to vision (Lytton, 2002, p. 20), inherited language assumptions that are unacknowledged and unexplored can deafen us to our habitual speech behavior. "... [L]anguage is as good as the assumptions underlying it" (Surindar, 1947).

Since language learning occurs long before we have the intellectual resources to question the fidelity of the content, we do not know to seek new perspectives that might plausibly lead to the discovery and correction of any errant cultural "facts" and the reasoning errors behind them (Jaynes, 2007, pp. 62, 133). Since we unknowingly inherit a prejudicial linguistic bias, we are deprived of an easy way to see our way out of the traps in our language, and we miss the global picture that surrounds us.

> Prejudice, the inability to carry out certain experiments, and the traps of language have all made it difficult to tease out the connections between mental events and events in the nervous system. (Edelman, 1992, p. 7)

The unexamined assumptions found in inherited language and variant cultural beliefs also represent a large and significant elephant, but its presence and any systematic distortions typically and inexplicably escape detection. Our human brain and the language we use can both lead us to imagine we see solutions and elephants when none are there, and then overlook solutions as large as elephants when they are in our presence, right in front of our nose. We can likely find better solutions if we take a new approach to language use that integrates and explains the relationship between information and knowledge by illuminating the mapping between the language system itself and the natural world (Christiansen & Dale, 2003, p. 606).

If we accept Alfred Korzybski's axiom that "the map is not the territory" but rather an abstract representation of it, then it follows that we will likely find our way around the territory more effectively if we learn how to make and maintain more accurate, more coherently correlated maps and take into account the dynamics of our changing environment (Korzybski, 1958, p. 750). When changes in the natural environment occur, organisms that fail to adjust their behavior to the new conditions generally do not survive and propagate (Johnston, 2008, p. 41). When changes in cultural and linguistic environments occur, failing to adjust our linguistic behavior effectively to the newer conditions may still suffice, and we may survive, but we may find our lives constrained, our options less than optimal, and our preferred outcomes thwarted.

We can consider the information we absorb as logically mapping to knowledge about our environment according to ecological validity (Brunswik, 1955). We can acknowledge the relevant value of information fidelity for enhancing our capacity to explicitly understand, explain, predict, and deliberately deal more effectively with the world's uncertainties. Effective use of language supports reliable information and knowledge, which enables us to maintain more coherently correlated maps. We can use this acquired knowledge as a foundation of prior information for acquiring even more knowledge (Browne & Keeley, 2007, p. 3).

How could we miss seeing an elephant in the room? Traditional wisdom claims that "seeing is believing." Given the ease with which we fall for magic and myths, we might alternatively say "believing is seeing." But what we see and what we believe are two sides of the same coin, with both affected by the amount of local distortion we incur internally from our perceptual processing bias. We see things that we have learned to expect and tend to ignore new data that are inconsistent with prior information (Jaynes, 2007, p. 133). We develop these habitual expectations in large part by what we say, hear, and learn, and by how we assume things work. Unthinking use of language can misinform us by distorting what we see and believe; how we think, feel, and behave; and how we perceive that the world works. Typical human behavior hardly requires a thinking brain. Our habitual behavior tends to automatically bypass the option of a deliberately thinking brain, at least until after the fact or when the wheels fall off.

Deliberate use of language that relies on accurately describing our world and our relationships can help us tune our perceptual processing bias to develop consistent internal models that focus on how the natural world works systematically. Systematic thinking is paramount to working toward a more harmonious coexistence.

Modeling Processing Fidelity

An organism's brain could be described as a world model, encoding many deduced rules about the regularities that govern the outside world and also represent that world's dynamic aspects (Braitenberg, 1984, p. 72). If we can develop a realistic understanding of how language maps our interactions to our belief, value, and logic systems in terms of learning, and how we obtain and process information into knowledge, we can better understand the human species "on its own terms" (Kirschner & Gerhart, 2005, p. 273). As Sean Carroll says (2006, p. 93), "If you want to get inside an animal's mind, it helps to see the world through its eyes."

We can model language and human behavior that applies global information, knowledge, and cognitive accuracy to recalibrate human language to represent a common scientifically generated referent for thinking, making decisions, and communicating as accurately as possible (Bailey, 2007b, 2009). Scientific culture supports a coevolutionary process for generating and referencing the fidelity and effectiveness of scientific knowledge. Fidelity and effectiveness come from how coherently a language corresponds to relations in nature. A natural referent establishes orderly systematic mechanisms and effective processing methods for sorting and matching information and knowledge to relations in nature, and it yields logical implications.

Choosing language and information fidelity that corresponds with a common natural referent in general terms provides a wide range of benefits individually, interpersonally, socially, and globally. Deliberate application of a natural referent can lay the groundwork for improving harmonious correspondence among information, variant beliefs, values, and biases. Such groundwork promotes policies that support optimizing effective communication, cooperation, and collaboration, thus increasing the possibility of a more peaceful and humane coexistence among disparate cultures.

A model with an accuracy-based referent incorporates logical statistical processing methods for identifying, correlating, and coherently representing systematic regularities found in nature as realities. These realized regularities represent our most efficacious understanding to date but are open to revision, as necessary, to account for new discoveries.

The value of a reliable referent lies in reducing the unnatural, unsupportable inferences that we might make about our observations and expectations of the external world, by establishing limits for evaluating how much we imagine about our experiences. In contrast to the fluctuating variability of accuracy for localized referents, invariant referencing enables generalization across the domain and limits acceptable information to reliable domain-dependent correspondence. These systematic limits constrain the evaluation of events and regularities as much as possible to deductions that can be reliably replicated or correlations that show relative invariance over time and across members of a class, category, or population. With relatively invariant external referencing, we can compare the accuracy of one brain's experience with that of another and more consistently evaluate discrepancies in correspondence among cognitive processing, information, knowledge, and nature. Natural referencing enables us to get a better notion of how effectively our knowledge relates to the outside world in a systematic manner.

Not all languages are created equal or at least not applied equally. Global language with invariant referencing to the physical world is the exception. Localized language constructions typically default to idiosyncratic cultural referencing with widely variable biasing of the world view. Cultural referents demonstrate an inherently subjective bias. Inferences based on localized language with variable and discernible differences in the fidelity of their information content create a range of individual and cultural biases (Prigogine, 1997, p. 2).

In the absence of a direct method for comparing individual experiences inferred from outside events, language can be applied to abstractly represent, encode, decode, and process those experiences. We are unable to touch abstract relations in the outside world directly, so we use language as an indirect processing method to help us figure out if what we experience logically corresponds with that world. The same language that biases our experience also provides an abstract medium for resolving the referencing problem in a relatively objective way. By establishing an explicit systematic bias that allows coherent calibration of our language, information, and knowledge, we can more reliably correlate what we experience to regularities in nature and natural physical laws. Deliberately applied systematic referencing can serve as a perceptual corrective lens for biasing the inferences we draw from observations of our experiences with that world. Coherently referenced language enables a reliably correlated history of domain regularities and experiences fit for explicitly representing systematic knowledge.

Language that accounts for external constraints on accuracy makes it possible to consider the effect of those constraints in our verbal deliberations and gives us a systematic tool for detecting evaluation errors. But rather than relying on accuracy, most predominant language models seem to ignore systematic efficacy and depend on localized language constructions to represent and define experiments and to evaluate the validity and relevance of experimental results. Local models tend to rely on simple relations between meaning and form, and omit referencing the language system to the world (Christiansen & Dale, 2003, p. 604). Such an omission creates isolated information distributions based on variant and transient regularities found in acquired cultural information and assumptions, which are generally irregular from a nonlocal perspective (Nicolis & Prigogine, 1989, p. 69).

Localized language neglects explicit pattern recognition of systematic information distributions. The localized constraints on explicit information can impede external risk assessments and error detection, and lead to ignoring actions that irreversibly harm the environment. We fail to notice the inherent risks associated with implicit thought processing that lacks explicit information. By default, local models exclude the identification and application of systematic language constructions that can extend our cognitive processing and analysis to statistically distributed information. For example, we tend to place more emphasis on our local risks and the "need" to consume more and more of a limited energy resource, while we ignore the larger systematic risks of the Earth's capacity to support a rapidly expanding population. We have a tendency to underweight what we perceive as highly unlikely or rare events we have not experienced, such as a tepid response to long-term environmental threats. "And when there is no overweighting, there will be neglect" (Kahneman, 2011, pp. 332–333).

We often hear about the importance of including various stakeholders in different ventures. Unfortunately, we forget about the primary stakeholder, Mother Nature. Taking into account relative harmony with nature allows us to consider deleterious effects of our actions, which have two-way ecological effects. When we take advantage of nature—as if its resources are a free gift and ours for the taking—such as by depleting fossil fuels, we can incur cost directly by destroying a part of nature and indirectly by polluting and contaminating our future and our children's future. For example, let's think of our world as a small spaceship that has a limited carrying capacity. We immediately notice all involved—we are all in it together. As the limited resources on a spaceship will eventually run out, this analogy exposes our immediate "needs" and local preferences that deal with our self-induced consumption and encourage conservation. In the same light, it reveals the concomitant risks of running out of energy that highlights the larger long-term risks of systemic neglect and abuse. We easily lose a systematic perspective when we reverse the analogy and extrapolate the size of the small spaceship to macroscopic world dimensions with which we can erroneously imagine unlimited resources and a vast world immune to physical insults. We often hear the affective decree "We *must* have more energy to meet our *needs*!"

> Living organisms create and maintain their essential orderliness at the expense of their environment, which they cause to become more disordered and random. (Lehninger, 1975, p. 8)

We are small parts of the environment rather than its master. In other words, "It's not all about me but about all of us." Whether or not we choose to ignore our own role as human beings that live there in the overall system, we are still responsible for our choices, deeds, and errors and are accountable for our part in creating those risks in the first place. When we borrow from the future, someone will eventually pick up the tab and pay the price, whether it is us, our progeny, or the grim reaper. Better resolution of this evolutionary dilemma might come from resetting our goals and expectations and prioritizing a systematically thoughtful relationship with nature. Even though we can conclude that it is unlikely that nature has feelings per se, we do best to consider mutually sensible interactions so that all of the constituents involved are as content as possible. Extending our language to a macroscopic perspective helps us to realize and consider the outcomes of our systematic errors.

How can we notice and decrease systematic errors if our language constructions are mutually exclusive of systematic information representation? Put another way, how can we inform ourselves systematically when we automatically and inadvertently ignore systematic information. Our language constrains the breath of our thinking and our ability to assess our own responsibility and accountability for our systematic errors and thus we fail to consider the long-term maladaptive effects.

Most of the language we use in everyday conversation and in many written forms inadvertently neglects the systematic representation of information and knowledge in dimensions of effective accuracy. This neglect compromises cognitive fitness, since representation influences not only the accuracy of the meaning we assign to environmental events but also the reliability of our thought-processing abilities, predictions, expectations, and thus our overall fitness. We also neglect overwhelming evidence that brains evolved systematically as general processing organs that operate probabilistically rather than absolutely. Ironically this predisposition for probabilistic neglect occurs even among researchers, whose scientific expressions about a complex world are typically constrained by localized language and absolute logic. These constraints seem to be compounded by the compartmentalization and attenuated communication within and across disparate scientific fields (Eliasmith & Anderson, 2003, p. 2).

We see communication difficulties related to word use, definition, and meaning across many scientific and academic areas. For example, disputes persist concerning whether something exists separate from the body that contributes to the body's experience—that is, whether mind has a separate existence from body. As with many other preconceived cultural arguments, at base mind–body concepts directly correlate to the dichotomous language structure that tends toward opposing or polarizing views. Of course, polarization does make it simpler to weight information, but at the expense of accuracy. Various isms typically describe the phenomena of affective experience in a polarized manner with poorly classified and locally defined semantics that presupposes the existence of the underlying phenomenon a priori.

Science attempts to mitigate such polarized thinking by minimizing the use of wording that presupposes the existence of referents without corroborating evidence. Take this sentence, for example: "The disastrous earthquake was the work of demons." We have physical evidence for earthquakes but none for demons. Arguing whether the demons caused the disaster remains moot until substantiation of their existence. In the absence of extraordinary evidence for supernatural forces, systematic knowledge of natural forces, processes, and relations offers more plausible explanations.

The mind and body conceptualizations demonstrated by the previous example rely on an insistence that the experience of seeing or believing that we sense something unusual that we cannot explain represents new knowledge not attributable to known physical facts. The notion of the independence of mind and body implicitly references these dichotomous concepts, definitions, and explanations to support the existence of a nonphysical dimension and underlying mysterious processes. Such an explanation requires unnatural contravention or suspension of known processes related to physical laws. Generally we would prefer more evidence for claims that exceed known physical relationships. Ambiguously referenced claims that exclude plausible information about what is known tend to lack redeeming scientific merit and are unlikely to contribute to our systematic understanding of nature. This will hardly poses a dilemma for those who believe that our experiences refer to the physical world (Miresco & Kirmayer, 2006; Mitchell, 2009, pp. 16–20). Carl Sagan concludes, "Extraordinary claims require extraordinary evidence" (Sagan, 1980).

If we consider the neurological phenomenon of sight, unusual sensations that we "feel" and causally relate to our experience can be more accurately described as new instances of physical, previously experienced inputs or, in some cases, spontaneously generated internal sensations. We know that input can incur affective variations in perceptual processing that take place internally in brains with genotypic and phenotypic variations. Also, in some cases variations in perceived input may relate to underlying, undiagnosed internal disturbances such as tumors, infarcts, and seizures. Different people see things differently, which is not surprising, since variability is a fundamental component of evolutionary processes. By deliberately modifying our choice of language constructions from affective to effective, we find that the perceptual dilemma evaporates. Changing the language changes the problem representation, available explanations, and the constraints for solutions. Orderly words with coherent domain equivalence reduce ambiguity, facilitate problem resolution in correspondence with nature, and increase effective communication.

> There is nothing wrong, of course, with using terms inherited from philosophy and psychology, as long as we do not forget that these are hypothetical constructs. After all, it is the verbal terms that allow for conversations among members of a discipline and that convey messages across the various scientific fields. However, this communication works best if we are able to create a structured vocabulary that restricts terms to unambiguous meaning that can be objectively communicated across laboratories, languages, and cultures without prior philosophical connotations. (Buzsáki, 2006 p. 22)

Locally referenced reasoning based on hand-me-down beliefs, logic, and predetermined values—used since antiquity with limited examination for natural fidelity—unduly constrains our systematic understanding of natural physical processes. While science has incorporated uncertainty and probability into its information processing methods, the application of uncertainty and probability principles outside hard science remains elusive. Not surprisingly, most people find uncertainty less appealing than certainty. The tendency to prefer what is perceived to be certain over the ambiguity of the merely probable has been termed the *certainty effect* (Kahneman & Tversky, 2000b, p. 20) or *ambiguity aversion* (Fox & Tversky, 2000, pp. 528–529). Constrained language paradigms based on certitudes that ignore systematic fidelity continue to be the general rule rather than the exception.

Even though deterministic language is logically insufficient for understanding relations in the world we live in (Bronowski, 1977, p. 61), few seem to recognize the value or even the possibility of extending language by applying probability; it seems that fewer still find reasons to attempt such a change. In somewhat the same way a novel species faces a struggle to survive, novel theories struggle to attract attention in the shadow of better-known, familiar, and widely accepted theories. However, as ecologist and entomologist Deborah Gordon points out, ignoring potentially useful theories for the reason that they do not fit a preexisting model makes little sense (Mitchell, 2009, pp. 293–295).

A linguistic model based on semantic fidelity makes logical sense, because by definition semantics represent the informational building blocks of language with which we relate to our world. Given the complexity of our world, we lack the resources to perceive everything around us, so we perceive some of the available inputs and miss others. While the phenotypic expression of our genes structures some of what we can and cannot perceive, experience plays a role as well. We encode, evaluate, process, and understand our experiences in large part with language. How accurately and reliably we reference, represent, and process our experiences will influence how we recognize those experiences in the future and how we think, feel, and act in the meantime.

Effective language and semantic processing enables our brain to reliably analyze and transform accumulated information into systematic knowledge. At least in part, perception consists of affective input constrained in scope by what we have learned or inferred in the past as subjects of our experiences. It follows that language influences behavior by biasing inferences we make about relationships experienced with the outside world, and those inferences interact circularly with what we see and attend to, perceive, and believe. How effectively we process our experiences influences our intended goals, the accuracy of conclusions we make about the intentions of others, our emotional state, and our subsequent responses.

> Although common wisdom has long recognized that emotion may alter sensation as well as reason, the recent insights from cognitive and affective neurosciences have now clearly shown that this notion is not just a metaphor. (Vuilleumier, 2009, p. 240)

Research on congruity effects suggests that there is overlap between the representations and attention processes involved in understanding language and the processes involved in interpreting visual images. It appears that representations created as a result of processing language can have broad effects on perception. In this case, processing unrelated language can change not only how quickly or accurately we can respond to a visual stimulus but also how qualitatively we change what we interpret the stimulus to represent (Dils & Boroditsky, 2010, p. 887).

Perception is not simply a feed-forward process of sensory inputs but rather an iterative interaction between exploratory/motor-output-dependent activity and the sensory stream (Buzsáki, 2006, p. 228). Language influences the perception of experience and informs the transformation to useful knowledge. The forming and storing of individual human experiences creates a knowledge base, a unique brain-based context that modifies the way the neocortex processes future sensory experiences and contingencies that affect subsequent actions (Buzsáki, 2006, p. 277). This interaction is unique in the sense of the expanded neocortex and language abilities in the human brain that enables abstract knowledge storage, which, in turn, influences our brain's ongoing perceptual processing and the adaptive effectiveness of the responses we produce.

The usefulness of transformed information for analyzing decision outcomes and evaluating constraints on solutions largely depends on the accuracy of the transformation (Schlosser, 2007, pp. 119–124). And efficient cognition relies on the accuracy of the available information (Anderson et al, 2002, p. 506). This analytical utility also applies to language constructions that influence the representation and processing of information and the efficient and effective transformation of that information to systematic knowledge. In turn, the fidelity of our linguistic representations biases our cognition and how accurately we measure the effectiveness of our reasoning process, if we consider it at all. Understanding the relevant relationship among effectiveness, systematic knowledge representation, and our affective value bias helps to clarify the modulating influence of language fidelity on logical thought processes. Language fidelity plays a role in the recurring loop between our sensorimotor system and nature that influences cognitive fitness.

When it comes to accuracy, what is measured, how it is measured, what referent it is measured against, and the measuring instrument itself all have significant ramifications for the results obtained (Lindley, 2007, p. 155); measurement is a means of communication (Prigogine, 1997, p. 150). As Poincaré (1905, p. 208) noted regarding the importance of finding systematic errors rather than averages, "the physicist [is] persuaded that one good measurement is worth many bad ones. . . ."

Fundamental language and semantic fidelity provide the basis for a reliable cognitive-processing model that includes systematic referencing for effectively representing, measuring, and processing globally sensible knowledge. A fidelity-based model enables the calibration of language, logic, and behavioral output correlated to the dynamics and constraints of the domain. Abstractly, this process follows the model of molecular biology, where the foundational rules for transferring information by DNA apply to organisms from simple single- and multicellular microbes to complex plants and animals, including humans. Stuart Kauffman has said (1993, p. 404), "We may find that *E. coli* and IBM do indeed know their worlds in much the same way." And Jacques Monod points out, "What is true for *E. coli* is true for the elephant" (Nüsslein-Volhard, 2006, p. 36).

In summary, the relevance of effective language processing may turn out to be the unnoticed elephant in the room. The elephant will remain invisible as long as we fail to understand how effective language informs our thought processes and overall cognitive fitness. We can conclude that the fidelity of our language directly influences how we perceive, value, and relate to our world. When we rely solely on language with constraints imposed by localized semantics, we not only ignore any local ones but "absolutely" fail to see the global elephant. In contrast, by deliberately choosing and using extended semantics, we can tune our language processing to explicitly represent, accumulate, and integrate higher-level global abstractions and systematic descriptions. We can then adjust and accommodate our expectations, perceptions, and attention to more likely notice and bring the elephant into better focus.

But what kinds of obstacles do we face when we attempt to search for reliable information that improves our understanding the nature of language and semantic processing on a systematic level? Are there other constraints that literally hamper our view? If so, can we overcome them? The next several chapters address these questions from a perspective of biases and systematic errors.

CHAPTER 5

Systematic Biases I

SYSTEMATIC ERRORS ARE KNOWN as biases, and they recur predictably in particular circumstances. The idea that our minds are susceptible to systematic errors is now generally accepted (Kahneman, 2011, pp. 1–10). Our behavior arises from the interplay among environment, experience, and the intrinsic properties of our neural circuits. The systematic interrelations of biasing constraints from brain, language, and environment make it difficult to artificially isolate one from the other. But looking at them individually in the context of correlated functional models allows us to take a system perspective of many parts influencing the whole. We can think about how the brain works from a perspective of the brain's architectural constraints, and how these constraints influence the types of language that can functionally overlap with cognition and brain structure from a brain's-eye view.

But we get a different perspective from the language software view as to how tuning the linguistic parameters can influence the effectiveness of the brain's output. We know that language and the brain co-evolved in the natural world. So it makes sense that by observing behaving organisms with brains and language operating within the physical constraints of natural physical laws from a bird's-eye view, we get a more global perspective.

The lower-order limbic system apparently evolved earlier in the progression to higher-order processing seen in the linguistic brain. We can view language functionally according to the different architectural constraints that influence the integration of the upper and lower brain structures and neural circuits in relation to interactions with the structure of nature. In other words, we can take a lower-order affective, higher-order effective, and integrated view of the brain from a functional systematic perspective. Then we can take a broader view to evaluate how our language, environment, and brain work together logically in the most effective manner.

In the next three chapters we will address several questions:

- What holds us back from understanding and behaving like we humans and our brains are all related?
- What holds us back from understanding that our behavior shows that we are all flawed and fallible humans?

- What holds us back from understanding and behaving like we are each responsible and accountable for our own thoughts, feelings, and behavior?
- What role, if any, does the evolution of the brain and the language that we inherited play in all of this?

As noted in Chapter 1 ("Relations and Processes"), we can think of the human brain as inherited hardware. And unlike computer hardware, which is hardwired to support certain operations, brains have the advantage of plasticity and learning during development and throughout the life of the owner. In other words, the neural wiring of the brain's hardware is to some extent modifiable and plastic. The nature of the hardware—anatomy, neuronal architecture, physiology, and chemistry—contributes to structural and functional constraints and biases that influence our cognitive development and thought-processing abilities, as does our limbic value system and our culturally inherited language constructions.

The biases covered in this chapter include

- Limbic bias
- Familiarity bias
- Status quo bias
- Developmental bias
- Comparison bias
- Blindness and confabulation bias

Biases can operate persistently, intermittently, and recurrently but, depending on the context of a situation, tend to differentially influence the brain's effectiveness as an interactive input-computation-output system. We can look at the functional aspects of individual biases and better understand their profound influence on how we think. But we run into a similar problem of using language to evaluate language as we do when we use the same brain to evaluate how the brain works, which can likewise bias what we discover. And who wants to admit their language or their brain is biased and imperfect? We will deal with language as a tool for understanding the brain and the biases we encounter by framing the discussion in reference to nature.

Limbic Bias

The implicit nature of lower-order limbic cognitive processing in the human brain presents an obstacle for recognizing liabilities associated with affective language constructions. The human brain evolved from less-complex organisms, and lower-order brain components tend to operate implicitly, automatically, and habitually. The absolute logic favored by our ancestors coexisted with and apparently produced language shaped by primitive limbic brain structures. Clearly language evolution was "shaped by the brain" (Chater & Christiansen, 2008, 2010), and languages seem to adapt themselves to a pattern dictated by human mental abilities, by the structure and organization of the human brain (Pierce, 1980, p. 245).

As a result, we inherited language constructions with absolute logic primitives that habitually pervade our thought processes. The force of habit and hidden assumptions embedded in our language constrain the transition to language constructions with updated logic to match the world in which we live. But our everyday language resists change; even change that can help us better understand our emotional limbic system remains thwarted and hidden from view, as if invisibly embedded in our limbic language habit.

> The core function of emotional systems is to coordinate many types of behavioral and physiological processes in the brain and body. In addition, arousals of these brain systems are accompanied by subjectively experienced feeling states that may provide efficient ways to guide and sustain behavior patterns, as well as to mediate certain types of learning. (Panksepp, 1998, p. 15)

We have an intrinsic emotional-affective system that has been well studied and described. The automatic lower-level survival-oriented limbic system exhibits commonalities between humans and other mammals (MacLean, 1990; Milgram, 2004; Panksepp, 1998). The limbic system includes the amygdala, which is involved in emotional processing, encoding salient memories, aggression, and response to danger (Mesulam, 2000, pp. 58, 90). Such ancient brain functions evolved long before the emergence of the human neocortex with its vast cognitive skills (Panksepp, 1998, p. 4). Many of our psychological constructs, such as emotions and feelings, boredom and passion, love and hate, attraction and disgust, joy and sadness, arise in the limbic system (Buzsáki, 2006, p. 280).

Stimuli are ordinarily evaluated subjectively in the limbic system, with assigned significance ranging from "feels good" to "feels bad," or they may be dismissed as neutral (Siegel, 1999, pp. 47, 124; Spitzer, 1999, pp. 300–301). The limbic system provides affective value tags from the organism's individual perspective that impart subjective meaning to incoming signals as afferent input from stimuli. These tags identify internal attractor sources for associating positive and negative values with stimuli, which results in movement toward or away from the stimulus, depending on state conditions (Siegel, 1999, p. 124). Positive and negative affective credits bias what we learn about the environment and how we modulate our behavior in that environment. Depending on experience, the limbic brain associates positive and negative value with meaning for stimuli and memories, intentions, behaviors, and outcomes, including perceived errors.

> ... [E]motional feelings not only sustain certain unconditional behavioral tendencies but also help guide new behaviors by providing simple value-coding

> mechanisms that provide self-referenced salience, thereby allowing organisms to categorize world events efficiently so as to control future behavior. (Panksepp, 1998, p. 14)

The most basic domain for communication in animals seems to be *emotional space* (Gärdenfors, 2010, pp. 407–410), or affective value space that includes positive and negative valences. In most simple interactions with our environment, we associate a space with position and the objects in that space to which we assign value. As notorious bank robber Willie Sutton said when asked by reporters why he robbed banks, "Because that's where the money is!" Willie was apparently attracted to the fundamental value that we associate with money and found it hard to resist that much value contained in one small space.

Emotions and values intersect where value corresponds as a fundamental affective component of emotion, modulating state changes for adaptive interactions such as avoiding dangerous or threatening situations and moving toward choices that enhance survival (Platt et al, 2008, pp. 135–136). Emotions can be classified according to positive and negative state changes that inhibit some behaviors and amplify others, and emotional states are associated with positive and negative valence (Platt et al, 2008, p. 136). Our emotional state influences perception and decision making by amplifying or dampening positive or negative affective attributes of stimuli input. The resulting value biases cognition, inference, and behavior, since emotions influence the relative "goodness" and "badness" afforded to stimuli.

"Goodness" or "badness" denotes the positive or negative significance of a stimulus relative to the organism, associated with the *affect heuristic*, in which we use our likes and dislikes to determine our beliefs about the world (Slovic et al, 2002, p. 397). In humans, stimuli values generally result from default cognitive processing by discrete absolute logic, in either–or, true-or-false, good-or-bad terms. Commonly, we have little if any knowledge about discrete logic or where feelings come from. Our emotional feelings are associated with the lower-order limbic brain and the evolution of complex higher-order interactions with prefrontal neocortex and language. The limbic system has a profound influence on our reasoning and regulation that includes inference, perception, predictions, decisions, and our emotional reactions to stimuli.

Many cultures still hold to the traditional analogy that the heart is the sentimental seat of emotions and that we feel with our hearts rather than with our heads. This popular view ignores ample evidence that the limbic brain produces emotions through the assignment and management of emotional valence (Buzsáki, 2006, p. 16). Cultures also ignore how the language we use influences our thought processing, feelings, and behavior.

> Men ought to know that from the brain and thence [from the brain] only arise our pleasures, joys, laughter and jests, as well as our sorrows, pains, griefs, and fears.

> Through it, we think, see, hear, and distinguish the ugly from the good, the pleasant from the unpleasant. (Hippocrates, quoted in Adams, 1972, pp. 344–345)

Without the benefit of understanding the goings-on inside our own heads and the multiple levels of cognitive processing that affect our values and behavior, our worldview can easily become distorted. Affective distortion can impair fitness because of spurious predictions, actions, and outcomes based on limbic reactive behaviors and their confusing consequences. Distorted values affect the reliability of our knowledge about the world we live in—and without reliable knowledge the conduct of nature appears merely a result of capricious events (Poincaré, 2007, pp. 77, 85).

The limbic-amygdala-related value system complicates our ability to understand our emotions, since limbic inputs have an implicit influence on our thinking. The limbic system operates as a general default system for affective processing that tends to operate automatically and implicitly with variable subjective influence on our actions. The lower frontal lobes, collectively termed the orbitofrontal cortex (OFC), automatically perform goal-related utility calculations via limbic system interaction with current input and our learning history, which includes prior reinforcement.

Situational evaluations result from computation of affective values in relation to our behavior and relative rewards, benefits or gains, and punishment, penalties, risks, costs, or losses. The values used in those computations derive from behavioral experience, which implies that at this affective level, the past determines the future. Accumulated positive-reinforcement values tend to ratchet upward, converging and normalizing to a mean value (Gilovich & Griffin, 2002, p. 5) called habituation, which is state-dependent according to deprivation and satiation (Skinner, 1953, pp. 72–81).

We tend to interpret negative or relatively decreasing positive values as losses or punishments depending on their magnitude relative to expectations. This predisposition has been described as *loss aversion* (Kahneman & Tversky, 2000b, p. 42, 2000a, p. 481; Loewenstein & Prelec, 2000, p. 568). These losses accumulate historically according to the magnitude of the negative values (Jennings, 1906, pp. 262–264, 293–297, 310, 338–339) and generally degrade gradually over time, a process often referred to as *graceful degradation*. Limbic value provides an inferential processing method with automatic logic that in the short term habituates to positives and in the long term, depending on availability, leads to a general accumulation and consumption of positives and avoidance or elimination of negatives.

Yet we recursively interact with the outside world, and eliminating negatives makes sense, but eliminating negative feedback leaves recurrent positive feedback unchecked, which can lead to runaway expansion, consumption, and catastrophic failure of the system. A general systematic bias for increasing rewards and decreasing penalties makes sense for an organism's safety and long-term survival but can increase susceptibility to overconsumption, such that we can probably add *consumption bias* to our list of systematic biases.

The limbic system operates fairly simply by routinely adjusting and resetting value references from prior norms and the normalized rate of change. Population norms, individual norms, and values change over time along with the expected rate and direction of change. The reset norms become the "entitlement" referent for expectations of behavioral outcomes "that must not be infringed" (Kahneman, 2011, p. 306). We see this in situations where cutbacks in entitlements to which people have become accustomed lead to discord and even violence. Depending on circumstances, animals commonly react unfavorably to another animal taking away or blocking access to a valued resource for reward or curtailing their range of actions (Panksepp, 1998, p. 54). Normalized values, and the behavioral expectations that accompany them, create an egocentric affective bias for risk–reward valuations. But norms, means, or averages fail to capture the extent of the parameter space, which can lead us to ignore crucial information about the dynamic range of fluctuations and variability in the system. In turn, averages constrain the effective range for assessing expectations and risks related to naturally fluctuating resources of rewards and penalties.

The limbic value system apparently evolved as a frugal way to measure and adapt to changes. With more highly evolved language and analytical processing abilities, we can measure change in relation to effective adaptation, which incurs a greater cost. We run into problems when we attempt to measure from static-like norms with commonly used absolute language. For example, a recent storm was predicted to result in an inordinate amount of rain on local communities in its path. It deposited up to five feet of water in some areas that were normally dry. A reporter standing in water well above the waist stated, "The water level isn't what it *should* be," which might have been a semantic faux pas, but the statement seems to assume that normal equals static.

If we were linguistically constrained and lacked any alternatives except for "shoulding" on nature, we might at least say that the water level is where it "should be" if we consider the amount of rainfall and the storm water system's drainage capacity. It seems more accurate to say that we sometimes face higher than average water levels in a world where change and instabilities represent the norm. We can safely conclude that over the history of the Earth, water levels change. Why should water levels always stay normal? Perhaps we at least might understand that no amount of wishing, pleading, chanting, or shoulding on nature can control the behavior of rainfall or water levels. But the limbic system tends to get affectively stuck in the static past and assumes that the norm predicts the future, consistent with routine limbic control. The limbic system apparently evolved as an automatic fast and frugal "smart enough" processing and behavioral control method for enhancing survival and reproduction with average sufficiency.

The limbic system operates in a world where one event precedes the other, producing a bias of linear causality that tacitly assumes the past predicts the future, which fails to account for coincidental occurrences except by retrospective rationalization. The *illusion of cause* bias arises because our brains are built to detect meanings in patterns, to infer causal relationships from coincidence, and to assume that earlier events cause later ones. This bias is related to the *gambler's fallacy*, the illusion that knowing the past endows the ability to predict the future (Macknik & Martinez-Conde, 2010, p. 195). We are more likely to see patterns when we think we understand what is "causing" them, and more likely to see patterns consistent with our intuitive beliefs (Chabris & Simons, 2010, pp. 153–155, 160). In other words, we tend to assume causality and intuitively find what we are looking for or expecting, while ignoring or denying nonconforming information and our systematic bias.

We see cases of blaming vaccines as the cause of unrelated illness that is in reality due to coincidental occurrences. Typically the blame relies on intuitive assumptions alone, without evidence or statistical correlation for calculating an association with risk. For example, an individual anecdotal report claims a child suffered from "mental retardation" after getting vaccinated, and others claim that vaccinations "cause autistic behavior." These claims persist along with ambiguous assumptions and emotionally appealing declarations that the vaccine "could potentially be a very dangerous drug," while ignoring the absence of any plausible scientific evidence for the claim and the overall benefits of the vaccine.

> ... [P]arents, the media, some high-profile celebrities, and even some doctors have fallen prey to the illusion of cause. More precisely, they perceive a pattern where none actually exists and confuse a coincidence of timing for a causal relationship.... Like the illusion of knowledge, the illusion of cause can only be revealed by systematically testing our understanding, exploring the logical basis of our beliefs, and acknowledging that inferences of causality might derive from evidence that cannot really support them. The level of self-examination is one that we seldom reach. (Chabris & Simons, 2010, pp. 174, 178)

The limbic system predisposes us to the *illusion of cause*. For a simple automatic system that tends to err on the side of caution, it suffices for the limbic system to operate as if caution were a correct assumption; after all, it has previously worked. When it comes to survival, the evolutionary cost of many false alarms that erroneously impute negative affect to causality of an event is relatively inexpensive compared with a miss resulting in a terminal error (LeDoux, 1996). The limbic system reflects the old adage "Better safe than sorry," which seems to work sufficiently well on average. Implicitly confusing chronological order with coincidental causality (Chabris & Simons, 2010, p. 167) makes for a fast and frugal automatic system in which the static past implies prediction of the future as a deterministic chain of causality. But the limbic system is oblivious to the implication of equating the past and future and assumptions of time flowing both in a forward and backward direction. As simply as before and after, the past and future are tacitly perceived as symmetrical, interchangeable, and equivalent. As a consequence we easily assume that the past predicts the future, where prior events appear to cause the effects that follow. Unfortunately, we are generally oblivious to the erroneous *illusion of cause* but apparently incur minimal immediate cost.

> It can be cheaper to occasionally make a mistake than to have more costly neural tissue and never err. (Heerebout, 2011, p. 148)

Implicit determination by causal assumptions may suffice in simple systems. Causal assumptions, though, fail to offer explanations for influences in time-dependent complex systems that display periodicity and oscillations (Buzsáki, 2006, pp. 5–10). Deducing causality is particularly difficult when the cause involves a reciprocal relationship between parts and wholes, as is often the case for neuronal oscillations and other properties of complex systems (Buzsáki, 2006, p. 10). The point made here is that the limbic system—oblivious to the direction of time—takes the inexpensive way out, which influences the compatibility of the language primitives we construct to explain relations we encounter in our world. We wind up with limbic-compatible affective language that is implicitly ignorant when it comes to applying one-way time. In this case, compatibility means the degree to which our primitive language software matches the architecture of the primitive limbic brain.

But the limbic system operates as a fairly simplistic processing system in which the accumulated history of events and coincidental affective values enable an inexpensive associative link for integrating action and value weights with the domain. Limbic integration operates automatically and subjectively and therefore is not readily amenable to deliberate analysis without an understanding of how our language, brain, and nature work together systematically. Higher-order analysis with effective language allows us to predict systematic influences and better understand brain and environmental interactions.

> To predict well [reliably] even when circumstances change we must use nonlinear prediction (Pierce, 1980, p. 213).

When we fail to account for probability, we easily succumb to a linear view that discounts context and blurs temporal distinctions between the past and future. Thus, we compromise our abilities to effectively evaluate, predict, and manage behavioral alternatives in response to change. In contrast, one-way time provides a natural temporal framework that distinguishes the static past from the dynamic future. Probability and forward-directed time naturally account for statistical predictions and behavioral risks associated with irreversibility, uncertainties, and instabilities that we experience from past to future.

> The messages of nature have to read in the direction of time which we experience. (Bronowski, 1977, p. 54)

The primitive limbic system provides a subjective platform that influences the evolution of language software used to represent information and reasoning (Lytton, 2002, pp. 81–82) and by default predisposes us toward two-valued either–or logic (Jaynes, 2007, p. 114). However, top-down regulation with extended language, logic, and systematic information facilitates effective rather than affective cognitive processing. Bottom-up affective processes, or *limbic symmetry,* include automatic lower-level limbic evaluations based on

- Beliefs and qualitative affective values
- Absolute logic with closed-ended processing
- Implicit two-way time

Top-down processes include the extended capability for *analytical symmetry* that enables deliberate higher-order linguistic evaluations based on

- Systematic knowledge and quantitative effective values
- Open-ended processing with probability as logic
- Explicit one-way time

Understanding how the brain works helps to dampen speculation about harmful effects we might incur from using analytical language and speech. Concerns that analytical thinking might turn us into emotionless robots or zombies are likely scientifically misinformed. It is unlikely that any amount of analytical language could annihilate the limbic system altogether. Language does not lobotomize our limbic system but instead establishes civilized internal policies and constraints and that can facilitate the humane modulation of our actions. In other words, analytical language can enable global knowledge for more effectively modulating our limbic-affective system. In turn, we broaden our understanding of imperfect human beings as a static class of *Homo sapiens* with divergent behaviors and the abilities to foster global cooperation.

Limbic symmetry automatically matches internal affective states to external reinforcement via absolute logic. The resulting assumptions typically defer to others' responsibility for actions and disturbances that we perceive as "causally" related. Analytical symmetry deliberately matches internal representations with external systematic relationships via nonabsolute logic. Global principles of influence replace idle speculation and rumors of causality. Language and global knowledge unique to humans enable analytical symmetry unavailable to other mammals. Symmetry breaking simplifies identifying information by distinctive differences of uniform features such as logical asymmetry between analytical values and limbic emotional values.

From a cognitive-processing view, language and logic are largely inseparable; language shapes thought, and thought shapes language (Spitzer, 1999, p. 235). Analytical asymmetry allows us to logically distinguish the asymmetrical forward flow of time from past to future and to live more effectively in the present. Language makes it possible to shift out affective limbic bias toward a more effective analytical bias.

Familiarity Bias

Familiarity bias presents another obstacle for understanding the nature of our brain and language behavior. By default, we sort information by comparing it with what the brain assumes we already "know." But if applied unthinkingly as a habitual heuristic, this familiarity-qualified information can trap us by suppressing the exploration for novel and potentially innovative alternative solutions that can help us deal more effectively with our everyday problems. There is a relevant distinction between habitual automatic thought and deliberate thought, where the former is generally implicit and the latter explicit. Language plays an influential role in explicitly mitigating or implicitly reinforcing our default biases. Deliberate use of effective language allows us to question the relevance of the familiar past and search for novel, creative solutions. Affective language reinforces the habit of familiar limbic solutions.

> The attraction of familiar experiences may have far-reaching consequences. Humans who have been exposed to certain ideas often begin to prefer those ideas. Perhaps a more poignant example of this is the ease with which people can defend ideas they have held dear since childhood or even for a short while, and how easy and natural it is to contradict the new ideas of others. (Panksepp, 1998, p. 259)

We see a distinction between the influence of the familiar past and that of the unfamiliar future on the logical input-to-output processing of information for predicting, deciding, and behaving. Distinguishing between the familiar and the unfamiliar allows us to differentiate between old and new information (Nicolis & Prigogine, 1989, p. 144). For resolving problems and comparing "what's going on in here" with "what's going on out there," it seems to make economical sense for the brain to first go effortlessly to existing information from past experience and passively default to habit and the familiarity of memories.

Passive automatic processing typically sorts new and old information as a habitual method that on average provides for sufficient adaptation. Understanding the distinction of passive biasing on our habitual perception enables deliberate analytical circumvention by actively suppressing or disassociating affect from incoming and prior familiar information in memory. The disassociation or suppression of affect enables a relatively neutral cognitive environment for analyzing the current incoming unfamiliar information in comparison with accumulated systematic knowledge. The effectiveness of the analytical processing and derived solutions proportionally correspond to the fidelity of the applied knowledge.

> Most of us grow more and more enslaved to the stock conception with which we have once become familiar, and less and less capable of assimilating impressions in any but the old ways.... Objects which violate our established habits of "apperception" are simply not taken account of at all.... Genius, in truth, means

little more than the faculty of perceiving in an unhabitual way. (James, 2007, v. 2, p. 110)

Prior information stored in our internal memory from experience can circularly bias reasoning, reasoning can bias incoming information, and both in turn bias perception, inference, decisions, and subsequent behavior. Deliberately applying language and semantic fidelity to upper-level regulatory processes enables explicit feedback for effectively evaluating and modulating our behavior and accumulating reliably informed knowledge. The deliberate logical processing of novel situational information with updated knowledge can override the default to habitual limbic regulation that favors familiarity. Deliberate upper-level cognitive processing can extricate us from lower-level limbic habits that perpetuate passively anchoring us to outdated beliefs from the past. We apply these familiar habits to the present and justify them with absolute language primitives. Kahneman and Frederick (2002, p. 51) and Sloman (2002, p. 383) separately describe upper- and lower-level cognitive models and reasoning systems in terms of *heuristics* and *biases*.

In a broad sense, the effectiveness of our upper-level cognitive processing depends on our choice of language constructions. Affective language supports lower-level regulation that typically binds us to the past through familiar emotional utility values automatically derived from our inherited learning and reinforcement history, sentiment, and habit. The upper analytical system relies on logic and semantics acquired from and embedded in our repertoire of language constructions. Extended logic constructions that match the dynamics of nature facilitate flexible interactions, alternative possibilities, and novel solutions for living adaptively in the present. Effectively matching language constructions to natural physical laws with an extended logic function facilitates analyzing and sorting information into more ecologically useful knowledge.

We easily mistake familiarity for reliable knowledge, a process described as the *illusion of knowledge*. We fall prey to the illusion of knowledge when we fail to recognize the importance of questioning our own knowledge (Chabris & Simons, 2010, pp. 121–123)—that is, what we think we know. Recognition of familiarity bias gives us an opportunity to question and evaluate the reliability of our knowledge, the logic behind our decisions, and the magnitude of our ignorance. We have an unprecedented linguistic ability for acquiring knowledge as well as for acknowledging our ignorance. But with the brain and affective language constraints we inherited, we seem especially adept at denying the latter.

We can mitigate ignorance by insightful thought processing with language, semantics, information, and knowledge based on systematic coherency and the value of extended logic rather than familiarity and absolute logic primitives. Language primitives support absolute logic processing with absolute contradiction and negation as the primary test of validity. By defaulting to absolute logic, we implicitly assume that information and beliefs can be identified as absolutely true or false. Without deliberate intervention, we tend to tacitly default to an absolute logic bias that routinely processes absolute solutions to our everyday problems. But we live in a constantly changing multidimensional world, where absolutes create one-dimensional decisions and solutions that constrain context and narrow our perception. Limiting context leads us to perceive that we live in a one-dimensional static world governed by the "absolutely true" or the "absolutely false."

The terms *true* and *false* represent polar opposites by definition. But this simple yes-or-no framework has no place for shades of doubt, ignorance, uncertainty, or probability (Jeffrey, 2004, p. 1). When structured by absolute logic, arguments to determine belief relevance can only converge to either "absolutely true" or "absolutely false." The familiarity of our internal frame of self-reference is typically reinforced by affective language primitives that contribute to shaping and perpetuating our individual value bias and behavior.

Furthermore, familiar value bias and absolute logic strongly resist new, unexpected, unusual, or unfamiliar views, which we perceive as strange and unknown. The resistance to unfamiliar views correlates with fear of uncertainty and dampened exploration for new ideas that imposes an anticreativity bias. Absolute logic augments the default tendency to negatively misperceive novelty as ambiguous, unfamiliar, dissimilar, or unsafe. This misperception distorts how we relate to novel objects and information, beliefs and values, and behaviors. We tend to relegate the unfamiliar to the status of irrelevant, threatening, or risky. We can then dismiss the unfamiliar by passively ignoring or actively rejecting it. This coarse method of using absolutes forces the classification and sorting of beliefs and values into either the familiar currently held "true" or the polar opposite unfamiliar "untrue." In effect, we tend to erroneously assign a probability of zero to any possibility that the unfamiliar might be "true" (Jaynes, 2007, pp. xxv, 343–355).

Such a situation is reminiscent of the story of the intoxicated fellow searching the ground under a streetlight on a dark night. A passerby asks what he is looking for there.

"I'm looking for my car keys."

"Where did you lose them?"

"I lost them about a half a block down the road."

"So why are you looking here?"

"Because this is where the light is."

It seems somewhat natural that we would first search locally where we have the best visibility under the spotlight of familiarity and certainty where we can see what we think we know. The alternative seems rather risky, searching aimlessly and blindly in distant darkness with our visibility further obscured by the unfamiliarity of the unknown, uncertain, and vague or imaginary shadows, even though better solutions might likely lie there.

Affinity for the familiar clearly limits our ability to discriminate among unfamiliar objects. In other words, our individual local connections tend to be the strongest and most familiar from our own perspective. Thus, the unfamiliar is subjected to amplified discrimination of differences and prone to weaker detection of similarities in comparison with the familiar. In turn, this default sorting typically subjects the unfamiliar to a higher probability of rejection. Language familiarity also trumps the unknown uncertainty and strangeness of the unfamiliar and by default blinds us to the realities of the value of extended language that could broaden our knowledge and enable a more insightful worldview. In effect, we evade reasonable introspection (Jaynes, 2007, p. 372).

When a language model finds itself on a local peak on the fitness landscape—where familiar absolute beliefs, values, and objectives suffice—it likely ends up trapped there by default. Absolute strategies fall prey to the fear of ignorance that constrains the ability to undertake open-ended exploration into the dark nonabsolute valley of the unknown and uncertain to search for possible novel solutions—at least until a catastrophe strikes. We easily become prisoners of our own subjective history. Both change and the accumulation of knowledge are difficult effort-wise and evolutionarily expensive. A shortage of knowledge structure marginalizes transitional scaffolding for building, informing, and innovating change that increases the probability of effortlessly staying close to home unless perturbed.

Evolutionary processes tend to create a distribution of possible solutions even though one may surpass the others under current conditions. Most utilitarian animals, including amoebas, ants, and humans, will relocate when the positive benefits of moving outweigh the negative cost of staying (Jennings, 1906, pp. 24, 38–39, 67; Nicolis & Prigogine, 1989, pp. 232–238). However, even though individuals or groups may move to other localities, they still face the constraints of their phenotype and familiar behaviors. Similarly, we humans face the constraints of localized language primitives, at least temporarily, wherever we go. Without open-ended options and updated language structure, we tend to default to the familiar and cycle back through the same rigidly ingrained processes at a different locality, sans knowledge of how the system works. But we have the capability of tuning our language to escape absolute constraints and break out of the familiar. Systematic knowledge and open-ended communication with preferences enables a creative approach with which we can mitigate the default bias of familiarity.

Default sorting of familiar and safe versus unfamiliar and dangerous serves to deftly preserve internally referenced familiar language, along with polarizing constraints that can deceptively trap us with localized cultural solutions. Linguistic behavior emanates from language absorbed into fundamental brain circuitry, which biases its default value. As a result, literacy and education alone do not necessarily improve our thinking. Rather, effective change benefits from deliberately shifting our preference globally toward systematic literacy by applying and reinforcing extended language for optimizing the analytical effectiveness of our thinking and behaving. Changing language behavior equates to action.

As imperfect humans, including those trained in research, all of us remain susceptible to our human condition of having an assuming brain prone to habit and familiarity bias. We can make an explicit effort to deliberately apply effective language constructions such as the ones we would apply to our analytical methods, experiments, and data. Still, we can easily overlook the potential value of applying those same methods to the familiar language and absolute logic with which we habitually think, speak, and exude confidence. Prior information, which we use to support and reason about our beliefs, values, data, the world, and relationships, and about our behavior and the behavior of others, does not necessarily receive the same screening as the methodologies we use in our experiments or the data we accumulate (Korzybski, 1958, p. 727; West-Eberhard, 2003, p. 6). In other words, we are easily deceived by familiarity.

In a changing world, the internal and external experiences to which we assign positive and negative values persistently fluctuate according to context. Value and how we assign it biases our thought, emotions, and behavior, whether we know it or not. Understanding how we implicitly incorporate and habitually use familiar values allows us to tune our language to explicitly mitigate those biases.

Status Quo Bias

Another obstacle for understanding the nature of language comes from the human brain's general tendency for affiliative attraction to localized beliefs and the adoption of cultural values held by the majority. Language plays a prominent role in status quo bias, since it transmits our affective cultural inheritance. The habitual human tendency for affiliation and reinforcement of mean beliefs and values held by the majority of a population encourages relative sentimental attraction to the mean and discourages drifting away from the status quo. We are generally predisposed to conform and implicitly go along with the group, but all the while we tend to profess our own autonomy (Milgram, 2004, pp. 114–115). Our brain's status quo bias, augmented by familiar, rigidly held beliefs, constrains learning and obediently binds us even more tightly to the static past.

Biological organisms tend to search for, find, adapt to, and become fixed on the system's favorable conditions, or bound to the sweet spots, and complacently resist inertial change from that status quo unless severely or chronically provoked (Churchland & Sejnowski, 1994, p. 341). A preference for the current state is sometimes referred to as *status quo bias* (Tversky & Kahneman, 2000, p. 146). This general principle makes some sense from the organism's perspective (Jennings, 1906, p, 311): Why risk change when the status of familiar things, which have acquired a positive affective value in part simply because of our history with them, seems "good enough"—especially when change and uncertainty carry an element of risk? Yet many less-complex organisms, such as ants, to some extent rely on exploring for alternative solutions through novel search, even when abundant resources are locally available for exploitation (Mitchell, 2009, p. 195).

Evolution doesn't necessarily produce the best solutions, but it generally produces enough possible solutions that some will suffice to satisfy the requirements of survival in a changing environment, since "satisficing" is usually good enough for survival and reproduction (Churchland & Sejnowski, 1994, p. 133). Humans and other mobile organisms generally flock to perceived positives and regard temporally noticeable, dramatic, rapid, or relative changes from the mean as aversive, and change their behavior by avoiding or pulling back as if from something injurious, or at least not beneficial (Jennings, 1906, pp. 23, 109, 311).

Such status quo behaviors become fixed over time, even though at some later point conditions may so change that the old responses are no longer appropriate, sometimes with disastrous effects on organisms trapped at the status quo (Jennings, 1906, p. 319). Status quo bias tends to bind us to the predetermined stepping-stones on the beaten path—and punish us for disobedience if we venture away. Depending on cultural norms and the distribution of phenotypic predispositions, we generally demonstrate an attraction to obediently climbing up the slippery status pole, attempting to achieve or surpass the norm, and competing to come out on top.

If we can come to understand how our attraction to the status quo can trap us there, we then have the opportunity to adjust our language behavior in anticipation of resolving the status quo conundrum. We can weigh the pros and cons of striving to achieve or competing to surpass the sentimental norm. Alternatively, we can decide by analyzing the plausibility of a choice according to what we think is in our best interest versus conforming to what others think is in our best interest. The human brain we inherited leaves us susceptible to the predetermined notions of status quo reinforced by the embedded beliefs and values we acquired from our rarely scientific and often superstitious ancestors. The localized language, belief, and value system we inherited may have been authoritatively or affectionately well intentioned, but much of it was clearly effectively misinformed.

We can try to patch a largely spontaneous and reactive status quo system by switching about language primitives and pretending to have a novel solution. We can switch as much as we like, until the cows come home, but no amount of switching likely extricates us from the simple fact that any combination of primitive linguistic constructs inevitably leaves us affectively polarized and statically mismatched with the dynamic physical world. We still have the option of using existing linguistic constructs based on effectiveness that offer alternatives to the status quo. Effective language facilitates open and consistent correspondence between our scientifically derived static human class and our imperfect status in a constantly changing imperfect world.

Developmental Bias

We incur a significant genotypic-related phenotypic bias during developmental learning according to the nature of our interactions with the environment and the language constructs we inherit. Developmentally acquired biases habituate our thought processing, limit insight, and complicate our explicit understanding of the language, beliefs, and values we implicitly learned in early childhood. These early beliefs establish scaffolding that influences discriminating the relevance of future beliefs and the accumulation of new information. Humans uniquely demonstrate an ability to learn, abstract, process, and express complex languages. Nüsslein-Volhard (2006, p. 115) noted that only humans exhibit this higher-level language skill, aided by an extended period of youth and an almost insatiable capacity for learning.

> Awareness of the role of early life experiences in shaping both *what* is processed and *how* information about the mind itself is handled can help us to negotiate our way through the complexities of the mind and social relationships. (Siegel, 1999, p. 63 [emphasis in original])

Typically, the formation of sensory organs and nerve connections continues after birth, heavily dependent on stimuli encountered in the environment. Initially many nerve connections are formed, but in most cases, only those that are used and activated will persist. To some extent, biological circuits wire themselves in response to internal guidance during gestation and to external sensory input after birth (Nüsslein-Volhard, 2006, p. 115). Without sensory stimuli during sensitive developmental periods, some relevant nerve connections may never form, creating deficiencies that constrain a variety of critical activities in later life (Wang et al, 2009).

Studies make it clear that anatomical wiring in early life is a critical developmental phase that influences the scope of functions that human brains can perform for the rest of their lives. Lenneberg describes this critical period as a window for language acquisition, similar to imprinting (Jackendoff, 2002, p. 96; Lenneberg, 1967). Early childhood experiences may affect complex processes such as the development of social communication and speech, since experience modifies connectivity in the developing brain (Buzsáki, 2006, pp. 220–221, 227). Observations of children neglected during this early developmental period suggest that outside stimuli are indispensable for the normal development of human cognitive capacities (Nüsslein-Volhard, 2006, p. 115).

> Although developing brains take care of their own input somewhat automatically, we may still assume that the better learning is synchronized with development, the more effective it will be. (Spitzer, 1999, p. 187)

A developmental relationship with learning further suggests that outside stimuli have a lasting influence long after early development (Bos et al, 2009; Korosi & Baram, 2009). In humans, the connection between language, beliefs, values, and reasoning is established early in childhood, forming the initial conditions of the system, which, once established, demonstrate resistance to change (Luria, 1981, p. 209; Panksepp, 1998, p. 245; Lane et al, 2000, p. 409; Milgram, 2004, pp. 114–115, 136–147; Tse et al, 2007; Munakata et al, 2008, p. 375).

Resilient automatic learning seems to provide sufficient evolutionary adaptation for most animals, with or without language. Humans can likewise do well enough using only language primitives. But language offers a higher-order resource for abstraction that influences cognitive processing and that we can deliberately fine-tune to go beyond good-enough solutions. Relationship experiences early in life shape the structure on which adult behavior is based (Sullivan et al, 2009, pp. 894–896). This resilient process seems to create a coherent worldview, but one easily confounded by the problems encountered in reality as the "knowledge" derived from these relationships biases how we construe reality (Siegel, 1999, p. 4).

> The use of language and interactions with the environment make up the sum of every person's experience and, throughout his or her entire life, provide the momentum for the formation and changing of high-level map-like representations of meaning. (Spitzer, 1999, p. 234)

As we develop as individuals, we normally transition from a condition of relative dependence to one of relative independence. Brain development parallels this transition as the frontal lobes, and the connections to them, mature to adult capabilities. In turn, the frontal lobes benefit from language that can effectively facilitate developmental transitioning to more independent thought processing and more flexible regulation of learning and behavior. But affective language traps us by default in an absolute, closed-minded developmental state dependent on rigid thought processing that constrains independent learning and behavioral flexibility. Developmentally acquired language, belief, value, and logic systems establish our initial state conditions that tend to bias our future learning and influence how we relate throughout our life.

> Relationship experiences have a dominant influence on the brain because the circuits responsible for social perception are the same as or tightly linked to those that integrate the important functions controlling the creation of meaning, the regulation of bodily states, the modulation of emotion, the organization of memory, and the capacity for interpersonal communication. (Siegel, 1999, p. 21)

Obviously the affective biases that suffice for modulating the brain and behavior of our mammalian cousins are not necessarily lethal to us. But understanding how the brain and language naively bias our learning during development can encourage us to tune the effectiveness of the language we use. It seems that taking advantage of our knowledge about how the human brain works and leveraging our systematic understanding of extended language in and of itself may be a large part of effectively dealing with the constraints and biases that we face. We describe our world with language that influences the fidelity of our understanding and the effectiveness of our adaptive behavior. Unthinking use of language can perpetuate static beliefs, misunderstandings, and needless upset that interfere with our happiness and our ability to adapt to the ever-changing conditions that we face throughout our lives.

Comparison Bias

Our brains are constantly making comparisons. Boroditsky (2007) has shown that comparing similar objects makes them appear more similar. Comparing dissimilar objects, on the other hand, may make them appear less similar. These results suggest that common cognitive processes such as comparison can introduce systematic biases into our evaluations of objects and conceptual understanding of relations by influencing how we treat their similarities and differences. Comparison may play a special role in how we partition bits of experience into categories by sharpening categorical boundaries and otherwise helping us to "create conceptual structure above and beyond that offered by the world" (p. 118).

Science sorts and correlates categorical classifications according to natural ordering. Science categorizes *mammals* in comparison with all other known animals in nature. We gained knowledge of mammals in the real world by laboriously evaluating the inclusion and exclusion of other groups of animals by observing for similarities and differences. Our brains tend to segregate by differences and integrate by similarities. Processing by similarities and differences and positive and negative value tags creates polarized perceptual value weights that simplify evaluations and decisions. We can deliberately modulate how consistently we order the sorting process, which often depends on our goal and purpose for making the evaluation, whether we want to find new information or keep old information, or whether we want to prove or disprove a theory, including proving our own theory.

Science values alternatives for updating theories. Science searches for novel information and theories, hypotheses, and explanations and evaluates them consistently in an orderly manner by comparable methodology, so as not to bias the resulting conclusions by inequitable treatment. In effect, this means that theories undergo scrutiny from many different angles, such as attempting to replicate results, looking for exceptions and errors, and even trying to disprove or falsify them. Falsifying a theory leads to further analysis of the method used for falsification. There are perhaps many ways to falsify a theory. Attempting to prove that a theory "will not work" by falsification provides limited evidence that it will not work in the specific manner in which it was tested, whereas testing under different conditions may demonstrate that the theory performs as expected.

General theories present a difficult case because they propose systematic solutions purported to be invariant across a domain. General theories may hold in a particular domain, such as the time domain, but tend to demonstrate frailness to testing or comparison in imaginary or hypothetical domains, or on the introduction of imaginary structures, methods, variables, parameters, constraints, and conditions that bias the systematic function space. In science we expect at best that relevant experiments test at least two plausible explanatory theories. In other words, there are two plausible competing theories rather than permutations of one theory that results in testing correctness by referring back to itself. We would expect that they both are evaluated consistently in the same orderly manner.

But in most of our everyday human experiences, our brains constantly compare incoming information with what we believe we already "know," or expect. We tend to measure experience against our internal reference of prior beliefs, sentiment, and assumptions that comprise and can compromise our worldview. The interactions we experience between our brain's senses and our behavior form our beliefs about the outside world that tend to lead to perceptual habituation. In other words, we become desensitized over time to our internal implicit processing so that the perceptions we construct become a slave to habit. Magicians take advantage of this brain habit when they perform their magic tricks. Perception is not a process of passive absorption but an active construction (Macknik & Martinez-Conde, 2010, pp. 1–6, 141, 142).

> "... [T]he general law of perception ... is this, that *whilst part of what we perceive comes through our senses from the object before us, another part* (and it may be the larger part) *always comes* (in Lazarus's phrase) *out of our own head*." (James, 2007 v. 2, p. 103 [emphasis added])

What is our purpose for the comparison? Evidence shows that two terms taken far apart appear to differ more than two terms taken together (James, 2007 v. 2, p. 645). A method for preserving information or theories can first contrast the differences and use them as reasons to discard the others. Thus if we wish to keep a theory intact or feel threatened by a novel theory, we can defend our familiar theory by emphasizing its positives, and attempt to destroy competing theories by emphasizing their unfamiliarity and hypothetical negatives. In effect, we cleverly resolved the competition by switching our methods—a common trick exploited by magicians (Macknik & Martinez-Conde, 2010, p. 35). By a bit of "slight of mind," we can magically switch the order of methods we use for the comparison by contrasting the competing theory and pointing out its hypothetical negatives as differences from the positives in our own. Deceptive switching of the ordering of our methods results in surreptitiously destroying the other theory and winning the competition. We can generally conclude that most theories demonstrate different levels of imperfection, which may go unnoticed in the heat of competition where our brains demonstrate a predilection for habitually resorting to deception. Regardless, we can leverage our advantage by pointing out hypothetical negatives and blowing any small discrepancies out of proportion.

If, on the other hand, we wish to look for new theories and information, we can first look for similarities of positives and get new ideas. Then we can eliminate or put aside alternatives on the basis of how well they match according to sameness and goodness of fit with the problem we intend to solve. Even in this process, we can trick ourselves, but how far we fool ourselves may depend on our choice of the methods, the consistency with which we apply them, how we order them and what we measure, and from where we measure. Ironically, if we do notice any slight of mind, we more than likely will notice the deception of others but seldom notice our own. When it comes to self-deception, we tend to remain oblivious to our folly.

> People exhibit a parallel tendency to focus on positive or confirming instances when they gather, rather than simply evaluate, information relevant to a given belief or hypothesis. When trying to assess whether a belief is valid, people tend to seek out information that would potentially confirm the belief, over information that might disconfirm it. In other words, people ask questions or seek information for which the equivalent of a "yes" response would lend credence to their hypothesis. (Gilovich, 1991, p. 33)

When we habitually reorder our sorting to pick the theory that confirms our familiar beliefs, our brain fools us into believing that we are "right" and others are "wrong." We tend to think of ourselves as normal and right while noticing the abnormality and wrongness of others and how "they just can't get it right. They just keep screwing things up." Since we are comparing our own ideas with our own subjective referent for relevance, we automatically find the positives and self-similarity in our own ideas. Our automatic limbic behaviors tend to keep us safely tied to what has sufficed in the past, but it also impedes us from exploring the unfamiliar. We prefer to go for safety first and examine the novel alternatives by suspiciously scrutinizing them for negatives, which typically leads to quickly discrediting and rejecting them (Gilovich, 1991, p. 50).

> People's preferences influence not only the *kind* of information they consider, but also the *amount* they examine. When the initial evidence supports our preferences, we are generally satisfied and terminate our search; when the initial evidence is hostile, however, we often dig deeper, hoping to find more comforting information, to uncover reasons to believe that the original evidence was flawed. (Gilovich, 1991, p. 82 [emphasis in original])

If our preference for the familiar to which we have committed much effort suffices, we find it too much work to switch and learn new tricks—at least until there is nothing left to explore or exploit, or disaster strikes. Of course we can claim that the uncertainty of new and unfamiliar ideas remains "untested" and could be very "dangerous" and lead to "disastrous calamities." But we forget to mention that our faulty assumptions and unacknowledged ignorance likely play the larger role in such calamities. We neglect to mention that shuffling the deck of absolute solutions will unlikely work any better now than it ever has, if it ever has. New ideas based on effectiveness unlikely will "cause the sky to fall," unless we believe that "only absolute pillars of *truth* can keep it supported." Embracing the pillars of truth invokes a theory lacking any known plausible evidence but in a pinch seems to comfort those who have developed the habit of certitudes. An immediate prediction of certain doom implicates the wary amygdala.

We inherited a brain that seems to automatically employ different sets of rules and sorting orders for ourselves and for others. We tend to habitually apply the rules that best suit our own interest and confirm our solution as the better or only way. Applying absolute rules to discredit others and to credit ourselves might seem tempting and hard to resist at times. But unless we pay more attention to our own behavior, we are unlikely to notice our own habitual tendencies and likely lack insight about our own deception. We can find 1% negative in the ideas that do not suit us, round up to 100% by overgeneralization, and simply discard the whole idea as absolute rubbish. We use these deceptive methods out of habit to feature our own positives while focusing on the negatives of others, and we cleverly win hands down.

> Understanding human nature is surely not as simple as understanding the nature of the subcortical emotional system we share with other mammals, even if they are the ancient centers of gravity for our affective value system. On top of these systems we also have strong intrinsic potentials for Machiavellian deceit. The brain of "the lizard" still broadcasts its selfish message widely throughout our brains. (Panksepp, 1998, p. 322)

Blindness and Confabulation Bias

Our habitual brain tricks us with a "staggering amount of outright *confabulation*—a fancy term for shamelessly making things up"—to construct our mental simulation of reality, which we seldom notice; our brain confabulates, we justify it, and then we stick to our guns (Macknik & Martinez-Conde, 2010, pp. 9, 171, 175). Confabulation bias distracts our spotlight of attention and distorts our understanding about the relationships we observe. We tend to automatically focus on the familiar. Neurons in the brain inhibit and suppress information outside that spotlight. In effect, we constrain our frame of reference and situational insight, which increases the chances of missing novel information and potentially relevant opportunities outside our spotlight. We actively suppress and ignore critical information, and the harder we focus, the more we miss, as if blinded by our own spotlight (Macknik & Martinez-Conde, 2010, pp. 64–65, 68, 84–85, 88–89, 157, 258).

Magicians have found that the longer the time between when a trick is seen and when it is reported, the more impressive the account. The delay apparently contributes to confabulation and the tendency for misinformed human memory "combining events seen with legends only heard." Our brains can reshape our memory with each recalling of them, and this propensity, combined with our tendency to be misdirected, leaves us an easy target for exploitation. We are susceptible to hoaxes, rumors, and assertions of truth that are repeated often enough. Our susceptibility results from a combination of source confusion and confabulation (Macknik & Martinez-Conde, 2010, pp. 112–113, 118–121).

We can relate such confabulation to the fisherman's story in which the fish grows proportionally larger with each retelling, until at last it swallows the fisherman. Since habitual brain behaviors tend to go unnoticed, similar to the nature of magic tricks, we easily fall victim to traps and remain ignorant or mystified about the deception. If insightful, we have the option of tracing the source of the hardware and software malfunction, perhaps with a bit of chagrin. But lacking insight, the option of traceability mysteriously escapes in our confusion. We apparently prefer to keep an idealized image of ourselves, including our own language, brain, thought processing, behavior, and knowledge, while resolving any confusion about our ignorance by simple denial or rationalization.

Memory source confusion is apparently a product of the brain's tendency to habitually assume and infer the validity of familiar information rather than deliberately seeking the information's source, plausibility, and reliability, especially where often-repeated and familiar information tends to be habituated and taken for granted. Once we have habituated to a feature of the world, we tend to safely assume it as a seemingly immutable part of our stable, reliable, and unchanging fabric of life, and tend to ignore that feature as an established "fact" (Macknik & Martinez-Conde, 2010, pp. 144, 154).

We sometimes exhibit *inattentional blindness* that limits attention to and perspective about our own expectations when we encounter something we were not expecting. We often blindly fail to see what we do not expect to see because it "should not" be there (Chabris & Simons, 2010, pp. 6–7, 20–22)—as if we expect things "should" stay the same. We easily habituate to pleasantries while desensitizing to and ignoring changes that do not result in egregious discomfort or that do not directly affect us. We often fail to compare what was in view moments before with what is in view now, a tendency called *change blindness* (Chabris & Simons, 2010, p. 55).

Our brains are constantly comparing incoming information with what it already "knows," or expects; new experiences are measured up against prior beliefs and a priori assumptions, which tends to give us "misplaced confidence" in our abilities to arrive at a "correct solution" (Macknik & Martinez-Conde, 2010, p. 142). We often assume that confidence and knowledge are related, but interestingly, research on the *illusion of confidence* shows that skill incompetence correlates with overconfidence as unwarranted certainty. Ironically, to overcome this means that we are challenged to better recognize our own limitations (Chabris & Simons, 2010, pp. 85–92).

Even though we can have a high sense of confidence in our sense of recollection for emotionally encoded memories, there is a behavioral dissociation between the influences of emotion on memory confidence versus accuracy (Hamann, 2009, p.192; Schacter & Slotnick, 2004). In other words, affective encoding produces subjective rather than objective memories. But our failure to notice the distinction between the two encodings tends to distort our objective perception. Research shows that those with lower competency skills tend to be the most overconfident, yet we still rely on confidence as an indicator of ability, and people who speak first and most forcibly are thought of as the most believable (Chabris & Simons, 2010, pp. 95–98). For example, patients found confident doctors who expressed certainty in their knowledge "more satisfying" and rated "least satisfying" those who expressed doubt by consulting a reference book (Chabris & Simons, 2010, pp. 106–107).

> The spooky truth is that your brain constructs reality, visual and otherwise. What you see, hear, feel, and think is based on what you expect to see, hear, feel, and think. In turn, your expectations are based on all your prior experiences and memories. What you see in the here and now is what proved useful to you in the past. (Macknik & Martinez-Conde, 2010, p. 8)

None of us with a human brain can claim immunity from its clandestine tricks. We all use a brain prone to making assumptions on average that when unattended automatically resorts to these habitual "good enough" fast and frugal processing methods for effortlessly predicting, remembering and adjusting our thinking about the past, and living in the present as best it can. The intuitive right side of our brain is unencumbered by analytical logic and appears to automatically detect and compute the emotional weight of solutions rather than deliberately elaborating and analyzing the global problem-and-solution search space.

If we deliberately engage the left side of our brain, we can use language to analyze and elaborate on the problem, identify potential benefits and pitfalls, and likely come up with more effective solutions. Effective language facilitates reliable decisions and predictable outcomes; it simply works better in general. It is not so much the decision making we do but the repercussions of those decisions that affect our lives. Researchers have concluded that our ability to detect and consider information that contradicts or challenges our established belief is crucial for learning about the world (Macknik & Martinez-Conde, 2010, pp. 176, 195). To achieve better accuracy, we tend to rely on effortless automatic strategies that are simple and work sufficiently on average, but the price we pay is the cost of ignoring systematic errors (Gilovich, 1991, p. 49). Our brains can automatically default to adaptive shortcuts that speed up or minimize processing and provide enough information to survive, but with compromised accuracy (Macknik & Martinez-Conde, 2010, p. 251). Where the accuracy matters most, deliberate analytical thinking helps us to

- Screen our beliefs about how the world works
- Adjust imperfections in those beliefs
- Effectively monitor and modulate our habitual thinking
- Explicitly work on decreasing systematic errors

If we understand the laws of value and action related to affective attraction and repulsion and how they tend to instigate habitual shortcuts, we can better understand our intentions and the intentions of others. We gain alternatives that offer a broader range of choices for adaptive behavior. We have the option of effective thinking that can extend our search to deliberately include the pros and cons of both the positives in the new and the negatives in the old. But the downside is that compared with habitual behavior, deliberate thinking consumes more time and energy (Macknik & Martinez-Conde, 2010, p. 143). The liabilities of automatic implicit thinking become more apparent when we consider how our spotlight of attention can selectively direct our energies toward preserving the old of the past at the expense of neglecting innovations, explicit creative thinking, and the possibility of constructing a more likely future.

> Even though we might conceive explicit thinking as more expensive than implicit automatic thinking, we might change our mind when we think of the cost of physically laborious tasks. In other words, planning ahead by thinking about efficient ways to move and arrange objects in a preferred manner can be much cheaper than real-world effort with trial and error, which can take considerably more energy to perform. We can exploit explicit thinking as a less costly virtual method for rotating and moving objects in space. (Casasanto & Boroditsky, 2008, p. 579)

There is ample evidence that our spotlight of attention can suppress surrounding information and constrain our ability to see the whole picture (Macknik & Martinez-Conde, 2010, pp. 59–89, 256–259). Constraints on attention, while economical in terms of energy, bias what we notice, attend to, and learn from while limiting the aspect of our perception from which we construct our worldview. Acknowledging our cognitive foibles affords us an opportunity to make allowances for effectively adjusting our language and our thinking.

> You'll recognize that the confidence people express often reflects their personalities rather than their knowledge, memory, or abilities. You'll be wary of thinking you know more about a topic than you do, and you will test your own understanding before mistaking familiarity for knowledge. You won't think you know the cause of something when all you know is what happened before it or what tended to accompany it. (Chabris & Simons, 2010, pp. 241–242)

Perhaps we can better deal with these obstacles and discover ways to bias our language's reliability and improve our cognitive effectiveness. Perhaps we can tune our language for effective performance within the reality of the time-related uncertainties we encounter in nature. It seems that understanding the value of applying extended language encourages improving our fitness, our adaptive performance, "how well we adjust to the realities of nature." Extended language and semantics support logically fit thought processing for more effectively navigating the dynamic world that we face in our daily lives. Incorporating language that effectively represents our knowledge of the world that we inhabit seems to represent a bargain when it comes to improving relationships all around.

Tuning our language can likely help us to mitigate the unseen systematic biases that may have previously escaped our attention and inadvertently deceived us with our own stealthy ignorance. Interestingly, children are notoriously difficult to deceive, as if their brains are not as yet sophisticated enough to be fooled (Macknik & Martinez-Conde, 2010, p. 153). The next chapter further expands on the role language plays in our processing biases and suggests ways to mitigate deceptive linguistics by applying effective language that can decrease systematic error rates and thus help maximize our cognitive behavior.

CHAPTER 6

Systematic Biases II

LANGUAGE, AND THE RULES that describe its common uses, has grown more or less organically from the largely habitual behavior of flawed and fallible humans. From the body of our inherited language, we construct our communication with others. Every construction embodies a bit of the user's perspective, a bias toward some particular meaning or application. When we speak of a linguistic bias, we refer to the inevitable predisposition that our assuming brain and our personal, educational, and social history imposes on the range of interpretations of meaning in a given communication.

Even though bias is fundamental in nature and in our everyday life, it has acquired a negative connotation, implying opposition or rejection of something. We could use the term *influence* instead, which has a more neutral connotation, but in science and engineering, *bias* generally seems to be the preferred term. Perhaps we can adapt to a more friendly perception of bias when we learn that a more effective bias can be our own best friend. On the other hand, common dysfunctional language biases can result in

- Deceptive classification
- Erroneous intuition
- Absolute thinking
- Overvalued beliefs

These biases tend to enhance rigid linguistic policies and inflexibility that hinder cooperative adaptive behavior.

Since the 1950s, research in rational emotive behavioral therapy (REBT) and cognitive behavioral therapy (CBT) by Aaron Beck, David Burns, Albert Ellis, Jesse Wright (Wright, 2004), and many others has demonstrated that changing the words and concepts with which we think significantly influences how we feel and how we behave. Because of that, these therapies have become a mainstay in scientifically based treatments for improving everyday thinking and reducing depression, anxiety, and obsessive thought, and they are often found beneficial in more severe thought disorders (LeDoux, 2002, p. 291). For example, adults and children who meet criteria for obsessive-compulsive disorder—considered a heritable physiological disorder—generally show significant improvement with CBT alone.

> Emotional self-regulation is presumably made possible through our higher cerebral endowments. Our symbol systems are especially effective in allowing us to negotiate such rough terrain. Language allows us to regulate our emotions. (Panksepp, 1998, p. 318)

If deliberately choosing different words to describe a problem can produce a different and more desirable emotional reaction, it seems clear that words can bias our thought processing and how we think, feel, and behave. Language can bias our perspective in many ways, between

- Static and dynamic
- Emotional affective and analytical effective
- Rigid and flexible
- Competitive and cooperative

Understanding that all living creatures, including humans, demonstrate biases makes it possible to uncover and explicitly adjust our own bias to reflect effective reasoning behavior.

Obedience-to-Authority Bias

We typically do not like the experience of being deceived or duped by others, but we often fail to acknowledge how easily we can deceive ourselves. Not only can we fool ourselves but it often happens as easily as breathing in and uttering words to describe our current situation. It is easy for us to ignore something and be naively deceived by default. But perhaps just as often, our own evaluations can mislead us, by fabricating reasons where none exist or by assuming and attributing failures to others rather than to our own misunderstanding.

Misunderstanding tends to arise from faulty information, and faulty information often springs from the assumptions and ambiguity hidden within our affective language. The better we understand how ambiguous language can bias our evaluations and lead us to inadvertently deceive ourselves and others, the better we can make adjustments and the less likely we are to fall prey to such deception.

We use declarative language and modal verbs to describe experience and convey meaning. Command declaratives, such as imperatives and prescriptives, attempt to leverage change or instruct the intentional state of others or compel obedient behavior. Other declaratives, such as informatives, convey information with an intention to change the informational state of others. Command terms tend to be discrete absolutes that constrain information and context by one-dimensional compression that converges to rigid decisions and closes the solution space. When phrased with open-ended language, informatives can propagate multidimensional solutions with contextual descriptions and alternative possibilities.

Recall that possibilities leave decisions open to tuning with the option of diverging to more effective alternative solutions. Words that carry more information tend to be more useful for thinking about and analyzing our different choices and preferences. Choosing language constructions that inform us rather than mandate can improve the fidelity of the information we use and increase our insight of the biasing role of language on the predictions we make, the behaviors we produce, and the outcomes we get.

Tomasello (2008, pp. 36–43) points out that our primate cousins communicate almost exclusively with imperatives and rarely, if at all, with informatives. Even when they have been taught relatively sophisticated communication means by humans, apes apparently continue to communicate with imperatives to get others to do things, and it appears that they may not comprehend cooperative informatives (Tomasello, 2008, p. 43). Rigid limbic-driven imperatives dominate the communications of most other animals as well (Mesulam, 2002, p. 19).

- How do we explain the imperative nature of our beastly cousins' communication?
- Do we understand that they operate with basically the same limbic brain as our own?
- Is it possible that we see the routine use of imperatives in each species as a result of our limbic commonalities?
- Why is affective language the language of choice among humans?

Imperatives enable the external expression of internally motivated intentions to control and instruct subordinates remotely. Associating language and word power with remote control implies some sort of force dynamics (Talmy, 2000, p. 441). It also suggests that vertical positions represent perceptual symbols of power (Schubert, 2005). Imperatives represent an extension of hierarchal regulation that operates somewhat like an authoritarian remote command-and-control strategy, allowing one individual to direct the local trajectories of other individuals or social systems, and they tacitly carry the local frame-of-reference bias of the speaker.

> Animal signals have the force of instruction rather than information. (Bronowski, 1977, p. 111)

As social creatures, nonhuman primates and humans employ a somewhat similar imperative hierarchal domination–subordination limbic axis for determining and controlling the status of other group members (Bauman & Amaral, 2008, p. 165; West-Eberhard, 2003, p. 350). We tend to express this dominance–subordination power hierarchy linguistically with imperative demands and intentions often related to competition, deception, coercion, force, demands, hostility, aggression, violence, and retroactive blame and punishment (Beck, 1999).

For example, Stanley Milgram's experiments demonstrated that people commanded by an authority figure will submit to the authority's expectations by doing cruel and dangerous things to study subjects, including administering severely painful shocks. He concluded that there is an association with the childhood learning of implicit imperatives—"And obey me!"—that is inseparable from the inculcation of an obedient attitude (Milgram, 2004, p. 136).

> The first twenty years of the young person's life are spent functioning as a subordinate element in an authority system.... [A]n underlying posture of submission is required for harmonious functioning with superiors.... [C]ompliance with authority has been generally rewarded, while failure to comply has most frequently been punished.... But the culture has failed, almost entirely, in inculcating internal controls on actions that have their origin in authority. For this reason, the latter constitutes a far greater danger to human survival. (Milgram, 2004, pp. 137–138)

It seems of interest that "the disappearance of a sense of responsibility" is the most far-reaching consequence of submission to authority (Milgram, 2004, p. 8). In other words, we relinquish responsibility for our behavior to the higher authority that we might imagine is pulling our strings, like a marionette with a basic "respect" for authority.

> We can conclude that in general people have no introspective access to the behavioral control mechanisms that are activated by the commands of someone in authority. (Vanderwolf, 2010, p. 5)

We imagine that we are not responsible for the behavior we produce, which introduces an imaginary control element void of feedback loops other than affective rewards and punishments. Such an imaginary control system might stump even the most astute engineer who attempted to draw or repair a control circuit that involves imaginary strings. Where could these invisible strings of obedience derive from? The following are prods used to encourage reluctant study subjects to continue to administer shocks (Milgram, 2004, p. 21):

- Please continue.
- Please go on.
- The experiment requires that you continue.
- It's absolutely essential that you continue.
- You have no other choice.
- You *must* go on.

> The key to the behavior of the subjects lies not in pent-up anger or aggression but in the nature of their relationship to authority. They have given themselves to

the authority; they see themselves as instruments for the execution of his wishes; once so defined, they [most] are unable to break free. (Milgram, 2004, p. 168)

"It's got to go on. It's got to go on," repeated one study subject (Milgram, 2004, p. 9). We can hardly miss noticing the explicit and implicit use of imperatives. Is it possible that these obligatory imperatives and curious aggressive behaviors are related to the invisible strings from which we seem unable to break free? Could we have possibly overlooked genotypic to phenotypic tendencies that are amplified by the biases embedded in the language we inherited? Are we as susceptible to obedience training as our canine friends? Is our proclivity for obedience to authority akin to the neurobiology of the tail wagging the dog? Language provides numerous affective terms to support the obligation to authority such as loyalty, duty, and discipline: We feel "pride" when obedient and "shame" when disobedient. And it is toward authority that the study subject turns for confirmation of his worth (Milgram, 2004, pp. 146–147). These are examples of localized words that deceptively rely on the affective power of imperatives. Consider the invention of the affective term *self-esteem* that represents a local artifact of the authoritarian system from which we have learned to rate our own "self-worth" in terms of obedience. In other words, the subjective nature of the "myth of self-esteem" derives from its subordinating social benefit (Ellis, 2005).

Even though imperatives and prescriptives tend to be associated with ordinate power hierarchies, vertical discrete regulation, and competition, ironically they are worded in such a way as to force obedient cooperation. In some cultures, the actual value in terms of costs and benefits is distorted by means of coercion and punishment to ensure maintenance of cooperative behavior (Floreano & Mattiussi, 2008, p. 551); for example, "You must cooperate or you will be punished." In contrast, descriptives and informatives tend to be associated with multilateral regulation, deliberate choices, and cooperation. In English and many other languages, local social communication tends to rely on imperatives, commands, and other absolutes (explicit or implicit) to deceptively leverage social ordering. We see examples in some societies such as the mysterious disappearance of people who disobey and the later recovery of their bodies, along with the bodies of others, in unmarked graves.

Constrained language environments imply a lack of deliberate choices that impose strict limits on alternatives and variability. In one-dimensional idealized spaces, imperatives may suffice and seem somewhat benign. A preponderance of imperatives suggests habitual subordinate communication with increasing liabilities for deception, as rank and power increase with positions of authority. Lest we forget we are talking about an interactive system in which language only plays a part, we will want to consider the role our brain plays in this hierarchy. It appears that the human brain and brains in general play a similar role in perpetuating obedience and subordination, as witnessed across many species with and without language.

Such dissociation tells us that our limbic brain is not as far removed from our fellow mammals and nonhuman primates as some might suppose. And we might expect that the limbic amygdala has an influential hand in our obedience to authority. Indeed, experimental evidence shows that the amygdala is involved in social conspecific behavior related to dominance-subordinate hierarchies. In social environments sensitivity to social information enables animals to distinguish the difference between acceptable and unacceptable behaviors. The amygdala is implicated in detecting and disambiguating threatening, fearful, and submissive behavior, including facial expression, eye contact, body posture, and direct and averted gaze (Buchanan et al, 2009, pp. 289–296).

Effective language can help us understand and modulate our subordinate mammalian predisposition for obedience to authority rather than rationalizing it by justification with affective language. Perhaps we have discovered a reason as to why we cling so obediently to affective language. There seems to be something in it for us that rewards domination and subordination and amplifies the intensity with which we support and defend our own behavior. Affective language appears to enable the fabrication of an absolute sense of "coordinated" direction for an obedient life, leaving us easy prey susceptible to authoritarian cultural indoctrination. Early cultural priming in childhood instills an almost fail-safe method for thwarting change, absolutely repudiating effective alternatives, and fostering our predisposition for obedience that tends to accompany us throughout our lives. The limbic sense of direction seen in humans is hard to differentiate from that of other socially sophisticated primates and mammals and to some extent even from insects such as ants that are obedient to an inherited social ordering of behavior.

Dominance–Subordination Bias

Hierarchical dominance can be seen in many animal species, including chimps, chickens, and humans. The policies of dominance systems reinforce top-down power, which emanates from the top and dictates solutions to the bottom. Among heads of state and in governments, businesses, and human society, hierarchical dominance with policies supporting obedient rule from the top tend to be the norm rather than the exception. Top-down power subordinates and constrains effective communication and impedes information flow by dampening unauthorized backflow of opposing information from the bottom, which is considered insubordination.

There are many stories where warnings from subordinates are ignored and their ability to take actions and make interventions are thwarted or suppressed; thus, even though the consequences are often predictable, solutions for avoiding the unpleasant outcomes fail to materialize until it is too late, if at all. Then, after the fact, the subordinates are criticized for not heeding their own warning even though they contend they spoke up but no one above them would listen. But as with a tree falling in a deserted forest where there is no one to hear, was there a sound? Of course there was a sound according to physical laws in nature, but apparently no one heard it, or at the least it went unattended. So there is a difference between making a sound, hearing a sound, and attending to and heeding a sound. People generally speak with the expectation of being heard, acknowledged, and, at least to some extent, heeded.

In status hierarchies, mandates rain down from the top, but the upward flow of information from subordinates at the bottom lacks status and struggles against the hierarchal gravity in the system. Bureaucracies typically have many layers in their hierarchy. Information coming up from the bottom to the top has to get through many nodes of middle managers. While some messages make it to the top, many are unheeded or get garbled along the way, and negative feedback is often rejected as "too negative," "unwarranted criticism," "controversial," or "not good for the institution." Typically, in a hierarchy criticism gravitates downhill; and in the end, the subordinated person at the bottom gets the blame and punishment that, in effect, covers up any systematic errors at the top. "If you don't understand that you work for your mislabeled *subordinates*, then you know nothing of leadership. You know only tyranny." (Attributed to Dee Hock, retired CEO, Visa International)

As Tomasello (2008, pp. 84–97) points out, even though imperatives relate to intentions, they require some form of joint attention, mutual assumptions, or common ground. But imperatives tend to carry little if any specific information and without qualification are fundamentally vacuous and nearly useless for effectively conveying information other than hidden assumptions. Prescriptive and imperative statements tend to converge directly to the conclusion or solution instead of explicitly addressing and elaborating on the problem space and considering supporting evidence, information, and plausible arguments. Thus, direct convergence to solutions can inadvertently lead to deception, whether intentional or unintentional. But even though we fail to notice our own use of imperatives, our imperative speech behavior speaks for itself—that is, if we care to listen to the words we use. Regardless, our speech behavior gives us away.

We define an argument as communication that presents evidence and inferences that allegedly lead to a specific conclusion, and we can run into difficulty when evaluating a conclusion without considering the supporting evidence (Browne & Keeley, 2007, pp. 18, 26). With the sensitivity to initial conditions found in complex systems, a small amount of assumed information unsupported by plausible evidence can introduce a large amount of deceptive noise that biases and distorts the accuracy of predicted outcomes. Even small informational errors are egregiously amplified by absolute logic and the nonlinear nature of the system (Nicolis & Prigogine, 1989, p. 179).

Like an oracle predicting the future, imperatives deny the noninterchangeability of the static past and the dynamic future. We tend to use imperatives to mandate predetermined solutions or conclusions to the exclusion of probabilistic predictions. Imperatives tend not to address probability, and their use reduces the likelihood of reliably predicting outcomes (Capaldi, 1987, p. 17). Research indicates that we would make more accurate predictions if we relied on statistical predictions rather than intuitive predictions (Kida, 2006, p. 196).

Prescriptives and imperatives declare control or forcing of others (Tomasello, 2008, pp. 117–118) and deceptively constrain problem resolution by omitting or ignoring plausible evidence that might have biased the reasoning behind the decision and the solution produced. The information vacuum created by the omission of plausible evidence confers constraints on our ability to evaluate risks and draw sensible conclusions about the reasonableness of our choices. Hence, we struggle to learn from and correct errors, thereby impeding useful knowledge for predicting outcomes of future choices. Without access to the source information, inferences, assumptions, and reasoning that led to the decision, we are less likely to learn from mistakes, detect further errors, and make appropriate adjustments to the decision-making process. Whether deliberately deceptive or not, absolute language tends to hide errors and perpetuate the risk for repetitive maladaptive behavior, reasoning, decision making, and learning.

Deliberately identifying and evaluating the plausibility of information, evidence, assumptions, and the reasoning process behind decisions enhances risk assessment and effective error management. Acknowledging the limitations and liabilities inherent in control language and the beneficial properties of informative language facilitates deliberately increasing information and reducing assumptions. Without making our assumptions explicit, we can hardly minimize them, but we instead incur an unknown systematic error burden that can impair the fidelity of our thought processing.

The choice to use descriptives, informatives, and interrogatives promotes cooperative, open-ended search for information and alternative solutions. Open-ended descriptives and informatives facilitate cooperative communication, information acquisition, information flow, and collaborative information sharing, while decreasing deception by improving learning through encouraging education (Bailey, 2007a). The promotion of interrogatives encourages questioning declarations—including imperatives from experts and those in authority—as to the plausibility and reliability of supporting information.

Competitive Status Hierarchy Bias

Absolute imperatives tend to leverage intentions unilaterally from the speaker's hierarchal local frame of reference, sometimes in a forceful, commanding, and occasionally even brutal way, to establish competitive, demanding dominance–subordination hierarchies. Lacking plausible evidence or logical relevance to support the reasonability of their demands, users of imperatives and prescriptives may refer to their authority and resort to applying oppressive physical or verbal force or subterfuge to leverage the accomplishment of their goals. Intentions typically include an expectation or anticipation of a certain outcome. Expectations can predispose to confirmation bias, and unmet or thwarted expectations can lead to arrogance, anger, and punishment.

In contrast, a policy that includes and deliberately reinforces cooperative language usage with informatives encourages collaborative communication horizontally across constituents. To induce and maintain the benefits of bottom-up problem resolution, an organization can establish a level communication structure that elicits cooperative problem solving and operates harmoniously alongside a structure for final decision approval as deemed appropriate.

The use of a systems approach that accounts for the influence of higher-order systematic interactions captures the importance of constructing policies that promote rather than inhibit distributed bottom-up problem-solving behavior. Similar to physical laws in nature, verbal policies induce limitations on behavior in the system, define domain constraints on the logical flow of information, and influence the cooperative tone of relationships. Whether explicit or implicit, the systematic organizational policies, or lack thereof, set the tone for communication and relationships within that system.

For example, across nature we see simple to more complex solutions evolving through the organization of life forms whose physical morphology, behavior, and policies naturally follow the laws of physics. A solution that cannot adapt to the functional constraints of the system or that violates the physical laws incurs punishment by impairment or selective elimination. In effect, evolution works by solving problems within top-down physical policies that allow for the bottom-up flow of cooperative solutions. Yet human behavior shows a propensity for inverting nature's favored system. We seem to prefer imagining an idealized top-down organization with obedience to authority, prefabricated monolithic policies, and solutions forced top to bottom in a take-it-or-leave-it fashion that meanwhile suppresses feedback.

> The meeting of a company president with his [or her] subordinates.... The subordinates respond with attentive concern to each word uttered by the president. Ideas originally mentioned by persons of *low status* will frequently not be heard, but when repeated by the president, they are greeted with enthusiasm.... Even when a subordinate possesses a greater degree of technical knowledge than his [or her] superior, he [or she] *must not* presume to override the authority's right to command but *must* present this knowledge to the superior to dispose of as he [or she] wishes. (Milgram, 2004, pp. 141, 144 [emphasis added])

Rather than emulating cooperative processes found to work in nature to deal with problems of adaptation, we tend to construct authoritarian policies that attempt to resolve problems by rank-ordered power hierarchies replete with imperative language. In nature the organization of solutions takes place by matching to goodness of fit of the problem. Incompatible with flexible thinking, imperative language instills arbitrary absolute constraints on communication that thwart the flow of self-organizing analytical solutions. Constrained language constructions fall short in supporting an understanding of higher-order analytical concepts such as open-ended thinking and information search, decreasing assumptions, elaborating on the problem space, and evolving cooperative solutions.

One-dimensional language may suffice for constructing imaginary monolithic solutions and absolute policies with mandates in simple or idealized systems with low information requirements. With limited acknowledgment of living in a dynamic physical world where heterogeneity and biological variation prevail, the application of absolute top-down policies necessarily amounts to reinforcing punitive wishful thinking about how the brain, language, regulatory policies, and the world work.

Higher-order policies can be extrapolated from bottom-up principles in complex systems and adopted as policy guidelines for deliberately encouraging insight into how these policies enhance effective communication behavior. Policies that fine-tune effective language behavior to the horizontal cooperative axis can improve information flow and reinforce cooperative behavior. Cooperation encourages the bottom-up flow of information and problem-oriented analytical solutions and generally enhances the ability to search for more robust alternative solutions.

The deliberate use of open-ended language as an information and communication source promotes the bottom-up flow of solutions by leveraging cooperative language that informs, requests, and reciprocates. A linguistic shift from vertical to horizontal tends to neutralize status-related hierarchal dominance by leveling the communication playing field. Bottom-up flow can produce informed alternative solutions that can be evaluated on the basis of possibilities weighted for the probability of effective problem resolution. Hence the problem itself imposes limits on the search space and allows solutions to organize as attractors to the problem. Domain knowledge further decreases the problem-and-solution search space. Alternative ideas are contrasted and weighted by explicitly comparing each solution's attraction with the problem on the basis of minimizing assumptions and maximizing plausible information, reliable evidence, and predictable effectiveness.

A cooperative strategy for problem resolution enables a redirection of energy and effort toward enhancing the fidelity of current information and generally considering the feedback process in a positive light. Both positive and negative feedback support the bottom-up organization of innovative ideas and cooperative problem resolution. Improved problem resolution corresponds to cognitive processing supported by effective language, information, and knowledge, since the language repetitively influences the fidelity of analytical processing. Informatives inform, and informed problem resolution relies on cooperation and collaborative communication and the accuracy of information. The sharing of information that informs rather than mandates tends to decrease assumptions and enhance analytical power for further developing knowledgeable solutions.

The distinct differences in the behavioral effects on communication between closed-ended imperatives and competition on the one hand and open-ended preferential informatives and cooperation on the other explain some relevant issues related to the evolution of language and communication. The development of verbal communication marks a major evolutionary transition; recent work using robot-based simulation shows that communication by signaling can arise spontaneously. Experimental studies show that a purely competitive setting with limited rewards gives rise to deceptive communication [signaling], while cooperative communication arises in settings with close group relationships and sufficient rewards (Lipson, 2007; Floreano et al, 2007).

Once established, the system's evolved communication style resists change (Floreano et al, 2007), potentially as a result of deceptive trapping on the local language fitness landscape. Inherited localized communication styles in humans tend to resist change in much the same way. This finding and others support the idea that the possible arbitrariness and imperfection of communication systems can be maintained despite their suboptimal nature and that once evolved, they constrain the evolution of more efficient communication systems (Floreano & Mattiussi, 2008, p. 581). Competitive communication behavior is found in chimpanzee societies, in the pecking order of chickens, and in the dominance hierarchies of human social structures and organizations. Dominance implies "I win because I'm above you" and often suggests "I will physically defeat you." Competitive statements, such as "I'm right and you're wrong," tend to constrain cooperation by operating on a noncooperative vertically regulated, top-down, winner-take-all hierarchy.

An employer who says, "I am the boss and it's my way or the highway!" forces employees to follow a single predetermined option regardless of its potential drawbacks and without specifically stating the problem or considering more cooperative alternatives. Employees may fail to meet goals when context changes, and doing as the boss says turns out to have unexpected adverse consequences. As the old saying goes, "They feel like they can't win for losing." Rather than encouraging open discussion of the systems influences on problem resolution, a competitive environment like this encourages deception, backpedaling, assignment of blame, and punishment. Such top-down systems are based on intuitively determining the objective a priori, then punishing unauthorized alternative solutions that do not fit the objective, while rewarding only those stepping-stones found on the predetermined path.

Choosing the objective ahead of time destroys alternative competing solutions by punishing them as noncooperative threats that impede the obedient progression to and the achievement of the prescribed objective. The boss at the top can take credit for successful solutions and blame the bottom for unsuccessful ones. Blame relates to a predetermined approach to problem solving where one part of a system is singled out as the disobedient absolute cause of the errors without taking account of multiple systematic influences. When we ask the cause of something, we typically mean "What [or who] is to blame?" (James, 2007, v. 2, p. 309)

Competitive speech tends to automatically default to habitual limbic tactics with affective rationalizations and justifications in lieu of more effective analytical solutions. Winning is the primary goal at any cost, including the use of brute force seen in humans, chimpanzees, and many other animals. When a being is lacking in brute-force abilities, deception is an obvious default, such as when an animal evolves a solution that can make itself look larger, like mammals whose hair stands up when threatened, or mimics a more dangerous animal, like spiders that are colored to look like a deadly species, or uses subterfuge, like squids that eject an ink-like substance as a smokescreen. Deception commonly occurs if "that's what it takes to win." Any combination of tactics that leads to winning by the predetermined solution defines acceptable behavior, especially when the option of losing offers the possibility of demise.

Solution, behavior, information, and *ideas* are equivocal terms from an evolutionary perspective. With humans, information and solutions incur similar competitive tactics to win, such as

- The brute force commonly seen in power hierarchies where the boss "always wins"
- Solutions that are spoofed up with hyperbole to make them look stronger, as in "Everybody knows this is the best solution"
- Smokescreens, such as "The fact of the matter is . . ."
- Insults and attempts to sabotage other solutions to demonstrate their "ridiculously obvious" inferiority

If our winner-take-all solution fails to out-compete the others, we can do and say whatever we want to defend ours and defeat the others. As highlighted in political discourse, if we have not displayed enough brute force to win, we can keep ratcheting up the force tit for tat and by increasing the deception with hyperbole, name calling, subterfuge, clandestine sabotage, and so on. We can do whatever it takes to win and continue spoofing up our solution and deflating the others until the bitter end: "The other solution is pathetically stupid." "It will cause a disaster and wreak havoc." "The sky will fall." If we haven't already, in a pinch we can resort to imperatives, such as "shoulds," "musts," and so forth. If nothing else works and we have enough brute force, we have the option of declaring all-out war, which usually starts with, "We *must* go to war because. . . ."

Some languages don't beat around the bush with imperative words but still rely on the implication of punishment: "Just do it." In other words, the receiver understands that disobedience can lead to banishment to a harsh environment. We typically learn imperative tactics from our early parental relationships even when they are implicit, such as a disgruntled look that tacitly conveys "Do it because I say so." We do not like to lose in retrospect, and out of habit we tend to continue to mimic parental authoritarian strategies that likely implicitly augmented our limbic brain's competitive circuits with imperative linguistic scaffolding and constraints and underlie our learning about "how we must deal with relationships."

The point here is not the hierarchal competition per se but the default deception that seems to rear its head when our limbic system takes charge. Limbic competition attempts to optimize problem resolution by affective leveraging of power. Top-down regulation by dominance can be regarded as noncooperative, with or without muted feedback. Rewarding only the predetermined absolute objective of winning is reminiscent of objective search algorithms and a tendency for local entrapment.

A horizontal axis rewards cooperative solution search with lateral inhibition that helps to contrast and converge multiple possible solutions in the context of the problem. Lateral inhibition, originally postulated in the 1950s by Horace Barlow, has become an important principle in neural computation (Barlow, 1953, 1959) and has been demonstrated to play a nontrivial self-organizing role in linguistic experiments via positive feedback for evolving coherent vocabularies in populations of autonomous robots (Steels, 1995; Steels et al, 2007; Steels & Kaplan, 2002). The apparent axial association with language and communication suggests that changing the semantic axis can noticeably change the communication space geometry. We have the option to improve cooperative communication by applying language that rewards effective collaboration and novel solutions.

We can shift our vocabulary from command to description and prediction (Bronowski, 1977, p. 153). Viewing competition and cooperation in terms of outcomes allows us to adjust the language priorities from command on the vertical competitive axis—prone to deception and penalties that inhibit the consideration of alternatives—to description and prediction on the horizontal cooperative axis, which relies on rewarding effective problem resolution. Cooperation is antithetical to deception, since deceit corrupts information.

Reinforcement paradigms have been studied by B.F. Skinner and others in animal and human experiments, with the general conclusion that to change behaviors, methods that reward the preferred behavior typically outperform methods that penalize the unwanted behavior (Skinner, 1953). Language that supports imperatives and deception and penalizes alternative solutions tends to suppress bottom-up cooperative behavior, so reconsidering our language preference makes sense. Effective language has relevant implications for improving cooperative communication.

Cooperative Bias

Because of the physiology and architecture of the brain, we cannot avoid bias in our behavior. *Bias* simply refers to a tendency toward a type of behavior, or perhaps the twisting of our perspective by such a tendency. We have been discussing biases that typically go unnoticed and unexamined and develop along with language acquisition in an already biased brain. Acknowledging the ubiquitous nature of bias and the association with language provides an opportunity to deliberately modulate that bias.

Research evidence indicates that human economic behavior perhaps evolved from, and demonstrates a close relationship to, the economic behaviors of our animal relatives. Studies by Glimcher et al (2006) suggest that other primates, such as monkeys, play mixed-strategy equilibrium games with relatively the same efficiency as humans; they tend to employ a strategy where each move matches the previous behavior of opponents—so-called tit for tat. As a reactive lower-level default mechanism, tit for tat has sufficed for less-complex organisms across the evolutionary scale and is found in the behavior of more complex organisms, including humans. Tit for tat apparently represents a more-or-less automatic, reactive, limbic default system for imperatively matching force with force (Mesulam, 2002, p. 19), analogous to the fundamental principle of Newton's laws of motion related to opposing forces, actions, and reactions.

Language radically alters the brain's online ability to compare, contrast, discriminate, and associate information in real time to deliberately guide thinking and problem solving (LeDoux, 2002, p. 197). Unlike organisms that lack sophisticated language abilities, we demonstrate a capacity for modulating our reactive mechanisms with deliberate inhibition (Panksepp, 1998, p. 322). In other words, our brain allows flexible regulation of our emotional system (Tranel, 2002, pp. 350–351). The language we inherit biases the flexible modulation of our community behavior. We incorporate cooperative and competitive biases from our brain architecture and early and later learning about how to act and communicate in relationships. The brain we inherit and our exposure to learning and practicing cooperation in various life situations play a part in our cooperative and competitive bias. We can enhance cooperation by modeling policies that support and reinforce cooperative behavior. Generally, the earlier we learn cooperation and the use of cooperative language, the greater the fluency of our cooperative skills.

Cooperation often fails to take hold in the absence of supportive policies. Policies create conducive linguistic environments by explicitly modeling cooperative language and providing oversight, feedback, and reinforcement. In situations lacking such support, we tend to operate on autopilot by default, in a sense blind, as if the policies do not exist or simply represent meaningless arbitrary dictates that have little bearing on our cooperative behavior. We easily default and revert to implicit habitual behaviors associated with familiar status quo policies (Tversky & Kahneman, 2000, p. 146; Kahneman et al, 2000, p. 163). Policies that rely on dominance-related hierarchies model and reinforce subordination and obedience to authority.

Game theorists and neuroeconomists sometimes face criticism for relying on highly constrained experiments and for producing allegedly trivial results or laboratory curiosities (Gilovich & Griffin, 2002, pp. 11–15), for conducting interdisciplinary arguments about how to reconcile prescriptive or descriptive approaches (Glimcher et al, 2006), and for their novel definition of rational behavior. Despite such criticism, their experimental results provide a radically novel and potentially enlightening perspective on processes of human reasoning and decision making. Their research has produced intriguing results, including the pioneering discoveries of Daniel Kahneman and Amos Tversky first published in 1982 with Paul Slovic (Kahneman et al, 1982) and elaborated in *Heuristics and Biases: The Psychology of Intuitive Judgment* (Tversky & Kahneman, 2002) and Robert Axelrod's elucidation of the prisoner's dilemma in *The Evolution of Cooperation* (Axelrod, 2006).

Axelrod points out that reciprocity and the "stability" supplied by the shadow of the future can nourish robust cooperation, and that the foundation of cooperation lies in the durability of relationships (Axelrod, 2006, pp. 173, 182). He offers the following pointers for improving cooperation:

- Enlarge the shadow of the future
- Change the payoffs
- Teach people to care about each other
- Teach reciprocity
- Improve recognition abilities for cooperation
- Avoid labels, stereotyping, and status hierarchies

> Perhaps if we understand this process better, we can use our foresight to speed up the evolution of cooperation. (Axelrod, 2006, p. 191)

Deliberate tuning of the language we use and the policies we construct can provide us with a fundamental resource for understanding, rewarding, and engendering cooperative interactions. We can deliberately apply our language-derived knowledge to increase effective communication and to bias policies that implement, reward, and accelerate the evolution of cooperation.

Policy Bias

Policies typically refer to a range of admissible actions or behaviors in a system. Recall that policies can be viewed as analogous to physical laws that describe admissible actions in a domain according to physical constraints. We find different styles of implicit and explicit policies that span a breadth from narrow to broad, rigid to flexible, local to global. Global policies are general and define the extent of systematic actions from a macroscopic perspective. We usually think of policies as organizational rules that govern conduct, but—thanks to language—policies are a routine part of our lives. The language-compatible brain we inherited can function implicitly. When we add languages, we enable the ability to define, represent, and express policies explicitly.

Our early abilities for thinking and communicating progress as we develop, along with learning that typically occurs automatically and implicitly with limbic modulation.

We learn to think and communicate in large part by default, but the influence of habits we incur usually lingers throughout our lives. We inherit language-related policies and values that bias our inferences and perception, how we inform our thought processing, and our actions for relating with the outside world. Deliberately choosing and asserting preferred policies on the basis of effectiveness allows us to ameliorate the affective nature of our limbic brain and cultural heredity. The many problems identified by game theory and neuroeconomics, related to competition, cooperation, and heuristics and biases, can in large part be mitigated by deliberately applying extended language and effective linguistic policies to improve our cooperative behavior, including our communication.

Cooperative communication relies on the ability to transition language constructions from vertical to horizontal, from closed-ended to open-ended, from absolute to relative. Such a transition allows us to extend our language from certitudes toward informed preferences, opinions, ideas, and possibilities. Policies that reinforce informed communication also support informed thinking.

Competitive policy-biasing problems arise when we default to absolute language, such as

- You always make stupid mistakes.
- You are a loser.
- You just don't get it. We need to finish the project today.
- Where is your sense of urgency?
- You screwed up again, and around here you're only as good as your last ball game.

We see a correspondence with policies that facilitate absolute thought processing. *Always* goes to infinity, which we understand is imaginary. The implication is that the mistake maker *is infinitely stupid*, which points to a systematic classification error between static classes and action. Mistakes we make are often related to implicit systematic errors from assumptions of certainty and the denial of ignorance that more than likely led to poorly informed planning and maladaptive problem resolution.

Just implies a single absolute *cause* for a problem that the person somehow lacks the ability to *sense* and deduce implicit information about. *Need* implies persuasion void of effective information by attempting to affectively leverage a preference to a matter of survival—for example, we *need* food, water, air, and shelter to survive—which seems disconnected from prioritizing the preferred completion of a task.

A sense of urgency implicitly assumes that we have special *extrasensory* receptors for detecting *urgency* produced by the perception of someone else's brain, to which we have no direct access. As far as we know, extrasensory transmission and reception has no scientifically documented physical basis. We expect that explicit communication is generally more effective than implicit assumptions and telepathy.

The last statement assumes the past *determines* the future, implies implausible cause and effect compatible with an idealized imaginary system and symmetrical time, and hardly fits the world in which we live and work. It also implies an authoritarian, top-down workplace policy bias that relies on hierarchal dominance and a winner-take-all competitive system in lieu of a bottom-up cooperative system.

In a cooperative environment we would expect benefits from effective language and a systematic perspective:

- We humans make mistakes. Let's go back and look at the system and see if there is a larger systematic problem.
- We can take a time out to elaborate on the problem space and consider alternatives for finding practical, adaptive solutions.
- Perhaps we can get the project finished today. We will make a note to review our planning and resource-management processes to check for systematic problems that we can adjust to enhance efficiencies and effectiveness.
- It seems that our system might be missing an explicit feedback loop for correlating time management and oversight. We can't go back and repair past mistakes, but we can probably learn something from this that will help with effectively tuning our system, decreasing our systematic error rate, and increasing the reliability of our predictions.

The former example of affective speech behavior demonstrates absolute policy-biasing constraints on systematic information and communication. It is difficult to get effective results with undetected ambiguous information fraught with unexamined implicit assumptions and coarse "true" or "false" processing. Further absolute constraints relate to incomplete logic forms, affective decision making, and a default to competitive behavior, including deceptive communication. As seen in the latter example of effective speech behavior, language changes the bias and the effectiveness of policies with which we resolve these problems.

For example, we can think about policies with affective language as the implicit default. Most cultures, including corporations, have implicit and explicit policies. They may have explicit written policies, but they typically lack the explicit language policies that enhance effective communication. If we think about language as information, we can construct and routinely apply explicit language policies that facilitate effectively informed communication. Similar to understanding the systematic policies (laws) of nature, we gain knowledge that informs us about how relationships in the system work. Knowledge of the policies does not mean that we can make predictions with certainty. It does give us a more effective understanding of natural systematic relationships and of the importance of expressing predictions as possibilities that we can extrapolate across other systems.

With effective language policies that offer a vocabulary based on explicit informatives, the fidelity of communication naturally improves. We can establish routine policies for stating preferences, opinions, and ideas as possibilities that replace absolutes, prescriptives, imperatives, and certitudes, such as *should, must, supposed to, have to,* and *need to.* Recall that prescriptives and imperatives are execute commands—"just do it"—that are pretty much void of information. They create power but also the inherent liability of an information vacuum. In the vacuum, communication and decisions default to implicit language habits as automatic language primitives.

Explicitly informed decisions generally enable more effective learning. Trial-and-error learning benefits from explicit analysis of the thought process, assumptions, and information that contributed to the decision and the systematic errors. When implicit decisions based on empty commands result in systematic errors, there is no way to go back in time and make corrections by informed analysis. Similar to derivatives in the mortgage crisis of 2008, when the derivatives became worthless, panic ensued when on finally looking in the bag, we found the bag to be empty. Unfortunately, in what initially seemed like a sure thing, the only surety was that somebody would be left holding the bag.

Policies that neglect or impede informed decisions can lead to undesirable outcomes and result in minimal learning. What did we learn at best? Not much. If any learning takes place from uninformed decisions, it is "coarse at best," such as these:

- Don't get left holding the bag.
- Don't buy empty bags.
- Shake the bag before you buy it.
- Don't get scammed.
- I knew better.
- I won't do that again.
- Why did that happen to me?
- How could I have done that?

For at least a couple of thousand years, we humans have demonstrated a tendency to persistently repeat mistakes, with no end in sight (origination often attributed to the Roman orator Cato). Perhaps eternally recurring errors boil down to a pretty simple problem in general, associated with the influence of an assuming brain and affective language, absolute information, and rigid thought processing. And perhaps a pretty simple solution: increase effective language that enables nonabsolute systematic information and flexible thought processing—put in place policies that are mutually exclusive with the default mode to absolutes. It may seem like faulty structure and anything but frugal, since effective thought processing involves a lot of work and practically guarantees "no free lunch."

But enhancing effective thinking looks more like a bargain when we consider the practical implications for decreasing systematic errors, and especially when we calculate the cost of those errors. Effective thinking is not necessarily easy, but it offers an information-processing method for producing and modulating more reliably adaptable solutions. When we consider potentially irreversible errors, such as those contributing to the destruction of our planet, effective thinking gains appeal. It seems in our best interest to increase our effective thinking abilities, and especially so for addressing concerns about unbridled pollution, greenhouse gases, and widespread destruction of forests. We often think about hindsight as 20/20, but hindsight applies to nonlethal errors from a population perspective. We can reevaluate the problem space and ask a few more questions in case we missed the systematic problem:

- Why do we often hear statements about humans repeating the same errors?
- Do we continue to repeat the same mistakes over and over?
- What keeps us from learning from our errors?
- Could the problem have anything to do with the effectiveness of the language and logic we use for thinking?

Perhaps the most relevant question is "Why do we choose and use primitive language constructions that support absolute policies and affective information to the exclusion of effective language and analytical information?" Of course, effective language hardly guarantees that we will never experience undesirable outcomes, but it does facilitate effective planning and learning from our errors. We expect effective language to more accurately inform our policies and hence our cognitive behavior. Cognitive fidelity—for information, information processing, and event-related application—integrates efficacy-based principles of REBT and CBT that support more coherent thought and behavior, including more effective speech behavior (Ellis & Harper, 1997; Beck, 1976). The policies for cognitive accuracy have been proposed in prior manuscripts and have been put in place as a pilot in a corporate setting, producing encouraging results in terms of cooperative behavior (Bailey, 2007a).

- We are all flawed and fallible humans
- Informatives and preferences with possibilities and probabilities generally enhance flexibility and work better than imperatives and prescriptives
- Each person is responsible for their own thoughts, feelings, and behavior

Accuracy-based policies—borrowed from bottom-up organizing principles found in complex systems—rely on orderly methods and procedures with

- Explicit referencing to and the application of forward-directed one-way time
- Extended systematic language constructions
- Probability as logic

- Plausible information and possibilities referenced to the domain
- Effective thought processing
- Open-ended language and communication conducive to open-minded solutions and cooperation

These systematic policies promote higher fidelity of information for effective learning and risk–reward assessment, and enhance the constructiveness of criticism with reliable positive and negative feedback. Reliable feedback decreases systematic error rates and contributes guidance for directing improvements in training and reinforcing cooperative language behavior. Accuracy-based cooperative policies facilitate flexible thought processes for effectively fine-tuning learning and correcting errors. The system is held responsible and accountable for the influence of explicit policies and oversight that reward information fidelity and the effective production and modulation of adaptive output rather than implicitly relying on subordinate blame and punishment.

An absolute strategy based on the imaginary notion of perfection and zero defects qualifies as particularly irrational because as flawed and fallible human beings, we can expect to err (Ellis & Harper, 1997). The nature of humans is to assume and err, and imagining otherwise offers little consolation. By accepting the notion of human systematic error, we have the option of rewarding cooperative optimization policies, persistent oversight, and training that can facilitate decreasing systematic error rates. Alternatively, we can continue defaulting to affective policies that try to punish errors away because of their disobedience in living up to imperatively mandated perfection. Destructive punishment in arrears hardly substitutes for constructive learning and rewarding cooperation in the present.

Our language-endowed speech abilities can enhance communication and cooperation (Cacioppo & Berntson, 2004, p. 978). Cooperation and informative communication can more likely evolve from policies that reinforce a safe, open, unintimidating environment conducive to collaboration and creativity. When we perceive danger or when anxiety is high, we tend to abandon such "lateral," creative thinking (Spitzer, 1999, pp. 271, 274–275). We can deliberately bias the direction of our effort toward language constructions that improve fine-tuning of efficient information flow, error management, and feedback. A policy bias that relies on linguistic fine-tuning tends to improve problem resolution, boost creativity, and increase the adaptive flexibility of the system.

An effective policy bias facilitates domain-correlated information and processing accuracy from a systematic frame of reference that defines plausible limits for elaborating the problem-and-solution search space. Since we live and operate in an imperfect, complex world with instabilities and uncertainties, it makes sense to reinforce thinking with linguistically coherent policies that correspond with the dynamics of that world. An effective policy bias facilitates thoughtfully informed decisions by rewarding the variability of ideas and encouraging alternative solutions, even though they may sometimes result in imperfect outcomes (errors). In other words, reward the decision-making process rather than punish how it turned out (Kahneman, 2011, p. 418). Effective policies facilitate cognitive accuracy that assumes individual responsibility and accountability for thoughts, feelings, and behavior by reinforcing the assumption—noted more than two thousand years ago by the Greek philosopher Epictetus— that our feelings and behavior can result from or be influenced by our own thoughts (Beck, 1976; Ellis & Harper, 1997; Hooker & Knight, 2006, p. 317; Ochsner, 2007, p. 107; Ochsner et al, 2005, p. 253).

Effective policies provide alternative possibilities beyond the either–or constraints of the "cooperate or defect" matrix described by Axelrod in the prisoner's dilemma. We have the option of extended language that enables analysis of multiple alternative solutions and enhances flexibility for more effective adaptation. An effective policy bias with accuracy-based referencing enhances cooperative communication by laterally correlating language to the domain rather than vertically and affectively to rank order. A policy bias based on effective language offers an adaptive tool for supporting, enhancing, and coordinating collaborative thinking that predictably supports cooperative communication. When it comes to cooperative policies, effective language matters.

Intuitive Bias

Intuition refers to the experience of knowing an answer in the absence of obvious deliberate analysis or insight into the processing, and lacking the benefit of explicit information. The results of intuition-based problem solving or decisions can range from desirable to less than desirable, and sometimes to disastrous. Most mammals seem to be adept at automatically solving problems using their limbic systems as guides. As mammals in general, we have a similar limbic brain that has been remarkably well conserved during the course of mammalian evolution, at least operationally (MacLean, 1990; Panksepp, 1998, p. 9). We would be hard pressed to argue against the appeal of intuition in comparison with analytical thinking for inexpensively solving problems; it seems to work "good enough."

In humans, as in other primates and mammals, bottom-up solutions are automatically generated by limbic-associated processes. But in humans, we often refer to these automatic cognitive behaviors as intuitions. We typically generate these intuitions on the basis of locally encoded information already learned within a subjective internal frame of reference. We also have a highly developed linguistic abstracting system, which extends our ability for deliberate analytical thought. Without language, we can still solve problems about as well as our mammalian cousins can, but with it, we can solve problems with a broader range of options for effectively adapting to situations that other mammals resolve affectively by fighting, freezing, or fleeing.

Our brain tends to silently address problems implicitly and automatically produce intuitive solutions that we interpret by the conventional encoding of our native language. Sometimes these intuitive solutions yield satisfactory or even surprisingly clever results. Our brain typically generates automatic solutions a step ahead of us, which allows for fast responses. We usually lack insight into how we automatically derive the solutions and make decisions. Nor do many of these situations require explicit insight—we could hardly function efficiently on a daily basis if we were left to deliberately decide how and when to move each foot as we walk about doing mundane tasks. For more complex situations, though, we have the advantage of applying language to help structure the problem, accumulate and process information, generate alternatives, and analytically evaluate possible outcomes.

Intuitively decided solutions are a separate step from reconstructing the thought process that produced them. We easily miss out on understanding the breadth of the problem, information, and processing involved. We might even overlook the option for explicit analytical deliberation. Although relying on intuition alone to address complex problems and decisions leaves us vulnerable to the linguistic inaccuracies embedded in our beliefs and assumptions, intuition can generally act as a good-enough filter as long as we know in advance what kind of contamination can occur (Jaynes, 2007, p. 536). In other words, acknowledging the constraints and liabilities of intuition encourages careful inspection of the derived solutions rather than carte blanche acceptance at face value.

Intuition and *making sense* are often used synonymously. The effective usefulness of intuition can depend on the fidelity of the acquired information, logic, beliefs, and values from which it was automatically derived. Intuitive solutions derive from prior learning and familiarity that tend to be biased and constrained by our brain and language. We can separate intuition from what we call insight depending on the reliability of the underlying knowledge and the effectiveness of the solutions. As noted earlier, intuition tends to depend on locally biased language and "common sense" rather than global sense.

For example, how would we answer when asked to explain why we prefer the language we currently use? Out intuitive answer might be "Because it works just fine." Where does the information for answering the question come from? Relying on arbitrary information without knowing the source, size, completeness, or representational and classification accuracy of the original data set can easily introduce noise and potentially induce noisy reasoning (Bishop, 2006, pp. 12, 32–39). With unexamined language and implicit reasoning, we likely run into difficulty coming up with a logical explanation for what we mean by "it works just fine."

We tend to automatically produce intuitive solutions and then apply upper-level cognition with affective language to explicitly validate them by justification, discredit alternatives, and explain and define the problem to conform to our intuition. Of course, we are more likely to take credit for positive outcomes and relegate the negative ones to others or caprice. We can always fall back on the ad hoc strategy of the "Texas sharpshooter adjustment." We cleverly move the target after the shot to wherever the bullet hit and make dubious proclamations of accuracy, certainty, and the perfect shot. This trick allows us to redefine the problems to fit our a priori intuitive solution. Acknowledging errors and learning from them takes a backseat to familiar choices and the certainty of being "right on target." We intuitively prefer winning to losing, even in retrospect, and are liable to deceptively turn back the clock to readjust the problem to fit the solution. Of course, turning the clock back and forth locally is hardly a problem with the imaginary flexibility offered by two-way time.

Without deliberate inspection, this implicit intuitive approach seems a bit backward and can lead to some awkward, complicated, and convoluted after-the-fact explanations for how the supposedly a priori calculations produced such absolutely certain accuracy and precision. This tactic might work nicely if we were walking backward with two-way time and relying on hindsight. After all, evaluations made through hindsight require less effort and may seem fairly accurate, since an event has already happened, but there are liabilities associated with walking backward.

Even hindsight runs into difficulties when the problem and solution spaces are not effectively explored initially. If noticed later, errors can appear as spurious coincidences when lacking initial assessment of the problem and explicit information related to the prediction. Learning, if any, is sparse without reliable prior information for measuring the systematic error signal on the basis of the difference between the predicted and realized outcome. Backward predictions or inferences "after the facts are in" still suffer from the a priori ignorance produced by intuition with previously uninformed certainty, unacknowledged assumptions, and the lack of alternative solutions.

There is a simple backward approach, by intuitively starting with an implicit solution, destroying alternatives, and then magically discovering the problem that matches inside the solution, or "I have the solution . . . what is the problem?" With deliberate reasoning, we can instead focus on understanding the problem before initiating a search for solutions and alternatives. We can first deliberately identify, explore, and elaborate the problem space. Then we can apply open-ended thinking to search for and consider possible solutions and alternatives that probably best fit the specific problem task. In this manner other alternatives can be put aside for future construction of solutions as required for different tasks.

Statistically predicting the most likely solution that best fits the problem a priori has more practical appeal than a backward approach of intuitively retrofitting the problem to the solution. Thanks to our clever language abilities and 20/20 hindsight, unique in the animal kingdom, we are perhaps the only animal capable of intuitively applying this reverse recipe. Hindsight often leads us to second-guess our intuitions. But, when possible, we can avoid second-guessing by relying on statistically informed solutions and resisting the temptation to guess in the first place. For example, when we intuitively guess at what we are "supposed to do," we lack explicit information for making the choice and correcting errors:

> I feel like my broker is an authority on stocks. He had a hunch that these stocks were a *sure* thing, and told me I *should* buy them. I'm *supposed to* trust my broker's intuition, so I bought them. But they keep going down, so my broker's hunch was wrong. I *shouldn't* have listened or maybe it was a coincidence. Maybe he *is* a *bad* broker? Now I am confused and still don't know what I was *supposed to* do in a situation like that.

After the fact, we often second-guess ourselves by saying, "I wasn't supposed to." More thoughtful decisions with statistically informed predictions typically improve outcomes and learning from errors.

- The stock market has inherent swings, cycles, bubbles, and oscillations, with various possibilities and difficult-to-predict risk–reward ratios.
- My broker recommends that I take my own risk tolerance into account and perform due diligence before making choices or authorizing purchases by others. After all, I am responsible and accountable for my decisions.
- It's not perfect and doesn't guarantee success, but staying as knowledgeable as possible about risks, not putting all our eggs in one basket, and deliberately informing choices seems to work better than relying on hunches and guessing.

In the stock market we see trends in which people seem to buy stocks on the basis of popularity. When stocks get positive comments, word of mouth grows and investors tend to intuitively pile on. In other words, we have trouble resisting our intuitive urge to rush in and feast on a good thing, especially when it seems like a sure thing and most agree. It is easy to think that future stock prices and financial markets are predictable because they reflect news events that are knowable. Think about this phenomenon in a natural kind of way. Stocks can advance or decline in two directions, either up or down. *Up or down* represents a competitive ordinal scale according to rank order, either above or below. In this case winning requires being at the top of the ranking order; winner takes all. That leaves room for only one winner at the top, similar to a power hierarchy. Think about vertical scales as trying to defy gravity so that we can climb higher and higher.

We know that humans, like most other animals, exhibit an emotional predisposition to accumulate positives; we follow the crowd because of our social herd tendency. Depending on our dominant–subordinate status, we intuitively try to stay up with or ahead of the status quo. But our limbic–amygdaloid system can only push stocks—or most any other thing we value—so far in one direction. We run or climb until the upward trend exhausts itself, gravitational forces take over, and the avalanche begins in the downward direction. We seem to forget the stock market mantra and nondeterministic disclaimer: "Past performance is not indicative of future returns." Do we intuitively believe that the past predicts the future and that they are equivalent? Our behavior gives us away and implies that we lack insight into what's going on with our limbic system. Why do we ignore knowledge about human mammalian behavior, claim humans act rationally, and blame our demise on bad luck or assumptions of randomness brought on by a capricious world? Perhaps assuming and blaming lets us take the easier way out without blemishing our status.

We think risks are low because we played it safe and let our neighbor go first. We intuitively assume that it's safe, since they told us how well they did and they obviously "didn't get eaten alive." There is a fairly simple rule as to how to make money in the stock market: "Buy low and sell high." Why does the alleged smartest animal on the planet buy high and sell low? We intuitively pile onto positives, and when at some point the rewards turn to penalties that we can no longer bear, we spin around and take off in another direction, similar to paramecium and many other single-cell organisms. Of course, we've known this for over a hundred years (Jennings, 1906), but apparently some news spreads slowly. At least we can still spuriously claim that we are rational and blame our economic misfortune on a capricious world.

Risks, penalties, rewards, and uncertainties are a part of our daily lives. We sometimes hear of people's fears about asteroids striking the Earth and causing a calamity. Great news—NASA claims to have discovered more than 90% of the larger ones and to know which way most of them are headed. Of course, we would like to know with certainty, but at least we have some knowledge of lower risks. What do we make of the instances of lawmakers passing special laws that benefit only themselves, padding their own salary, double-dipping, and accelerating their pension? Are they random acts? Perhaps they are intuitively following what is known as Sutton's law: "Because that's where the money is!" Of course, we prefer the risk–reward ratio to be in our own favor.

Do we believe people intuitively behave in such a way that suggests there is something in it for them? Why would some groups be against reducing toxic emissions? Do they say, "We are biased and looking out for ourselves"? It is unlikely. When we make emotional decisions, we tend to intuitively increase our proclamations of "unbiased" and give a litany of reasons awful things will happen if we act otherwise: "we can't afford it; it will ruin our economy; it will kill jobs; no one can meet those expectations." When we see these polemic arguments going on, we might suspect that the limbic–amygdaloid system is likely at work, along with an intuitive emotional attraction and reaction to some locally polarized beliefs. On the other hand, sometimes we are in a quandary about how to behave and are left guessing as to what we are "supposed to do."

Depending on the supposed authority to which we answer, "supposed to" typically can vary quite a bit in the moment and the context of a situation. Understanding that we have choices leads to the importance of maximizing information prior to reasoning and behaving, and making effective rather than affective predictions and intuitive decisions. Predictions based on different possibilities enable us to rely on explicit information that helps us to weigh choices and risks analytically rather than emotionally. Rather than ignoring information and habitually relying on implicit intuitive reasoning, we can deliberately base our predictions and choices on thoughtfully processed plausible information. The point here is not that everyone "should" keep a calculator in their pocket and make endless calculations for every decision. But thinking and decision making generally benefit from deliberately considering many possibilities and elaborating on the problem space, gathering plausible information, and checking underlying assumptions. We can expect plausibly informed decisions to typically result in more effective outcomes.

The more we know about the how the world works, the more plausible options we have available, and the more likely we are to think in terms of various possibilities. Thinking about the world in possibilities, opinions, and ideas means we can think more open-mindedly. Intuitions based on closed-minded absolutes leave no room for uncertainty. Rather than accepting automatic intuitive decisions and rationalizing them later, cultivating analytical methods by practicing deliberately thinking things through can lead to replacing the habit of spontaneous decision generation with plausibly informed thoughtful reasoning. Trading intuitive "supposed to" for deliberate thinking enables the consideration of alternative choices and various possibilities for solutions that better correspond to adapting to the world in which we live. We own the choices we make and the actions we take, for which we alone remain responsible and accountable.

Generally, effective approaches consist of deliberately exploring and defining the problem space a priori, then searching for different alternative solution possibilities based on plausible information and choosing a solution that probably best fits the problem. This iterative open search process continues by deliberate and persistent feedback, comparing the results with the predicted solution and looking for omissions, assumptions, or other contributing factors that increase the probability of errors. We can recalibrate our computations to reflect effective error assessment, ongoing oversight, and constructive learning.

Trial-and-error learning benefits from a process that acknowledges and anticipates mistakes, in which discovering and learning from the mistakes takes the front seat. A simple improvement can be made by deliberately shifting from intuitive certitudes and discrete two-valued logic to open-ended computations with possibilities. Intuitive decision making can start with a spontaneously discovered solution, then define the problem in terms of that solution, and sometimes offer amazingly good results. In the long run, conducting a priori evaluation of the problem, developing multiple alternatives, and converging to a solution with analytical reasoning generally facilitates effective outcomes and adaptive learning. The solution can then be evaluated and adjusted to tasks through ongoing feedback.

We tend to habitually base our decisions on the most accessible information, which often emerges intuitively with the first thought that occurs, without further reflection (Schwarz & Vaughn, 2002, pp. 104–105). The most familiar thoughts tend to automatically pop up first. The intuitively derived bottom-up solution can then be set as the top-down absolute static objective. The resulting stepping-stones to the static objective are rewarded, while alternatives and novelty are punished and dismissed. But we have already seen how with static objectives we can miss out on the discovery of alternatives that can lead to innovative solutions. Searching the problem and solution space for more information and alternative solutions can increase the discovery of previously unknown information, assumptions, risks, and opportunities.

The ease with which we intuitively reach a yes-or-no decision can seduce us into ignoring the behind-the-scenes limbic processing that produced the decision. We can benefit from acknowledging the variability and constraints of intuitive thought, which suggest that we avoid submitting intuitions or any other automatic thought as certitudes. Rather it encourages us to deliberately and carefully evaluate intuitions according to effectiveness. We routinely apply intuitive thinking for everyday problems, but confusing intuitions with chiseled-in-stone truths can mislead us. The solutions that our brains automatically construct may seem like certainties, but the typical either–or solutions that intuition produces result from our limbic architecture, physiology, and discrete linguistic bias rather than from the effective fidelity of informed thought. The limbic system is tuned to the processing of localized affective information. Habitually depending on limbic intuitions can lead us to believe we know a lot. But the limbic brain's area of expertise comes from knowing what feels good and what feels bad. In other words, the intuitive limbic system has subjective expertise in feelings, and affective rather than effective value, of which it remains ignorant.

Ignoring the absolute nature of intuition can automatically trap us in a world of black-and-white certainties and constrain our search for more effective information and knowledge. In turn, intuitive traps can also prevent us from discovering hidden faulty assumptions and relevant information. When we lack insight, our ignorance can hamper the effectiveness of our day-to-day reasoning. Acknowledging the potential liabilities and deception associated with intuition can encourage deliberate thought that leads us to reconsider the reliability of our certitudes. We might also consider why we intuitively believe that certitudes have much of anything to do with the world in which we live. Analyzing the habitual intuitive nature of our brain to automatically accept the assumptions that it produces as facts encourages us to work on increasing our explicitly informed reasoning before acting—including thinking before speaking.

Intuitive limbic thought processing knows little if anything at all about effective dimensions, analysis, probability, or multiple alternative solutions. The limbic brain simply does what it does best: it naively divides the input space into two regular cells, which tends to discretize continuous information and produce either–or solutions. Discrete limbic compression of information tends to result in intuitions with one-dimensional problem resolution. Implicitly hiding assumptions and limiting the consideration of possible multiple solutions and conclusions results in the wasting of information (Jaynes, 2007, p. 174).

Even though intuition sometimes yields desirable insightful outcomes, when it fails, its opaque nature can lead us into self-deception by implicitly concealing the information, assumptions, and systematic error bias that contributed to the errant decision. Thus deceived, we can find no easy explanation to account for or correct our erroneous predictions. The intuitive process remains hidden from our view behind a tangled wall of implicit assumptions and biases that limit our insightfulness. These implicit biases can be logically deduced by carefully examining the information given about the problem and evaluating the resulting solution. By examining the *fingerprint* of the output product, inferences can be made about the logic most likely involved in the reasoning that led to a particular solution from a particular input. In effect, we ask: "What situational logic explicitly maps output prediction y to input stimulus x?"

We typically benefit from taking the time to deliberately evaluate intuitively derived solutions before enacting them, including decisions that seem simple but where we might be ignorant of the greater systematic risks. In the next chapter we will evaluate further liabilities associated with systematic biases.

CHAPTER 7

Systematic Biases III

THE DISTRIBUTION OF DIFFERENT human traits tends to produce a bell-shaped curve. Most of us fall somewhere in the middle on most of these traits, depending on the circumstances. For instance, a trait distribution of flexibility and inflexibility (rigidity) that places each trait at one end of the curve with a 50/50 mix in the peak at the middle indicates that the majority is in between the extremes. Although we can attribute some part of the traits to genetics and development, language also has an influence. It has been shown that REBT and CBT can help to decrease inflexibility by teaching people to apply more effective language to their thinking and communication that better accommodates flexibility and dealing with uncertainty.

Inflexibility Bias

Why is flexibility a relevant issue? Usually, more complex animals communicate more flexibly (Spitzer, 1999, p. 39). Signaling complexity tends to increase from single-cell organisms to vertebrates and human primates. *Signaling* is defined as the transmission or communication of information, and global signaling entails the unfettered flow of information across the entire system. Recall that a global system requires generalizations and open-ended processes that leave the solution space open. Effective language provides extended open-ended construction and generalizations for representing systematic information. Generalizations increase flexibility and broaden the adaptive range for information transmission as signaling complexity increases in dynamic domains. Extended language facilitates flexible signal tuning related to cognition and behavior. Situational contexts change, and flexible thinking increases adaptive performance over a broader range of changing conditions.

Humans who tend to think in absolute terms often display rigid and intolerant beliefs and behavior and have difficulty understanding the contextual significance of a particular conclusion or solution in a specific situation (Browne & Keeley, 2007, p. 183). For example, when we discuss issues as true or false without considering context, we can rigidly argue for and defend one side of an issue as the absolute truth. Context typically involves multiple variables that allow many possible configurations of perception in different situations. If we include a consideration of context in our discussion, we find it more difficult to reduce the possible options to one dimension: yes or no, true or false. For example, think about a two-sided coin with heads on one side and tails on the other representing either true or false. We are limited to viewing only one side at a time that confirms either true or false. The dynamics we face in the real world present a greater challenge, since information continuously changes and hardly resembles the example of a two-sided coin.

A rigid, absolute perspective allows one view. Rigidly focusing on one perspective can limit us to that view, and multiple other views are suppressed by default. A change in context then requires ad hoc explanations to shore up the rigidly fixated perspective. A rigid perspective that only allows a single view, under which nothing changes, limits options in a complex world where things change continuously. But context inevitably changes—even in the face of rigid beliefs that we refuse to change—requiring us to constantly change our explanations. Cognitive inflexibility with one rigid perspective may sometimes suffice but works best in an imaginary idealized world where context converges to static absolute perception: "Take it or leave it." Children exhibit a similar cognitive inflexibility when they do not understand that a single stimulus may look different and reveal unfamiliar features when seen from different perspectives (Zelazo et al, 2008, p. 556). From research in normal psychology, we know that people do not readily change their views, even when faced with substantial contradictory evidence (Spitzer, 1999, p. 291).

If we evaluate adults by the principles of cognitive flexibility, we find a remarkable persistence of the rigid perspectives of childhood (Pronin et al, 2002, pp. 641–642), even in experts trained in experimentation. Science continuously struggles to integrate the rapidly growing accumulation of information about the human brain, cognition, and behavior. But we seem to find it much easier to closed-mindedly and expertly argue "absolute right" or "absolute wrong," and to criticize the absolute imperfection of unfamiliar hypotheses, theories, beliefs, opinions, or ideas that are not self-similar, and often disapprove of and belittle the people who hold them.

> We acquire and retain material possessions because of the functions they serve and the value they offer. To some extent, the same can be said of our beliefs: We may be particularly inclined to acquire and retain beliefs that make us feel good. (Gilovich, 1991, p. 86)

As a practical matter, the constructive effectiveness of criticism seems to benefit from an open-minded effort to understand the object of the criticism, which requires cognitive flexibility. Why would we argue against an idea of which we know very little or virtually nothing at all? Saying "I do not believe in it" hardly excuses us from being a victim of our own ignorance. Criticizing poorly understood, unfamiliar ideas suggests defensive or destructive rather than constructive intentions. Inflexibility tends to freeze perspectives and shift the focus of attention to irrelevant details that obscure the big picture. We can rigorously scrutinize those details, no matter how irrelevant, to find a perceived error in some subpart, leveraging any flaw to dismiss the whole as irrelevant or absolutely wrong. Such a tactic, though, results in absolutely ignoring the idea in total. In other words, we have rigidly trapped ourselves in a local subspace and miss out on seeing the global picture because of inflexibility and self-imposed ignorance.

Inflexible people tend to be highly critical of others' beliefs. Ironically, they are often the ones least likely to accept critique of their own beliefs, but instead rigidly defend them while blatantly rejecting any criticism from others. Unfortunately, inflexibility seems to keep us from engaging in the kind of open-ended cooperative communication that science relies on to explore and logically understand different possibilities in the natural world.

This problem calls to mind the story of the blind men who encounter an elephant: by their other senses, each examines a different part and makes different inferences on the basis of their local sensory perception (Lytton, 2002, p. 224). One feels an ear, one feels a trunk, one feels a tail, and each has discovered only a small part of reality. They fail to realize that reaching the most accurate conclusion depends on taking a collaborative approach and integrating their isolated information. Each blind man believes he has discovered the absolute truth and thinks that the others are babbling nonsensically. It never occurs to them to pool their information and take a systematic approach. Rather than rigidly cling to an idealized concept of an absolute truth, they could have discovered a world filled with novel information, ideas, and possibilities. They are blinded not so much by lack of vision but by faulty reasoning with rigid absolute logic. Individual opinions are important—indeed we often employ them—but they tend to be counterproductive when absolute, uncritical devotion to a single opinion obscures the view of the larger system.

But even with all of our senses intact, we still struggle to see the whole. As the saying goes, sometimes "we can't see the forest for the trees." When we narrow our focus to an individual tree, we lose sight of the dimensions of the whole. Of course, we can still discover valuable information about the individual tree. Locally constructed language with discrete measurements can segregate and sometimes make sense of parts of the system down to the microscopic level. But we lack a globally extended sense and the flexibility that could give us a macroscopic perspective along with the micro.

To see the big picture, we widen our focus to take in more dimensions and use global sense to perceive the larger context in generalities. Open-ended language can serve as a lens for flexibly tuning our generalized sense that lets us see the possibilities at a systems level. Systematic sensing improves our possible options for flexibly changing our point of view between convergence and divergence, and it facilitates the discovery of novel alternative information for updating our knowledge.

We can measure inflexibility by the amount of information it would take to change a belief or opinion. Increasing requirements for the amount of information to change an opinion is associated with higher levels of rigidity. When only absolutely true information is admissible, changing even imaginary beliefs can require an infinite amount of "absolutely true" information. When the answer is that no amount of information would change our belief, then rigidity goes to infinity. But "not even the Bureau of Standards [currently named the National Institute of Standards and Technology] can give us evidence that good" (Jaynes, 2007, p. 574). Measuring from an implausible referent of idealized perfection typically leaves us with an idealized static snapshot at best.

Absolute Logic Bias

In a complex environment, few if any absolutes exist. The system and its information components change continually. In our daily life, absolute logic often seems to suffice, but we would do well to remember that it tends to obviate a macroscopic perspective. Absolute logic generally places performance constraints on effective cognition and adaptation.

Practically speaking and somewhat analogous to the constraints of a fixed fitness function with true or false objectives, an absolute logic bias sets the stage for maladaptive behaviors. Localized language constrained to absolute true-or-false logic lacks extended dynamic parameters for computing generalized solutions to the complex problems we encounter under real-world conditions. Absolute reasoning can deceptively obscure information that could improve coherent understanding and solve problems that we face on a daily basis (Langer, 2000; Langer & Piper, 1987; McInerny, 2005, pp. 94–95; Browne & Keeley, 2007, pp. 54, 182–183). Does this deception happen routinely, and if so, how?

Complex organisms naturally inherit default brain biases that suffice for survival and reproduction, and they learn other biases during their lifetime depending on their environment and level of neural complexity. Humans as a species are biased by our biological brains, language, and conventional localized learning and cultural belief systems. Our inherited beliefs and the embedded logic and values contribute to biasing our perception and inferences. Our biases tend to hide silently in our day-to-day habits, and we take our own worldview for granted as "normal." This invisible logic bias operates implicitly as the self-centered referent we routinely use for evaluating the bias of others.

Habituation to our own logic can make it "absolutely" invisible to us. We can see our behavior to some extent but usually lack insight into the logic by which it was processed and produced. Operating from our own locally biased frame of self-reference makes us more apt to notice the biases of those around us. We notice and critique the behavior of others when it does not make sense to us or "does not seem logical." We routinely notice how others differ from our "normal," but we just as routinely ignore or deny our own biases, as if "unaware of our own unawareness" (Pronin et al, 2002, pp. 638, 662).

Bias and ignorance of our bias affect our thought processes, perceptions, emotions, and behavior. We can spend time arguing, all the while trying to disprove any biases on our part, but fail to realize we all have biases. We are all susceptible simply for having brains with habitual tendencies. We have gotten by for a long time with local true-or-false limbic logic, but it becomes a liability as systematic errors accumulate. Limbic logic with its intuitive nature implicitly constrains us to coarse-grained convergent thinking, poor generalizability, and an absolute bias, which resists flexibly switching to alternatives. By exclusionary default, absolute logic bias creates mutual exclusivity with effective language and extended logic—incompatible with escaping its own self-referenced bias. Extended logic confers a sense of continuity across the entire system, which is global sense. We have the option of extended logic unavailable to our animal cousins. Extended logic enables divergent open-ended thinking that gives us more possible options to choose from, including equivalent solutions and convergence to probabilities.

It seems clear why evolutionary processes in a dynamic domain are open-ended rather than closed-ended. Open-endedness facilitates adaptation. Even though absolute logic doesn't fit the domain dynamics, it is relatively inexpensive and works sufficiently on average. The same criteria apply to language, with the caveat that language offers analytical tuning for enhanced adaptability and the ability to systematically decode information and effectively update knowledge. Absolute logic biases the way we learn and update our information and knowledge. Perhaps it is worth repeating: When we consider the updating of old, familiar information, absolute reasoning stodgily requires that any new information "must" be absolutely true to be allowed to replace or modify old information, even while our own familiar information goes unexamined. Is it possible that we accept our old familiar information without reasonable inspection? But on the other hand we can scrutinize new unfamiliar information to death and discard it for the slightest blemish. Does this sound like equitable treatment? Furthermore, we tend to require absolute proof to falsify our old information before we ever consider discarding it. We perpetuate our own absolute bias that anchors us to the static past and obstructs adaptive learning of new information.

As Sir Francis Bacon once said, "We prefer to believe what we prefer to be true" (Jones, 1998, p. 45). Absolute logic allows us to confirm the "truth" of whatever information, belief, or value we choose or find comfortable. Research on confirmation bias, which was previously discussed, has shown that we tend to thoughtlessly search for confirming data and reject contrary data (Chapman & Johnson, 2002, pp. 133–135; Gilovich, 1991, p. 37; Kida, 2006, pp. 155–162). Confirmation bias reinforces the features creating the bias (Siegel, 1999, p. 305), and denying all the while, we become "absolutely stuck" on our self-defeating predetermined path.

Organisms benefit from updating predictive competence in light of new information, and we benefit from logically updating our inference processes. Logical inferences can yield more complete and coherent information and multiple equivalent solutions. Dichotomous true-or-false thinking creates impediments to multiple conclusions (Browne & Keeley, 2007, pp. 182–183).

> Rigid dichotomous thinking limits the range of your decisions and opinions. Even worse, it over simplifies complex situations. As a consequence, dichotomous thinkers are high-risk candidates for confusion. (Browne & Keeley, 2007, p. 183)

Constrained coherence created by an absolute logic bias limits reasoning abilities and effective communication. Constraints on coherence reduce representation, transformation, and discrimination accuracy, resulting in lower feedback reliability and compromised predictability (Braver & Ruge, 2006, p. 321). Absolute reasoning limits flexible learning and our ability to logically evaluate alternative conclusions, which places adaptive constraints on sensible relationships. There is a liberating effect of recognizing alternative conclusions (Browne & Keeley, 2007, p. 188).

In an unfortunate and ironic sense, we may be the only animal species "absolutely guilty" of withholding evidence from ourselves, but we usually lack insight into our own deception. While an absolute reasoning bias may suffice in more simple contexts, most of the problems we face exceed the applicability of absolute logic, whether we recognize it or not. An absolute logic bias becomes a liability when it supersedes the use of extended logic that offers effective tuning of adaptive performance.

Classification Bias

Classification is a constructive process that is critical to understanding relations in a dynamic environment. We form classifications to model our environment by specifying and separating observed regularities into a logical categorical ordering of information. For example, animate objects are classified as living organisms, whereas inanimate objects are not. Human beings are classified as a static class within vertebrates and separate from the static class of invertebrates. We tend to construct classes in large part on the basis of what we feel or observe in our experiences with objects and actions. How we perceive and describe our experiences depends on our language and what we notice and deem relevant, which in turn biases how we structure our classifications.

For instance, when early scholars conjectured that all matter fell into the four element categories of air, earth, water, and fire, subsequent researchers spent considerable time trying to explain phenomena that did not fit into that structure. Those hard categories gave way to today's expanded table of elements that is based on atomic structure, which led to solving many problems simply by reclassifying and redefining them in more ecologically effective terms. The updated periodic table gave us a natural reference that expanded our processing perspective by tuning our focus and attentiveness to relevant structural patterns in nature, which in turn led to a broadened perspective and more logical understanding of how nature works.

Our language enables us to observe relationships and define classes that fit with the logical structure in nature, such as the periodic table. We constructed the table to classify fundamental microscopic elements in nature. Therefore, we can think about the logical ordering of elements that make up nature's fundamental microscopic structure. Yet our implicit limbic system automatically classifies stimuli according to true-or-false logic as either good or bad, which better fits a static, idealized world. With language, we can construct arbitrary classifications of an idealized world that offers anything from simplistic correspondence with nature to almost purely imaginary relationships. How we construct systematic classifications of fundamental macroscopic processes plays a key role in how well we understand the logical ordering of relationships in nature. We can construct what seem like realistic classifications that fail to capture nature's fundamental macroscopic dynamics.

For example, Newtonian physics provides a systems perspective, but with idealized classifications of processes and relationships inconsistent with nature in a general sense—that is, closed-ended physics, deterministic, conservative, two-way time, and so on. Nonequilibrium physics provides an extended systems perspective with generalized classifications of processes and relationships consistent with those observed in nature—that is, open-ended physics, nondeterministic, nonconservative, irreversible processes, one-way time, and so on. We can see how the physical classifications we construct to correlate the ordering of nature have a fundamental relation to our language structure that we apply to understand the very same natural processes. If our language refers back on itself, we can see a conflict of interest between locally referenced language and the fidelity of our understanding. We expect effective classifications to demonstrate fidelity corresponding across language and the physical world. In this case, we can mitigate the conflict of interest with externally referenced global language and systematic classification.

With inherited localized language, we create "perceived categories" and classifications that mostly reflect our a priori beliefs about that which we attempt to classify. As already noted, we have a predisposition to confirmation bias. For example, cultural beliefs often violate the static classification of human beings by labeling with internally referenced affective values or behaviors that change, such as labeling a human being as a "bad person" instead of labeling the person's behavior as "We think that person sometimes behaves badly." We can no more avoid a tendency for classification bias than any of the other biases we have discussed here. But classification bias can be tuned to the global environment, as demonstrated by science. Classification is a product of language and brain correlations that work most effectively when in correspondence with external physical relationships rather than internal affective values.

In other words, effectively classifying relationships about the external world requires external referencing to that world, which establishes a natural physical bias. As with most all biases, we can work to gain insights into the affects of underlying beliefs and look for ways to tune and redirect biases that produce less-effective outcomes. Language that coherently corresponds to the physical domain refines our constructions for the natural classification of objects and logically admissible actions, values, processes, and relationships. Domain correspondence between language and classifications confers a natural bias that facilitates the production of reliably informed knowledge and the modulation of effective adaptation—and a logical understanding of regulation in biological organisms.

It follows that improving the effectiveness of natural language constructions facilitates effective classifications and correlations consistent with the observable physical ordering of nature. Some things move and some do not. Some things move actively on their own accord, such as living animals. Other inanimate things, such as falling rocks, move passively because of the forces of nature. Once established, our classifications bias how consistently we perceive and relate to our world where physical laws prevail. How well we manage to construct our classification bias will make a difference in the reliability of our subsequent perception of relationships and how effectively we predict and adapt as things change.

Internally referenced absolute reasoning reduces the problem space to coarse-grained true-or-false categories, which biases discrimination of feature differences for effective classifications. We tend to aggregate similar things into categories—trees, cars, banks—and then treat them as identical (Holland, 1995, p. 10), and we tend to discard "junk" items or ideas as a form of random noise, especially when the external idea interferes with our internal familiar, self-similar ideas.

We often classify something we do not understand, do not agree with, or "do not believe in" as junk. We tend to make uninformed assumptions on the basis of the belief that if some detail is false, that justifies classifying the whole as irrelevant. This strategy makes it easy to ignore the whole (Holland, 1995, p. 11). Any belief, idea, or piece of information can be arbitrarily labeled either totally true or totally false. In many cases, when we say

- "I do not believe in it."
- "I am against it."
- "It's not good for anything."
- "It doesn't make any sense."

We more likely mean

- "It doesn't fit with what I *believe*."
- "I am unfamiliar with it."

- "I do not understand it."
- "So I will call it junk and discard it."

And perhaps contrary to conventional wisdom, unfamiliarity and contempt often go hand in hand.

By tagging objects or ideas with arbitrary local labels, it is easy to focus on the details we deem relevant and ignore the details we consider irrelevant (Holland, 1995, p. 13). We find it fairly easy to rail against something when only a few details are known and poorly understood. But it is much more challenging to first identify, evaluate, and classify the relevant variables and dimensions of the problem and seek additional information to identify systematic meaning before offering an opinion. It is harder still to refocus energy toward finding more viable global solutions and formulating new alternatives that better fit the facts (Jaynes, 2007, p. 139).

Overgeneralized classifications with tags or labels create locally biased opinions, inferred by generalizing in an arbitrary direction starting from an individual frame of self-reference. Tagging or labeling also allows for construction of vertical hierarchal organizations (Holland, 1995, p. 15) and can produce implicitly biased dominance interactions with arbitrary values. Arbitrary hierarchal values likely evolved largely along with social-labeling processes associating different patterns of primitive feelings in various social contexts, and eventually accepted as distinct entities (Panksepp, 1998, p. 301). Arbitrary social labeling tends to propagate polarization and maladaptive classification by ordinate value-ranking of people, ideas, and beliefs. This polarization leads to disdain, discrimination, and discordant relationships.

Ordinate hierarchal ranking reflects a rather simple linear function that represents the value of the whole as the weighted sum of arbitrary values attributed to the parts. Understanding nonlinear interactions generally requires going beyond simply averaging the sum of parts generated with arbitrary labels (Holland, 1995, pp. 15, 166). Linear true-or-false discrete logic imposes strict limits on a system's information processing capacity. "To a system with limited information processing capacity, complex input is nothing but noise and leaves no traces in the system" (Spitzer, 1999, p. 188). Classifying by applying idiosyncratic labels tends to produce overgeneralized classes by associating arbitrary features with coincidental behavior and inaccurately assuming cause and effect (Skinner, 1953, pp. 84–85; Browne & Keeley, 2007, p. 147). We often support our existing beliefs and expectations by selectively embracing coincident correlations that give the illusion of causation, without discriminating as to the accuracy of the information or the logic behind the illusory extrapolation (Kida, 2006, p. 156). For example, we can arbitrarily tag a person or group with a prefabricated affective label such that we coincidentally associate causation with negative emotions.

Word labels have been shown to activate and modulate emotional brain regions, emphasizing the importance of their impact on appropriate cognitive regulation of emotions (Rolls, 2008, pp. 166–171). This emotional correlation suggests that labels play a relevant role in understanding human behavior. By arbitrarily applying local either–or value rankings, labels bias what and how we think and how we habitually classify and rate the worthiness of ourselves and others. Understanding how our brain creates these primitive subjective feelings helps us appreciate how the nature of our values and inner affective states are stitched together by our more recently evolved cognitive apparatus (Panksepp, 1998, pp. 301–302). Our recently evolved prefrontal cortex enables language with analytical abilities useful for accomplishing this understanding and classifying nature more objectively.

Without language constructions that enable classifications better correlated with the ordering of nature, we fall back on our ancestors' idiosyncratic classifications, subjective associations, and affective rationalizations. When our classifications do not match well with our subsequent observations, we conclude that the world is random, chaotic, confusing, and disordered, but the error instead results from our ineffective modeling of that world. In the brain or in computational models, classification bias influences the efficient processing of effective solutions. Put more simply, effective classifications matter.

Overvalued Belief Bias

As members of a linguistic species, we inherit our local beliefs, values, and logic from previous generations, along with the language we learn as children. Our linguistic inheritance biases

- Our brain's cognitive-processing abilities
- How we classify and process information and knowledge
- How we perceive and infer values
- How we predict and behave

Overvalued beliefs disproportionately amplify affect and magnify our isolated and proportionally small, localized worldview, along with our distorted perceptions of that world. Our inherited beliefs and values play a role in biasing how we see the world, how we think it works, and how effectively we relate to it.

When faced with multiple stimuli, we tend to discriminate among them with value-laden labels, generating overvalued beliefs that bias our thought processing, emotions, expectations, and actions. Global labeling distinguishes among static classes, dynamic classes of actions, and values that functionally bias the effectiveness of our information processing. Value biases cognition, and affective language supports an emotional value bias whereas effective language supports an extension to an analytical value bias. Overvalued beliefs amplify our affective value bias, lead to emotional distortions in our perception and behavior, and preclude a systematic perspective.

Perception biases input sensitivity and stimulus-correlated signal detection and feedback. The bias can amplify the positive or negative influence of the perceived magnitude of an input stimulus. In turn, perception biases the magnitude of the output response (Jennings, 1906, pp. 302–305). We readily notice locally biased value variance, when one person's signal is another person's noise. Unfortunately we seem to turn down the feedback gain when we evaluate ourselves or our actions toward others and turn up the feedback gain when evaluating others, what they do, and how it feels to us. This "me, not me" distinction may account for perceptual discrepancies seen in the escalating force of tit-for-tat responses (Shergill et al, 2003). The inconsistent influence of a self-absorbed brain, overvalued beliefs, and absolute truths across individuals and disparate cultures tends to exaggerate "me, not me" effects. We easily notice the bias where one person's treasured, overvalued "should" represents another person's vehement, overvalued "should not," or when a "should" of today arbitrarily turns into tomorrow's "should not."

Overvalued beliefs lead to misinformed discrimination that typically involves arbitrarily classifying, tagging, or affective labeling of individuals, entire groups, or static classes of human beings according to a transient behavior based on superficial features perceived locally as having a negative value. Inaccurate discrimination adversely affects sensitivity accuracy (Spitzer, 1999, p. 149) and distorts how we associate value with stimuli, contingencies, credit assignments, and risk–reward assessments.

We inherit preestablished categories and values from our ancestors and then pass them along to our heirs. Once the arbitrary classification tags have been applied, the stereotyped group can be devalued, disregarded, ostracized, or even severely punished simply by association with the arbitrary value tag that classifies a person's worth according to a disliked behavior feature, implying that a static class *is* its behavior (Jaynes, 2007, p. 21). Liking or disliking, valuing or devaluing any particular behavior is different than arbitrarily creating a local class bias by statically encapsulating an individual or a whole group to invariantly represent a transient behavior, as if they are the behavior. The brain encodes information as durable knowledge, but overvalued beliefs and static negative classification of individuals or discrete groups according to behavior in a simplistic true-or-false manner can artificially and permanently devalue them from human to unworthy less-than-human status by implying that they are their behavior (Bloom, 2002, pp. 93–94, 254; Milgram, 2004, pp. 37, 161; Mitchell et al, 2006; Phelps & LaBar, 2006, p. 440).

Language allows us to arbitrarily construct these abstract value systems with absolute overvalued beliefs and status hierarchies. Driven by overvalued beliefs and unfamiliar ideas, the beliefs of others assume a negative polarity (Pronin et al, 2002, pp. 651–653) and can be misperceived as foolish, illogical, and ill-conceived. In some cases, the verbalizations of others can be perceived with such polarity that we deem them physical threats against which we believe we *need to* or *must* defend ourselves or retaliate with force. Rather than a difference of opinion, we perceive criticism of our overvalued beliefs as absolute insults and often retaliate, sometimes harshly, even brutally. We can even upset and anger ourselves in response to symbolic gestures, reflecting how past learning and current appraisals can influence emotional arousal (Panksepp, 1998, p. 190). Of course, if we are uninformed about the local meaning of the symbolic gesture, we see it but simply ignore it; the same applies to alleged insults.

The emotional response systems of animals can be rather inflexible (Panksepp, 1998, p. 37). When we react automatically to the verbalizations of others, our behavior does not differ significantly from that of other mammals that operate without the benefit of language and the higher-level cortical regulation it offers. Mammals typically have limited affective reactions in a threatening situation—to freeze, fight, or flee. Those affective impulses come from deep in the limbic–amygdala system. When faced with what we deem a threatening situation, rather than taking the time to apply higher-level cortical analysis, we tend to automatically inject whatever we might have learned and react affectively on intuitive subcortical "gut feelings" alone.

The brain has numerous survival mechanisms, and when threatened, we may find it difficult to subdue this reaction. But we do have an alternative not available to other mammals (Panksepp, 1998, p. 190). We can stop and think analytically about what we classify as a threat. We can assess our emotionally laden descriptions and compare them with the analytical descriptions. We can challenge and change what we think. With words alone, we can transform an imagined threat into something that merely concerns us or does not concern us, or into something that we can observe with some annoyance while we modulate our behavior more effectively and dampen our affective reaction. Simply put, there is no physical law that says we *must* feel insulted and infuriate or enrage ourselves and resort to outrageous behavior regardless of the context.

Knowing how nature works helps us by enabling us to put behavior in context. For example, "You cannot *make* me angry, because there are no physical strings attached. But if I am bonded to you and your behavior distances you from me, the bond is stretched and I will feel bad. I am angry because I feel bad and attribute the cause of those feelings to you." When this happens we tend to protect ourselves or retaliate with punishment, similar to many other mammals. When we understand how the world works and the physical limits that constrain actions within that world, we can realize other alternatives. We can take context into account and deliberately modulate habitual behaviors and decide not to automatically upset ourselves or blame others. We can realize that in relationships bonding feels good going in and unbonding feels bad going out, and typical relationships are rarely one or the other. In effect, we can think about how we relate to others and take responsibility for our own thinking, emotional feelings, and behaving. We have the option of treating others as we prefer to be treated.

Rather than assume that the negative feelings we experience were caused by the people around us, we can carefully observe, and a reliable bias will reveal that our own limbic reaction intervened between what we think was the cause and what we think was the effect. We can retrospectively justify our perception, which hardly overcomes the affective bias that influenced our misperception of "no strings attached" reality. When we interact with someone, we implicitly construct the meaning of the interaction from our belief system and then react to our construction. If we construct an inaccurate meaning, we expend energy reacting to something that did not happen except in our limbic system but that still can lead us to develop overvalued beliefs associated with that person or group. The closer to real-world systematic referents we construct our meaning, the less likely we are to misclassify and assign blame to others, and the more likely we are to accept responsibility for our own behavior and construction of reality, inherited or otherwise. In other words, we are each responsible for our own thoughts, feelings, and behaviors with no strings attached.

The power to modify our behavior by modifying the words we use to describe events is rooted in the way words influence our perceptions, our beliefs, what we value, and our cognition and emotions. In CBT, a patient learns to use different words to describe an event, moving from emotionally leveraged absolute words and phrases such as *hate* and *awful* and *I can't stand it* to lower-intensity, less-rigid words and phrases such as *dislike* and *unpleasant* and *I find it annoying*. The same situation with different words becomes less threatening, less unpleasant, and thus more manageable. The patient still may not like how it feels but now has a better opportunity to deal with it more reasonably. The rule of thumb is this: "The good news is that you don't *have to* like it, but it seems to be in your best interest to minimize upsetting yourself and resolve the problem as best you can." By shifting the problem and responsibility internally, we can maximize our resources for resolution. The patient learns the importance of self-talk and that each of us is responsible for our own thoughts, feelings, and behavior (Ellis & Harper, 1997).

Our perceptions are greatly influenced by what we expect to see, so we often inadvertently anticipate, see, and react to the irrational, unfounded fears, beliefs, and values assigned by and inherited from our ancestors (Kida, 2006, p. 20). We can relate emotionally to the past experiences of others through inherited beliefs, values, and the hand-me-down stories that accompany them. Even though we never witnessed the experiences, we feel the emotional connection that seems to reach across time and vicariously transfer to our own beliefs and values. Our limbic system could not care less whether we directly experienced the emotion or whether our belief is overvalued. If we feel angry, the easier it is for us to engage in reactive tit-for-tat.

Overvalued beliefs, noticeably incompatible with a sense of humor, can sometimes lock us into a sense of self-importance instead. Our subjective confidence in the certainty of our overvalued beliefs can give us a sense of omnipotence, which can sometimes lead us to do the unthinkable. We kill people over their beliefs as if beliefs alone can threaten us. As far as we know, we are the only animal to do so.

We see many examples of emotionally labeled beliefs across most societies. For instance, children are superstitiously labeled as witches in Africa, with the justification being the common belief that the alleged witches can communicate with the dead and cause misfortune, illness, and death. An article on stereotypes titled "Schizophrenics Battle the Disease, the Myths" fails to mention that labeling people by a disease engenders negative connotations by creating a stigmatizing stereotype. The outcome goes against the professed intentions of eliminating stereotypes. Despite the misinformed labeling used by the less educated, people diagnosed with schizophrenia are human beings who unfortunately have a recognized chronic psychiatric disorder. They are not evil, and they are not a disease. They are *Homo sapiens* like the rest of us.

Stigmas label people as less than human, which results in dehumanizing them. A similar phenomenon happens when a scientific organization professes an intention to fight the stigmas of mental disorders but continues to apply colloquialisms and other antiquated lay labels to humans who misuse or abuse alcohol or substances, labeling them as alcoholics and addicts or alleging them to have a disease colloquially labeled as alcoholism. The prejudicial labels continue to be habitually used even when the terms have long since been dropped from scientific nomenclature because of their colloquial nature and emotionally laden stigmatizing effects. Once popularized, old habits seem to die slowly even among professionals.

Reasoning with a reliable value bias allows us to separately evaluate and more coherently correlate human thought, feelings, and behavior across humans as a species. A reliable value bias allows us to see ourselves as members of a static class of probabilistically imperfect beings, *Homo sapiens*, or perhaps *Homo imperfectus*. Fine-tuning our classification and value bias enables us to relate to others as members of the same class, despite different ranges of behavior and preferences. Once we acknowledge our joint membership in the same general class, we can choose to relate to each other as human beings. Even though we may like or dislike particular beliefs or behaviors, we then have the option of deflating our own overvalued beliefs about them.

A systematic value bias consistent with such insight enables us to separate a person, group, or static class from beliefs or behavior and explicitly apply what we know about relationships and context to adjusting our own beliefs. When we understand the logical interrelations among objects, action, and value, we can distinguish transient actions and values from static classes. By avoiding overvalued beliefs, coherent classification facilitates thought processes and solutions that reliably correspond with how the world works. Coherent classification can engender more consistently humane interactions.

Overvalued beliefs and emotional labeling lead us to separate humans into arbitrary groups and assign ordinate values to their worth on the basis of culturally biased goodness or badness. In doing so, we inherently promote polarizing prejudices and even violence. Our value bias influences how intensely we feel, think, and behave and how harmoniously we relate with others.

Microscopic Bias and Systematic Blindness

When we view brain and language biases as a group from a macroscopic perspective, we encompass a broad systematic category that includes a microscopic bias, which results in ignoring or neglecting the systematic space. A microscopic bias discretely converges to smaller and smaller dimensions, inhibits divergence, and defaults to systematic blindness. It is no surprise that systematic biases go undetected with localized language that also demonstrates a microscopic bias. On the other hand, it is no surprise that biases come into focus with systematic language that flexibly extends to divergence and macroscopic dimensions. As far as we know, we are the only animal in the animal kingdom capable of both systematic knowledge and systematic ignorance. We incur this distinction by ignoring and thus failing to acknowledge the relevance of coherent systematic relations among language, brains, and how the world works. Thus, we are the only animal known to have the dubious distinction of self-imposed systematic blindness, which does not bode well for our idealized vision of human superiority in the animal kingdom. But what is the nature of the processing behind these biases?

Can extended language enable divergence and open our eyes to our systematic biases, better inform our thinking, and help to make practical improvements in our actions? The answer is yes, we can see that language has a lot to do with biasing our beliefs and values and providing an adaptive mechanism for improving our lot in life. We can learn from effective language and in turn better understand our responsibility for advancing harmonious adaptation with our world, which is addressed next.

CHAPTER 8

Effective Language

WE CAN LIKELY EXPECT that more accurately matching our language and cognitive processes with a reliable understanding of nature can lead to more effective adaptation.

Better knowledge about our language and how we use it to relate can potentially contribute to thinking about the world more realistically and making better predictions. Knowing the limitations of language helps us to fit our language behavior to the physical constraints of the world in which we live. In this chapter we consider systematic referencing and the processes that integrate relationships among language, cognition, and the physical world with the idea of cognitive fidelity and fitness landscapes. The discussion emphasizes the influence of effectively represented time on information-processing models.

A language model with systematic referencing and real-time tuning abilities facilitates coherent evaluations in correspondence with our observations of relationships in nature. Coherent evaluations highlight the connection between cooperation and fitness for optimizing adaptive relations. Furthermore, language that fits the domain structure offers insight for realizing practical solutions that go beyond rhetoric.

The Nature of Language Fitness

We are statistically integrated with our world because we use the information we obtain from the environment to form our internal model of that world (Buzsáki, 2006, pp. 47, 50). Adaptation to change depends on flexible cognitive regulation and the extent to which we can develop and maintain more accurate, more precise, and more up-to-date sensorimotor representations in our brains.

We use sensorimotor representations of our experiences to modulate and calibrate our perceptual interpretation of immediate sensory inputs, to predict future events, and to plan actions in anticipation of their potential consequences. Representations that coherently reflect the statistical regularities in the environment are more likely to correlate with effective predictions and performance.

> "Representation" of external reality is therefore a continual adjustment of the brain's self-generated patterns by outside influences, a process called "experience"

> by psychologists. From the above perspective, therefore, the engineering term "calibration" is synonymous with experience. (Buzsáki, 2006, p. 11 [emphasis in original])

Effective language contributes to reliable cognitive modulation and time-sensitive calibration by providing a practical method for representing and processing information and knowledge. As a result, we see more effective modulation of cognition, enhanced error tuning, decreased errors, and increased learning rate. Language that supports flexible tuning enables adjusting cognitive processing to match external change and temporal context.

Information is relevant in time, since both are concerned with aspects of change and the time context within which the change occurred (Buzsáki, 2006, p. 13). Understanding experiences in the context of time and space in which they occur enables us to accumulate a history of reliable correlations of regularities as systematic knowledge. In turn, systematic knowledge enables statistical predictions for developing and adaptively tuning responsive solutions going forward in time according to effectiveness.

In a dynamic system, predetermination of only one solution prior to an event ignores the statistical distribution of information, constrains flexibility, and increases the probability of maladaptive spur-of-the-moment responses when the future dares to be indifferent to our solution. Dynamic systems deal with a priori constraints by relying on alternative possibilities. Rigidly predicting and relying on one predetermined solution inadvertently excludes alternatives. The dynamic future is under construction, and things change. It makes sense to seek out language constructions amenable to tuning and capable of correlating information concurrently with change. Open-ended language enables flexibility for considering alternatives and accepting new information from the current input that enhances ongoing tuning and adaptation of solutions to fit changing problems. Language that facilitates flexibility presents us with value-added adaptive benefits in a world of change and instabilities.

> The presupposition of a stable environment, which was a given during the evolution of human beings, is no longer granted. In many areas of human activity, stability is no longer achieved, desired, or even perceived. People can therefore easily get into a situation in which their acquired knowledge and skills are no longer needed, or in which they continue to filter from the environment rules and values of parameters that no longer hold true. (Spitzer, 1999, p. 56)

From an evolutionary view, history shows that as circumstances changed, the fittest species of many eras gave way to new species better fitted to new conditions (Carroll, 2006, p. 39). When calibrated to a reliable external systematic referent, language can play an influential role in understanding these changes and in evaluating implications for human fitness. Language appears to have evolved from more primitive forms of cognition by trial-and-error learning in various local environments (Chater et al, 2009). Language without systematic referencing predisposes us to more trials, errors, and tribulations than we might prefer and, at best, likely leaves us locally trapped and struggling to adapt on a subjectively distorted landscape of our own making.

> So the critical step in the conception of an open plan is certainly this: that "the survival of the fittest" must be understood as the selection of those fitted for change as part of the total concept of fitness to a changing environment. Adaptation has to match the changes in the environment, but adaptability has to match the rate of change. (Bronowski, 1977, p. 171)

Fitness Landscapes

Consider the fitness landscape of genetic material. *Genotype* refers to the genetic makeup of the DNA of an organism, while *phenotype* refers to the organism's chemistry, physiology, shape or physical morphology, and behavior. The genotype can be mapped to the phenotype, and the phenotype mapped to fitness; the fitness landscape is a convolution of these two mappings, which directly maps the genotype to fitness (Nowak, 2006, pp. 30–31). A fitness landscape abstractly connects genes to behavioral fitness as a logical process of adaptation to that landscape. Logic couples the elements of a system and changes the properties of the landscape and search parameters that potentially optimize adaptive evolutionary processes. Changing the logic function changes the interactive rules and the shape of the fitness landscape. Thus, adaptive evolution favors adaptable complex systems (Kauffman, 1993, pp. 95, 121, 191, 232, 235).

Think about standing on a mountaintop and peering out over the distance at a broad vista of different elevations with a vast array of peaks and valleys. We can think of a fitness landscape as an array of multiple local peaks inhabited by individual subpopulations of solutions that climb up those peaks by incremental genetic changes. Some subpopulations ascend to the top position at the peak, continue to compete for survival, and become statically trapped there since moving to lower elevations implies lower fitness. We can visualize this distribution of local landscapes and competing solutions as individual cultural entities perched and trapped at the top of each peak, touting their local solution as the winner and the one and only absolutely true solution. Yet if we subscribe to the "no free lunch" theory, we might suspect that they have not yet discovered the utility of novel alternative solutions and open-ended search processes. Even though each peak has its own vantage point, it is a narrow, static perspective and by definition incurs localized constraints that obscure a dynamic systematic view.

Regardless of the variety and expanse of the landscape, a horse looking outward with blinders incurs a narrow field of vision that restricts attention within a narrow range in the direction of travel. The horse plods along, focused on the road ahead, oblivious to what's going on in the rest of the world outside the line of sight. We see similar localized limitations with closed-ended language that limits the breadth of our worldview to a small surrounding area that directly concerns us, and we ignore the rest. We become statically trapped at the top inside our own little world, boxed in by and oblivious to our narrow-minded thinking.

Think of each peak with individuals distributed at different levels up and down, across, and around the mountainside. Every individual has a unique perspective, biased by the evolution of their individual history, with variable genotypic and phenotypic influences from brain, language, culture, and learning that each encountered. In turn, each has a unique linguistic vantage point, making it difficult for one person to understand why others fail to see the obvious "my way—they just don't get it." If we think about a panoramic systematic view, which tends to neutralize or dampen the emotional landscape, the polarized local perspectives come to light, and it is easy to understand why people see things differently from where they stand and think, and why they hold their view so adamantly. A global perspective levels the fitness landscape by enhancing systematic learning and smoothing the discrete true-or-false, right-or-wrong solutions that constrain learning. Smoothing enables a phase transition to dynamic populations of solutions that can continuously oscillate among alternative possibilities. With a systematic perspective, it seems obvious as to why and how these local views are trapped at a particular location by the polarizing language, brains, and beliefs and the systematic processing errors that biased the trajectories that led them there. They are trapped by and stuck in their own rigid polarity.

From a global perspective, it is of little surprise that each local view is locked in at a particular position and competes to prove it is the right one, since each view seems obvious from that particular location. Indeed, anyone standing there would see roughly the same picture if they had incurred the exact initial conditions, learning history, and predetermined path to that view. But we understand that beginning from the exact same starting point and arriving at the exact same position is unlikely in complex systems with inherent sensitivity to initial conditions, ongoing nonlinear influences, and the lack of prescient knowledge for exact predetermination of future disturbances. Each history reflects experience and learning encountered within variable context and constraints that bias perception of current input and lead to a range of different actions rather than the one and only "right" action. Recall that with one-way time we calculate correctness in arrears by observation rather than by feelings of premonition for the future.

It turns out that "right" is a matter of absolute linguistic constraints on perspective. It comes as little surprise that the local perspectives hardly have a clue about the existence of a global perspective. The only clues available are local ones with localized constraints that recursively converge back to the same locality on the landscape. In other words, we understand why and how we got there but cannot figure out why others cannot stand in our shoes, look through our eyes, and see what we see. We can become so caught up in being right that we fail to notice our own inability to see the world as others do. From a global perspective, we see convergent limitations, incurred by the exclusive use of closed-ended language, that trap us by impeding divergence to a broader perspective and dampening systematic exploration.

In a changing environment, statically constrained language and semantics distort perception and deceptively compromise information and domain knowledge. A static perspective inherently incurs a systematic deficiency that biases the order of the language-fitness landscape by imposing conflicting processing constraints on the dynamic landscape. With a static perspective, the landscape seems more rugged and multipeaked, creating isolated traps because of the compromised correspondence between information and domain knowledge. To enhance escape from localized portions of the fitness landscape, we benefit from having flexible regulation, alternative solutions, and open-ended search processes. Open-ended regulatory processes enable alternatives that can lead to novel solutions and potentially allow us to escape blind alleys and dead ends at local peaks, and safely navigate the valley to extend our search across the global landscape.

Open-ended language and semantics enable processing compatibility with abstract descriptions of phase transitions that correlate the statistical structure of dynamic fitness landscapes to the global flow of populations. In other words, the language fits the domain dynamics. Systematically calibrated language supports nonlinear processing and time-dependent fine-tuning consistent with evolvability and adaptive fitness in self-organizing systems with inherent uncertainties and dynamic instabilities. Language constructions that coherently match the structure of the fitness landscape facilitate effective modulation of cognitive processing and adaptive responsiveness to change.

Language-fitness landscapes can be defined as the distribution of the capacity to perform some processing operation over language space by connecting the structure of the language to behavioral fitness. Generally, the more coherently the language and logic fit the environment, the more effective the cognitive processing, learning, and adaptation. Even without an a priori fitness function, fitness can be compared ad hoc by evaluating differences in adaptive effectiveness between a predetermined absolute solution and the best likely choice among alternative possibilities. Simply, we can measure effectiveness as the difference between predicted and realized outcomes and make error corrections. We expect that knowledge learned from decreasing systematic errors can facilitate global smoothing and generally enhance predictability. Learning makes the fitness surface smoother and simplifies evolutionary search (Nolfi & Floreano, 2000, p. 157).

Similar to novelty search, with extended language we expect the rugged landscape to evaporate into an "intricate web of paths leading from one idea to another . . . [where]

the concepts of higher and lower ground are replaced by an agnostic landscape that points along the gradient of novelty" (Lehman & Stanley, 2008). The high-dimensional global landscape is functionally compressed by generalized language that facilitates the flow of populations unencumbered by local idealized constraints. Dynamic undulations are dampened by neutralizing emotional value polarities and analytically generating a statistical distribution of open-ended possibilities. Open-ended language enhances learning that smoothes the global language-fitness landscape and extends search capabilities for discovering novelty and creatively resolving more complex problems.

We live in a world with a landscape of continual change. If we attempt to use static language to interpret and describe what we observe, we wind up with a distorted, uncorrelated image. By using a static system that operates independent of time, our static image mismatches solutions with the domain dynamics, since the context keeps changing in relation to what we see as the problem resolution. Context adds relevant information about experiences that improves learning rates in a dynamic environment, but absolute language leaves little room for context.

To derive more accurate information in the context of a complex world, we can acquiesce to a statistical opinion about possible states that is based on a systematic processing perspective: "Give me your current best opinion of how the world generally operates as a whole." This statement describes a globally referenced view with systematic context constraints that fit with the population dynamics we see on a natural landscape such as Earth.

A transition in the landscape enables a shift from a localized, individual processing view to global processing and a population view of ensembles. There is also a shift in the statistical method for processing information at the global level as probability distributions, which correspond to continual change, uncertainties, and instabilities in the domain. The global transition enables a central distribution of information with alternative solutions that tend to enhance problem resolution across the landscape.

Language fitness describes how well the use of language correlates with effective cognitive processing and adaptation on a given fitness landscape in light of long-term survival and reproduction. In this case, language is evaluated by its logical contribution to adaptation on a global language-fitness landscape and by how effectively it correlates our cognitive processing, understanding of the domain, and predictability for adaptive responses to changing conditions. We expect that the more logically our applied language represents and corresponds to the dynamics of the domain, the more effectively we can predict the probable effects of our behavior and make adaptive adjustments over time. Effective language extends and references information, behavior, solutions, and adaptation to the external system, which in turn enables systematic measurement of effective fitness. We expect that tuning the language fitness increases the probability of transitioning from local sufficiency to approximating global optimization.

Calibrating Language

Computational cognitive models link the symbols used in the model, such as words, with their semantic referent in the external environment (Cangelosi, 2001). With language we acquired the capacity to correlate symbols to the physical world through the intermediary of internal representations (Steels et al, 2007), a process described by Stevan Harnad as symbol grounding (Harnad, 1990). (For review see Barsalou, 2008). *Grounding* is a widely used term in electrical engineering for describing an electrical circuit that establishes a reliable direct connection to the earth. Similarly, grounding language with a reliable symbolic correlation to the physical world establishes a communication circuit for reliable information flow. Even though language provides an indirect connection to the physical world, it enables a higher level of abstract expression as systematic correlations (i.e., systematic grounding by correlation). Language confers a means for producing and processing lower-order concrete and higher-order abstract descriptions of nature corresponding to the events we experience. Thus language helps with understanding and expressing direct and indirect relationships.

Humans have the option of a delayed shunt for deliberately inhibiting reflex-like impulses (Bronowski, 1977, p. 144). The delay allows for initiating higher-level cognition, and thinking by simulating the possible consequences of behavior before acting. A delay allows time for deliberate higher-level analysis of the input information by the language faculty for logically processing, predicting, and modulating behavior. Effective language includes flexible tuning parameters that permit logically processing and adjusting the input-to-output product for adaptive real-time responses to change according to feedback experienced from the domain. Optimizing the benefits of the feedback requires deliberate recognition of the error signal, understanding domain relations, and calibrating internal-to-external correspondence to decrease errors.

We can internally calibrate language and semantics with the external environment via external referencing and symmetry breaking that enables consistent evaluations of events across time and space. In effect, symmetry breaking syncs and separates the dynamic future from the static past. Symmetry breaking enables separating and storing past experiences as memories and distinguishing simulations of future events that have not yet occurred. We struggle to redeem ourselves from errors after they have occurred and absolve ourselves from future errors. Understanding symmetry breaking facilitates noticing and learning from past errors. Effective learning improves planning and predictability that can decrease the systematic error rate. The deliberate application of effective language extends linguistic parameters and expands the dynamic tuning range for correcting systematic errors. Education, in particular higher-level education that introduces principles of global referencing, symmetry breaking, and extended language, encourages the pursuit of effective information processing and systematic knowledge, if we are so inclined.

> Education is very important, also permanent education. The world is changing so fast, that we need to update continuously our knowledge in order to adapt. (Prigogine, 2003, p. 74)

The intrinsic hidden nature of affective bias and static absolute local beliefs dampens flexibility and the creative search for novel solutions and systematic knowledge. Furthermore, static methods with short-term, narrow-minded planning directly extrapolated from past experience threaten society with fossilization or, in the long term, with collapse by dampening the adaptive possibilities of society for innovation and originality (Nicolis & Prigogine, 1989, p. 242).

Language structure logically calibrated to one-way time and regularities in nature enables fundamental improvements in insight, adaptability, cooperation, and robustness. Robustness can be thought of as skills or behaviors that offer a high level of effective adaptability. "A biological system is robust if it continues to function in the face of perturbations" (Wagner, 2005). Robustness in evolution has been equated with cooperation, since cooperation can enhance adaptability (von Dassow & Meir, 2004).

The plasticity and robustness of language acquisition indicate its central evolutionary importance to *Homo sapiens* (Wimsatt & Griesemer, 2007, p. 281). Accurately calibrating language to information, information processing, and timely knowledge in the moment has been proposed as a method for enhancing the robustness of decisions with a higher probability for adaptive outcomes (Bailey, 2009). Language calibrated for fidelity and logical consistency provides a naturally efficient interactive loop with the environment for enhanced predictability and adaptive adjustments to feedback.

Furthermore, a bias toward accuracy offers improved error detection and correction with enhanced risk prediction. Adequately monitoring and regulating behavioral output depends on feedback (Jeannerod, 2003, pp. 83–85). The inverse relationship between increasing fidelity and decreasing errors leverages a system's feedback ability such that small changes in accuracy parameters tend to produce significant gains in predictability, flexibility, adaptability, and robustness (Camazine et al, 2001, pp. 32–36; Kirschner & Gerhart, 2005, p. 225).

According to Daniel Siegel, small changes in a person's perspective, beliefs, or associations with particular forms of information processing can rather quickly lead to large changes in state of mind and behavior (Siegel, 1999, p. 221)—suggesting that calibrated changes in language fidelity might predictably produce improvements in cognitive fitness. The relation among processing fidelity, predictability, and adaptability implies that logically mapping language, cognition, and action to changing object positions and value states outside the brain can enhance fitness.

To an extent, the accuracy of our thinking depends on how well we calibrate our language to the dynamics of our world. Aristotle's idea that language, thought, and reality correspond with the natural world can indeed make sense, if applied probabilistically

(Johnston, 2008, p. 325). In other words, systematic calibration enables logical consistency and a new coherence with nature. Is it worth considering the notion of calibrating internal-to-external correspondence between language, brains, and nature? Symmetry breaking and temporal systematic referencing support logically correlating cognitive regulatory processing with regularities and physical processes in nature. Taking a broader holistic view can potentially illuminate the hidden secrets of the system's macroscopic behavior.

Systematic Referencing

Who cares about referencing? And why would anybody care? If we acknowledge that we care about the meaning our speech conveys, we can ask, "Meaning in relation to what?" What are we measuring and from where, and is it tangible? If we think about referencing meaning to something invisible with which we are unfamiliar, such as one-way time, it does seem a bit strange at first and likely presents a challenge to argue otherwise. We are searching for plausible solutions, and sometimes plausible solutions are not necessarily ones with which we are familiar. In this light, even if we come to agree that referencing in general and external systematic referencing in particular is plausible, it still may seem a bit eccentric. We will address the former questions first and then continue to follow up on the latter issue of plausibility throughout the book.

What happens when we use unnatural referencing to establish meaning, such as a reference lacking known observable physical effects in the domain? Unnatural referencing assumes physical properties and structural–functional relationships unsupported by scientific evidence that requires yet undiscovered sensors. Science establishes evidence by applying logical processing methods to measure sensible effects corresponding to known physical processes and relationships found in nature. For example, about 65% of people apparently believe that if someone is staring at the back of their head, they can sense that person looking at them as if they have eyes in the back of their head, even though science has verified only five physical senses and none in the back of the head other than touch. This belief continues despite the scientific debunking by an experiment reported in the journal *Science* by psychologist Edward Titchener; it is of interest that the article was published in 1898 (Chabris & Simons, 2010, pp. 199–200).

Approximately 150 years of concerted efforts for substantiation without confirming evidence for its existence points to the likely dubious nature of extrasensory perception. Perhaps the more important question is, even without evidence, "why so many people nevertheless believe in its existence" (Gilovich, 1991, pp. 169–170). We might think that all of these efforts have been in vain, but the one outcome of note is that it gives us overwhelming evidence confirming that people are willing to accept their intuitive thoughts about whatever they imagine or believe as factual without any factual physical evidence whatsoever. Confirmation rests on the circular argument equating perception and reality, inclusive of imagination, such that since it can be perceived, it therefore *must* exist and *must* be real.

Yet we have evidence that the tendency to believe without evidence is systematic rather than coincidental and that the confounding independent variable is the failure to separate imagination from physical reality. We see two types of logical arguments: One argument is referenced to a local frame for simulating a relation between action and affect translated to an imaginary internal world, and the other is referenced to a systematic frame for simulating a relation between observable action and effect that translates to the external physical world.

In the example of belief in the ability to sense someone's stares, as far as we know the sensory relationship expresses affective imagination rather than sensible referencing to natural laws. We might also consider a possible relationship between this extrasensory phenomenon and coincidental exposure, or cultural memory from childhood when children are told that their parents have eyes in back of their heads and always see and know what they are doing. Perhaps this peculiar-sounding child-rearing method is thought to work because children might be less likely to engage in forbidden behavior when they think someone with supernatural powers is always watching over their every move. Of course, as we mature, this primitive notion of someone constantly watching is somewhat foreign to the educated mind. These persistent beliefs of old likely were passed through many generations by narratives that were subjected to further contortions along the way, sans reliable referencing. But that hardly explains the whole story. Some fairly common beliefs of today correspond with quite ancient "wisdoms," suggesting that the beliefs apparently derive from the brain's local affective bias with a proclivity to overlook systematic errors due to systematic blindness.

> There is no good reason to assume that the brain is organized in accordance with the concepts of folk psychology.
> —Cornelius H. Vanderwolf (Buzsáki, 2006, p. 3)

We see many examples of implausible referencing across cultural boundaries even in societies perceived to have more sophisticated thought and wisdom. We see a common thread suggesting the burning human desire to know how and why things happen. But our efforts, driven by this burning desire to discover any "how" and "why" receptors, have failed. We have found much evidence that how and why are produced by internal perceptual processing associated with our language, logic, learning, beliefs, and knowledge about the external world. Of course, such evidence does not stop us from searching to confirm our cherished beliefs and trying to prove that what we "know" exists in the real world—nor lead us to disregard our illogical contortions.

Telepathy, sometimes called precognition or clairvoyance, is the supposed ability to transfer thoughts, feelings, desires, or images directly from the mind of one person to the mind of another by extrasensory channels and without using known physical means; it was at one time considered by many as a mysterious form of occult communication between brains (Buzsáki, 2006, pp. 3–4). The brain does emit electromagnetic waves, but these are known to be measurable only with direct skin contact, corroborated by physical evidence almost 100 years ago by neurologist and psychiatrist Hans Berger (1929). Perhaps systematic referencing offers a plausible method for addressing the problem of measuring the effects of invisible processes?

> We recognise that our language, illogical though it often is, is a cherished product of a long history but, nonetheless, we must take care that it does not interfere with our ability to understand the natural world. (Vanderwolf, 2010, pp. 72–73)

Recall that the premise is that with effective language referencing, we can potentially identify and accumulate accurate information about how the world works. Language and thought are inherently about how information is acquired, used, and transmitted (Feldman, 2006, p. xv). We expect that with effective language, thought processing, and information, we can build knowledge that accurately represents nature in a reliable manner and facilitates finding more effective solutions to problems we face. We see many examples in scientific fields such as medicine and engineering where education, learning, and the search for information lead to innovative solutions. It follows that enhancing the reliability and accuracy of information and knowledge that we think with can possibly contribute to our long-term survival. We can further develop the notion of modeling the brain as an information-processing organ for integrating knowledge about natural relationships into effective predictions. We are now left to identify and elaborate on what we mean by external referencing and demonstrate a level of correspondence with systematic processes. *Systematic* and *global* are considered roughly equivalent terms from a macroscopic perspective.

In theory, reliable external referencing can provide a working domain-dependent framework or frame of reference. For example, a systematic framework takes in the breadth of the system that includes the Earth and its atmosphere. Think about this frame of reference as the platform from which we can take in the largest macroscopic perspective that can converge to lower-level microscopic perspectives while maintaining logical consistency. A systems view is somewhat like the analogy of opening nested Russian dolls from the outside in. We open the outer doll to find another that looks the same, only smaller. Each time we open the outer doll, there is another and another, each one smaller and smaller. The difference here is that we are searching for something we cannot see directly, but we can accumulate only indirect evidence by looking at the effects of the global processes. We are looking for relatively invariant referencing that corresponds with and offers general explanations for lower- and higher-order processes and relationships across the breadth of the system. We expect effective generalizations to offer reliable explanations that hold up all the way down to smaller subsystems.

The similarity to the Russian dolls comes from the way in which system-wide general processes maintain self-similarity of scale from larger dimensional spaces to smaller microscopic spaces. Each doll incurs a ratio made to the same scale, so we can say they are relatively scale invariant. For our purposes we can think of these as fractal-like processes such that they have self-similar features. For example, when we stand between two mirrors, we see repeating self-similar images that appear to decrease in size according to distance from our position. The processes we are searching for work relatively the same at lower or higher dimensions. In other words, they generalize throughout. We cannot see these processes directly, but we might suspect that they form the invisible thread that weaves the fabric of life. We now understand the problem as establishing a reference by which to measure the effects of general processes that follow relatively invariant systematic rules. How well we chose the reference will become obvious when we measure predictions by the difference in correspondence between predicted effects and realized results.

As already noted, we expect effective referencing to generalize throughout the system, so we expect the referencing to help with systematically decoding locally referenced descriptions and in defining sensible limits in the domain. Systematic referencing takes precedence throughout the domain and establishes logical constraints on arbitrary cognitive processing and the production of nonsensible locally variable meanings and solutions. Language that corresponds to reliable external referencing sensibly correlates systematic meaning to domain regularities. Reliably correlated meaning generally enhances predictability and effective adaptation to external change.

Effective natural language facilitates the logical extension and expansion from absolute certitudes to relative possibilities and global domain knowledge. We know that the information we perceive in the world meets statistical criteria because it comes and goes, waxes and wanes, and demonstrates anything but certainty. For example, consider an informal conversation about the "sunrise," a topic discussed in Chapter 1 ("Relations and Processes"):

"The example about the sunrise is silly. Who cares what you call it. I know the sun will rise for certain in the morning. As a matter of fact, according to the almanac it will rise at exactly 7:00 AM. That is about 15 hours from now."

"Are you certain the sun will appear in the morning?"

"You bet I am certain. It has risen every day since recorded history."

"You do realize that the sun does not have an infinite amount of energy."

"Of course, everybody knows that, so what is there to quibble about?"

"Sometime in the future it will run out of energy and we won't have anything to argue about."

"Okay, so what's the big deal?"

"According to what you agreed to, it is not a certainty. But tomorrow it is highly *likely* that the sun will appear in the morning."

"No, I am certain it will rise tomorrow!"

"Do you mean that you are so certain that you would bet it will appear tomorrow?"

"That's right!"

"How much would you wager?"

"I will bet as much as you want to bet. What about a million dollars?" (Chuckles to self.)

"Okay, I'll take that bet. We have a deal?"

"It's a deal!"

Both shake hands.

"Come with me."

We get on a plane and travel to the South Pole. The sun does not appear the next morning as predicted, or even by the end of the day. The sun is a no-show; there is total darkness.

"I win the bet."

"You tricked me." (Grumbles.)

"No, you tricked yourself. But I will call off the wager if you agree that the sun does not *rise* each morning. The appearance of the sun rising in the sky is due to the Earth's rotation. The appearance of the sun *rising* is simply an illusion, and a statistical one at that. The illusion is perhaps highly likely but probabilistic nonetheless."

"It still seems like the idea of a sunrise is a simpler way to describe what is going on."

"Perhaps so, but it's not so much that the analogy lacks technical correctness. The larger problem is that you were so convinced by your habitual thinking that you were willing to bet on a sure thing. Viewing the world in terms of absolute certitudes seems simpler, but those idealizations reflect your habitual idealized way of thinking. Now what is so difficult about changing one word, from *certain* to *probable*?"

"Well, I guess it is pretty simple, but it's not so easy, because I've been thinking the other way for a long time."

"Habits are inherently difficult to change, some more so than others. Think about it. With some practice, you'll have it down pat in short order."

We see that our common vernacular associates certainty with the analogy of a "sunrise," which is akin to "believing what we think" and routinely "thinking what we believe," to the exclusion of other possibilities. We can add, "Of course we know that. Everyone knows about possibilities and probability," but relegating knowing about probability to an ad hoc afterthought hardly corresponds to actively applying probability to our thought processes and speech. We have the option of believing in, relying on, and applying statistical processes to our thinking and speaking, but certitudes eliminate possibilities and freeze us right in our tracks. Whether habitual or otherwise, absolute certitudes are mutually exclusive of alternative possibilities. Understanding the changing world we live in and applying what we know adds clarity that takes the form of effective thinking and alternative possibilities and probabilities.

But what has probability got to do with invariants and referencing? And how is probability an invariant when probabilities can change from moment to moment? Recall that we are looking for processes. More specifically we are searching for processes that are relatively invariant across the system, and time is involved. Continual change and statistical information distributions throughout the system as alternative possibilities define the hallmark of complex systems. It is difficult to find an invariant in a continuously changing world, but we know that change is a matter of time. The universal invariant that immediately stands out is the unalterable forward direction of time that makes possible a symmetry-breaking phase transition from static certitudes to dynamic possibilities and probabilities. It is difficult or close to impossible to measure and understand change in dynamic systems without applying symmetry breaking, one-way time, and possibilities and probability. The persistent process of change is relatively invariant throughout the system, at least on some scale, from top to bottom and bottom to top. Therefore, effectively recognizing information patterns in a changing environment relies on symmetry breaking and statistical processing that matches the constantly changing distribution of information. We have identified what appears to be a relatively invariant statistical process for referencing systematic information in dynamic domains, such that our ruler is probabilistic and leads to probabilistic information processing.

Probability defines a statistical processing method for measuring and making predictions of time-dependent changes. It is calculated by logical processing that enables measurement of results according to correlations of history-dependent output reliability and differences in correspondence between predictions of the expected and the realized output product, where output is defined according to effective adaptation in response to change.

We can expect that the fidelity of the language we apply to our search for information and knowledge corresponds to reliable referencing and processing within the domain of our search. We operate in the domain of nature, and it makes logical sense to use a frame of reference that demonstrates a bias toward the physical and structural constraints of that domain. A systematic bias places sensible domain-dependent limits on the application and performance measurement of language and semantics. To approximate this systematic bias, we can choose generalized language constructions that coherently correlate information across the extent of that particular domain.

Language referenced to the domain restricts the universe of discourse to problems and solutions that logically operate within that domain's physical constraints. In effect, changing to extended language shifts our world perspective from static idealizations to dynamic realizations and we come to the end of certainty (Prigogine, 1997).

We use words to represent and communicate sensory evidence about actions and correlations we abstract from nature. External domain-dependent referencing enables calibrating our brain's perceptual and conceptual sense connection between the natural world and our language, words, and ideas. Then the expected effects probabilistically correspond to our internal cognition and predictions. We have a coherently correlated loop between our brain, words, ideas, information, and knowledge that reliably corresponds to our world as possibilities and probabilities. Coherent correspondence among language, brain, thought, behavior, and the environment helps us measure how effectively we can understand and predict relationships in relative accord with nature.

> ... [T]hat the mind should have the power of reacting in just that duplicate way [mirror reality] can only be stated as a *harmony* between its nature and the nature of the truth outside of it, a harmony whereby it follows that the qualities of both parties match. (James, 2007, v. 2, p. 618 [emphasis in original])

We use language and words to describe the limits of sensible domain interactions by defining the logical correlations and orderly connective properties and processes among structure, function, and behavior in the overall system. Functional domain-dependent correlations provide plausible descriptions and definitions for limits of sensible search methods. The limits rely on coherently correlated physical relations with nature that help to identify and correct imaginary ideas and beliefs that tend to result in infinite searches.

> I protest against the use of infinite magnitude as something accomplished, which is never permissible in mathematics. Infinity is merely a figure of speech, the true meaning being a limit.
> —C.F. Gauss (Jaynes, 2007, p. 451)

Depending on our reference frame, absolute or relative, and the effectiveness of our language, imagination can allow us to assume an unbounded world we can never search or understand, since it goes to infinity, or optionally, perceive a physical world with finite constraints. A reliable domain-dependent referent limits the search space and dampens the tendency for blind or exhaustive searches by promoting plausible search policies that operate within the physical constraints of the domain and the physical limitations of the organism's body morphology. In systems with open-ended processing, accepting domain-dependent limitations relegates plausible measurements to finite dimensions in lieu of infinite. In effect, finite limits support generalizations, reduce the problem-and-solution search space, and enable logical information processing and measurement with a tractable number of variables and dimensions.

We expect that language acts as a representational intermediary in cognition, and prefer a systematic language-processing model with a logical coupling to reliable external referencing against which to measure predictability and effectiveness. *External referencing* enables specifying the upper limits of domain dynamics, structural and functional dimensions, measurement parameters, where to measure from, and how to measure domain information. A *global referent* accounts for symmetry breaking, domain-dependent time, open-ended logic, orderly classification structure, and statistical representation of information.

Global referencing enables systematic

- Predictions that are based on probabilities
- Correspondence with open-ended evolutionary processes
- Standardization of categorical definitions
- Standardization of information classifications according to natural source ordering
- Extension to and support for models of nonequilibrium dynamics, including
    - Domain-dependent time
    - Irreversible changes
    - Continuous processes
    - Persistent interactions
    - Multidimensional influences
    - Domain instabilities

We expect a natural referent to define the general systematic ordering and dynamics in the domain with nonequilibrium constraints. In other words, nonequilibrium dynamics apply where a nonconservative system operates in an open-ended manner with finite variables, irreversible processes, and domain-dependent time. These constraints define physical limits, variables, and parameters involved in effectively modulating instabilities within evolving natural systems where life forms operate in conditions that are far from equilibrium. Nature's dynamic structure is established by the invariant directional asymmetry of time that points toward the future and forms the event thread correlating and connecting context, actions, and events in space. We seem to have come across a fundamental invariant that defines the nature of change and phase transitions from the static past to possibilities and probabilities of the future. Could this be the mysterious invisible thread behind the curtain that weaves the fabric of life itself?

- Time is invisible.
- It defies our senses and cannot be touched, felt, seen, heard, smelled, or tasted.

- It spans the global domain.
- It connects relational experience contextually as actions and events across space.
- It plays a definitive role in the process of change and probability.

Asymmetrical time represents the essence of biological processes and relationships. We have identified external invariant referencing. We can enter *time*, *symmetry breaking*, and *asymmetrical time* in our reference list of general systematic processing and relational principles:

- Time, symmetry breaking, asymmetrical-time referencing

Asymmetry and *symmetry*, and especially *asymmetrical time*, aren't exactly household words, but we will try to address them along with the process of symmetry breaking in a palatable manner, formally and informally. Formally, *symmetry* in physics includes all features of a physical system that exhibit the property of invariance or equivalence. According to a particular observation, the aspects of the systems are unchanged under certain transformations. Time is asymmetrical and invariant in the forward direction and noninvertible. In other words, time is irreversible. This means that time translation in a physical system has the same features over a specified interval of time. Even though the past and future are different and naturally asymmetrical in the forward direction of time, each is symmetrical across all observers for the same time interval, as the following example and that of Maxwell's demon demonstrates.

As evidenced by the asymmetrical direction of evolution in nature, the static past and the dynamic future are not interchangeable, despite common absolute wisdoms implied by local linguistic conventions that support two-way time. Attempts at testing, tuning, and optimizing a two-way time-dependent system in a dynamic domain are likely to produce an entangled hairball rather than a solution. Testing language for two-way time simply requires explicitly searching for absolutes, imperatives, prescriptives, and so on. Time-symmetrical absolutes and imperatives indicate guilt by association. Simply assuming the use of one-way time or denying the use of two-way time fails to provide adequate confirmation or a reliable test when our speech behavior exposes its own affective fingerprint with entangled past and future. When we predict the future with certainty, we act as if we visited the future, informed ourselves of future context, and came back to prescribe a current or future behavior. Not only have we violated one-way time constraints but we also are feigning having more information than we actually have. In other words, one-way time eliminates the current possession of future information, with the exception of possibilities and probabilities. But with systematic blindness the norm, this constraint of reality is unlikely to prevent routine misuse by the general population and charlatans alike.

We can compare this charade to a confidence scheme that leverages the tendency for human gullibility by selling guarantees of the future to those affected by trust. We are guilty of pretending to steal information from the future to defraud the present with information that has not yet been constructed, similar to the imaginary Maxwell's demon—introduced to a public audience by James Clerk Maxwell in his 1871 book, *Theory of Heat* (Leff & Rex, 2003, p. 3)—who implicitly stole information from the future. Ironically, the demon's apparent violation of irreversible one-way time (Prigogine & Stengers, 1985, p. 239) tends to go unnoticed by those attempting to solve the puzzle even today.

> Maxwell's demon is no more than a simple idea. Yet it has challenged some of the best scientific minds, and its extensive literature spans thermodynamics, statistical physics, quantum mechanics, information theory, cybernetics, the limits of computing, biological sciences, and the history and philosophy of science. (Leff & Rex, 2003, p. 2)

"Of course we believe in one-way time," we say, but apparently we demonstrate a failure of commitment and application of one-way time to our everyday thinking. How can we claim to believe in something that verbal evidence shows we continue to ignore? If we assume the future is already constructed and knowable, we imply certainty and a static world. Implicit or not, pleading ignorance hardly excuses the crime. Implicitly stealing information from the future would hardly stand up to legal scrutiny in a court of law that typically judges ignorance as an unacceptable excuse for any violation.

Language tests that find explicit evidence of possibilities, probabilities, preferences, and opinions confirm support for one-way time. Even though nature provides abundant evidence supporting the direction of one-way time, everyday language and semantic conventions across the world commonly use symmetrical time, perhaps implicitly. But no one so far has provided plausible evidence supporting two-way time. Simply put, if we fail to explicitly apply language tuned to the invariant forward direction of time to our cognition, we tend to tacitly default and succumb to imaginary thinking based on two-way time.

> There is a relation between the direction of evolution and the direction of time. In a history of three thousand million years, evolution has not run backward. (Bronowski, 1977, p. 170)

Research findings show that conceptions of even such fundamental domains as time can differ dramatically across cultures (Boroditsky & Gaby, 2010, p. 1635).

While we have ample language constructions that can explicitly support the invariant forward direction of time, our conventional language use overwhelmingly favors bidirectional time, even language used in scientific communication. For instance, classical physics uses symmetrical time, calculus generally relies on time-symmetrical deterministic formulas, and most cultures routinely use absolute beliefs and predetermined values that implicitly assume time symmetry, where the past predicts the future and implies interchangeability (Bronowski, 1977; Prigogine, 1997). Implicit symmetrical time is currently the prevalent temporal referent for language, beliefs, communication, and policy making around the world. Perhaps the appeal of symmetrical time is that it tempts us to revisit and revise or extricate blame from past deeds and relish thoughts of infinite rewards? Symmetrical time supports beliefs in an idealized imaginary world, far from the realty in which we live.

Symmetrical time is tacitly included in localized language, which allows it to inexplicably evade detection because localized language refers back to itself and hides its encoding. The most obvious example is classical equilibrium physics with symmetrical time, where the past and future are indistinguishable. In contrast, nonequilibrium physics explicitly tuned to one-way time makes a clear distinction between the past and future. Bidirectional time presents as a "negative symptom" problem where time symmetry is implicitly silent. We incur a stealthy linguistic problem for recognizing the covert negative symptoms. Our familiarity and comfort level with our own habituated "normal" leads us to ignore alternatives and keeps us in the dark, out of touch, intellectually stationary, and systematically blind.

Regardless of status, heads of state, politicians, perennial experts, and the rest of us remain at risk for unwittingly relying on language locked in to symmetrical time. If asked, we would say, "Of course we know that the results of our behavior cannot be reversed in time and we cannot change the past." We also know that predicting the future with certainty works no better than reprimanding the past.

"We should not have done that."
"That should not have happened."

"But we did do that and that did happen."
"We might suspect that it should have happened because it did."

"Okay, we should not have done that yesterday."
"But I know what we should do tomorrow."

In these examples the now, past, and future are tacitly interchangeable. When we choose and use absolute linguistic primitives that create static policies as if nothing ever changes, we imply time equivalence that distorts our thinking. Language tuned to symmetrical time induces static policies in a dynamic world by binding our future to the past. But why is time so important, and why is it so puzzling? When discussing vision, Marr states:

> Perhaps the key to the puzzle is the analysis of motion—more so, perhaps, than any other aspect of vision—time is of the essence. This is not only because moving things can be harmful, but also because like yesterday's weather forecast, old

> descriptions of the state of a moving body soon becomes useless. . . . Perhaps one of the most primitive types of motion analysis is concerned with noticing something has changed. . . . (Marr, 2010, p. 162)

Timing may not be everything, but naturally tuning language to one-way time and applying it to our everyday language use plays a critical role in our understanding of nature, communicating reliably with others, and increasing our sense of possibilities for a cooperative future.

Symmetry defines equivalence of the past among all observers for that time period and separately defines equivalence of the future among all observers for that distinct period. Symmetrical time assumes implausible two-way time that can variably flow backward or forward. Breaking time symmetry distinguishes real-world time as one-way time that invariantly flows only forward. Symmetry breaking establishes explicit incompatibility between real-world systems with open-ended processes and uncertain futures and idealized systems with closed-ended processes, certain futures, and implicit assumptions that the past determines the future. *Symmetry breaking* is a process of choices where instabilities encountered by the system lead to decision points and irreversible transitions that create novel structure for solutions. These novel solutions inform adaptive action and the modulation of the instabilities.

The arrow of time, preferred asymmetry, and irreversibility play an informational role in evolution and entropy, in language, and in communication (Nicolis & Prigogine, 1989, pp. 144–145). In complex systems, representing information as statistical distributions allows better understanding of systematic influences. In large nonlocal entropy-producing systems, we find dynamic instabilities at transitional decision points that lead to novelty-producing bifurcations (Nicolis & Prigogine, 1989, pp. 147–148, 164–171). Systematic language supports immutable sequential descriptions of experience consistent with the forward-directed natural ordering of events in time.

> Unlike the adjacent order of a series of objects, the sequential order of a series of events cannot be rearranged. Some of the *individual* events that compose the grand sequence of a day can go backward, displacements for example, but the sequential order of their occurrence is immutable. This is why "time" is said to have an "arrow," I believe, and this is why "time travel" is a myth. (Gibson, 1986, p. 109 [emphasis in original])

We have inherited affective language that correlates to how our limbic brain processes and integrates past experiences in memory to set expectations for behavior and predict future experiences and outcomes. We know that limbic brains act as if the past predicts the future and imply symmetrical time—that is, time that flows symmetrically in both directions, backward and forward, as opposed to asymmetrical time, which flows in only one direction, toward the future. But what kind of person other than an imaginary time traveler would believe in symmetrical time? And why would we or anyone else care about the direction of time, changing our language, or our thinking? Why would we make the effort when we can simply conclude, "Who cares which way time is pointing, when fast and frugal intuition works *just fine*." We may or may not be impressed by the knowledge that evolutionary processes operate on one-way time and rely on open-ended capabilities and alternative possibilities for innovating creative solutions. Of course, evolutionary processes operate toward the future where the irreversibility incorporated into the description of one-way time places constraints on the direction one can travel in real time.

Symmetries may be generally classified as global or local. Global symmetry holds relatively invariant at all points of space and time. Local symmetry has variant symmetry transformation at different points of space and time that arbitrarily allows forward or backward time and confounds the distinction between the past and future. Local time varies implicitly with arbitrarily referenced linguistic conventions and habits, whereas global time explicitly correlates an invariant natural referent that standardizes time and events systematically for all observers (Edelman, 1992, p. 203). As in Einstein's theory of relativity, local time remained a reversible time (Prigogine & Stengers, 1985, p. 17). Einstein once said, "Time [as associated with irreversibility] is an illusion," and "for us convinced physicists, the distinction between past, present, and future is an illusion, although a persistent one" (Prigogine, 1997, pp. 58, 165).

Symmetry breaking governs the separation of events that happened in the past from future events that have not yet occurred. Memories are static and orthogonal (at right angles) to time such that we can make marks on a timeline that represent individual memory episodes for specific events. If we were marking the events in real time, each one of our marks would break time symmetry by separating the past and future at the time of the event. In effect, symmetry principles enable memory formation of past experiences in evolving organisms (Edelman, 1992, p. 207). No matter where a person or groups of people are located, the past has the same general symmetry for all observers, is static, and is somewhat similar to a picture or notepad that carries information as past memories that are suitable for storage.

We typically lack complete knowledge of an event, and memories tend to fade over time. Even though memories can be fragile, whatever happened cannot be changed by any amount of forgetting, rituals, or wishful thinking. The same general symmetry applies to predicting the future for all observers. The future is dynamic, probabilistic, and nonabsolute—uncertainty prevails no matter how much we imagine or wish for infinite certainty. After the predicted event happens, it is relegated to the static past. The process of symmetry breaking enables separation of time-invariant information descriptions of past and future events—that is, memories and predictions. Thus, information can be defined and identified according to symmetry-breaking criteria and a time-dependent probabilistic ruler. We can add probabilistic information processing to our reference list of general systematic processing and relational principles:

- Time, symmetry breaking, asymmetrical-time referencing, and probabilistic information processing

When we concomitantly account for both past and future views, the difference between the static past and dynamic future is generally asymmetrical for all observers. In other words, regardless of our point of view for our observations, what we believe or perceive, or how we speak, all humans live by the same systematic time standard. When our language and speech match with the general laws of nature, our observations confirm the invariant direction of time pointing toward the future, wherein the past and future are neither equivalent nor interchangeable. Explicit recognition of this irreversible asymmetry provides a coherent invariant reference for understanding our world, encourages us to take responsibility and accountability for thoughtful planning, and increases the probability of improving our adaptive behavior. An invariant referent informed by nature establishes a source for symmetry breaking and making measurements and predictions of action in the system that coherently correlate with the extended laws of physics. Then, our language and other behaviors we produce will likely correspond as adaptively as possible with how our world works.

Describing referents and transient discrete interactions locally might suffice in small spaces, but effectively describing continuous interactions in large multidimensional spaces across the domain, such as the atmosphere, requires delocalized descriptions and referents (Prigogine, 1997, pp. 114, 141). Nonlocal models provide a continuous-state action space for representing and describing persistent relations in high-dimensional systems. Localized linguistic models lack sufficient parameters to scale to nonlocal dimensions for dynamic measurements and predictions in evolving systems.

Table 2 shows the extended linguistic parameters and characteristics that enable systematic representations and processing descriptions. The column on the left lists variables and parameters expressed by language that impart localized constraints on information representation and information processing. The column on the right lists *extended language* with global constructions that contribute to efficient information compression and effective exploration and exploitation of high-dimensional domains. In complex domains, extended language facilitates optimal representation, processing, and application of information in real time. From left to right we see increased computational expense and effort as the "cost of doing business" that correlates with optimizing systematic fidelity. Variables from left to right columns increase task-dependent efficiency, corresponding to a range of computational output effects between local sufficiency and global optimization.

A global natural referent enables dynamic logic and measurements that can extend across the entire system. The invariant laws of nature provide consistent referencing that supports systematic knowledge. Nature presents us with irreversible time-dependent events and constant change, where knowledge and predictions are best represented as possibilities according to natural laws and evolutionary processes. Knowledge derives according to a systematic referent existing in and informed by the physical laws of nature rather than from outside of nature—that is, rather than according to being unnatural, paranatural, supernatural.

Table 2. Extended Language and Parameter Space

Local	*Extended*
Variant	Relatively invariant
Stability	Modulated instabilities
Rigid	Flexible
Static	Dynamic
Equilibrium constraints	Nonequilibrium constraints
Nondissipative	Dissipative
Reversible processes	Irreversible processes
Conservative	Nonconservative
Symmetrical time	Asymmetrical time
Low-dimensional	High-dimensional
Cause and effect	Systematic influences
Transient	Persistent
Linear	Nonlinear
Closed-ended	Open-ended
Infinite limits	Finite limits
Discrete two-valued logic	Continuous probabilistic logic
Perfection	Range of imperfection
Absolute	Relative
Certainties	Possibilities and probabilities
Bottom-up	Top-down
Automatic	Deliberate
Affective	Effective
Implicit synthetic	Explicit analytic
Simple to complicated	Complex to complexification
Specific	General
Irregularities	Regularities
Idealized physical laws	Realized physical laws
Nonevolvable	Evolvable
Deterministic	Nondeterministic

For example, referents outside of nature influence how we refer to external relationships, which affects our inferences and perception, beliefs, and imaginary simulations, including

- Superstitions
- Illusions of perfection and stability
- Certainty of future events
- Infinite worlds
- Influence of mystical forces
- Communication with invisible people

It is somewhat common for children to have imaginary relationships, but these habits can carry over to adulthood. Understanding how things work in nature enables the application of global knowledge to inform local beliefs and predictions about physical relationships that we experience in our everyday lives, naturally augmenting effective real-world communication with others.

Reliable referencing can help to correlate and calibrate internal and external information with relative equivalence to nature. Correlations expressed with globally reliable language allow us to compare the coherency of our internal cognition, information, values, behavior, and predictions with our external knowledge. A narrower range of systematic errors and fewer discrepancies between our predictions and our results support coherency that corresponds to more reliable predictions with less variance between expectations and realized outcomes. Better predictive abilities enhance the possibility for effective adaptation to change within that system. As an information form, extended language can be tuned systematically to enhance the effective utility of our thought processing, logic, and predictions, and thus our fitness. The next section elaborates on the relationship between language and the physics of nature.

SECTION II

Language and Physics

CHAPTER 9

It's a Physical World

IN THIS SECTION WE explore the development of perspectives on physical laws and their general historical relevance. We also explore the divergence to newer models of physics that appear to resolve compatibility issues between conservative physics and nonconservative evolutionary processes and physics of today's world. We have already seen how conservative systems with closed-ended deterministic processes lack the means to support open-ended nondeterministic evolutionary processes found in nature.

The Nondeterministic Nature of Physics

At any given time, our world perspective likely depends on our individual beliefs or current knowledge that may or may not include the evolution of scientific thinking in relation to nature and general physical laws. Our understanding of nature and how it works has evolved from a deterministic perspective to a nondeterministic one. We now have the option of applying nondeterministic thought processes that correspond to what we know about the open-ended processes of evolution. But science and thinking in general both struggle as popular pastimes, especially when it comes to thinking off the beaten path. It is hardly surprising that *evolution* and *nonequilibrium physics* are not necessarily household words or commonly accepted knowledge about nature. Historically, we know that scientific knowledge about the world slowly evolves to more widespread popular acceptance over time that correlates to some extent with higher-level education.

Such an evolution in our thinking allows us to take a macroscopic view that significantly changes what we see. Nondeterministic processes have been around for a long time and were identified by Henri Poincaré at the beginning of the twentieth century when he discovered resonant interactions in large high-dimensional spaces. We see evidence of nondeterministic processes in evolution and oscillatory brain rhythms, but only recently has physics extended and codified the details. According to this extended perspective, we live in a time-dependent, perpetually changing environment with far-from-equilibrium dynamics and a difficult-to-predict future.

In a difficult-to-predict environment, it makes sense to have the flexibility offered by alternative solutions. Evolution operates on nondeterministic processes that enable adaptation, which generally suffices in face of uncertainties. Not surprisingly, evolutionary processes are generally open-ended and produce multiple alternative solutions. Open-ended logical processes enhance flexibility for correlating solutions to the complex dynamics of nature. It is no less surprising that open-ended, nondeterministic language also matches these systematic dynamics.

In physics, the idea that we can make accurate long-term predictions about complex systems' behavior has evolved to the probabilistic view of complexity theory and dynamic systems of today. Over the centuries, the laws of physics represented idealized simplifications that assumed a deterministic relation between cause and effect in nature, leading to the belief that it is possible to make accurate long-term predictions of a physical system as long as the starting conditions and causal relationships are known.

Deterministic physics assumes that the universe unfolds in time without deviating from predetermined laws, like the workings of a perfect machine. In a deterministic worldview, every cause has a unique effect, and every event or action inevitably results from identifiable preceding events and actions. Thus the workings of the system can be predicted in advance, or in retrospect, given sufficient information and processing power.

In other words, deterministic theory tacitly assumes that if we can know the initial conditions of a system with sufficient accuracy, and we understand the laws of cause and effect, we can predict future states exactly. If inaccuracies in our measurements of the initial conditions lead to errors in prediction, we can simply reduce the errors in our measurements to reduce variance in the outcomes, until we achieve the accuracy that produces the desired level of precision in our predictions. Deterministic theory implies that with perfect knowledge we can produce perfect predictions.

Newton's universe is deterministic and reversible, and sometimes depicted as working like a game of billiards, where the trajectories of molecules and celestial bodies can be likened to colliding billiard balls. Since the results are mathematically predetermined from the initial conditions, the interactions can be run forward or backward in time like a movie. Unexpected variations in the results are attributed to a lack of precision in the measurement of those initial conditions. Newtonian physicists assumed that if they could shrink the magnitude of imprecision, they could similarly shrink the uncertainty of the predicted behavior. Deterministic assumptions and certainties persist today in the physical sciences and in society as well, even though Newton's laws were superseded by a larger set of nondeterministic physical laws based on uncertainty found in dynamic systems.

The uncertainty principle has been interpreted in the past as a constraint on precision that limits our ability to make infinitely accurate measurements; it assumes if we cannot achieve perfect knowledge, then our predictions will likewise be imperfect to some extent. But this deterministic view assumes that nondeterministic descriptions of uncertainty are probabilistic only due to our imperfect measurements and our imperfect knowledge, rather than the inherently nondeterministic nature of the physical world. These assumptions represent a basic flaw of deterministic physics.

In the early twentieth century, Poincaré identified another basic flaw in deterministic theory when he observed that some complex systems continue to diverge from predicted behavior despite improvements in the measurement of their starting conditions. He showed that for these systems even the smallest differences in starting conditions will produce large, unpredictable differences in outcomes.

Since we are unable to perfectly eliminate differences in initial conditions, prediction accuracy in these systems will inevitably vary. These variations came to be known as dynamic instabilities, or simply as chaos, and the theory that defines them is known as chaos theory. Theoretically, with perfect and complete knowledge we could accurately and completely explain or predict the results of change in complex systems with dynamic instabilities, but such perfect knowledge is unachievable with real-world measuring devices. The information vacuum created by this inability to produce perfect knowledge leads us to question the deterministic notion that we can reliably predict outcomes with perfect accuracy when we live in a nondeterministic world where dynamic instabilities prevail.

Some 50 years after Poincaré's discovery, Edward Lorenz (1963) published an article in the *Journal of the Atmospheric Sciences*, "Deterministic Nonperiodic Flow." He had written a basic mathematic software program to study a simplified model of how an air current would rise and fall when heated by the sun. His computer code contained the established deterministic mathematic equations, which governed the flow of the air currents. Since the computer code was deterministic, he expected to get the same result each time he entered the same initial values into the program. Instead, he got different results every time with what he thought were the same initial values. He concluded that while he considered his initial values "identical," the computer interpreted them as variable enough to produce significant differences in the ending states of different runs of the model, which is the signature of dynamic instabilities in complex systems.

The sensitivity to initial conditions discovery by Lorenz's mathematic weather model is amusingly referred to as the *butterfly effect*, where a small change in initial conditions, such as a butterfly flapping its wings in the Amazon, might, in principle, ultimately alter the weather in Kansas (Kauffman, 1993, p. 178). Or the beating of the wings of a fly in Massachusetts could well be at the origin of a major climate change in the Indian subcontinent (Nicolis & Prigogine, 1989, p. 124). This effect is considered the basic reason that long-term forecasting in dynamic systems is notoriously weak. Ever so slight or unseen initial differences can lead to large changes in the predictability of a high-dimensional nonlinear system, such as the global weather system.

At the time of its discovery, the phenomenon of chaotic systems was considered a mathematic oddity. In the decades since then, physicists have found that dynamic instabilities tend to be the norm in the universe. The presence and prevalence of dynamic instabilities in complex natural systems seem to further limit our ability to determine predictions with certainty. Ilya Prigogine and other scientists have questioned the relevance and reasonableness of a deterministic description of the universe as well as deterministic beliefs. Assumptions expressed as certitudes about the universe are being replaced by theories of physics that operate on possibilities and probabilities (Prigogine, 1997). The notion of randomness is evolving into nondeterministic descriptions of nonrandom regularities in the natural world associated with dynamic systems, complexity theory, and self-organization.

We notice a difference between physical interactions and self-organizing relations in living biological systems that add a level of complexity and the ability to adapt to changing conditions. In biological systems, living organisms carry their history along with them in the form of genetic information, such as DNA. The brain functions as a complex adaptive system (Buzsáki, 2006, pp. 15–17), and the continuing study of dynamic instabilities in complex adaptive systems is important for neuroscience in the study of brain and behavior. In these complex adaptive systems, nondeterministic physics prevails.

Deterministic descriptions perhaps suffice when confined to relatively simple, idealized, low-dimensional interactions, and the dynamics of individual trajectories, but they typically face increasing constraints as the complexity of the system increases. In entropy-producing complex natural systems, we encounter

- Bundles of trajectories as ensembles
- Higher-dimensional interactions
- Multivariate nonlinear influences
- Instabilities
- Irreversible processes
- Asymmetrical time, where the term *dynamic* is extended to descriptions of time-dependent *dynamical* systems

In natural systems, we see dynamic instabilities and asymmetrical time, which irreversibly flows from the past to the future. In such systems, physical law would then assume an entirely new aspect; it would no longer be solely a differential equation but would take on the character of a statistical law (Poincaré 2007, pp. 110–111). Time asymmetry allows a global statistical description of physical laws that is based on probability distributions of information that correlate with domain regularities and the flow of correlations ordered in time as a history of those correlations (Prigogine, 1997, pp. 78–81, 122).

Deterministic physics, language, and mathematic equations work fairly well for describing simple idealized systems and averages, and when used in hindsight can describe some version of the static past. But they fall short for reliably describing or predicting the dynamic future. With symmetrical time, it comes as no surprise that the wave function collapses in theoretical quantum physics (Prigogine, 1997, pp. 48–50, 53–55, 130–132). Deterministic language describes the mathematic formulas, which incorporate symmetrically reversible time, and both lack the dynamic parameters to support nondeterministic descriptions.

> But the fundamental object of quantum mechanics, the wave function, satisfies a deterministic, time-reversible equation. (Prigogine, 1997, p. 5)

It is also not surprising that language correlates to physics, since we use language to describe physical and mathematic relationships. Of course, language itself does not collapse the wave function. But by understanding the difference between localized and global language constructions, we get clues about what may be occurring. Localized language constructions isolate the solution space and hide the global problem we are trying to solve, where part of the solution is inside the problem. Nonequilibrium systems with dynamic instabilities rely on asymmetrical time (Prigogine, 1997, p. 157). Even though quantum theory incorporates probability, the probabilities are "time symmetric" (Prigogine, 1997, p. 137).

> Figuratively speaking, matter at equilibrium, with no arrow of time, is "blind," but with the arrow of time, it begins to "see." (Prigogine, 1997, p. 3)

Language and mathematic formulas based exclusively on deterministic principles fall short for "seeing" and describing or understanding instabilities found in nonequilibrium systems. In effect, deterministic principles create systematic blindness by default and, by ignoring the irreversible asymmetrical arrow of time, place constraints on the search for novel solutions, creativity, and innovation. The term *arrow of time*, coined by Arthur Eddington (Nicolis & Prigogine, 1989, p. 124), denotes the privileged direction of time. Extended language with nondeterministic characteristics corresponds with changes we observe in our evolving world. Without extended linguistics and one-way time, our dynamic perception easily becomes temporally distorted and relegated to a static view.

Time and Space: Where the Action Is

Time stands out as a paramount feature of a complex environment, and to account for one-way time—the forward noninvertible direction of invariant time—we can distinguish language that correlates with the dynamics of the physical world. Extended language supports relevant temporal dimensions for coherently correlating events with one-way time, cognition, and external change far from equilibrium.

> In a broad sense, *time is a measure of change*, a metric of succession, a parameter that distinguishes separate events. (Buzsáki, 2006, p. 6 [emphasis in original])

The natural world is a global stage buffeted by the winds of persistent change, on which physical evolutionary processes unfold and flow forward in time and space. Science seeks to observe and describe nature and the open-ended global processes of entropy and complexification in terms of time-irreversible evolutionary processes that point in the same invariant direction, as does the arrow of time, toward the future.

> Exactly as biological evolution cannot be defined at the level of individuals, the flow of time is also a global property. (Prigogine, 1997, p. 20)

Representing and describing these higher-order global processes requires language with an extended descriptive capability that enlightens our perspective. Language is typically viewed in terms of grammar, syntax, and semantics. Grammar provides descriptive rules for using words to construct sentences in a given language. Syntax addresses the output of preferred linear sequences of communications ordered in time. Semantics describes word and sentence meanings relative to their use. While each contributes overall meaning, semantics supports definitions and contextual usage that apply to past, current, and future experiences across time. Words satisfy static constraints for storing memories, for dynamically connecting our thoughts to the world, and directing those thoughts constructively toward the future. Semantics that provide extended higher-order abstractions and explicit systematic meaning empower grammar and syntax to a global level.

Without higher-order semantic representational capability, language is limited to localized descriptions by default. Global descriptions of the arrow of evolution, the arrow of entropy, and the arrow of complexity become possible when language has the extended semantic capability to represent, process, and explicitly synchronize these concepts with the arrow of time. When we apply the extended "arrow of semantics," we can escape from the static certitudes of localized representations and evolve to global descriptions of possibilities and probabilities. Global descriptions enable a new coherence with the natural world and an understanding of the uncertainties and dynamic instabilities we encounter there.

Deliberately correlating semantics to nature on a global level and aligning language with nature's arrow of time produces descriptions that distinguish the static past from the dynamic future by semantically breaking time symmetry and synchronizing the nature of language with the language of nature. The arrow of semantics provides language abstractions with which we can represent the nondeterministic ordering of the space where life exists, between the static certitudes of absolute rigidity and the tumultuous randomness of absolute chaos—the fluid open-ended land of possibilities and probabilities in which we live.

In entropy-producing nonequilibrium systems, instabilities lead to oscillations and choices among possible alternative solutions. During oscillations, the event of choice produces a bifurcation to a preferred state for improving the adaptive modulation of instabilities. Bifurcations at decision points add information that increases complexity in the system. The decision reflects the influence of current inputs, the current state, and domain knowledge accumulated from the evolutionary history of correlations. Information about the system prior to the event is static and no longer changeable, while the system's future is undetermined, changeable, and probabilistic. The fluctuations occurring at the bifurcation and choices made by the system will appear arbitrary and noisy to an outside observer lacking systematic knowledge. As previously noted, this process of choice is called *symmetry breaking*, where irreversible space–time transitions create novel structure that enables adaptive modulation of behavior within the instabilities encountered by the system. We see somewhat similar novel structure created by the scaffolding of memories when new learning from experience takes place.

For example, an organism's DNA represents knowledge accumulated as the history correlating the organism's long-term survival as adaptations to disturbances it encounters. DNA carries variations as complexity added to the system at reproduction, which adds alternative information as possible phenotypic solutions for novel behavior. In humans, we have language that can also lead to novel solutions. When things change and what worked previously fails to resolve a problem, rather than increasing rigid constraints, we can apply language to process and flexibly reconfigure what we know into developing new solutions with increasing complexity.

Language with systematic correspondence enables systematic knowledge for building more complex solutions when encountering disturbances. Innovation comes from processing different variations and integrating novel configurations of our accumulated systematic knowledge. We can use the accumulated knowledge to decrease systematic errors, increase effective choices for resolving the everyday problems we face, and add our new learning from the outcomes to continue building on our history of systematic knowledge.

According to Prigogine (1995), "complex systems carry their history on their backs." In other words, DNA carries the genetic histories of evolving creatures analogous to the backbone of complexification. Operational knowledge of an evolving system and the progressive increase in knowledge it contains represent understanding of irreversible evolutionary processes over time. The systematic knowledge we gain from understanding evolutionary processes can improve how we make predictions corresponding to natural physical processes—how we communicate, cooperate, and adapt in correspondence with nature.

> Communication is at the base of what probably is the most irreversible process accessible to the human mind, the progressive increase in knowledge. (Prigogine & Stengers, 1985, p. 295)

We might arguably assume that most languages have some capacity for representing, processing, and understanding physical relations within the natural world. The application of extended language expands the linguistic spectrum and enables information processing and descriptions of physics at a systematic level far from equilibrium. Language structure that correlates with nonequilibrium physics enables a practical macroscopic perspective and extended processing for understanding change within dynamic natural constraints. Life is not merely a structure but is instead a changing structure, a process (Bronowski, 1977, p. 164).

Even though language provides us with a resource for unlimited imagination, we operate in one world only—the real world of physics (Skinner, 1953, p. 139). Humans, other living organisms, and the natural world represent one large interacting physical system. Practically, it makes sense to focus attention on where the action is and the subset of parameters and assumptions that most strongly affect the system's behavior (Jennings, 1906, p. 107; Ellner & Guckenheimer, 2006, p. 261).

> In studying the behavior of any organism, the first requisite to understand is working out the action system. (Jennings, 1906, p. 300)

In a dynamic system, the first assumption is that time represents the foremost, overriding, and all encompassing feature; time and reality are irreversibly linked (Prigogine, 1997, p. 187). Since biological systems are dynamic systems, models that ignore time altogether may inadequately explain or fail to account for overall system complexities compared to models that integrate time into their methods (Eliasmith & Anderson, 2003, p. 219; Spitzer, 1999, p. 87).

Static snapshots, or queries isolated in time, can provide valuable information, but they are less likely to capture relevant statistical information about persistent dynamic processes and relations in the overall system. Models that assume bidirectional time will more likely converge to local average descriptions and conclusions and are less likely to capture the dynamic range of divergent processes in evolving systems with unidirectional time, including oscillations, instabilities, and irreversibility. We know that averages compress information and hide the variability on which dynamic systems thrive.

> The direction of evolution is an important and indeed crucial phenomenon, which singles it out among statistical processes. For so far as statistical processes have a direction at all, it is usually a movement toward the average—and that is exactly what evolution is not. (Bronowski, 1977, p. 167)

Physical science that includes non-Newtonian irreversible entropy-producing processes views time as proceeding independently in one direction only, by simply sliding continuously into the future (Prigogine, 1997, p. 18). The arrow of time flies irreversibly forward along with the situational spatial occurrence of physical events in the natural world. The flow of time represents a global property, and all arrows of time in nature have the same preferential orientation, producing entropy in the same direction as time, which is by definition into the future (Prigogine, 1997, pp. 20, 102). Irreversible physical processes are defined by the second law of thermodynamics that distinguishes between past and future by the forward direction of time (Mitchell, 2009, p. 43; Prigogine, 1997, p. 19). A dynamic temporal correlation suggests that a failure to effectively integrate time and space into problem solving consequently reduces the probability of finding the best solution at any given moment (Fuster, 2003, pp. 62, 109).

Characterizing information processing at the event level defines a special contextual correspondence between time and situational events, in effect collapsing time and space into event space, or Minkowski world (Minkowski, 1952, p. 75–76, 79–80). Event-level descriptions establish a relatively objective referent for evaluating situational interactions within the natural world at a given location in time and space (Einstein, 1961, p. 161), and for correlating relevant nonlocal information about relationships into reliable systematic knowledge referenced to reality in the context of that time and space. Event-level specificity provides a specific time-related referent for establishing an independent reality (Minkowski, 1952, p. 76) and provides a nonlocal method for the ordering of dynamic events according to a single time sequence (Prigogine, 1997, p. 147).

> For these [nonlocal] systems there is a privileged direction of time, exactly as there is a privileged direction of time in our perception of nature. It is this *common* arrow of time that is the necessary condition of our communication with the physical world; it is the basis of our communication with our fellow human beings. (Prigogine, 1997, p. 54 [emphasis in original])

In this space and time context, relative objectivity refers to the application of a natural external referent that relies on one-way time and event-related symmetry breaking. Breaking time symmetry synchronizes the temporal aspects of language and cognitive processing with actions in response to changing external relations in space and time. Leibniz pointed out long ago that space and time are not things but relations (Bronowski, 1977, p. 40).

> Science is a system of relations.... [I]t is relations alone which can be regarded as objective. (Poincaré 2007, pp. 135–138)

The term *objective* takes on a new meaning when applied to processes of dynamic change, action, and event relations of objects in space that correlate knowledge structure with the updated nondeterministic laws of physics. Thus knowledge of relations serves us as rule of action (Poincaré, 2007, pp. 112–115, 128).

> The only link between the verbal and objective world is exclusively structural, necessitating the conclusion that the only content of all "knowledge" is structural. (Korzybski, 1958, p. 20)

The idea of *reality* represents an abstract linguistic construct referenced to one-way time, by which we distinguish the independence between past and future. One-way time provides a probabilistic processing method for parsing time-dependent events and actions. Distinguishing the static past from the probable future at the event level enables distinctions between descriptions of identity as "being" or passively belonging to a static class, "being" as a persistent process within living individuals that make up that class, and "becoming" as passive change over time. For instance, humans belong to a static class of human beings, where "being" allows for changes in living systems over time. Passive verbs act as copulas or connectors that make it possible to arbitrarily mislabel, misconnect, and bias naturally ordered categories of static classes, dynamic behavior, and value. Linking verbs, such as *to be*, *is*, and *are*, can extinguish context by arbitrarily creating a category of one dimension, somewhat similar to one-dimensional imperative and prescriptive modal verbs.

> Let us notice that initial conditions, as summarized in a state of the system, are associated with being; in contrast, the laws involving temporal changes are associated with becoming. In our view, Being and Becoming are not to be opposed one to the other; they express two related aspects of reality. A state with broken time symmetry arises from a law with broken time symmetry, which propagates it into a state belonging to the same category. (Prigogine & Stengers, 1985, p. 310)

A given event takes place within the context of location and time. At the moment of an event, the flow of time is partitioned into past and future. Relative to the event, the past is now static and unchanging, while the future is undetermined and probabilistic. In a sense, we are living in the future as our choices and outcomes take on a statistical meaning in a dynamic world. Imagine that we are on safari, hiding in the bush watching wildlife interacting around the local watering hole. We have our cameras handy for taking photos and our video recorders for action videos. We can record or take a snapshot at any moment we like. We see a herd of wild water buffalo drinking their fill and decide to take a photo. The instant we click the button and take the picture, we break time symmetry and produce a static photo even though the memory is still reverberating in our mind. The same applies when we film the whole experience over a period of time, except that we can watch the natural extravaganza over and over again as much as we like. Similarly, we can recall the episode in our memory and retell the story again and again. Of course we know that playing the video or repeatedly experiencing our memories does not change the fact that the actual experience is static and in the past, even though we might feel like we are reliving it.

We can naturally partition and classify experience according to operational descriptions of static existence of classes passively denoted by *is*, the static past denoted by *was*, and dynamic behavior as an active process of action, change, and probabilities. Passively misappropriating static and dynamic class characteristics confounds temporal context, which in turn confuses temporal separation of information, misinforms prediction and systematic error correction, and disrupts effective adaptation to changing states. Separating the static past from the probabilistic future reflects the evolutionary development of the universe described within the fundamental laws of physics (Prigogine, 1997, p. 29). The dynamic laws of nature now deal with possibilities, rather than the certitudes of the old deterministic laws, and thus overrule the age-old dichotomy between being and becoming (Prigogine, 1997, p. 155).

Even though language provides an abstract rather than a concrete representation of the external world, it provides a reliable extensional substrate for describing and establishing a relatively objective external referent. When we wish to compare observations about the external environment, whether within one brain or among groups of communicating brains, an invariant referent to nature, such as an atomic clock, oscillations produced in cesium atoms exposed to electromagnetic radiation, or to some extent even a simple calendar, establishes temporal context with the invariant forward direction of time. One-way time reduces the amount of unreliable temporal inferences that might otherwise complicate effective cognitive processing and communication. With an invariant referent established, explicitly correlating language to one-way time in nature turns a complicated task into a fairly simple but nontrivial one that enhances effective thought processing and predictions.

Communication operates according to common time as a referent for measuring, processing, and synchronizing perceptions with the internal language constructions with which we represent nature. Communication that follows the extended laws of dynamics, including the breaking of time symmetry, enables us to transform information received from the outside world into actions on a human scale to function more reliably in that world (Prigogine, 1997, pp. 150–151).

Abstractly, we can indirectly reference reality, systematic knowledge, and probability with the logical correspondence among our cognitive processes, our experiences, our world observations, and our linguistically constructed worldview. In other words, the structure of our language influences the structure of our knowledge. Systematic referencing enables performance measurement and evaluation of the effectiveness of our cognitive processing and our knowledge by weighing how reliably observations and predictions at one instance in time correspond with other instances across different time scales. Functional evaluations provide logical evidence for how effectively we have modeled the relationship between cognitive processing and linguistic structure in correspondence with regularities observed in the outside world and with the structure of natural physical laws that govern those regularities and that world.

> A physical theory, we have said, is by so much the more true, as it puts in evidence more true relations. (Poincaré, 2007, pp. 140–141)

When we deliberately construct and calibrate our language and semantic representations to fit the behavior of relations in the world in which we live, we naturally process and produce consistent structure for effective speech. The essential distinction here is that behavior is a physical event that can be observed externally or detected by a recording device of some sort (Vanderwolf, 2010, p. 16). In other words, behavior provides an observable physical connection with the reference environment.

Without systematic referencing we are vulnerable to the variable beliefs and whims of our own locally constructed referencing, two-way time, and absolute certitudes. We are left implicitly anchored and bound to a history of coincidences from the static past rather than correlated to the dynamic realities in the present and possibilities in the future—blind to the global nature of the arrow of time.

Orderly Classes, Objects, and Actions

Computational processing models rely on orderly classes of objects and permissible actions to compute solutions. Reliable classes tend to correlate with effective solutions. To make sense of our experiences, we use language to identify classes to model the structure of the environment and specify and separate the entities into a logical ordering. We make logical sense of the world with classes that coherently correspond with the systematic ordering of nature.

For instance, recall that the early classification of elements into air, earth, fire, and water yielded inconsistent results and was of little help to us for processing misclassified information into orderly explanations about how the world works. Over time and with better instruments, techniques, and a lot of hard work, this earlier system eventually evolved into a more complete table of the elements that offered explanations which better corresponded to the systematic ordering of natural processes and relationships. Without improved elemental classifications representing coherent relationships, we would still be trying to solve the processes of nature in terms of air, earth, fire, and water. Language, coherent classifications, and effective problem resolution go hand in hand.

The natural world includes matter and objects, which are structurally equivalent at some level. Observing the physical relation between matter and identifiable objects provides an informational description for classifying matter in nature into animate and inanimate objects. Biologically living humans obviously fall into the former class.

This rather simple biological classification system has applications for computing naturally preferred information representing contextual relationships among time and space, objects and positions, energy, and actions. Classification allows groups of like things to be treated similarly (James, 2007, v. 2, pp. 646–647). In effect, classification is a method of symmetry breaking.

John Maynard Smith has advocated the importance of understanding biological systems from an information-processing perspective (Maynard Smith, 2000). The biological description of living systems as information processing or computational networks is common in many areas of science (Mitchell, 2009, p. 169; Siegel, 1999, p. 209). The physics of the world the body inhabits, the physics of the body the brain inhabits, and indeed the physics of the nervous system itself have shaped the evolution, and hence the computational style, of nervous systems (Churchland & Sejnowski, 1994, p. 416).

Classifying matter and objects systematically according to symmetry breaking correlated to their natural physical relationships provides an information-processing framework for representing and describing the complex behavior of dynamic systems in computational terms. Complex systems adapt to the environment by processing information learned about that environment (Mitchell, 2009, p. 170). The ability to maximize the systematic fidelity of information, along with the quantity of information available for computational processing, along with how that information is represented, classified, defined, and valuated, will tend to bias the partitioning of information in the system. The partitioning influences the processing of solutions and observations made about the system's effectiveness. Ineffective partitioning can misinform our understanding of the system's state, leaving us with faulty and often confusing conclusions.

> A central part of this problem considers the criteria for natural classification. The capacity to know a world requires that sufficiently similar states of the world be able to be classified as "the same." (Kauffman, 1993, p. 233)

Science relies on a fidelity-based classification system to describe the systematic ordering of processes and relationships in correspondence with nature. With a scientific classification, we can represent and organize information in a logical manner that correlates the processing of that information with the natural ordering of objects, actions, and events according to their structure and functional physical relationships in the environment. Classification fidelity enables effective processing and understanding of regularities observed in natural relationships. In turn, we improve our ability to make predictions, eventually yielding what is collectively known as scientific knowledge. Scientific language is deliberately constructed to represent, classify, and process information into effective knowledge that corresponds to the time-dependent systematic structure of nature.

Breaking the Static Language Barrier

We use language to describe and understand the world. So it follows that we can more likely improve our understanding of the world if we have a systematic understanding of how language works. The catch-22 is that we can only use language to talk about language. It's kind of like trying to see our own eyes—we can only do that by looking in a mirror. So what can we use as a language mirror to study the effects of language on our understanding of the world?

Recall that scientific methods allow us to study and discover regularities in nature, which enhance our understanding of how the many complexities in the physical world operate according to physical laws. We find systematic information distributed statistically in the environment as probabilities. Reliable systematic knowledge depends on a logically coherent relationship with an external referent for effectively evaluating, processing, and correlating information according to probability distributions. Science relies on probability-based logical processing for accumulating systematic knowledge.

Natural systematic representation and processing methods provide an operational understanding corresponding to the structure of the physical world we inhabit. When our thinking corresponds with structural relations and actions in nature from a systematic perspective, we can think more responsibly about how we treat others and our world. An external referent to the physical world shifts the information distribution from local to global (Prigogine, 1997, p. 96). Global information yields a better understanding of the behavior of the organisms that live in the world and the natural ordering of the physical relations among all involved, including the relations among members of the species *Homo sapiens*.

Living organisms represent a complex of many processes, of chemical change, growth, and movement proceeding with a certain energy requirement. Organisms are composed of matter, subject to the usual laws of physics and chemistry, and they operate within the limits of natural internal and external physical constraints.

> All the potency of behavior, and of everything else that exists, must lie in the laws of matter and energy—physical and chemical.... (Jennings, 1906, p. 326)

These physical correlations have relevant implications for complex systems. A central issue in complex systems concerns communication and information processing in various forms (Mitchell, 2009, p. 41) that operate within natural constraints of the physical domain. The scientific study of information began with thermodynamics and physicists' notions of energy, work, and entropy (Mitchell, 2009, p. 41). Entropy and information are closely related (Prigogine, 1997, p. 24). These physical relationships provide valuable insights about the world and also demonstrate that for dynamic biological entities, acquiring information and modulating entropy requires work (Mitchell, 2009, pp. 46, 72).

Work implies costs for any given biological strategy, algorithm, or recipe, with sustained effort to solve a given problem. Of these costs, entropy plays the major role as the price paid by nonequilibrium structures found in nonconservative open systems (Prigogine, 1997, p. 175). No algorithm is sure to succeed, and as a result there is no free lunch (Wolpert & Macready, 1995, 1997; Rogers et al, 2003). Conservative systems—which are closed by definition—are idealized energy-conserving systems, which correspond to a free lunch. But in complex systems, we can expect to pay for the modulation of increasing entropy and complexification.

Interactions in physical systems include relationships between mass and energy in the universe stated by Einstein's equation, $E = mc^2$ (Einstein, 1961). In light of this formula, the universe and nature begin to make sense in a scientific rather than a mystical way. We live in a physical world operating under natural laws—a dynamic world composed of mass, energy, time, space, and action. Herman Minkowski (1952) proposed four-dimensional space by unifying time with three-dimensional space, and Einstein's relativity theory complements Minkowski's revelation by integrating space–time and the relation between mass and energy in a "conservative" universe. We now understand that conservative laws formed the stepping-stones and a jumping-off place, leading to the current laws of nonequilibrium physics that confirm the nonconservative nature of our environment. For instance, a conservative system represents an idealized closed system that is self-contained such that energy and information is confined within the system rather than freely exchangeable with a perpetually changing external environment.

In Einstein's formula, the single constant *c* accounts for both space and time, leaving mass and energy proportionally interchangeable (Mazur, 2007, p. 206). The age-old mysterious motion paradox popularized by Zeno, and dating back to the time of Socrates and Aristotle, was conceptually solved by measuring time relative to distance in terms of the constant (*c*) representing the speed of light in a vacuum (Mazur, 2007, pp. 3–9, 32–37, 110). Even though time and motion are different, they are effectively inseparable (Mazur, 2007, p. 37). Time and the forward direction of time are invariant (Prigogine, 1997, p. 168), an essential feature of complex systems in that the past and future are distinguishable and not interchangeable.

> The descriptions of nature as presented by biology and physics now begin to converge. (Prigogine, 1997, p. 162)

We can more readily learn to adapt and use our language to accurately describe the world if we understand how convergent and divergent processes correlate our thought and behavior with nature and the invariant forward direction of time. Language that temporally corresponds with our experience allows us to make predictions and behavioral adjustments that fit in with the natural forward flow of time. For example, notice time-related distinctions and differences, and the statistical expression of time for past and future events in the following sentences:

- I made plans for today, and that is not what I expected to happen. It *should* not have happened.
- I *should* have gotten what I expected.
- I *know* that I will get what I want tomorrow.

- I can make plans for today and tomorrow, but it is unlikely that I can predict and expect future events with certainty.
- Today, tomorrow, and thereafter, I will likely have some surprises.
- I may get surprises that I prefer and those that I do not prefer.

These examples point out the hazards of neglecting the forward direction of time in a dynamic environment where future events and predictions take a statistical form as possibilities and probabilities. In a pinch, we can resort to the old standby, "Of course I use one-way time," which typically means "I assume I use one-way time." Our words expose the story told behind our thoughts. For example, when we say, "Do what you are supposed to do," we imply that "supposed to do" has some meaning in the real world. If that were the case, and since what we are "supposed to do" is one-dimensional, we imply static time constraints that extend what we are supposed to do to infinity in either direction as if we believe in two-way time. So we follow up:

"Do you believe that time only goes in one direction toward the future?"

"Of course. Everybody knows that!"

"So how does a person figure out what they are supposed to do?"

"You should know what you are supposed to do."

"Okay, what if I do what I am supposed to do and it results in a mistake?"

"Well, that's simple; you didn't do what you were supposed to do."

"But I would have to have some occult knowledge or to time travel to the future to find out what I am supposed to do."

"Oh no, that's not necessary. Just do the right thing. Do what's right."

"I thought you said you believe in one-way time. But it appears that you are using one-dimensional language with two-way time that limits temporal context."

"What, are you implying I am a liar?"

"I am telling you that you are a flawed and fallible human who assumes that because you know time only goes in the forward direction, you also assume that you actually *use* one-way time when you think and speak. What you say you believe is incongruent with your language behavior."

"You're not implying that I am a liar?"

"No, of course I'm not. But a bit of education might be helpful if you would like to understand how and why you violated one-way time."

"You mean I traveled the wrong way on the one-way street of time. I hope it's not a felony."

"Yes, that's about it, and no, it's only a misdemeanor."

"Whew, so what kind of remediation do you recommend?"

"Relax, it's a common mistake. I don't usually make recommendations, but I can give you some information and alternative solutions to consider that might be helpful."

"Okay."

"Listen to the words you use to talk to yourself in private and the speech you use in public. Look for absolute words and imperatives that *freeze* your solutions. Absolutes easily extend to informatives and preferences by adding words to convert the expression to possibilities that correspond with one-way time. Select the words you would like to replace. When you are thinking or speaking, identify them, challenge them, and replace them with the words you prefer. Our speech is by and large implicit, habitual, and automatic. Changing speech habits takes deliberate effort and a lot of practice."

"You mean the same way you get to Carnegie Hall?"

"That's pretty much it—practice, lots of practice."

"So I'm not incorrigible?"

"Hardly so. You only demonstrated that you are flawed and fallible like the rest of us, along with our imperfect human brains and outdated language habits, replete with absolute prescriptives and imperatives."

We see how absolute thought processing effortlessly neglects time context, which leads to confusion about past and present states, implying two-way time. In comparison, relative thought processing includes context that corresponds with one-way time in the natural physical world and distinguishes past and future states. We can see how absolute logic allows us to effortlessly amplify our affective state and easily upset ourselves because the world fails to behave as it should. Absolute logical habits effortlessly compute a fusion of absolute goals and expectations that exist in idealized imaginary worlds. We live in a changing world where making the effort to explicitly incorporate one-way time into our language improves the correspondence between our thought processes and the physical realities of the world in which we live.

It is fairly simple to find better words, such as preferences and informatives. But it will likely take some effort to identify, challenge, and replace our old familiar words with newer and more effective words (Ellis, 2001, pp. 144–145). When a person begins to work on improving their word usage, we often hear bemoaning of the effort: "I *have to* think before I speak." More formally, "Before, my words flowed effortlessly. But now, searching for, finding, and choosing better words requires a deliberate effort to explicitly intervene before I speak." We call this deliberately effortful process *thinking*.

The words *should* and *must* turn up in many examples in this book, not in order to create boredom but because they are common examples of frequently used control words prevalent in spoken English. *Must* directly exerts control as imperative behavior, and *should* indirectly exerts control as prescriptive expectations for behavior. We can take notice of the frequency of control words by monitoring our own private self-talk and public speech, listening to the speech of others, and observing the written word in the news, books, and other media.

Regardless of medium, control words are often applied for persuasion and appeal to emotion. Yet we would expect words used for persuasion to correspond to a level of temporal fidelity that effectively informs our thought processing and expect explicit expression regarding the plausibility and reliability of the information, the logic form leading to the solutions, and the opportunity for alternatives. Persuasive speech is held to the same standards of fidelity as any other source of information. Analytical persuasiveness is measured by the quantity and quality of plausible, reliable information and evidence it contains and the possible or probable temporal effectiveness for a given choice, decision, or solution. The difference between informing words and controlling words seems related to alternative choices and direct or indirect implication of obligatory obedience from a position of authority.

We can say that command words (imperatives)

- Are fast and frugal
- Suffice locally
- Often work to get others to behave in the way we prefer
- Are effortless and inexpensive information-wise
- Are closed-ended but flexible in the sense that we can fill them with all sorts of implicit assumptions and vagaries
- Are one-dimensional, which confounds time context, so we can simply transport them backward and forward in time
- Implicitly leverage affective thought processing and amplify affective tone

Misinformation, whether implicit in persuasion or otherwise, and if we accept it, can compromise our choice toward less-effective solutions.

We are each responsible for gathering evidence and processing sufficient information for our choices. Control words are empty command shells that some speakers use to teleport commands to the past and future. The take-home message here is that teleportation implies time travel, which as far as we know and according to one-way time is imaginary and incompatible with known physical laws in complex systems. In upcoming chapters we will further explore and explain the source of this mysterious time traveler.

We can say that informative words

- Are effortful, slow, and expensive
- Can optimize globally
- Supply information to help others consider choices and possible outcomes
- Are rich information-wise
- Are open-ended and flexible but require work to enhance fidelity
- Are multidimensional with time context that distinguishes past and future
- Explicitly leverage effective thought processing and modulate affective tone

Most languages can simply clarify the distinctions between the past and the future states of the physical world. The past remains static and does not foretell the future with certainty. The future harbors possibilities. The past and future are neither equivalent nor interchangeable, or negotiable. We can choose and use language that acknowledges the physical realities of our changing world. We can choose and use language as if we live in a static world. It is up to us to choose and use language that gets the results we prefer. The key word here is *use*.

The previous examples in this chapter have pointed out the relationship among time constraints in the physical world, time-coherent language, and effective thought processing. As awkward and effortful as it may seem, time-coherent language usage is fundamental to understanding effective brain processes for thinking, feeling, and relating and behaving adaptively in response to change. We have seen how exposing and breaking the static language barrier puts language on par with the general systematic structure of our dynamic world. We see benefits from changing our language habits and learning to think and speak about the world dynamically.

The next chapter addresses complex systems. We will develop an understanding of these complex systems, and find that in a sense, they seem simpler than we might expect from their name alone. Since complex systems are systematic in nature, they can be explained fairly simply according to their general structural and functional dynamic features. *Dynamic system* seems like a more palatable nomenclature.

CHAPTER 10

Complexity

WE HAVE AN OPPORTUNITY for understanding nature when we acquire a basic understanding of complex systems, since nature itself operates as a complex system. Organisms operate as complex systems that can learn, engage in adaptive responses to change, and improve their lot in life. For example, evolving agents might develop a range of variable behaviors that act as alternative solutions in a changing imperfect environment. The wider range of behavior enables increased options for adaptively responding to unexpected change.

In a perfect world, one static solution might work, but novel alternatives can increase the options for finding effective solutions within the uncertainties and instabilities of the complex dynamics of nature. New solutions usually build on variations of old solutions that worked, and thus they increase the options for the creation of more complex higher-level innovations. For humans, language stands out as the newest and arguably one of the most sophisticated solutions produced by evolutionary processes. Indeed, natural language represents an innovative neophyte on the evolutionary landscape.

Language is considered by many authors to be another example of a complex system (Steels, 2000b) that likely evolved from more simple forms (Corballis, 2002). Simple language works well in a static environment and can suffice on average with limitations in a complex environment. Language that matches the dynamics of a complex world can likely produce more optimally effective solutions. Simple, absolute language allows simple solutions but relegates us to simply imagining how our complex world works as an idealization. Alternatively, language that fits the complex system's general organizational structure offers a simpler perspective and a sensibly correlated worldview.

> Some of the complexity we intuitively ascribe to internal cognitive states may actually be a matter of complexity in the world with which a relatively simple organization has evolved to mesh. (Churchland & Sejnowski, 1994, p. 423)

A sensible language structure that corresponds with general systematic structure facilitates effective thought processing, predicting, and adapting. In the long run, sensible language generally enables simpler and more optimally effective solutions. But trying to understand the solutions without effective language can easily overwhelm us where the world's complexity seems complicated far beyond anything we humans can grasp. We can see where ambitious efforts in pursuit of such extensive systematic knowledge could easily lead us down many blind alleys and quickly get expensive. But what is the value of language and knowledge? One answer is that together they help us make sense of the complexity of nature, increase the reliability of predictions, and enhance our understanding of adaptive behavior.

A Sensibly Correlated World

Effective language usage is instrumental in sensibly correlating our worldview, thought processing, and behavior. Primates appeared an estimated 120 million years ago, *Homo sapiens* evolved approximately 2.5 million years ago, symbolic thought developed around 50,000 to 100,000 years ago, and an acceleration of increasingly complex behavior and possibly the rudiments of language evolved over the last 40,000 to 50,000 years (Striedter, 2005, pp. 313–321). Scientists think that language likely developed during that time frame along with unique human brain development. Writing appeared less than 5,000 years ago, with evidence evolving from primitive markings and cave drawings to sophisticated text. We extended and expanded many of the rudiments of today's science over the last several hundred years on the stepping-stones of knowledge and thought dating back thousands of years to early Greek and Roman scholars. We can relate the evolution of explicit knowledge and thought with our linguistic brain.

> It would be fair to say that the [language-related] lateral prefrontal cortex probably helped early primates to exhibit more complex and flexible behavior. (Striedter, 2005, p. 309)

We can speculate that language structure evolved from early building blocks similar to ecosystems and organisms, where speech behavior likely evolved from simplistic utterances by early *Homo sapiens* that were ignorant of how the world works—that is, except for the default to rudimentary affective assumptions of implicit causality and positive and negative limbic associations. Regardless of the origin, affective language can misinform our understanding of the external world. Localized language tends to default to and rely on *affective input* and processing of event-related feelings from an egocentric perspective—how the world treats us. "Events in the default mode lead to automatic, predetermined, and obligatory responses without allowing much neural space for thought, foresight, choice, innovation, or interpretation" (Mesulam, 2002, p. 25).

Our brains, like the brains of other mammals, demonstrated adeptness for affective cognitive processing long before we developed the ability for language in general and effective language in particular. The language structure that we inherited has apparently developed over a long period and correlates affectively with the structure and function of the limbic system. But our language structure has evolved largely without the benefit of modern science. Understanding science and its orderly search methods for making novel discoveries seems fundamental to sensibly understanding language structure and thought processing that corresponds to the structure of our world.

Extended language enables processing of interpretations of events on the basis of *effective output* and adaptive responses to external change—how we treat the world. In a recurring circular manner, effective processing relies on understanding the temporal relationship among experience, sensory input, cognition, adaptive behavioral output, and the effects of actions in the environment. We gain such an understanding by open-ended exploration with temporally reliable language, such that a search for information and knowledge enables exploiting that knowledge systematically for enhancing the reliability of predictions. Effective language improves learning efficiency by modulating our sensory information processing in relation to effective action output and adjusting for errors in prediction. The resultant feedback from the environment reveals how effectively we interact and adapt in response to change and informs us as to how cooperatively we treat nature. Effective language relies on higher-order abstractions with greater output correspondence to the structure of the physical world. Put differently, effective language enables informed, systematic correspondence of real-time brain processing to external relationships.

Science is a dialogue between humankind and nature (Prigogine, 1997, p. 153) that helps us make sense of the physical world we live in by advancing our thought processing from absolute to nonabsolute, from localized to globalized, and from certitudes to possibilities. Chris Eliasmith and Charles Anderson contend that by measuring physical properties such as velocity, force, acceleration, and position, physicists can explain and make sensible predictions about the ever-changing universe. The dynamic nature of the system indicates the relevance of accurately modeling the physical properties of the system's current state to predict its possible future states, and why understanding physical facts about the environment enables more reliable predictions overall (Eliasmith & Anderson, 2003, p. 29).

We interact with physical laws and principles every day: time, position, distance, acceleration, force, momentum, matter, velocity, space, work, charge, energy, and so on. Such temporal interactions pervade the daily life experience of both individuals and entire societies. Our interpretations of these interactions, from local to global perspectives, influence our worldview regardless of our acknowledgment. Individuals and society alike are integral parts of the interactions within the macroscopic system.

> A dynamical model of a human society begins with the realization that in addition to its internal structure, the system is firmly embedded in an environment

> with which it exchanges matter, energy, and information. (Nicolis & Prigogine, 1989, p. 238)

Energy enables living organisms to interact with objects in the environment. Living organisms derive energy from the conversion of matter from one form to another. Energy provides the organism with the potential to do work (Mitchell, 2009, pp. 41–42), such that energy represents a force potential for local interactions within complex systems. These physical interactions contribute information and knowledge about processes that we gather from observing the effects of the relationships among environmental events, objects and position, energy and value, time and space, and action. If we find general rules that explain the effects we observe from these relationships, we can use the rules to understand processing and action regulation across the entire system. When we see correlations among theories, hypotheses, explanations, and solutions at the systematic level, it suggests an extension to general global laws.

When we extend language from localized to global, we enable a macroscopic linguistic perspective that facilitates the discovery of general theories. Extended language enables generalized cognitive processing suitable for exploring, evaluating, and exploiting global theories according to systematic effectiveness. General theories encompass systematic inclusion of relationships and processes in lieu of defaulting to convergence and exclusionary localized partitioning. Extended language can represent relationships across an open system and account for divergence, open-ended processes, and physical constraints throughout that system. We can conclude that extended language facilitates general descriptions that correspond to the systematic definition of open-ended processing in global systems.

In contrast, recall that localized language discretely partitions an open global system into smaller subsystems by default and creates variant representations and descriptions that define closed-ended processing typically found in isolated localities. Developing a sensible systematic understanding requires general language structure that corresponds structurally and functionally with regulatory processes in the global system. We expect that regulation in complex systems follows the same macroscopic structural and functional relationships in correspondence with the physical laws of nature—from brains to the computational biology of living cells.

Systematic Regulation

> Why does the organism choose certain conditions and reject others? This selection of the favorable and rejection of the unfavorable ones presented by the movements is perhaps the fundamental point in regulation. (Jennings, 1906, p. 40)

With a macroscopic perspective we can view the global organization and ordering of structural interactions, processes, and functional regulatory principles in a systematic manner. That is, the regulatory principles are reflected in the actions of the organisms living in the system. Organisms evolved as dynamic systems corresponding to their domain (Eliasmith & Anderson, 2003, p. 14). The natural world plays an essential role in the system's regulation, where the environmental output represents an input to the organism and the organism's output acts as input back into the environment (Nolfi et al, 2008, p. 1430).

At the most basic level as integral parts of the system, autonomous organisms shaped by natural selection know and interface with the natural world by navigating from their default position or localized frame of reference. Localized frames operate within global environmental constraints defined by the physical laws of nature. Understanding regulation systematically within this dynamic input–output relation relies on a sensibly referenced language. Effective language frames event-related interactions in the context of their occurrence. Contextual frames facilitate understanding regulation logically in terms of processing, orienting, and directing the control of position, internal states, and action in response to instabilities and change encountered within the overall system.

As an interactive product of the system, the organism's dynamic existence could be said to define the system as dynamic rather than static (Jennings, 1906, p. 289). An integrated system relies on sensible feedback processing for modulating the internal milieu within the limits of internal and external constraints. Reliable pragmatic feedback processing improves effective internal regulation by activation, inhibition, and switching that influences behavioral adaptation in response to external change including change of position and state values.

Structure, function, and regulation are found across the spectrum of biological systems. *Regulation* refers to the operation of various adaptive biological control mechanisms that play a large role in dynamic systematic modulation and robustness (Callebaut et al, 2007, pp. 31–37). Regulation involves the persistent monitoring and modulation of overall structural and functional interactions within and between organisms and the environment (Jennings, 1906, pp. 339–350). Biological organisms rely on feedback-control mechanisms similar to those found useful in engineering practices (Shadmehr & Wise, 2005, p. 5; Pierce, 1980, p. 218). Regulation via feedback control was introduced to systems biology by Norbert Wiener (1948) in his classic book *Cybernetics*, and the principles have now become an important component of computational neuroscience (Swanson, 2003, p. 90).

Developed in the 1860s by French physician Claude Bernard and later popularized by Walter Cannon, the concept of homeostasis describes the ability of biological systems to maintain their physiological state within acceptable limits (Shadmehr & Wise, 2005, p. 5). Homeostasis, a form of symmetry or dynamic balance, describes a simple adaptive principle motivating the regulation of an articulating system (Ashby, 1960, pp. 85, 209–214; Floreano & Mattiussi, 2008, pp. 263–264). From this perspective, organisms modulate homeostasis through regulatory control using positive and negative feedback from the environment (Mitchell, 2009, pp. 296–297).

Homeostasis implies steady-state regulation, which is often associated with local descriptions of stability in deterministic systems with reversible processes that operate around an equilibrium point, where stasis implies a static state. Yet static states and stability hardly address the complex nonequilibrium dynamics found in nature associated with irreversible processes, persistent change, and instabilities. Regulation in complex dynamic systems encompasses nonequilibrium modulation far from equilibrium. Nonequilibrium modulation in complex systems enables the creation of novel solutions and learning in response to instabilities incurred by disturbances. The novel solutions enable adjustments for adapting the system's modulation of instabilities within the limits of life-compatible physiological boundaries. A global perspective highlights the importance of understanding systematic regulation in biological organisms in terms of effectively modulating responses to instabilities and tuning in to the dynamic reality of nature. In other words, global nonequilibrium explanations provide dynamic solutions that match natural changes, in contrast to equilibrium descriptions and solutions that imply stasis, or idealized static conditions.

> Complex systems are open, and information can be constantly exchanged across boundaries. Despite the appearance of tranquility and stability over long periods, perpetual change is a defining feature of complex systems. . . . [T]hings are different in open, complex systems that exist far from a state of equilibrium. (Buzsáki, 2006, pp. 11, 12–13)

Complex biological systems operate nondeterministically under continuously changing conditions, and their regulation is better described as the dynamic modulation of instabilities and irreversible choices leading to bifurcations and solutions away from equilibrium (Prigogine, 1997, p. 126). More simply, complex systems live in a world with an undetermined future, marked by disturbances, instabilities, and choice-making bifurcations; they learn by correcting errors, accumulating knowledge, and modulating adaptive responses. The organism changes its behavior as a result of interference or disturbance in its physiological processes. When the interference abates, the organism selects and retains the favorable condition reached by ceasing to change its behavior (Jennings, 1906, pp. 342-343). The bifurcations represent behavioral choices where new solutions are added to the systems memory. This option of action choices for new solutions enables more elaborate complex adaptive behaviors. Operating in an environment that continually changes and where information is expensive, complex systems tend to learn general rather than absolute solutions.

> The fundamental fact must be remembered that the life processes depend upon internal and external conditions, and are favored by conditions that are rather

> generally distributed throughout the environment of organisms. (Jennings, 1906, p. 342)

With our access to extended language skills and a brain that can exploit analytical cognition, we can accept the concept of systematically augmented thought processing. A language code that offers a systematic understanding of the physical world translates to a systematic understanding of the brain, emotions, and behavior. Language commensurate with providing that systematic understanding appears to be fundamental to accomplishing that end. We expect as much from our human brain, in which information gained from evaluating the effectiveness of our actions can circularly lead to the accumulation of knowledge that enhances effective regulation and predictions—after all, we are prototypical complex biological systems.

Predictability and Probability

Dynamic models describe how system properties change over time (Ellner & Guckenheimer, 2006, p. 1). A dynamically interacting system generally operates in a state of continual flux (Mitchell, 2009, p. 15). Predictability becomes an issue for understanding the system. The system's dynamic complexity and instabilities inevitably complicate predictability, since multiple variables persistently fluctuate. Thus, dynamically regulated nonequilibrium systems are best described functionally in terms of possibilities rather than certitudes (Prigogine, 1997, p. 126).

As Heisenberg discovered, an observer cannot predict with 100% certainty the conditions and behaviors of a complex system operating in an uncertain environment, even at the subatomic level (Lindley, 2007). With no way to turn back the clock to check the accuracy of observations, one can improve predictability by logically factoring in probability, on the basis of all available reliable information (Jaynes, 2007, p. 87).

> The prospects for making accurate predictions in complex, uncertain, and ever-changing environments appears to be rather limited, and people may use newly obtained information to evaluate their outcomes and to challenge or question the validity of their initial predictions.... All predictions are made under varying degrees of uncertainty, with no meaningful prediction ever being completely certain. (Armor & Taylor, 2002, p. 347)

Statistical approaches give the probable behavior of the system (Mitchell, 2009, p. 48). The system's prediction accuracy in large part depends on the effectiveness of the statistical approach and the reliability of source information and systematic knowledge. Statistical dependence in complex systems highlights the relevance of systematic referencing, which confers logical processing methods and statistical measures that are based on symmetry breaking for effectively parsing and processing information and knowledge, evaluating alternative possibilities and probabilities, and making reliable predictions.

In evolving systems, reliable parsing of events enables a distinction between information of the static past and the prediction of probable future states. Statistical dependence in complex systems benefits from reliable information, information processing, and learning from errors. Symmetry breaking leads to a statistical method for logically referencing, integrating, and processing systematic source information from correlated regularities into global knowledge. A reliable global referent facilitates logical evaluation, correlation, and calibration of information about regularities found in nature (Jaynes, 2007, pp. 252, 279, 292, 411).

Furthermore, we can think of evolution as a practical open-ended process, in which organisms evolve according to probabilities and where individuals and populations meet certain natural conditions, rather than viewing evolution as a random process in response to blind laws or arbitrary events (Prigogine, 1997, p. 189). A natural physical view supports the notion of characterizing evolutionary processes as likely resulting from logical interactions of positive and negative forces among energy, matter, attraction, repulsion, and approach–avoidance that we see throughout nature.

Understanding the systematic relations among physical processes in the world enlightens our predictions, but systematic ignorance leaves us groping blindly in the dark. Paramecia seem to have learned as much about the importance of relations and making predictions, as they continuously sample possibilities in the environment and change behavior when conditions change in an unfavorable manner. This sampling behavior leads these animals to predictably collect in certain regions of favorable conditions and avoid unfavorable ones (Jennings, 1906, p. 38). Prediction implies learning and planning, but are we to expect that the "lowly" paramecia can plan ahead?

> Thus Paramecium is continuously receiving "samples" of the water in front of it. . . . By thus receiving samples of the environment for a certain distance in advance, it is enabled to react with reference to any new conditions which it is approaching, before it actually entered these conditions. (Jennings, 1906, p. 47 [emphasis in original])

We have the advantage of extended language for informing us of logical alternative possibilities prior to predicting and acting. Approaches that ignore alternative possibilities constrain our understanding of evolutionary processes and complex systems, the evolving nature of language, and life itself, all of which operate by exploiting alternative possibilities. We can choose to replace absolute certitudes with possibilities and probabilities that coherently match our cognition with the dynamic distribution of information in nature. With informed possibilities, we incur enhanced plasticity that we can apply to thinking, predicting, error learning, and flexibly regulating adaptation. Exploiting our ability for plasticity confers an adaptive regulatory advantage when we are faced with the challenges of persistently evolving change in a complex world.

Adaptive Plasticity

Complex systems rely on plasticity and flexibility. As a plastic organ, the brain responds to experience and facilitates adaptive learning (Coolidge & Wynn, 2009, p. 50). Santiago Ramón y Cajal (1852–1934) theorized that learning might be related to changes in the strength of connections between nerve cells (Kandel, 2006, pp. 158–159). This idea was elaborated by Donald Hebb, who proposed that the brain has a significant ability to connect, reconnect, and change weight values [connection strength] and associations in neuronal circuits in response to the organism's experience with the world outside the brain. He also suggested a seldom-quoted dual-trace mechanism with the addition of a second learning mechanism operating independent of any structural change related to transient, unstable reverberatory traces [oscillations] (Hebb, 1949, p. 61).

Numerous studies (e.g., Kandel, 2006) have generally confirmed Hebb's conjectures about how plasticity and oscillations demonstrate a correlation to learning and working memory (Buzsáki, 2006, pp. 349–353). In effect, we can think of learning as corresponding to modification of connection strengths by experience (Burgess, 2007, p. 716) that influences memory via the process of plasticity (Siegel, 1999, p. 216).

> The word *neuroplasticity* denotes the remarkable capacity of the brain to adapt continually to the demands placed on it by experience. (Spitzer, 1999, p. 137 [emphasis in original])

Neuroplasticity is often used to explain learning in terms of experience and cognitive fitness. Language corresponds to plasticity as an intermediate substrate for learning that facilitates

- The construction of environmentally coherent higher-level abstractions
- Abstractions that effectively represent and process environmental relations and learning from experience
- The application of that learning to effectively regulate adaptive fitness in a flexible manner according to probabilities

The outcomes of our actions are generally probabilistic and uncertain, at least to some extent, regardless of our acknowledgment. Though probabilistic descriptions enhance cognitive flexibility and plasticity, they can easily be viewed by some as weak, wishy-washy, uncertain, or indecisive compared with the reassuring certainty of rigid, absolute descriptions.

From a narrow perspective flexibility implies weakness, but it demonstrates a remarkable strength, especially in complex systems where plasticity enhances flexible learning and adaptive responses to change. Excessive rigidity does not necessarily represent toughness and can often exhibit low stress tolerance (Shadmehr & Wise, 2005, p. 28). *Rigidity* is another word for stiffness; in a dynamic environment, increasing rigidity and stiffness can increase brittleness, and the more rigid the structure, the more likely it will fail when stressed. Conversely, flexibility implies malleability, plasticity, resilience, and a tendency to bend rather than break under stress.

One of the essential features of complex behavior is the ability to transition between different states (Nicolis & Prigogine, 1989, p. 36). Rigidly ingrained states reduce variability in the system, which diminishes its adaptability in the environment (Siegel, 1999, p. 236). In terms of suitability for survival in the face of instabilities and uncertainties found in nature, flexibility generally makes for greater strength with a higher tolerance for stress. If survival is mapped to usefulness, then fitness corresponds closely to strength as the counterpart of fitness (Holland, 1995, p. 65).

Effectively regulated flexibility provides organisms with a stronger fitness that relies on neuroplasticity-related cognitive processing for learning ability, adaptability, and survivability. In dynamic systems rigid regulatory processing incurs a fault-intolerant brittleness. Flexible regulation produces a stronger, fault-tolerant buffering mechanism that facilitates adjustment to the instabilities and uncertainties encountered in nature.

Over time, dynamic buffering mechanisms enable organisms to adapt to the environment by acquiring new skills that enhance survival. The importance of flexibility appears early in evolution when eukaryotes that shed their rigid cell walls were able to form multicellular organisms (Nüsslein-Volhard, 2006, p. 122). Plasticity has been found to play a major role in high-level cognition (Spitzer, 1999, p. 167) that enables a flexible process for calibrating input sensitivity and tuning of the output gain in response to stimuli by learning from experience.

Flexible learning and reliable feedback contribute to decreasing errors and fine-tuning the dynamic range of the output in response to change. As an evolutionary innovation for adaptive flexibility, plasticity has logical implications for rigid versus flexible language applied to learning and the processing of information and the accumulation of knowledge. Flexibility plays an important role in cognitive processing for thinking about choices, making decisions, learning from experience, correcting systematic errors, and effectively adjusting responses to a changing world. Rigid language arbitrarily imposes strict constraints on cognition that impedes plasticity-related learning and adaptation.

At the most basic level, the brain can be considered a flexible, open, and dynamic system (Siegel, 1999, p. 17). *Flexibility* refers to the system's dynamic range and sensitivity to changing environmental conditions and involves the capacity for variability, innovation, and creatively dealing with novelty and uncertainty. On the other hand, rigid adherence to previously ingrained states produces excessive continuity and minimizes the system's ability to adapt and change (Siegel, 1999, p. 219).

An adaptive agent shows increasing goal-directed competence over time based on experience and adapts by changing rules [beliefs or policies] as experience accumulates (Holland, 1995, p. 10; Johnston, 2008, p. 353). An adaptive agent applies learned information to calibrate input and regulate output to facilitate effective interactions. The agent learns to deconstrain or adapt to shifts in input–output relationships, rather than rigidly attempting to control them (Kirschner & Gerhart, 2005, pp. 142–143). Simply put, rigid language, beliefs, values, and regulatory control generally dampen behavioral adaptability.

> Experience modifies human behavior. And experience modifies human beliefs. These two things are not independent of each other. Behavior often results from beliefs; beliefs are potential behavior. (Polya, 1954, p. 10)

The flexibility to adaptively modify information and input–output correlations in an effective manner both depends on and boosts brain plasticity. We learn through interactions with the environment, which change the connections in our biological brains (Spitzer, 1999, p. 38). Plasticity provides the mechanism for enriched learning and flexible fine-tuning that enhances the effectiveness of a system's knowledge base over time. An organism's effective output and adaptive responsiveness in a complex environment rest on its ability to reliably learn from experience.

Learning provides a decisive evolutionary advantage, and evolution seems to favor genes that increase fitness by enabling learning and communication (Murre, 1992, pp. 107–108). With language, we acquired the tools to flexibly modulate learning and tune input–output processing to augment adaptive levels of behavior. The caveat is that the range of flexibility correlates with the language constructions used, such that language with a broader range of flexibility can circumvent the rigid closed-ended constraints imposed by absolute semantics and either–or logic. Plasticity allows for adaptively fine-tuning cognitive processing in response to experiential feedback. With extended language, we can tune relevant meaning to enhance effective information processing and adaptive regulation, improve error learning, and reduce systematic errors.

Language allows us to explicitly fine-tune our cognition and adjust our language software by deliberately modeling language that promotes adaptive learning in nature (Lytton, 2002, p. 78). We can ask to what extent our model resembles what we have learned about the real [natural] world (Jaynes, 2007, p. 75). One of many possible solutions entails extending language parameters to confer flexible fine-tuning of cognitive processing amenable to promoting more robust adaptive learning and cooperation—extending the effective reach of cooperation across many individuals and societies. Extended language can facilitate greater flexibility and cooperation, which has far-reaching social implications for all of us, today and for the future.

> Our everyday experience teaches us that adaptability and plasticity of behavior, two basic features of nonlinear dynamical systems capable of performing transitions in far-from-equilibrium conditions, rank among the most conspicuous characteristics of human societies. It is therefore natural to expect that dynamical modes allowing for evolution and change should be the most adequate ones for social systems. (Nicolis & Prigogine, 1989, p. 238)

In short, language that augments flexibility and fits the changing world we live in can facilitate the advance of cooperative, adaptive social systems.

Complex Systems

The possibility of readily transitioning among various modes of behavior is the principal fingerprint of complexity (Nicolis & Prigogine, 1989, p. 232). György Buzsáki describes complexity as nonlinearity with the emergence of unexpected solutions, where the evolution of complex systems is not predictable by the sum of local interactions, and causes are not "one or the other" but instead are embedded in the configuration of relations (Buzsáki, 2006, pp. 13–14). For the biological complexity discussed in this section, the term *complex* does not simply mean *complicated*; it implies a nonlinear relationship between constituent components, history dependence, nonabsolute boundaries, and the presence of amplifying–dampening feedback loops (Buzsáki, 2006, p. 11). All of these are common characteristics of complex living systems.

> It is more natural, or at least less ambiguous, to speak of *complex behavior* rather than complex systems. (Nicolis & Prigogine, 1989, p. 8 [emphasis in original])

In contrast to complex systems, systems we think of as complicated may have some feature similarities but typically display random organization that defies analysis. We can also define complexity and complex behavior in the statistical terms of global or nonlocal information distributions. This nonlocal definition encompasses complex structure and the functional modulation of choices and probabalistic behavioral transitions in response to disturbances, instabilities, and change. In complex systems nonlocal descriptions enable understanding the systematic dynamics that include multiple variables and persistent nonlinear interactions of populations or ensembles. According to Prigogine, complex dynamical interactions take place far from equilibrium within the instabilities of nonconservative, time-irreversible [dissipative] systems (Nicolis & Prigogine, 1989, p. 197; Prigogine, 1997, p. 66).

The arrow of time conveys irreversibility and nonequilibrium entropic processes that make the evolution of complexity possible. Thus, the arrow of time plays an essential role in the formation of structure in both the physical sciences and biology (Prigogine, 1997, p. 71). Life is associated with entropy production and therefore with irreversible process, and complexity has consistently been associated with irreversibility (Prigogine, 1997, pp. 63–64). Complexification depends on symmetry breaking that adds structure to the system under conditions of far-from-equilibrium dynamics, where the distance from equilibrium becomes an essential order parameter in describing nature, and bifurcations can be considered the source of diversification and innovation (Prigogine, 1997, pp. 63–64, 70). Far-from-equilibrium conditions enable the evolution of new order in the open system, within systematic limits, where the system modulates adaptation to instabilities by adding complexity.

> ... [T]he passage toward complexity is intimately related to the *bifurcation* of new branches of solutions following the *instability* of a reference state, caused by the nonlinearities and the constraints acting on an open system. (Nicolis & Prigogine, 1989 [emphasis in original])

Prigogine (1997, p, 126) summed up this definition quite simply: "Again, *complex* means time symmetry is broken." In irreversible systems with dynamic instabilities, symmetry breaking produces new structure and alternative solutions that lead to complexification. Dynamic instabilities in complex systems, current event-related inputs, and history-dependent systematic correlations take precedence over arbitrary predetermined local biases for influencing decisions. Hence, complex systems display nondeterministic dynamics.

We have developed a heightened understanding of complex systems that has improved weather forecasting through the processing and integration of knowledge of atmospheric conditions, the careful correlation of past patterns with current observations, and the making of statistical predictions on the basis of the given atmospheric conditions. Predictions are probabilities computed from accumulated knowledge as a history of reliable correlations and thus relate to uncertainty. Predictions typically state an expected outcome, while a forecast may cover a range of possible outcomes. Meteorologists routinely incorporate probability into their forecasts. On the basis of feedback and learning from previous predictions, they continually adjust their predictions, mathematical and statistical models, and computer programs that process and generate those predictions. Since we have weather every day, they receive immediate feedback about their predictions, and their knowledge of probabilities accumulates over time (Chabris & Simons, 2010, p. 144–145).

Despite our typical preference for certainty, weather predictions are short term and general rather than absolute. Interestingly, most forecasts are very accurate except for days with a forecast of 100% chance of rain. On those days the accuracy of the predictions drops to 90%. Despite this rather stellar record, we are more likely to notice prediction errors, especially when the forecast ruins our plans and expectations. Among the general public, a thorough understanding of probabilistic weather predictions to some extent depends on the level of scientific literacy, which varies but typically remains constrained even on the driest of days. Scientific literacy tends to appear as the exception rather than the rule, and some cultures and societies even harbor disdain and contempt for scientific literacy and the uncertainty it implies.

Unfortunately, those with rudimentary scientific knowledge and limited experience with complex systems can interpret the uncertainty implied by probabilities as a lack of skill. Therefore, meteorologists frequently make a point of explaining the limitations of their work, and through advanced computer modeling, they have demonstrated the relative accuracy of short-term forecasts within the acknowledged constraints of complex global weather systems.

Weather operates as a complex system with inherent dynamic instabilities, making it unlikely that any amount of precise weather history can predetermine long-range forecasts or consistently produce 100% accurate short-term forecasts. Dynamic instabilities and nonlinearities amplify small differences in the initial conditions of the system into new structures in weather patterns that preclude absolute predictions. Recall Lorenz's butterfly effect described in Chapter 8 ("Effective Language"), where a small disturbance such as a butterfly flapping its wings can theoretically influence atmospheric disturbances and weather on the other side of the globe.

The capability for increased complexity allows biological systems to produce novel innovations and inventions by open-ended evolutionary processes that correspond to self-organizing principles and the physical laws of nature—and not surprisingly, all have dynamic properties correlated with nonlinearity and one-way time. The term *self-organization* was introduced by British psychiatrist W. Ross Ashby, who applied it to studying neuroscience, but somewhat ironically it has since migrated to other disciplines and generated much interest in the field of physics (Buzsáki, 2006, p. 12). Prigogine defines self-organization as the choice between solutions appearing at a bifurcation point, determined by probabilistic laws, where far-from-equilibrium self-organization leads to increased complexity (Prigogine, 1997, p. 205).

Rather than arbitrary arrangements, self-organized structures select from systematically ordered patterns characterized by inherent order such as self-similarity, repetition, and regularities found in nature, and in order for a brain to guide an organism safely through life, we would expect it to be in tune with the environment—that is, a brain that is in harmony with nature (von der Malsburg, 2003, p. 1002–1003). Self-organizing structures involved with self-replication, self-regulation, and self-sustaining complex behavior, such as cognitive processing in the brain, rely on correlating adaptation to the dynamics of the environment in which it evolved.

Understanding which functions a system can realistically perform in the natural world as the logical consequences of the constraints of one-way time characterizes the system's dynamics. The world operates in the same forward direction of time, such that explicitly incorporating the natural direction of time into models improves ordered correspondence among processing, representation, description, and understanding the systematic dynamics in relation to complex systems. We have evolved in this world, and we would expect that our dynamics would match the dynamics of that world (Eliasmith & Anderson, 2003, p. 219).

Modeling a system with complexity theory allows discovery, examination, and feature characterization for mechanisms, parameters, and behavior involved in interactions within and between system components that generate particular system-level properties (Camazine et al, 2001, p. 72). Identifying the mechanisms that underlie a behavior is a primary step toward understanding how that behavior evolved (Camazine et al, 2001, p. 490). Complexity theory provides a widely relied-on scientific model for observing complex dynamic systems, also referred to as complex adaptive systems, or dynamical systems. "The word dynamic means changing, and dynamical systems are systems that change over time in some way" (Mitchell, 2009, pp. 15–16). Complexity theory provides a systematic approach for modeling biological processes and dynamic relationships in the world as a self-organizing system.

Self-Organization

Viewed as a dynamical model that changes over time, evolutionary processes of natural selection, survival, and reproduction operate on populations of organisms derived from self-organizing processes. Under natural selection, adaptive responses reflect the local rules of interaction among the self-organizing components. The natural environment shapes the rules that produce adaptive regulation of the state interactions among the components of the living system and the environment (Camazine et al, 2001, p. 494).

In other words, local rules are a product of local self-organizing principles and local activation of survival-related adaptive behaviors within the natural constraints under which they evolved. Even though self-organization is sometimes described as operating without a central controller, it is reasonable to assume that molecular interaction and self-organizing processes of attraction and repulsion correlate behavior to local reinforcement and the "invisible strings" of short- and long-range forces of physical laws. The organism's genetic instructions and phenotype evolved under those laws, and the local rules and attractor dynamics (approach positives and avoid negatives) respond to and operate under the global influence of the same physical laws and non-equilibrium constraints. Hence, we find compatibility across perspectives of local self-organization, local activation that operates under physical constraints of global laws, and global solutions correlated to domain knowledge. In other words, global laws explain local self-organizing interactions.

> External fields such as the gravitational field of the Earth, as well as the magnetic field, may play an essential role in the selection mechanism of self-organization. (Prigogine & Stengers, 1985, p. 14)

Local self-organization works through positives and negatives related to short-range attraction and repulsion and short-range plastic learning, but within long-range and long-time-scale macroscopic influences over evolutionary time. We see evidence for these forces across a range that includes both atoms and the biochemistry of molecular processes in living cells, which are replete with positive and negative polarities throughout (Lehninger, 1975). Close relationships among entities increase the intensity of attraction and repulsion. Even on a human level, the intensity of bonding in close relationships tends to automatically increase our focus on local behaviors that affectively feel more intense and thus more meaningful at short range. Self-organization demonstrates the evolutionary accumulation of structure operating at a local portion of the landscape. Even though short range, behavioral activation at this local level continues to operate under the auspices of global laws that define global inhibition and regulation within domain limits.

Consider that life evolved by self-organization. Fish evolved in the ocean, to amphibians, reptiles, birds, and mammals. Different animals evolved in different localities but all under the same natural laws. They live and operate within a local region but are still influenced by natural global laws that apply to all organisms. In nature, physics is physics that we cannot hide from regardless of our location. Global laws and constraints do not interfere with or degrade perceptions of local self-organization but instead enable globally informed knowledge about how self-organization contributes to the evolution and regulation of living creatures.

As a general phenomenon, we see self-organization across the whole of living, evolving species, where the total complexity of the organism's behavior exceeds the sum of the parts. Frequently used examples include ant colonies and the immune system. In almost every case we see a distribution of variable behaviors that enable general alternative solutions. Ants can shift between exploration and exploitation depending on the available food resources. In some species a portion of the ant colony continues novel exploration even with an abundance of resources. In our immune system we see systems that modulate between exploration and exploitation while keeping alert to novel invading pathogens. The immune system has generalized cells that on detection of a foreign body can trigger a chain reaction of cells with more specific binding properties and antibodies that better match the actual properties of the invader (Mitchell, 2009, p. 195; Nicolis & Prigogine, 1989, pp. 232–238). The cooperative behavior of the whole group accomplishes more than possible by the sum of individuals functioning alone. In both cases we see generalization where complex systems rely on exploration and exploitation of novel alternative solutions while modulating attractor relationships on the basis of approach-and-avoidance behaviors.

Furthermore, we see similar behavioral relationship across mobile organisms in general. For instance, what general relation can we observe among humans, immune cells, and puppy dog tails? In the immune system we see analogous negative reactions to repulse or destroy foreign invaders, as we see with humans and across societies. In other words, the immune system recognizes symmetry-breaking differences between familiar and unfamiliar objects and evaluates whether they represent a threat. How does the immune system know if an object is familiar? Essentially the same way we humans know an object is familiar: we compare it with our memory.

If we come in contact with an object that is marked in our memory as negative, we can react by

- Eliminating the problem of avoiding an object by initiating an offensive response and fighting if we think we have an advantage or lack other alternatives
- Protecting ourselves by a defensive response if we are attacked
- Freezing and taking a chance that we will be ignored
- Avoiding the object with an evasive fleeing response if we think we are disadvantaged

In the immune system we expect that the memory helper T cells will inform us of the presence of a negative object, and we react by

- Protecting ourselves by eliminating the object with an inflammatory response and ingestion by killer T cells and macrophages
- Ignoring the object if it is confused as familiar and self-similar, which is seen in cancer cells that often avoid detection by evolving deceptive defenses that "fool" the immune system (Xue and Stauss, 2007, pp. 173, 179)

If puppy dog tails are unfamiliar, we usually know right away that the dog is friendly by evaluating its characteristic symmetry-breaking tail wagging or unfriendly by a snarling display of teeth and vicious, threatening barks. Depending on circumstances, we are inclined to approach the former and avoid the latter. We perceive the event of an approaching dog as negative if it is snarling, but if it turns and avoids us, we perceive the event as positive. We perceive the event of an approaching dog as positive if it is wagging its tail, but if it turns and avoids us, we perceive the event as negative. These are essentially the basic tenets of behavioral reinforcement. Among all of these, the similarity with change of position, positive and negative values, approach and avoid, familiarity and unfamiliarity, and the role of symmetry breaking and memory is striking. These are local actions that evolved universally across organisms in the system and tend to operate between local minimum and maximum within the global constraints; and applicable to a theory of cancer including an exaggerated competitive bias.

A systematic optimum correlated to domain knowledge involves accounting for the global constraints of the system. Global constraints acknowledge domain-dependent referencing that supports reliable statistical measurements across a range of possibilities for evaluating predictions and effectiveness of outcomes. Natural domain limits can be abstractly defined in relation to global referencing according to the probable optimal limits of global information and knowledge accuracy. A global optimum does not replace local self-organization or activation but defines systematic boundaries for those processes as statistical limits of general knowledge. In systems with inherent instabilities, statistical boundaries reduce the range of local speculation of certainties to manageable levels of possibilities and probabilities. The take-home message in complex systems: Possibilities and probabilities that correlate with the laws of nature replace certitudes (Prigogine, 1997, p. 4).

Knowledge of distinctive systematic properties depends on a macroscopic perspective, which suggests that no matter how infinitesimally small the microscopic focus or how intensely we study the local properties of self-organization in an isolated system, the general global structure, regulatory processes, and action relationships most likely cannot be inferred or discovered from knowledge of a part of the system. A part, no matter how large the proportion, constitutes a subsystem by definition. In many of these local isolated subsystems, the relevant evolutionary variables involved are not known to any degree of detail, and the pertinent variable may be a part of the problem that is trying to be solved (Nicolis & Prigogine, 1989, pp. 186, 217–219). In other words, the global solution remains impossibly hidden in the problem, which has been subjected to piecemeal dissection and thus locally trapped. Furthermore, with local search we are more likely to deceive ourselves and find and confirm what we were looking for in the first place.

> ... [O]ur brains construct the feeling of a global percept by sewing together multiple local percepts. As long as the local relation between surfaces and objects follows the rules of nature [as we understand them], our brains don't seem to mind that [obtaining] the global percept [in this manner] is impossible. (Macknik et al, 2010, p. 8)

Not surprisingly, we process and develop global descriptions and solutions by adopting global perspectives to satisfy the constraints imposed by the very nature of the problem. We are more likely to discover macroscopic perspectives through open-ended search processes that diverge into the larger systematic space. Since predetermined objectives can automatically circumscribe and close the global search space by definition, they will likely result in futile attempts to diverge to higher-level macroscopic descriptions and converge by default to local dead ends. Without higher-level descriptions, the complex behavioral dynamics of a complex system cannot easily be predicted or deduced from the behavior of individual lower-level entities (Buzsáki, 2006, p. 13).

> The upward direction is the local-to-global causation, through which novel dynamics emerge. The downward direction is a global-to-local determination, whereby a global order parameter "enslaves" the constituents and effectively

> governs local interactions.... If the component relationships within the system become optimized for a particular task as a result of external perturbations, the system is called adaptive. (Buzsáki, 2006, pp. 14–15 [emphasis in original])

Global descriptions allow for local self-organization and activation while matching the statistical correlation structure of the fitness landscape to a systematically referenced optimum. A globally referenced description explicates local behavioral performance in higher-order terms of systematic information and knowledge correlated to the global landscape. Defining the statistical limits of the global optimum facilitates the orderly description and interpretation of problem and solution spaces and helps identify logic parameters for effectively fine-tuning solutions within the systematic limits of global knowledge. In effect, by hyperfocusing on self-organization as a strictly local phenomenon to the exclusion of global correspondence, we also throw the baby out with the bathwater and extrude the self-organizing principles that relate to general systematic principles throughout the domain.

Global descriptions develop general knowledge of how the entire system works, which extrapolates to enhancing optimal reasoning in local regions. It is precisely the global analysis that permits the interpretation of the local changes so that an explanation of how the system works is available (Churchland & Sejnowski, 1994, p. 377). In an abstract sense, global knowledge can correlate sensorimotor integration between local motor behaviors and globally decoded processing of sensory information. Hence, global knowledge can correlate local search with higher-dimensional constraints and knowledge by processing information systematically, thinking globally, and acting locally. Thus it is possible to redefine the bottom-up self-organizing limits constraining the solutions at local optimum in terms of effectiveness, efficiency, and top-down systematic constraints imposed by probabilistic global laws.

> ... [T]op-down data constrain bottom-up hypotheses, thus narrowing the search space. Quite simply, it is more efficient to have some idea of what you are looking for than not. (Churchland & Sejnowski, 1994, p. 148)

It also helps to have some idea of what you probably will not find, including absolute certainties and symmetrical time. Understanding the global laws of nature starts with understanding the arrow of complexity—how complexity evolves and increases from local self-organization—which depends on arrow-of-time irreversible descriptions that break time symmetry. Extended language enables one-way time-dependent descriptions and representations that improve the coding fidelity of systematic information in correspondence with the ordering of time-dependent domain dynamics. Information fidelity based on extended language with one-way time-dependent reliability has relevant analytical value for enhancing effective thought processing. Time-directed information and globally ordered knowledge typically trump locally informed beliefs for effective cognition. We benefit from information and understanding that corresponds with the consistent ordering of the forward-directed flow of nature's time. In turn, we can direct our thought processes and behavior to more efficiently navigate and effectively adapt to the instabilities and uncertainties we encounter in our complex world.

It is possible to acknowledge the evolutionary principles of local self-organization—which seem to weigh heavily on the polarity of localized language—without being trapped by affective language constraints and imprisoned between a local minimum and maximum. Even though self-organization evolves in isolated local portions of the fitness landscape, it is shaped and defined by general physical laws. Descriptions of local self-organization, somewhat like the evolution of language and global scientific theories, evolve from a linguistic phase transition from deterministic to dynamic nondeterministic processes. As a part of evolutionary processes and nature, self-organization remains subservient to global laws and probability. Local self-organization, although an isolated phenomenon, still operates under the global laws of nature and is best described by globally coherent language that extends past the affective limitations of localized language.

We can miss the big picture while railing about how self-organization happens locally and is therefore only amenable to interpretation as a local phenomenon. Optionally, we can learn about global influences in complex adaptive systems. Nature consists of invisible processes among short- and long-range interactions of high and low energies and short- and long-time scales. If we ignore the macroscopic interactions of forces and temporal scales that self-organization evolves under, we can miss seeing the profound global influences associated with local periodic and aperiodic behavior. Invisible periodic processes related to oscillations and hidden under relations of approach-and-avoidance behavior can leave us to easily dismiss them as random phenomena. In contrast, in the absence of a visible controller, the nonrandom behavioral regulation of organisms may appear to take on mysterious operational qualities. Likewise, an organism's spontaneous movement without the attachment of visible wires or strings can appear as the work of mysterious unknown forces that at times somehow magically correlate to the local landscape. Misinterpretation of the obvious effects of invisible regulatory processes can easily lead to confusion and local speculation, while the not-so-obvious systematic processes go ignored and unexplored.

To some extent, all known organisms evolved in a periodic world. Across nature, in organisms from small to large, we see local attractor relationships and periodic oscillations that extend across individual cells, brain oscillations, and diurnal and circadian rhythms (Buzsáki, 2006, pp. 117–120). We see changes in physical relationships of the Earth's axis, gravitational forces, and magnetic fields and global variations in semi-daily, daily, monthly, quarterly, and annual periodicity. We see symmetry breaking, periodicity, and attraction and repulsion among attractor relationships operating from local subatomic microscopic to a global macroscopic level. The inclusion of macroscopic influences on local processes seems like a prudent consideration when describing localized self-organizing relationships on global landscapes. Attention to macroscopic features helps to remind us that the world operates on global symmetry-breaking principles and irreversible one-way time according to possibilities, rather than locally on two-way time and certitudes—and encourages us to adjust our language behavior accordingly.

Self-organization complements Darwin's evolutionary theory and plays an important role in complexity theory by modeling biological systems that are based on functional features of dynamic attraction and repulsion and positive and negative feedback (Floreano & Mattiussi, 2008, pp. 516–518; Camazine et al, 2001, pp. 15–23). The second law of thermodynamics describes entropy in an isolated system as driving expansion toward its maximum value (Lindley, 2007, pp. 23–29; Mitchell, 2009, p. 71; Prigogine, 1997, p. 202). Put another way, the second law states that the disorder in a closed system mostly increases and at best stays the same. Modulation of attractor relations can potentially mitigate the effects of expansion for a price paid in energy and entropy. Self-organizing dynamics, attributed to symmetry-breaking bifurcations, irreversibility, and attraction and repulsion, represent common components influencing interactions within complex living systems.

Open-ended complex systems irreversibly point toward the future, deal with instabilities, and produce complex structures, innovative solutions, and creative novel behaviors. We use language to represent, describe, and understand these complexities. Simple, discrete language parameters with certitudes that work fairly well in an idealized static environment may suffice on average in a complex environment. But averages decrease and hide variability such that we pay a price for this sufficiency by limiting knowledge about alternative possibilities for higher-order cognitive processing, which places further constraints on the search for novel solutions and stifles creativity. To facilitate dealing effectively with the uncertainties and instabilities in nature, dynamic language solutions include statistical descriptions parameterized to generate populations of alternative possibilities rather than one-dimensional certitudes.

Complex systems and self-organization are best described and understood by language constructions that correspond to the domain dynamics. Language that matches the domain dynamics offers an option for extended cognitive processing that can facilitate the efficient production and effective optimization of solutions. Effective language generally produces more optimal results than affectively constrained language, especially when addressing the continuous dynamics of open-ended systematic processes. Recall that affective language, exclusive of continuous logic, computes with discrete two-valued logic and produces rigid, absolute solutions that close the solution space.

In contrast, effective language includes probability as logic that enables continuous processing and open-ended solutions as possibilities that leave the solution space open.

In higher-dimensional domains, probability as logic enhances flexibility and output optimization in response to change, which is consistent with the continuous domain dynamics and one-way time. Hence, effective language enables continuous logical processing that facilitates open-ended solutions and effective adaptation in response to a fluctuating environment. Simply put, we live in a world of many possibilities, and we expect language that scales to the world in which we live, to work with greater effectiveness.

Understanding our world with an open mind helps us to get on the same page with our world and interact coherently therein. Language and science play a key role in this open-minded understanding that can reap widespread benefits for all involved.

> Our belief is that our own age can be seen as one of a quest for a new type of unity in our vision of the world, and that science must play an important role in defining this new coherence. (Prigogine, 1997, p. 186)

In complex biological systems open-ended regulatory processes operate probabilistically on fairly simple information coding principles related to symmetry breaking, such as positive and negative input values and approach-and-avoidance output reactions. We see fundamental statistical relationships among position, action, value, charge and energy, and internal–external processes of attraction and repulsion related to approaching positive and avoiding negative states. Value, charge, and change are general features of the complex system we call nature. The next two chapters integrate extended language and the intentional brain with cognitive processing, predictions, and external relationships, and deal with the operational aspects of affective and effective language in relation to value and action.

CHAPTER 11

Extended Language and the Intentional Brain

NATURE, EVOLUTION, AND THE brain operate as open-ended processes. Language that enables macroscopic perspective and processing lets us see the big picture that we miss with constrained language. Language that corresponds with the open-ended processing of information within confines of the structure of the world in which we live and the brain with which we interact conveys a more coherent understanding of relations we experience.

Understanding relationships and what makes them tick has a lot to do with language. Even though we might expect some sort of understanding even without language, it is language that enables explicit expression of our thought and knowledge about the world. A world without language would likely leave us somewhat on par with many other creatures that appear to lack the sophisticated cognition and communication that language makes possible. We might think it would be very different from our world, but our primitive limbic cognition would likely automatically prevail. How can we extend beyond primitive limbic reactions? Does language hold the key to resolving this conundrum? We can look at various language forms and decide which we think works the best, including affectively constrained and effectively extended language.

Constrained Versus Extended Language

For example, constrained versus extended language deals with

- The certain versus the possible and the probable
- Static versus dynamic
- The idealized versus the realized
- *Always* versus *sometimes*
- *Must* versus preference
- *Me* versus *them*
- Individual versus population

- How the input feels affectively
- How effectively the output works

Extended language carries a greater amount of information than language with closed-ended constraints, such as prescriptives and imperatives. Extending language clarifies distinctions between the static past and probabilistic future, and possibilities provide open-endedness that matches with continuous processes and change. And linguistic extensions facilitate focusing on informed choices with responsibility and accountability for actions and separating affect from effect. Consider, for example,

- "It will *certainly* happen" versus "it will *possibly* happen"
- "It will *possibly* happen" versus "it will *probably* happen"
- "It will *always* happen" versus "it will *sometimes* happen"
- "I *must* go" versus "I *choose* to go"
- "*You* upset me" versus "*I* upset myself"

Extended language enables us to understand that we live in a dynamic world of possibilities rather than a static, idealized world with imaginary certainties and invisible strings of obedience. The conversion of possibilities to probabilities depends on language constructions that fit the domain and provide sufficient quantities of current information and world knowledge to make reliable predictions. Generalities, such as *sometimes*, facilitate correspondence with open-ended processes and relations in the finite world in which we live. Absolutes, such as *always* and *never*, correspond to an imaginary infinite world with closed-ended processes. *Often* and *seldom* correspond to the occurrence of context-dependent change and allow flexibility for making adjustments, and choosing alternative preferences consistent with dynamic interactions generally found in evolutionary systems.

Obligatory imperatives, such as *must,* exclude choices and freeze perspective. "You must stop upsetting me" demonstrates affect-laden reactive cognition. Our safety oriented limbic system implicitly assumes an external threat "caused" the negative sensation "effect" and automatically protects us with an imperative defensive reaction. The implicit misperception of affective responsibility for "upset feelings" responds to explicit remediation with effective language. "I do not like your behavior, but I refuse to upset myself about it" reflects effectively tempered thought processing and response. The former affective reaction attempts external behavioral control and deflects responsibility and accountability. The latter effective response takes responsibility and accountability for internal behavioral modulation. From these choices we can ask

- What level of regulation do we think would generally produce the best results between affective and effective language?
- Which language delegates "self" responsibility and accountability, affective or effective?

- If we could choose language for our own personal use, which would we choose, affective or effective?
- If we could choose everyone's language, which would we choose, affective or effective?

Of course, we can and do choose the language constructions we prefer. But we might wonder why we would construct an absolute language that fits an idealized world that hardly resembles the world in which we live. We might also wonder why we would consider using affective language for our everyday speech. We have the option of using effective language that fits with the real world. We likely can see affective language use in others but have difficulty recognizing that we use it in our own everyday speech. Perhaps exploring language from several different perspectives will shed more light on how we got the language we use, why we use it, and why we are oblivious to its affective nature.

Theories, explanations, and descriptions are often put forth to try to illuminate the correspondence between language, the brain, and general physical relations. These include event-level correlations between classes, objects, actions, and methods found in higher-level object-oriented computer science programming language (Seldon, 2010, pp. 275–276; Steels, 2008), the processing of nouns and verbs in artificial neural networks (Cangelosi & Parisi, 2004), and analyses of the way we use nouns and verbs in language (Bailey, 2009). These descriptions correlate with objects, behaviors, and intentions in usage-based language theory, and with the mirror neuron system first described by Giacomo Rizzolatti and his colleagues (Arbib, 2003, pp. 606–611; Tomasello, 2005, 2008; Shadmehr & Wise, 2005, pp. 172–175; Rizzolatti et al, 2009, pp. 632–636; Rizzolatti & Sinigaglia, 2008, pp. 39–48, 76–79, 127). These theories can help us to extend and expand our perspective to develop a systematic understanding of affective and effective language, local-to-global processes and physical relationships, and language's profound influence on our cognition—our thoughts, feelings, and behavior.

Relevance

Relevance predictably influences thought processing proportional to our affective or effective linguistic bias. An affective linguistic bias automatically increases the sensitivity and affective tone for emotionally relevant stimuli, which tends to amplify the input intensity and either the perceived pleasantness or unpleasantness of stimuli. The increased emotional intensity biases the predictability of the cognitive processing and behavioral output. As a form of linguistic symmetry breaking, *relevance* means having either affective or effective value asymmetry applicable to the matter at hand, such as positive or negative weight allotted to information that influences our attention, choices, and decisions.

We have variable histories, so what one person thinks is relevant may seem irrelevant to another. Even though our brains are roughly the same, exclusive of more severe dysfunctions, some brains come predisposed to different levels of sensitivity to stimuli with underdampened or overamplified reactivity. Affective language tends to exacerbate these predispositions and amplify or dampen the perceived emotional intensity associated with stimuli relevance that in turn biases behavioral reactions. Therefore, behavioral predictability is at the mercy of how we perceive the environment at any given moment, depending on our current affective state. Effective language enhances the analytical modulation of cognition, perception of stimuli relevance, and choice of behavior; thus, behavior is typically more predictably adaptive and less reactive. We see both genotypic and cultural influences on our phenotypic perception of relevance, which tend to positively respond to increasing linguistic effectiveness and education.

An effective linguistic bias deliberately addresses input sensitivity and feedback reliability for effectively adjusting behavior. A bias based on effectiveness enables analytical recalibration and modulation of input sensitivity, perceived stimulus values, and output responses as an alternative to affectively biased processing and emotionally driven reactions. Assessing the effective value enables information processing and decisions that can account for the pleasantness or unpleasantness of stimuli while making analytical predictions that are based on possibilities and probabilities for maximizing adaptive behavior. The effectiveness of our language enhances analytical sensitivity that in turn biases the effective fidelity of our cognitive processing.

With a brain that comes with systematic biases and uses language, we are susceptible to affective and effective linguistic biases. The fidelity and bias of the linguistic constructions we use influences sensitivity to relevant stimuli and the reliability of the predictions we make. Differential contributions from genetic inheritance, learning, and language usage can influence how we perceive and respond to relevant stimuli that we deem pleasant or unpleasant. The perceived relevance of stimuli begins in early stages of development when we are sensitive to learning how things feel, what to approach and avoid, and how our feelings and behaviors correspond with our predictions of how the world works.

Usage-based language theory presumes a functional meaning in the informational context that the organism finds specifically relevant. Infants learn to correlate certain speech stream segments with objects and actions in the world around them and their spatial positions (Friederici, 2008, p. 117; Iliescu & Dannemiller, 2008, p. 138). Associating affective value with objects and actions likely co-evolved with regulation and reinforcement that eventually led to affording affective meaning to the correlated speech on the basis of the perceived pleasantness or unpleasantness of input from the environment. Our mammalian limbic system apparently operated sufficiently prior to the invention of natural language. Long before our ancestors possessed the capacity for language or abstract reasoning, they moved in relation to objects and places in their environment. This relationship presumably led them to naturally parse the world into objects, groups of objects, and actions (Nowak, 2006, pp. 251–252).

Human brains apparently evolved higher-level cognitive-processing capacity on top of, or in parallel with, already evolved motor learning pathways, and, like other vertebrates, we first learn how, when, and where to move in response to our perceptions of observed objects. This sensorimotor relation suggests that the ability to estimate and predict the intentions of environmental objects evolved long before language (Shadmehr & Wise, 2005, pp. 1, 9).

Early sensorimotor pathways provided a local mechanism (Rizzolatti & Sinigaglia, 2008, pp. 21–52; Arbib, 2003, p. 609) for

- Reaching, pointing, and grasping
- A natural platform for motor learning
- Adaptive motor behavior regulation
- Identifying and understanding intentional relationships
- Eventual language development

Broadly speaking, our nervous systems experience environmental stimuli and make predictions that are based on the perceived relevance we derive from those experiences. A great deal of learning, including motor learning, depends on several key mechanisms, such as predictions, error signals that indicate the accuracy of those predictions, and error correction that improves the accuracy of future predictions. All play an important role in modulating instabilities, fitness, and decision making (Shadmehr & Wise, 2005, p. 41).

Motor activity also interrelates with several other domains, including perception, cognition, and emotion (Angulo-Barroso & Tiernan, 2008, pp. 155–156). Experience in neurorobotics has advanced the argument that the motor system strongly influences learning via exploration of the environment, such that recognition [relevance] and action correspond to the effectiveness of perception (Arbib et al, 2008, p. 1470). In turn, we can expect that linguistic effectiveness influences cognitive thought processing and related perception of stimulus relevance, predictions for adaptive behavior, and expected actions and reactions in the environment.

We use language to aid in categorization (Koziol & Budding, 2010, p. 74) and to represent relevant features we perceive in the environment. Language provides a representational system that facilitates the organization and influence of motor plans (Grafton et al, 2009, p. 648). The evolution of language provided us with internal symbols and semantic referents that helped to optimize sensorimotor coherency between our internal operations and the external environment (Cangelosi, 2001, p. 94; Steels et al, 2007). Linguistic coherence can improve statistical correlations with environmental regularities in terms of understanding relevant meaning and predictability.

Extended language coherently correlates and flexibly associates our experiences to the natural ordering and contextual meaning of events in time and space via

- Words
- Concepts
- Percepts
- Inferences
- Values
- Information
- Knowledge

Correlating cognitive processing to globally preferred information asymmetry establishes reliable meaning for effectively analyzing, understanding, and communicating experiences in correspondence with external physical reality. The affective language we use in our typical day-to-day speech for interpersonal communication relies on brittle emotional connections formed by absolute logic processes that tend to arbitrarily distort meaning. In turn, distorted meaning limits our ability to analyze, understand, and predict the relevance of future experiences. How do we decide what is relevant? How can we make sense of all this and improve our predictions?

Like other physicists before him, Heisenberg saw that we can attach meaning to position and momentum by measuring them (Lindley, 2007, p. 146). An object's direction, mass, and velocity form a vector that can be interpreted as intention. Meaning, as relevant environmental information, can be abstracted from intentionality as the interrelationship between goals, perceived states of the world, and actions taken in pursuit of those goals (Moisl, 2001, p. 450). Simply put, meaning results from sensorimotor integration of experiences corresponding to sensory values, goals, and actions.

Conceptually, intention integrates and conveys two independent variables: action and value (Skinner, 1953, p. 36). Action conveys a change in position, and value associates relevant affective meaning with directed action and change of state. The intent of an action can be interpreted positively as beneficial or negatively as harmful. For example, consider these statements:

- "What have you done? What is the meaning of all of this? You did this on purpose."
- "I didn't mean to do it. I mean, it wasn't intentional. I didn't mean to cause any harm."

Intentions provide meaningful and usable information for processing and understanding environmental reinforcement value via feedback (Mitchell, 2009, p. 296). By estimating the intentions of an animate object, an organism can attribute or infer relevance for predicting the general salience of the current interaction—that is, the gist or general essence of the event (Siegel, 1999, p. 28; Spitzer, 1999, p. 65).

Gist

The gist of an unfamiliar, unexpected, or novel object can be assessed fairly quickly by symmetry breaking. Inference from an internal reference frame yields an estimated positive or negative value weighting associated with the object's position, change of position, and presumed intention. The idea of an intentional representation of behavior addresses a central problem facing neuroscientists: how neurobiological systems represent the world and how organisms employ representations, via transformations, to guide behavior (Eliasmith & Anderson, 2003, p. 5).

The *gist* is a fast and frugal information form that conveys limbic-related affective meaning, purpose, and intention. Understanding how the gist is automatically computed and inferred by our brain points to the importance of analyzing the influence of language on how we perceive our own intentions and the intentions we attribute to others. Since we compute the gist automatically, making the effort to deliberately access and analyze our intentions allows us to think about the long-term effects of our behavior from a system's perspective. In effect, we can compare the risks and benefits from a local, self-centered emotional perspective against a systematic, world-centered analytical perspective that is based on effective adaptive behavior.

Language represents knowledge in the sense that it provides categories that structure and define abstractions mapped from space and time, multiple influences, objects, intention, and possession (Tomasello, 2005, p. 54). When a neural system computes a transformation from an input variable to an output variable, the relationship produced contributes to generalized mapping patterns that encode the input variable (Shadmehr & Wise, 2005, p. 403). This mapping also reflects the generalized categorization of intentions and behaviors that predict potential utility as value associated with actions and their expected outcomes. Relevant spatiotemporal information abstracted and mapped from an object's relative identity, position, directional heading, velocity, and inferred value yields the gist or general contextual representation. The gist represents the general meaningfulness of a particular event that influences inference and prediction of possible purposeful intentions.

By getting to the gist, we quickly infer transformation of stimulus-relevant affective values to identify purposeful actions and anticipate an object's possible positive or negative intention. From an organism's internal frame of reference, having a fast and frugal processing method that is based on intention allows for quick and simple identification of relevant value for orienting current action, initiating automatic responses, and recording the outcomes in memory. The association of memories with positive and negative values according to experience facilitates long-term memory storage and retrieval. Our memories provide us with an updatable information source for integrating past and current experiences into our thought processing. Thus memories play a role in evaluating relevant stimulus values and predicting the outcomes of future actions. Memories that include analytical processing with systematic information provide logically informed assessments of the current state of the environment, consideration of the probable effectiveness of various alternative solutions, and selective tuning of predictions before choosing and executing our actions.

Our memories play a significant role in how we recognize and categorize similar objects or events we encounter in the future (Shadmehr & Wise, 2005, p. 41). Intentions provide a fairly inexpensive and efficient source of information for quick evaluations and responses. Affective language relies on implicit assumptions of intentions, automatic limbic processing, and reactive behavior. Alternatively, with effective language we can deliberately analyze intentions, gather information, consider relevance, overrule our limbic system, and make effectively informed responses.

Affective values associating memories and experiences from the past can influence how effectively we think and behave in the here and now. Understanding what is relevant for thinking and behaving effectively in the present looms large for reliable predictions. We can gain knowledge from past behavior for understanding our current behavior. Past perceptions of relevance tend to accumulate and create affective interference in the present. Even though no longer relevant, they can influence our current thinking and planning for the future. Affective interference encourages considering effective cognition. We have effective options for recalling, revaluating, and editing affective relevance from past learning. Effective language plays a pivotal role in coding our understanding of relevance in the here-and-now context of our own actions.

Using language that is conducive to analyzing and correlating relevant current information with our thought processes, predictions, and experiences facilitates effectively fine-tuning our knowledge and cooperative responses. Effective language can extend our thinking beyond affective gist, convey relevant meaning in the here and now, and enhance our understanding of relationships that we encounter in our daily lives.

Cognition, Meaning, and Purpose

> ... [A] theory of language is woefully incomplete without a serious account of meaning.... [W]e would like to be able to account for the way that (more or less) the same thought can be mapped into expressions of different languages, allowing for the possibility of reasonably good translation. (Jackendoff, 2002, pp. 271–273)

Meaning can be defined according to its significance, purpose, or intended purpose that can be described in relation to systematic information. Meaning and purpose play an interactive role in cognitive processing and communication for sharing information, planning actions, and understanding the intentions of others (Risberg, 2006, p. 10). The language we use influences how meaning is conveyed and perceived as information.

We can understand meaning as information produced by language usage that affects and effects our communication. *Affective meaning* refers to locally referenced usage that obscures context and omits or disregards systematic perspectives. Affect corresponds to input and how received information feels to us. Affect-laden language tends to saturate the input, leading to corrupted processing that can distort the output gain and the interpreted meaning of messages. Effective language facilitates the modulation of emotional responses. *Effective meaning* extends the definition in context to significance across the system. Effect corresponds to output and how the analytical tuning of cognitive processing enhances reliably informed meaning and predictions that correspond with adaptive actions.

For example, loud, angry, belligerent speech perturbs the environment. Generally, persons in this environment who receive this emotionally intense display will likely perceive threatening behavior. The intensity of the affective input stimulates the limbic system's direct emotionally reactive pathways that can result in freezing, fleeing from, or attacking the source of the disturbance; any one of the responses may possibly suffice, but fleeing appears to lessen the risk of harm. The other alternative is to pause, suppress reactive behavior, separate the affective meaning, and process an analytical solution. Thoughtful solutions without an affective burden can enhance effective resolution, but sometimes quick, reactive solutions save us from harm. Reacting by matching affective intensity tit for tat with loud, angry, belligerent speech typically leads to escalation. We often see what appears to be purposeful tit-for-tat behavior in power struggles that reward the person with the most aggressive behavior as the winner.

Effective language helps us to distinguish between affective and effective solutions and question any role that purpose might play in how we respond to disturbances. When we incur disturbances, affective purpose tends to occur effortlessly and reactively, and without thoughtful intervention tends to emotionally bias our behavior. In a sense, we observe some benefit on average from affective purpose as a sufficient quick fix, but affective purpose comes up short for maximizing effective adaptation in the long run. Where does meaning and purpose come from?

Although we have ample evidence supporting the mechanisms for the evolution of biological organisms, it seems that evolutionary processes are somewhat neutral to purpose. In other words, we see no predetermined objective. In light of this, we can assume that purpose is defined and attributed by humans. But we still ask how and why things happen and wonder about the meaning of it all. The biological organisms that we observe seem to exhibit purposeful goal-directed behavior attributable to survival and reproduction, as witnessed by a predisposition for attraction and approach to positives and by repulsion and avoidance of negatives. So it is not surprising that animals are attracted to rewards such as food sources and sexual activity, and generally avoid unpleasant stimuli. Purpose seems to operate largely on the local level, which we can infer and interpret retrospectively as an intrinsic systematic bias in the service of propagating the species through survival and reproduction.

The notion of purpose brings up the question of how our behaviors come about and how they relate to affective and effective processes of brains, cognition, language, and information. The connection appears to operate through the local-to-global development of meaning. If our behavior is goal-directed, what is our goal, our purpose? If our brain is idly waiting around for a goal in order to know how to proceed, we could wind up in a pickle, so to speak. With affective language, though, we can rationalize imaginary purpose and meaning that covers up for our limbic-like behavior in lieu of considering analytical assessment of our goal-directed behavior that is based on cognition with effective language. But we often seem left in a quandary: "Where does meaning come from, and how does our brain make use of it?"

Somewhat similar to the effect of mutations in DNA, meaning produces changes in the organism that creates that meaning (Carroll, 2006, p. 80). In humans, the meanings we assign to words and learn by our linguistic abilities produce functional changes in our brains and our cognition.

The right upper frontal lobe helps modulate visual-spatial meaning in more generalized information terms, identifying the gist of the situation (Schmalhofer & Perfetti, 2007, p. 184; Long et al, 2007, p. 330; Tapiero & Fillon, 2007, p. 365; Siegel, 1999, p. 331). The left upper frontal lobe, or dorsolateral prefrontal cortex (DLPFC), abstracts more specific details as meaningful information via language and also helps integrate internal and external information in response to changing conditions (Kurzban, 2008, pp. 159–160; Gazzaley & D'Esposito, 2007, p. 188; Risberg, 2006, p. 6; Siegel, 1999, p. 326; Wagner et al, 2004, p. 714). Arguably to some, our left DLPFC supports linguistic modulation of executive cognitive activity, explicit analytical cognition, and deliberate purposeful behavior. *Executive function* has been described as the capacity to deliberately generate adaptive behavior in the absence of external direction, support, or guidance (Koziol & Budding, 2010, p. 71). Of course, we know that even though we can act deliberately, we are still a part of the environment and operate within physical constraints, so we acknowledge the mutual interactive influences between us and nature. Executive functioning enables sculpting a view of the world that is realistic and consistent (Stuss et al, 2001, p. 107).

The lower orbitofrontal cortex (OFC) primarily modulates emotion and object–affect association related to obtaining rewards and avoiding penalties, and it is involved in implicit, rapid, emotion-related decision making (Rolls, 2008, pp. 52, 123–124, 136, 525–526). The OFC via limbic-related interactions provides important value-based utility calculations (Padoa-Schioppa & Assad, 2006; Glimcher et al, 2006). The OFC contributes to social and emotional cognition and purposeful utilization behavior (Koziol & Budding, 2010, pp. 75–76). With knowledge about objects, actions, and value relations, the frontal lobes influence human executive function via routine monitoring of higher- and lower-order information integration, reasoning, decision making, direction of attention, and inhibition of inappropriate responses (Goldman-Rakic, 1995; Lichter & Cummings, 2001; Middleton & Strick, 2001; D'Esposito & Postle, 2002, p. 177; Salloway & Blitz, 2002; Tranel, 2002, p. 351; Striedter, 2005, p. 334; Curtis & D'Esposito, 2006, p. 295, Figure 9.8; Lee & Seo, 2007, p. 108; Chow & Cummings, 2007, p. 29).

Higher-order inhibition in nervous systems apparently evolved as a sophisticated regulatory mechanism that provides more complex computational ability for resolving higher-dimensional problems in complex domains. We can think of an upper-level system involved in more abstract, higher-order analytical processing operations, and the lower-level system involved in more concrete first-order emotional processing. In humans, the left DLPFC plays a principal role in deliberate execution of higher-order inhibition and selection of language and semantic processing in large part via working memory. The left DLPFC is part of a functional language module with the ability to manage information and knowledge resources related to verbal skills, and via the OFC implicitly influences the engaging and disengaging, switching and dampening, and editing of lower-level subcortical loops (Stuss, 2007, p. 293; Alexander et al, 1986; Middleton & Strick, 2001; Chow & Cummings, 2007, p. 29; Cummings & Miller, 2007, p. 15; Poldrack & Willingham, 2006, pp. 130–131).

The left DLPFC's ability for explicit language in humans plays an instrumental role in cognitive processing through working memory and top-down regulation of information (Knight & Stuss, 2002, p. 577; Mesulam, 2002, pp. 19–21; Petrides & Pandya, 2002, p. 39). Language enables explicit logical modulation of cognitive processing and deliberately tuning cognition in relation to statistical regularities in the environment. The DLPFC plays a prominent role in regulating planning and implicitly biasing mammalian motor output in general, and biasing motor output for speech behavior in humans, augmented by the explicit language (Rolls, 2008, pp. 525–527).

Explicit modulation provides enhanced sensitivity and more reliably biased inhibitory mechanisms to suppress irrelevant stimuli, reduce processing of irrelevant information, and mitigate irrelevant habitual or maladaptive motor responses (Braver & Ruge, 2006, p. 320). Suppressing motor output—similar to pressing the clutch and disengaging the drive train in an automobile—allows extended offline time to think or simulate alternative solutions. By disengaging the motor output, the DLPFC can suspend behavior, which allows extended cognitive processing and an opportunity to inhibit and edit misdirected or misinformed plans related to incidental errors, unexpected external changes, changes of state, and so on.

Recall that humans have an increased ability to disengage, inhibit, and delay motor output that allows us to go offline and think about the past, reflect on the present, make and test premotor output copies for plans, and think about appropriate responses before acting (Bronowski, 1977, p. 144). We can take advantage of the left DLPFC, which provides us with a resource for long-term cost–benefit planning related to analytical decision making, whereas the OFC typically displays immediate emotion-related, reward-based decision making (Rolls, 2008, p. 526).

> We have plenty of disreputable impulses but, compared to other species, we can inhibit them far more effectively. This gives our minds freedom to contemplate the long-term consequences of our acts and to behave accordingly. All that freedom and perspective naturally comes at a price, for it also means we have become responsible for our deeds. This unprecedented power and responsibility of choices makes us a terribly conflicted "paragon of animals." (Striedter, 2005, p. 344)

We might be tempted to focus only on higher-level, top-down brain structures. A strategy with a top-down focus can easily blind us to relevant parallel interactions with lower-level subcortical cognitive structures. Leonard Koziol and Deborah Budding describe the tendency to focus on higher-level structures as a "cortico-centric bias" (Koziol & Budding, 2010, p. 5). No part of the brain, however interesting or seemingly important, likely operates entirely independently. Cognition involves reentrant subcortical loops involved in affective processing and integrating motor functions by the limbic system, basal ganglia (BG), and cerebellum.

> In Latin, *limbus* means a surrounding ring. The French neurologist Paul Broca introduced the term *la grand lobe limbique* in 1878. Papez (1937) went further by postulating that emotions reverberate in the ring, often referred to as the Papez circle. (Buzsáki, 2006, p. 281 [italics in original])

The term *limbic system* has been introduced since and generally includes the amygdala, hippocampus, entorhinal cortex, and hypothalamus (Buzsáki, 2006, p. 281). The BG are sometimes described as being part of the limbic system because of their connection with automatic output processing, and in humans the left DLPFC typically has an opportunity to have the "last word" on the behavior they produce. An integrated cognitive view that includes cortical and subcortical processing enhances our understanding of the brain's overall functional modulation. The flexibility to selectively integrate information between more computationally expensive deliberate, higher-level processes explicitly and less-expensive automatic, lower-level, implicit processes increases our options and our ability to operate efficiently in difficult-to-predict complex environments.

> The cortex, basal ganglia, and cerebellum are three brain systems that run in parallel. Each region makes a unique contribution to behavior, whether that behavior is motor, cognitive, or affective. (Koziol & Budding, 2010, p. 22)

Reliable computation depends on inhibiting irrelevant information while maintaining attention (Koziol & Budding, 2010, p. 71). We process current information, compare it with recalled information stored during previous experiences according to differences and similarities, and correlate options and predictions for choosing action in response to current events. We then integrate further learning from the outcomes that our actions produced. Reliable information and feedback facilitates our ability to deliberately edit working memory, effectively correct mistakes that contribute to systematic error variance, and suppress prepotent affective biases from habitual limbic responses. We can see where our language plays a role in cognition at this level and how words, beliefs, and values can influence relevance and what we attend to, what we perceive as errors, and how we make predictions and error corrections. The effectiveness of our language places constraints on how we process information, such as affective language that defaults to supporting either–or limbic processing and biases perceptual relevance.

We use our vocabulary to represent and remember the semantic categories we develop through cortical sensory perception. This perceptual overlap suggests mediation through the medial temporal lobe memory system. According to this view, language involves interplay between subcortical and cortical systems (Koziol & Budding, 2010, p. 169). The reliability of our language and semantics is thus instrumental in explicitly monitoring and modulating the fidelity and effectiveness of our brain's computations and output. Effective thinking improves decision making related to our choices and the irreversible consequences we can face in an uncertain world with one-way time. Effective language broadens the dynamic range available for top-down modulation of tuning in relation to sensory input, processing, and behavioral output.

Bottom-up activation can be inhibited by the prefrontal cortex (PFC), and top-down activation can be accomplished indirectly by disengaging an inhibitor, which releases the systematic inhibition that normally suppresses unintended and random bottom-up signal transmission. Deliberate PFC inhibition can suppress and recalibrate automatic limbic-associated affective responses in favor of analytical modulation based on current information and accumulated systematic knowledge. In effect, lower-level fear-based affective responses are susceptible to upper-level modulation. For example, studies in the regulation of fear have shown that the amygdala response is decreased as a result of inhibition from the PFC (Phelps, 2009, p. 211). The concept of the PFC as a significant contributor to executive regulation by inhibition has a long history dating back to the pioneering efforts of Alexander Luria (1902–1977).

DLPFC executive functions play a role in inhibition that provides a regulatory link between sensory evidence and (medial) mPFC–OFC–limbic pathways. Deliberate DLPFC inhibitory modulation of the OFC enables editing of working memory and indirect regulation of limbic value-related computation via suppression of habitual pathways involved in affect-associated choices and actions. Thoughts and memories placed in working memory can influence what we attend to, the way we see things, and the way we act (LeDoux, 2002, p. 319). Executive functioning plays a role in modulating perception related to hindsight, planning, anticipation, and a sense of time. Integrating information in the time domain is a critical PFC operation for goal-directed actions (Fuster, 2002, p. 99).

Executive inhibition enables suppression or the switching off of motor output, giving us an opportunity to think, plan, and imagine (Barkley, 1997, pp. 86, 163; Bronowski, 1977, pp. 142–145). Suppression is flexible and typically described as relative tonic or phasic inhibition rather than absolute (Koch, 1999, pp. 118–119). We can think of the language-related DLPFC as playing an executive role in higher-order inhibition and regulation analogous to a computational input-processing-output model. The DLPFC allows for deliberate top-down regulation of alternative solution preference by flexible inhibition according to current stimulus input and the monitoring of cognitive computations and memory associations in relation to the effectiveness of outcomes.

Top-down DLPFC executive processes can provide deliberate recall and editing of working memory, and regulation of effort, alertness, processing speed, and attention, by inhibiting irrelevant stimuli. Inhibition influences the focusing, sustaining, and shifting of attention according to stimulus relevance and enables the prioritization, organization, initiation, and completion of tasks. The significance of executive regulation and explicit language becomes more apparent when we consider that the limbic value system can implicitly affect almost all of our cognitive processes. In short, executive functions make practical thinking and reasoning possible (LeDoux, 2002, p. 178).

The frontal lobes provide a site where association cortices and the limbic system interact (Mesulam, 2000, p. 45). The hippocampo-entorhinal complex and the association neocortex are continually reconstructing, updating, and elaborating associations that collectively lead to the consolidation of new memories. Associative memories form a part of long-term knowledge, where new information can be written on top of existing items. Old information is protected by the transient storage of limbic-dependent information until behavioral relevance is established, before being transferred to long-term association cortices. These organizing processes limit the indiscriminate influx of new information, while allowing adaptive learning under changing conditions (Mesulam, 2000, pp. 62–63). We can see where we might incur incidences when new learning might not fit into the structural scaffolding that underlies what we "know." In such cases, novel information can easily be tossed aside and discarded depending on the affective value we infer and associate with the experience.

The DLPFC can suppress limbic-driven behavior and contribute to behavioral self-regulation by monitoring and mitigating bottom-up impulses, managing frustration, and modulating emotions (Brown, 2005). In humans, DLPFC inhibition can delay motor response and allow for separation of the affective emotional charge from value associated with events, in order to allow a more appropriate response (Barkley,

1997; Bronowski, 1977). For instance, we get involved in a heated argument with an individal after someone we hardly know informs us that the person made disparging remarks about us. We inadvertently accept misinformation from an unreliable source, jump to conclusions, and loose our cool. But we have the option of counting to 10, calming down, and thinking practically. We have the option of taking the time to elaborate the problem space, search for additional evidence, and investigate the reliability of the accumulated information and the source. Thus, we could avoid the altercation altogether. Of course, we could also say, "What they think about me is none of my business," and then walk away. But without PFC dampening of the emotional solution, we might be hard pressed to avoid the altercation.

The frontal lobes play a deliberate executive role in the explicit regulation of flexibility, insight, planning, and the modulation of instabilities encountered between internal and external states. Language and the frontal lobes can help us dampen the impulsive tendency to needlessly upset ourselves and endanger ourselves and others; and they increase the probability of realizing a practical resolution.

The upper-level DLPFC allows memory editing via language and working memory, with which we can fine-tune learning and domain knowledge. Working memory allows us to plan and organize a higher-order behavior instead of automatically responding to the immediacy of the environment (Koziol & Budding, 2010, p. 49). DLPFC regulation influences the registration, storage, and recall of recent experience, which includes

- Sorting
- Associative search
- Recombination
- Selection
- Online reintegration

These processes are generally attributed to executive functioning and working memory. Furthermore, frontal lobe processes play a role (Mesulam, 2000, pp. 48, 63) in

- Disassociating appearance from significance
- Grasping changes of context
- Shifting mental sets
- Assuming multiple perspectives
- Comparing potential outcomes of contemplated actions

Lesion studies provide useful information about the effects of structural and functional disturbances on brain processing. Prefrontal lesions disrupt the flexible modulation and transition between state functions by shifting the emphasis toward stimulus-bound utilization behaviors (Mesulam, 2000, pp. 45, 47). The PFC likely plays a crucial role in abstract thinking and resisting default to limbic imperatives and literal (stimulus-bound) association in favor of a less obvious inference implied by the context (Mesulam, 2002, pp. 19–21). Frontal lobe damage undermines the effectiveness of encoding and retrieval and contributes to impoverishment of the associations we use to reconstruct context and temporal order. Damage also decreases the speed with which internal data stores are searched, increases the tendency to confabulate, and impairs strategic thinking and risk management (Mesulam, 2000, pp. 48, 63).

Furthermore, frontal lobe damage has been shown to disrupt cognitive flexibility, especially when we would benefit from modifying our behavior to fit a novel context. Disruption can lead to stimulus-bound and concrete behaviors. Then, during reasoning tasks we have a tendency to reach closure prematurely, jump to conclusions on the basis of incomplete information, and perseverate, which dampens our willingness and ability to consider or explore alternative solutions (Mesulam, 2000, p. 45). The frontal lobes interact with subcortical structures that can also contribute directly or indirectly to these dysfunctions (Koziol & Budding, 2010, pp. 13–16, 83).

The PFC appears to sit at the apex of behavioral hierarchies (Mesulam, 2002, p. 25). Generally frontal lobe executive processes are, at least in part, sensitive to linguistic biases for reliably inhibiting irrelevant input, computation, and output and for optimizing effective adaptation in complex environments. In effect, when we add language we extend the functional top, or upper, limit of top-down regulation by providing a higher level of abstraction and extended linguistic parameters suited for analytical thought in a complex world. Effective language enhances the reliable inhibition and modulation of bottom-up, limbic-related affective processing. The trade-off is between more-expensive analytical cognition and less-expensive limbic cognition and affect-related tone of the experience, the benefits of which can depend on the linguistic effectiveness of the available information and knowledge. Kahneman (2011, pp. 20–30) describes two systems (previously proposed by Stanovich & West, 2000): system 2 (upper), which is slower, deliberate, orderly, and effortful; and system 1 (lower), which is quicker, automatic, impulsive, and effortless or nearly so.

> System 1 is gullible and biased to believe, system 2 is in charge of doubting and unbelieving, but system 2 is sometimes busy, and often lazy. (Kahneman, 2011, p. 81)

We can think of the lower limbic system as analogous to System 1. The automatic limbic-OFC system provides a lower-level perceptual default for rapidly processing information and also for subjectively filling in, or inferring, missing information from historical state information. For example, we fill in parts of visual scenes that our brain cannot process, which affects our perception (Macknik & Martinez-Conde, 2010, pp. 13–14). The lower-level bottom-up default system contrasts with the deliberate but slower and effortful upper-level analytical processing of the language-associated left DLPFC. The DLPFC demonstrates the additional ability to explicitly search for and more knowledgeably fill in missing information within the constraints imposed by the effectiveness of the available linguistic coding and representations. Albeit with more time, effort, and additional energy cost than implicit automatic cognition, the explicit DLPFC contributes adaptive benefits by providing alternative routes to action (Rolls, 2008, p. 525).

> Indeed, it would be adaptive for the explicit system to be regularly assessing performance by the more automatic system, and to switch itself in to control behavior quite frequently, as otherwise the adaptive value of having the explicit system would be less than optimal. (Rolls, 2008, p. 527)

The brain's fast and frugal automatic limbic system computes output predictions on the basis of the perceived stimulus strength of the input in comparison with similarities and differences between previously experienced outcomes and reinforcement. The slower, effortful, and more expensive deliberate system provides a task-dependent option for trade-offs between speed and accuracy. Such trade-offs allow an organism to continuously modulate and smoothly switch responses to events occurring or recurring in different contexts. Continuous recurrent modulation allows for attentional switching at initial observation according to the relevant salience inferred from the transformation of visual, auditory, and other sensory inputs.

Value Perspective

How do we develop our worldview, and what influences that development? The simple answer points to the interplay among our inherited brain hardware, our inherited language and beliefs, and the reinforcement learning we accumulate from experience. All of these revolve around symmetry breaking and value, which biases cognition, perception, behavior, and learning. In other words, value symmetry breaking is the common denominator. Value asymmetry influences our worldview and how we perceive and relate to others around us. We can consider value from the inside out and from the outside in, and gain insight into how our values form and profoundly influence our habitual cognitive behavior.

Imagine standing in the middle of a busy city block and looking outward and upward from a self-centered individual perspective. We call this a localized view. Imagine floating above the city and viewing the same city block and watching all the people, including ourselves, hustling and bustling around. We often call this a bird's-eye view. These different views represent our perspective and relate to how we see and value ourselves and others, as well as the objects and actions, in the world around us. We can continue to zoom farther out. Think about being an astronaut rocketing toward outer space and looking back toward an immense perspective and quite spectacular view of the magnificent blue globe we call Earth—as captured in previously transmitted images from space. At this distance there is no sign of city blocks or humans. From farther and farther away, Earth becomes a pale blue dot and eventually only a mere speck of dust in the cosmos among billions and billions of stars (Sagan, 1994, p. 7).

Changes in how we view things enable different perspectives about essentially the same issue, object, or being. Opening our view by shifting between a narrow and a broad perspective lets us see things differently. Sometimes we gain insight into things to which we may have become habituated and that we previously overlooked, but more often, the thought of looking simply doesn't occur to us. And, as discussed earlier, habits become implicit, taken for granted, and invisible. Even though out of sight, the invisible force of habit biases and impinges our perspective. With a closed perspective we can miss a lot, possibly spectacular and relevant things. We tend to shift focus similar to shifting our position by zooming in and zooming out according to what we value and deem relevant. But our limbic system imposes an affectively charged bias on cognition and the position we take.

We typically rely on our own locally cultivated affective values, but we also have the option of looking at value analytically from a systematic perspective. Values take on different meaning when we view people interacting, including ourselves, from a distanced position above. A bird's-eye view separates us from the local emotional interactions going on down below and can give us a less polarized, more analytical perspective. From this distance we might wonder why people interact with each other the way they do and with so much emotional intensity, and how their reasoning processes might work. They each seem to act as if the world revolves around only them.

Copernicus provided evidence that the Earth circles around the sun, freeing humans from the primitive illusion that this planet sits at the center of the universe (Swanson, 2003, p. 41; Skinner, 1953, p. 7; Spitzer, 1999, pp. 9–10). But since we typically navigate from our own internal coordinates or frame of reference, most of us still can operate as if the universe revolves around us (Laplace, 2009, p. 164). Our internal referencing easily confuses us as to the importance of our own beliefs and values. We tend to place ourselves in an unduly positive light, overestimate our skills, and fail to recognize our inadequacies (Dunning et al, 2002, p. 324; Kruger & Dunning, 1999). For example, a survey of one million high school seniors found that 70% thought they were above average in leadership ability, and only 2% below average; a survey of university professors found that 94% thought they were better at their job than their average colleague (Gilovich, 1991, p. 77). Perhaps this self-centered bias explains why we generally seem to learn more from our own experience than from the experiences of others, and why we tend to operate to our own advantage and favor our own internal values, directly or indirectly, even when we cannot reasonably explain why. And similar to other animals, we tend to choose favorable conditions and reject unfavorable ones (Jennings, 1906, p. 340).

Most organisms routinely navigate from internal coordinates that provide a default reference as an internal perspective for simple, intentional, and attentional discrimination related to position, place, space, external objects, and memory integration (Kandel, 2006, pp. 307–316). Changes in positional coordinates convey information that can be understood in terms of the transformation of internal neural representations (Eliasmith & Anderson, 2003, p. 5).

In effect, relative coordinates define positional representations correlating behavior with external task geometry. We call *task geometry* the external task representation of goal-directed output in relation to internal coordinates and stimulus representation. "The geometry of the brain is often organized to reflect and exploit the geometry of the physical world" (Stanley et al, 2009). By taking advantage of intrinsic–extrinsic symmetry and asymmetry, this geometrical input–output relationship can be exploited computationally as an agent's regulatory controller, as demonstrated in artificial intelligence research (Risi & Stanley, 2010; Stanley et al, 2009; Gauci & Stanley, 2008; D'Ambrosio & Stanley, 2007). Simply put, there is a correspondence between what goes on inside and outside of the brain:

> . . . [T]he system of coordinate axes to which we naturally refer all exterior objects is . . . invariably bound to our body, and carried around with us. . . . To localize an object simply means to represent to oneself the movements that would be necessary to reach it. . . . Thence comes a new distinction among external changes: those . . . we call changes of position; and the others, changes of state. (Poincaré, 2007, pp. 47–48)

As in the physics of relativity, recognition of an object and its existence depends on the dimensionality of space, internal and external (Braitenberg, 1984, pp. 42). The features of routine objects are easily compared with memories for correlations that are based on familiarity and affective value and can be promptly processed by the perirhinal–parahippocampal cortex and hippocampal formation in the automatic lower-level limbic brain (Price, 2006, p. 49).

The hippocampal complex makes significant contributions to the larger theme of information processing expressed in this book; it plays an instrumental role in cognition and value that includes

- Linguistic and nonlinguistic information processing
- Short- and long-term information storage and retrieval
- Information association
- Information affect association
- Information integration and segregation
- Internal and external information search
- A frame of reference for novel search and goal-related navigation
- Assumptions of linear cause and effect
- Brain oscillations

- Event-related symmetry breaking that separates the past and future and establishes spatial, temporal, and value context
- The mapping of internal-to-external position and change in position that contributes to valuating objects and states and modulating approach-and-avoidance behavior
- Support of the statistical distribution of information by providing an event-related platform to deal with the flow of constantly changing information

The hippocampus occupies the crossroads of associations, correlating language with cognitive processing, value and internal system states, and the state of the external world. The crossroads provide an interchange between internal affective state, place, position, action, and processes of change that correspond to external processes of change and action among state values, places, and positions. We can think of the hippocampus as an integrator of relationships, including linguistic representations stored in memory from our experiences with the external world. Memories that intersect there represent relationships stored according to their content.

Relational encoding provides flexible access to content-addressable information in situations quite different from those of the original learning (Morris, 2007, p. 662; Burgess, 2007, p. 721). Content-addressable information is accessible as a whole or by parts of the content (Buzsáki, 2006, p. 289; Rolls, 2008, p. 560), as David Marr originally theorized some 40 years ago (Marr, 1971). Accessing only a part of the information out of context can tacitly transpose its dynamic context to a static description of the part and limit perception of its systematic relational value. Extended language provides explicit temporal context to resolve the problem of implicitly confusing the static past and dynamic future. It seems that limbic reactions "couldn't care less" about such lofty concepts as one-way time, but remember, limbic skills were hewed for fast and frugal responses when we do not have the luxury of analysis.

> When implicit memory is reactivated in the future ("retrieved"), it does *not* have a sense of self, time, or of something being recalled. It merely creates the mental experience of behavior, emotion, or perception. (Siegel, 1999, p. 65 [emphasis in original])

The hippocampal formation, via the perirhinal–entorhinal cortex, provides a strategic convergence region for information, including salience features, from nearly all high-order cortical areas, brainstem nuclei, and sensory modalities, most of which have reciprocal projections from the hippocampus as well (Churchland & Sejnowski, 1994, pp. 248–249, 282–287; Moisl, 2001, p. 454). Serial and parallel pathways through the hippocampal formation allow information processing by divergent and convergent integration and routing of multiple information streams (Amaral & Lavenex, 2007, pp. 107–109, & Figs. 3-56, 3-57).

A large part of the modularly organized neocortex is tuned to identify statistical regularities in the input streams conveyed by sensors. Virtually all neocortical regions project to the perirhinal and entorhinal cortices, and neocortical information is funneled to the hippocampus by these structures, making the hippocampus the ultimate association structure, receiving the highest-order neuronal information via multiple recurrent loops (Buzsáki, 2006, pp. 278, 281–282, p. 52, & see Fig. 2.6). The components of the hippocampal complex play functional roles in the storage, recall, and integration of information (Mesulam, 2000, pp. 59, 65, & see Fig. 1-13). A functional representation that simplifies memory systems allows us to take an object-based view of the brain. The output of the visual system (Rolls, 2008, pp. 19–22, 269)

- Represents invariant objects independent of position on the retina [enables association and integration of the object by the visual "what" pathway—the temporal lobe and hippocampus]
- Associates the object with reward [OFC and amygdaloid-accumbens complex]
- Associates the object with its position in the environment [hippocampus via the parietal "where" pathway]
- Recognizes the object as familiar [perirhinal cortex]
- Associates the object with a motor response in habit memory [BG and association networks]

The BG automatically modulate familiar object relations and intention with reward-driven instrumental learning of internal and external stimuli associations, which leads to selecting actions deterministically as if past reinforcement predicts the future (Koziol & Budding, 2010, pp. 58–60, 80–81). The BG play a role in the automatic integration of procedural habit memory with value and motor output by interfacing with numerous recurrent cortical loops that exchange information with the

- Supplementary motor area
- Frontal eye fields
- DLPFC
- OFC
- Inferior temporal lobe
- Parietal cortex
- Anterior cingulate cortex (ACC)

The ACC integrates paralimbic value processing with the nucleus accumbens reward–approach and amygdala punishment–avoid circuits (Koziol & Budding, 2010).

The brain's structural and operational organization allows the functional integration of language, which provides another level of description for symmetry breaking and evaluating object relations. Language influences regulation and evaluation of relevance, attention, and action choice. Top-down language processes can play an influential role in modulating affective representations in the brain, such as reward–approach and punishment–avoid. The top-down processes correlate cognition and attention that can have beneficial effects in directing sensory and emotional processing toward stimuli and events that the cognitive system has deemed relevant (Rolls, 2008, pp. 167–171, 289). Relevance, sometimes referred to as salience, tells us what to attend to and what to ignore.

The association between relevance and ignorance elevates the significance of ignorance. But do we possibly misunderstand the term *ignorance*? The word has a plethora of colloquial negative connotations that can discredit someone (or something) by labeling them in such a way that implies they are "dumb" or "stupid." For example, "They are all ignorant. They're just a bunch of ignoramuses." There seem to be few, if any, benefits for feigning ignorance, denying ignorance, or deliberately remaining ignorant about how nature works. This view holds especially when we come to the far-fetched conclusion that everyone is ignorant except us, which requires denial of our own ignorance. It is hard to believe that ignorant is an either–or phenomena.

But it seems that ignorance is important and has a positive side that we tend to overlook. When we acknowledge our ignorance, it can inspire learning and the search for knowledge, and, better yet, ignoring saves energy. Ignorance comes with our status as human language beneficiaries. Arguably, ignorance would not exist, at least not explicitly, without some type of language capable of expressing abstractions relating us to the external world. Yet, thanks to the limbic nature of animal behavior, even we humans will carry on the theme of focusing on what we believe or perceive as relevant, and by parsimonious default ignoring or dismissing the rest to the irrelevant.

We have a limbic system that plays an automatic processing role in implicitly directing and modulating our attention and our ignorance. Extended language provides a tool for deliberately overriding the limbic system and enabling higher-ordered cognition with explicit information and knowledge. Typically, we can analyze relevance on the basis of effectiveness, make effectively informed decisions, and deliberately modulate what we prefer to attend to or ignore. Of course, we live in a complex world with a lot going on, so it's unlikely that we know it all. Thus none of us knows everything that there is to know, and by default, like it or not, we all have some amount of ignorance. At any given moment there are so many things going on that it is difficult to choose the best value when it comes to paying attention. The term *paying attention* takes on special meaning as the cost of doing business and practically guarantees no free lunch in a high-dimensional dynamic world. It is not surprising that the frugality of evolutionary processes flushed out the fact that deliberate attention is an effortful activity and selectively cuts corners by leveraging affective value as a simpler automatic default that conserves energy by promoting effortless ignorance.

> The often-used phrase "pay attention" is apt: You dispose of a limited budget of attention that you can allocate to activities, and if you try to go beyond your budget, you will fail. It is the mark of effortful activities that they interfere with each other, which is why it is difficult or impossible to conduct several at once. (Kahneman, 2011, p. 23)

Discriminating between what to attend to and what to ignore, with thoughtful ignorance, has practical benefits for encouraging learning and saving valuable energy. We can theorize that learning is born of ignorance. Accepting our ignorance can encourage the option of searching for more information and knowledge, increasing learning, and putting forth a greater effort to identify, analyze, and separate the relevant and the irrelevant. We can attribute relevance and ignorance to our automatic limbic system as affective ignorance and to our deliberate analytical system as effective ignorance. With effective language, we expect the deliberate analytical system to facilitate the acknowledgment of ignorance and to exploit the importance of learning.

We can see another dilemma brewing in the trade-off of fast and frugal versus slow and accurate, because the affective value of ignorance is generally inexpensive compared with the effective value that is the cost we pay for deliberate attention, learning, and the accumulation of systematic knowledge. We are left to vacillate between affective and effective value and efficiency, caught up in the battle of fast and frugal versus slow and expensive. It seems that the answer might be: "It depends on the context and the goal." So in the meantime, we can make do with the vacillations and the oscillations we incur when "it depends," and inhibition can increase oscillations according to goal-related feedback that biases the regulation of brain rhythms when we "can't make up our mind."

The Nefarious Amygdala

When we take on a positive bias, we tend to disproportionately ignore negative feedback. If we have a negative bias, we tend to disproportionately ignore the positive. Brains generally have a built-in negative bias that leans toward survival. We can automatically avoid or eliminate what we perceive as unpleasant and potentially harmful, and approach and embrace what we perceive as pleasant and helpful.

The brain processes physical pleasure and pain on both an emotional level and a physiological level that can make physical feelings difficult to differentiate from psychological descriptions of internal mood states. For example, *depression* describes extended negative mood states. For practical purposes, positive and negative mood states influence perception and our behavior associated with approach and avoidance. Mood changes often accompany changes in bonding and social affiliations, such as when positive moods accompany gains associated with acceptance, approval, or promotion, or when negative moods accompany losses such as companionship deprivation, rejection, or criticism. We see that positive and negative biases can have a profound influence on our behavior.

Since we prefer to feel happy, it may seem strange to us that brains would evolve with a negative bias that leads to negative mood states. The evolutionary benefits apparently offer functional, state features for shifting behavioral patterns. On an analytical level it seems that seeing both the positives and negatives as accurately as possible helps us to better adapt our behavior in consonance with the world around us. Effective language therefore likely offers practical adaptive benefits for tuning our bias and sensitivity, increasing accuracy, and decreasing stress and upset. It seems that the emotional price we pay for feeling good when obtaining positives going in is paid back by feeling bad when losing them as we incur negative emotional repercussions on the way out. There's no free lunch in nature; we can pay dearly for the collection of pleasant and unpleasant memories we acquire. We can chalk up the expense to the local cost of doing business with trial-and-error learning of what best to approach and avoid.

Overall, the local price we pay for attraction to positives, repulsion from negatives, and a negative bias hardly feels like a bargain, unless we consider the benefits of survival. Perhaps even though it is not free, the local cost seems to offer sufficient benefits on average for decreasing some risks and rewarding survival and reproduction. Yet we often struggle with the challenge of dealing with our emotionally laden predispositions. Local positive and negative rewards and penalties have a profound affective influence on our behavior, as if gravity was pulling our head down and keeping us concretely focused on the local surface or as if some other strange force was nefariously leading us around with our nose to the ground. Is there evidence for the existence of such a strange and mystical force? And is there a more scientific explanation?

We might ask what grounds us to the local landscape and also enhances our survival by automatically affording value symmetry breaking, affective meaning, and relevance to our experiences. Can we find the brain's mysterious sentinel responsible for contributing the affective component to safely modulating and integrating our brain's automatic processing behavior? The answer may come as a bit of a surprise. Overwhelming evidence for culpability points to a limbic structure near our olfactory center in the rhinal (nose-related) cortex. The particular bilateral structure is small and almond-shaped and is found in most mammals. The culprit is called the amygdala, a limbic structure that implicitly affords and associates functional affective meaning to mammalian cognition. The amygdala automatically facilitates risk avoidance and safe reward-seeking behavior and tends to passionately lead us around by our nose, so to speak, as if carrying on a natural local tradition. Is the amygdala implicated in understanding or misunderstanding cognitive control and the mysterious invisible processes hiding behind the curtain—in the how and why of behavior? The amygdalae lie nestled deep in our temporal lobes, snuggled right in the thick of our cognition; they appear to be involved in modulating [almost] all of cognition (Buchanan et al, 2009, p. 312).

The amygdaloid complex is thought to act as the danger detector of the brain. In evolutionary terms, it seems obvious that danger detection can make a large contribution to an organism's long-term survival by simply alerting our attention to the relevant as something not to be ignored. The amygdala is highly conserved in mammals and plays an instrumental role in keeping us and other mammals safe from external threats. It apparently plays a valuable functional role in the brain and in evolution. We can speculate that the amygdala may be close to the affective root of what makes us emotional human beings and contributes to our civilized and uncivilized behavior. If so, understanding the amygdala's modus operandi may give us clues that help us discover the relations among evolution, the physical world, the human brain, emotions, behavior, and even affective and effective language. We would expect brain structures that are important for integrating information to receive many inputs. The amygdaloid–hippocampal complex seems to be at the center of it all with dense neuronal connections to and between both and most of the rest of the brain (Freese & Amaral, 2009, pp. 3–42).

> Not only is [the amygdala] associated with the ability to detect and retain the motivational value of environmental events, but it is crucial for orchestrating a wide range of physiological reactions that allow the organism to adjust to these events. Thus the amygdala not only receives and integrates various inputs from external and internal sources, but also projects to many output systems that can then modulate autonomic, motor, memory, cognitive, and perceptual processes. (Vuilleumier, 2009, p. 220)

The correlation of language, brain, internal affective state, and external rewards and punishments is an integral theme in cognition and central to this book. As an extended part of the hippocampal complex, the amygdaloid complex lies near the crossroads of information flow and is well positioned for imposing an affective influence on behavior while allowing monitoring by the PFC. The left DLPFC has access to analytical modulation via language and in general can deliberately overrule limbic control and regulate behavior more effectively. There is a trade-off between deliberate left DLPFC and automatic information processing by the mPFC–OFC that is reminiscent of the fast-and-frugal versus slower-with-accuracy trade-off. The left DLPFC can deliberately modulate systematic error tuning and the effectiveness of behavioral output, dependent on the effectiveness of the language resource. The mPFC–OFC implicitly modulates local error tuning according to current and past affective inputs. The deliberate effective and automatic affective systems modulate, switch, and integrate roles depending on task demands for speed and fidelity. Cooperation and competition between the systems enables balancing risk versus reward and the costs versus the benefits of exploration and exploitation, taking into account effort, energy resources, and time constraints. The ability to modulate between exploration and exploitation enhances efficient switching between foraging and consumptive behavior depending on resource availability.

The amygdala is well positioned to exert a strong polarizing influence on stimuli relevance related to attraction and repulsion and the brain's struggle between intellectual finesse and emotional brawn. Simply put, we are looking at the emotional engine behind our automatic thinking, feeling, and behaving that permeates our existence. Evidence indicates that the amygdala plays a role in the enhanced processing of emotional stimuli within distant cortical sites (Vuilleumier & Brosch, 2009, p. 929). A brief summary is provided in Table 3, which shows some of the structural and functional relationships and neocortical and subcortical limbic connections with the amygdala (Freese & Amaral, 2009, p. 16).

We see that the amygdala is indeed well connected. If we suppose that being well connected has something to do with structural and functional relevance, then it looks like the amygdala has evolved its way to a highly valuable evolutionary stature. We have a left and right amygdala, and they demonstrate widespread involvement in

Table 3. The Functions of Various Brain Structures

Structure	*Function*
Hippocampal formation	Learning and memory
mPFC	Inhibitory modulation of appetitive behavior
OFC	Modulation of incentive value for expected outcomes
DLPFC	Sparse and indirect modulatory connections
Insular cortex	Recognition of the affective states of others
ACC	Somatosensory integration and error signal
Ventral tegmental area	Reinforcement and reward signals
Bed nucleus of the stria terminalis	Generalized anticipatory affective tone
Brainstem	Autonomic and visceral functions and defensive reactions
Thalamus	Sensory integration
BG/striatum	Automatic integration of motor response, reward, and goal-directed behavior
Hypothalamus	Endocrine, autonomic and visceral, and defense reactions
Olfactory system	Odor valence
Basal forebrain	Aggression, motivation
Inferior posterior parietal lobe	Position or location, "where" input
Temporal cortex	Memory association areas and visual cortex, "what" input

ACC, anterior cingulate cortex; BG, basal ganglia; DLPFC, dorsolateral prefrontal cortex; mPFC, medial prefrontal cortex; OFC, orbitofrontal cortex.

emotion and affective linguistic processing. Emotion is central to human experience. It colors how we perceive the world and influences our decisions, actions, and memories (Morris & Dolan, 2004, p. 365).

Elisabeth Murray and colleagues suggest that the amygdala not only contributes to negative reinforcement and negative affect but also appears to contribute to some aspects of positive reinforcement, hedonic liking, and incentive values related to wanting and seeking. Their work extends the role of the amygdala to include the application of emotional valence to cognitive constructs, including words, rules, concepts, and conclusions. The amygdala also participates in associations related to survival, including food-seeking, eating, and reproductive, parental, and defensive behaviors, and it is involved in the orientation of surprise responses to unexpected stimuli (Murray et al, 2009, pp. 83–85), including predator odors, and neuromodulator release (LeDoux & Schiller, 2009, p. 54).

We can likely add to this list a role in the processing of ambiguity, uncertainty, novelty, social relevance, dominance–subordination, and negative stereotypes (Buchanan et al, 2009, pp. 290–293, 304), modulating vigilance related to learning (Whalen et al, 2009, p. 279), and providing a centrally connected multidimensional scaling representation (Buchanan et al, 2009, p. 312). Studies and data reinforce the view that amygdala processing tags personally relevant emotional information, regardless of valence (Canli, 2009, p. 254). It also likely influences other affective states, including maternal, sexual, rejection, criticism, jealousy, blame, and outrage.

Murray and colleagues describe the amygdala as linking information processed by the higher-order association cortex to "instinctive" behavior and value assignment; it links objects to value and updates or otherwise alters representation of value, and contributes to modulating performance rules such as "approach stimuli of positive valence." They propose that the amygdala provides a key link between emotions and cognitive constructs, words, images, ideas, abstract thoughts, preferences, and abstract goals (Murray et al, 2009, pp. 97–100). Rather than encoding and storing emotional memory itself, the amygdala apparently facilitates ongoing memory-encoding processes in other memory systems and performs an evolutionary adaptive role by enhancing relevant memory encoding for emotionally salient events stored for future occasions (Hamann, 2009, p. 178).

In effect, the amygdalae occupy a strategic position, acting functionally like computational context nodes for biasing thoughts, sentimental feelings, and behavior correlated to affective value. The influence extends to time and place, orientation and direction, and familiarity. The amygdala contributes to value symmetry breaking at the time of an event by associating asymmetrical positive and negative valence with memories of our experiences. Symmetry breaking, memories, spatial context, events, and event-related rewards and penalties are all orthogonal to time. We experience events as distinct entities with time-dependent differences from past to future.

Relegating our experiences to the past creates memories that are essentially static, which are amenable to updating when and if new information becomes available. If projected into simulations of the future, our memories remain static and not directly accountable or responsible for future events. But the memories provide information for indirectly calculating predictions as possibilities and probabilities.

> If the amygdala is important for linking object representations with value... then why would it not also be important for linking other types of representations with value? In our species, with our profoundly derived capacities for abstract thought and language, perhaps the amygdala provides the key link between ideas and emotion.... [T]he ability of the amygdala to link the products of cognition to value may serve as the basis for images of ourselves as positive (or negative) entities moving through time. (Murray et al, 2009, p. 99)

The amygdala is possibly the most densely interconnected region of the primate forebrain and is widely accepted as receiving inputs from many cortical and subcortical structures. It enables an individual to ascribe emotional meaning to events, affectively coordinate adaptive behavior, and modulate cognitive processing (LaBar & Warren, 2009, p.157), and it is involved in cognitive reappraisal strategies (Ochsner et al, 2005; Phelps, 2006). The amygdala also influences emotional arousal, attention, perception, and context related to encoding, storage, and retrieval of memories (Hamann, 2009, pp. 179–183, 193–195).

> Thus the amygdala plays a key role in connecting external sensory information to the instinctive processes that underlie the most fundamental aspects of vertebrate behavior. (Murray et al, 2009, p. 85)

Studies have shown that amygdala recruitment plays a role for processing cues that signal safety in addition to those that signal danger. It helps individuals assign the emotional significance of stimuli as their social and emotional contexts change during development (Tottenham et al, 2009, pp. 108–109). The brain's upper *where* pathway locates an object's position, and the lower *what* pathway identifies the object's relevant contextual features according to value. Functional imaging data of the amygdala suggest that lower-level object identification *what* pathway associations are formed between a stimulus and its emotional significance, which changes as the emotional or social environment changes (Tottenham et al, 2009, p. 113). In effect the amygdala acts like a relevance detector for tuning affective affordance to our cognitive and social processes that gives priority to relevant events (Phelps, 2009, p. 216).

The amygdala has been described by M.-Marsel Mesulam as the gateway into the neurology of value. The amygdaloid-accumbens-septal complex provides hippocampal valence inputs (Mesulam, 2000, pp. 2, 56, 58–63) for learning negative and positive stimuli associations, and allows for simple situational object integration by association with reward and place (Rolls, 2008, pp. 188, 269). Also in human primates, the amygdala plays a prominent role in salient facial feature recognition relevant for social behavior, including eye contact, affiliative and aggressive behavior, and social emotions and their communication (Mesulam, 2000, p. 58). The outputs of the hippocampal formation and the amygdala are basically the same as their inputs: the neocortex (Buzsáki, 2006, p. 282). The amygdaloid complex is connected with the entire hippocampal formation (Freese & Amaral, 2009, p. 22) and displays oscillations associated with paralimbic and hippocampal rhythms (Buzsáki, 2006, pp. 308–313). Effective modulation of these oscillations can potentially produce harmonious interactions and constructive resonances with the environment.

The amygdala not only is well connected in the brain but also seems to play an instrumental role in keeping us affectively connected to the local landscape. With safety in mind, it automatically activates, inhibits, and biases navigation toward favorable conditions associated with what can be described as attraction and approach to positive valence and away from unfavorable conditions associated with repulsion and avoidance of negative valence. We can draw an evolutionary analogy to ignorance-regulated behavior based on associations and learning that reinforce an affective bias as what to ignore and what not to ignore (Johnston, 2003).

> Positive and negative affect indeed appear to be generated by the nervous system as a neural code "to those aspects of the environment that were a consistent benefit or threat to gene survival in ancestral environments" (Johnston, 2003, p. 173).... Appetitive-approach tendencies towards positive stimuli and aversive-avoidance tendencies away from negative stimuli have likely developed in evolution to maximize fitness benefits and to minimize fitness costs, respectively. (Heerebout, 2011, pp. 77–78)

More than one hundred years ago, Jennings (1906) described similar action tendencies of attraction and repulsion and movement toward favorable conditions in experiments with paramecium and other simple organisms. He described approach-and-avoidance phenomena as a general principle of action regulation produced throughout the spectrum of organisms:

> Behavior can thus be largely classified into two great classes: "positive and negative" reactions; movements of "attraction and repulsion," of approach and retreat. (p. 265)

We can add polarity—*positives* and *negatives*, *attraction* and *repulsion*, and *approach* and *avoidance*—along with *relative change of position* and *state* to the list of general systematic processing and relational principles:

- Time, symmetry breaking, asymmetrical-time referencing, and probabilistic information processing
- Polarity: positives and negatives, attraction and repulsion, approach and avoidance
- Relative change, change of position, change of state

We see a distinct functional correlation between symmetry breaking, the limbic amygdala, and automatic behavioral regulation by activation and inhibition that supports survival and reproduction. Next we will explore the hippocampus, which is at the crossroads of relationships with time, space, and value.

CHAPTER 12

Hippocampal Crossroads: Time, Space, and Value

THE HIPPOCAMPUS CAN BE viewed as a single, supersized cortical module with largely random connections that create a vast searchable multidimensional space, making it the ultimate search engine for the retrieval of archived information, somewhat like the librarian for the huge neocortical library (Buzsáki, 2006, pp. 288, 331). Influenced by language, the librarian manages the file system in charge of filing, retrieving, and monitoring the mapping of information, working with the amygdala to facilitate tagging event-related information with affective value according to its contextual relevance, and both segregating and integrating input and output across the brain's memory storage system with internal state information. In effect, the hippocampus facilitates memory formation by separating the past and future by event-related symmetry breaking. The information available for retrieval from memory storage typically results from the accumulation of local learning over time.

Context, Maps, and Position

The hippocampus–entorhinal complex influences mental content via idiosyncratic experience-derived affective associations, which endow percepts and events with contextual anchors and subjective personal significance. The complex also plays a critical role in the long-term storage and explicit recall of such arbitrary associations (Mesulam, 2000, pp. 58–65). Individual organisms have different histories of positive and negative experiences that create variant affective associations. The ability of emotional stimuli to enhance the storage and retrieval of explicit memories is likely dependent on the consequences of amygdala activity (LeDoux & Schiller, 2009, p. 53). Assuming that humans produce mental representations of objects in the world, we would expect those representations to influence behavior (Kauffman, 1993, p. 232; Cheney & Seyfarth, 2008, p. 236).

Suzanna Becker (2007, pp. 1, 14) has proposed a global optimization principle for learning in the hippocampal region, based on the goal of accurate input reconstruction, combined with neuroanatomical constraints, leading to simple, biologically plausible learning rules for regions within the hippocampal circuits, including ventral stream *what* and dorsal stream *where* inputs. Both of the streams carry information from visual inputs; the *what* pathway proceeds from primary and secondary visual areas to the memory-related temporal lobe, while the *where* pathway leads to the visual-spatial–related parietal lobe (Spitzer, 1999, pp. 112–113).

> The map of our environment is composed of a set of place representations connected together according to rules which represent the distance and directions which connect the places in the map of an environment derived from the animal's movements in that environment. (O'Keefe & Nadel, 1979, pp. 488–489)

The hippocampus could be considered a *cognitive mapper* (Morris, 2007, pp. 617–619; O'Keefe & Nadel, 1978) in that it provides an event-level auto-association or autocorrelation mechanism for context by connecting situational space and time, regulating the order of perceptual categorizations and mapping mental representations to emotional appraisal centers (Siegel, 1999, p. 330). By its computational definition, an *auto-associator* is a self-correcting network that can re-create the previously stored pattern that most closely resembles the current input pattern, even if the pattern is only a fragment of the stored version (Buzsáki, 2006, p. 289). Auto-association involves generalized rather than absolute pattern matching.

In effect, the hippocampus plays a key role in contextual learning, mapping, segregating, and integrating internal state relationships among neocortical cognitive maps in memory of objects, locations, and affect associations to external goals, objects, and places. The operational internal and internal-to-external mappings enable explicit declarative linguistic associations to narratives and spatial navigation that correlate internal historical mapping to past experiences, external mapping to present interactions, and simulated mapping to future states. The putative mapping system also encodes where landmarks are located in relation to each other in some kind of geometric framework (Morris, 2007, p. 619). The mappings take on a statistical character because information continuously changes, with many moving objects and various positional locations encountered across time.

The hippocampus also acts somewhat like an internal compass that integrates head direction or heading involved in spatial navigation, as demonstrated by James Ranck in experimental results (Ranck, 1985; Taube, 2005; Buzsáki, 2006, pp. 305–306; Morris, 2007, p. 618). Spatial and directional templates already incorporated into the hippocampo-entorhinal complex contribute to the representation and integration of external spatiotemporal information with context. The hippocampus correlates associations and inference to intention (action and goal-related value) and integrates contextually coded information at the event level with

- What (animate or inanimate)
- Who (friend or foe)
- Where (location/position)
- When (now, before, after)
- Why (process)
- How (coincident cause–effect)

> Contextual encoding is clearly an effective, often automatic associative mechanism for binding events in memory. . . . As with the formation of stimulus configuration or relational associations, contexts can influence the attention paid to stimulus, "set the occasion" for a stimulus to predict one outcome or another, disambiguate stimuli and their predictive significance in other ways, or provide an incidental or deliberate way of organizing attended information. (Morris, 2007, p. 669)

Context, as a neural construct of space, could be considered a major driving force in the evolution of the distinctive cognitive systems that mediate it (Morris, 2007, p. 622; Jeffrey et al, 2004). The contextual representation takes place as a result of automatic integration in the hippocampal complex that coincidentally breaks topographical spatial-mapping symmetry and time symmetry (Buzsáki, 2006, p. 302). The information naturally generalizes and synchronizes spatiotemporal coherency between the organism and the domain and in turn provides information crucial for processing and modulating adaptive interactions that correspond to external changes.

In effect, the hippocampus provides a general structure for the breaking of spatiotemporal symmetry via situational integration of event information correlated in time and space. The relationship with the amygdala enables the breaking of value symmetry as negative and positive valence. Value-added integration provides an affective attentional mechanism for automatic salience detection and safety assessments of stimuli in relation to distance and external-to-internal positional and state changes. Change conveys temporal information about the environment for predicting actions according to relevant positive and negative values. In turn, external changes correlated to internal position and change of state values provide real-time information for direct arousal and indirect processing of predictions, and the modulation and coordination of adaptive approach-and-avoidance reactions. Hippocampal integration allows the modulation of an organism's direction, position, and affective state in contextual relation to a changing environment.

With all of these time-related changes and things to remember, we might think we would run out of brain cells long before adolescence. And up until 1998 it was thought that we had only whatever number of brain cells we were born with to last throughout our lifetime. But Fred Gage's revolutionary research demonstrated the generation of new brain cells in the adult brain. The neurogenesis takes place in the dentate gyrus of the hippocampus and supports his proposed model of pattern separation and "time-stamping" of memories (Gage, 1998; Gage, 2002).

> Today, it is widely recognized that the brain's ability to generate and sense temporal information is a prerequisite for both action and cognition. (Buzsáki, 2006, p. 6)

In humans, the right hippocampus is generally thought to be more involved with spatial information and navigation and the left hippocampus more with temporal and linguistic information (Morris, 2007, p. 618; Burgess & O'Keefe, 2003, pp. 540, 543; Burgess et al, 2002).

Hippocampal integration allows the modulation of an organism's direction, position, and affective state in contextual relation to a changing environment. At least in part, linguistic information is also integrated at the hippocampal level. The reliability of the language and how accurately it represents information play an influential role in cognitive processing of spatiotemporal context, perceptual acuity, and behavioral effectiveness. Constrained language obscures context and counterproductively limits executive functioning.

A Sense of Direction

A sense of direction can be found in single-celled to complex multicellular organisms, including rodents, primates, and human primates (Jennings, 1906). But without an external source of reference, we may think we are going in a forward direction but not know where we are headed. An external referent gives us a way to orient and adjust our bearings, which can help us avoid getting lost in the expanse of a difficult-to-predict, changing environment.

Language, beliefs, and values play an abstract and literal role in our sense of direction, especially when we are navigating or looking toward the future, imagining, and predicting. For example, language-related beliefs, as memories, enable mimicking traveling backward in time for reminiscing that influences the direction we take today in pursuit of goals and the predictions we make. Beliefs influence the direction for imagining, simulating, and planning in pursuit of future goals. In the same light, culturally inherited language conveys predetermined affective biases that literally influence our navigational direction toward or away from stimuli, including people, places, things, and ideas. The predetermined influence is linguistically encoded by our cultural values, which prescribe positives to approach and hold dearly and negatives to avoid or defeat at all costs.

These examples of inherited biases lend support to the thesis that effectively referenced language can beneficially influence our spatial and temporal sense of direction—how and what we value across space and time, past to future. Language influences our perceptions, our sense of where we have been in the past and where we are now; it has a profound influence on our sense of the future (Bronowski, 1977). The hippocampal complex plays a fundamental role in integrating our sense of direction with affect, time and space, action and change, and the outside world.

The architecture and functions attributed to the hippocampal complex lead us to describe cognition and behavior in terms of vectors and vector space (Burgess, 2007, pp. 737–738; Buzsáki, 2006, p. 307). Idealized local vector descriptions have limitations when referring to the behavior of nonlocal ensembles and persistent interactions that operate according to statistical descriptions as probabilistic processes extending over the entire system (Prigogine, 1997, pp. 37, 114–117). Systematically correlated information and semantic abstractions referenced at the macroscopic level enable extended linguistic ability beyond local trajectory descriptions to global descriptions of populations. Extended language

- Logically describes the processes, relationship dynamics, and evolving nature of the world with which we interact
- Orients us in time and space
- Establishes our direction for thinking about the world from past to the present to the future
- Provides reference for statistical methods for describing our direction and change, including changes related to choices and bifurcations at decision points

There is a large body of evidence supporting the role of the hippocampus in learning, navigation, and cognition related to the integration, storage, and retrieval of information, and also in natural language. Generally, studies in animals other than humans make up the bulk of our knowledge, since invasive experimentation cannot be performed with brains of human subjects (Churchland & Sejnowski, 1994, pp. 429–430). Much of the human information comes from evaluation of brain injuries, observations accompanying neurosurgery, and other noninvasive techniques such as electroencephalography and functional scanning. Taken together, they reflect a wealth of knowledge about the workings of the hippocampus and the brain in general.

The human capability for language and the differences found in the human neocortex make for difficult comparison between studies of less-complex animals and humans. It may be that fundamental principles can be discovered in animal models, and that knowing these will provide the scaffolding for answering questions concerning those aspects of the human brain that make it unique (Churchland & Sejnowski, 1994, p. 430). Since we have analytical language abilities, studies of less-complex animals can provide valuable insights into parallel cognitive processes in humans that can also contribute to understanding the adaptive advantages of linguistics.

In other words, we can add language to or subtract it from the equation. We can observe the differences in adaptive effects on behavior when comparing both brains as animal brains. Then we can draw conclusions about how language influences the tuning of affective and effective changes to input, computational processing, and output—individually and taken together:

- Can differences in language enable tuning cognitive behavior beyond the level of merely sufficient?
- How does extended tuning influence human cognition and linguistic behavior?
- How do differences between constrained language and extended language influence our nonlinguistic behavior?
- How do these differences compare with the behavior of our mammalian cousins and other less-complex organisms?
- Is there a measurable difference between affective tuning and effective tuning on our sense of direction?

We can predict that there is no statistical difference in adaptive behavior for animals with language versus without language—that in comparison with nonlinguistic animals, humans with affective language demonstrate no difference in the direction of nonlinguistic approach-and-avoidance behavior or effective adaptation in response to change. In other words, we can predict that affective language in comparison with effective language will show no significant difference in changing behavioral direction or effective adaptation.

We use language constructed by humans for their own use to predict behavior across different animals, and we run into the previously mentioned circular bias problem with a language that directly refers back on itself. Comparing internally referenced affective language and externally referenced effective language will likely demonstrate biases with noticeable differences in the reliability of predictions for adaptive responses. Because we want useful results, we prefer to carry out the experiments or observations under continuous real-world conditions rather than in a laboratory idealization with discrete constraints. The language and methods we choose to use—local or global—impose constraints that can significantly influence our results.

Furthermore, in complex risk–reward situations under real-world conditions, we expect extended language to modulate affective limbic reactions and enhance the fine-tuning of analytical decisions. Of course, it is somewhat unfair to the other mammals with rudimentary communication abilities that do not have the advantage of selecting between affective imperative or effective analytic language. We find little consolation in knowing that other mammals apparently lack the linguistic ability to rationalize or blame others for their directed aggressive behavior. But what are the differences in adaptive behavior, and what role does language play?

The future is under construction, which requires thinking in a forward direction. How could we ever consider destroying our future? Or—perhaps a more puzzling question—why do we ignore the destruction of our future? And how much can we borrow from the future without compunction—as if we are certain that the future goes to infinity with an infinite amount of resources and we can do no harm? If we take a linguistic approach and quantitatively compare aggressive destruction of the environment as negative adaptation, humans may show a remarkable superiority for anti-adaptation. Would such results enlighten or confuse us? Could we accept results that do not confirm our beliefs about human superiority, or would we diligently search to find justifications for rejecting such results? The answer appears to reside with our ignorance of systematic biases, aided by affective language.

> In short, many of the ancient, evolutionary derived brain systems all mammalians share still serve as the foundations for the deeply experienced affective proclivities of the human mind. (Panksepp, 1998, p. 4)

For example, we see directed aggression as a common agonistic behavior across animal species, including fruit flies, and from less-complex mammals to humans (Maxon & Canastar, 2006, p. 3). We see other common features in primates of command-like vocal displays and imperative gestures (Tomasello, 2008, pp. 36–43), and many mammals display threatening behavior and vocalizations for threats, warnings, and sexual calls (Panksepp, 1998, pp. 188–191). Perhaps one day comparative experiments including humans will come into vogue. In the meantime there is a wealth of evidence from REBT and CBT that changing language habits can confer flexible tuning that enhances effectively modulating the beneficial direction of our thinking, behaving, and emotional well-being (Ellis & Harper, 1997; Beck, 1976). There is probably even more evidence that education in general and higher education in particular offer improvement for more effectively directing higher-order thought processes and analytical abilities.

We can evaluate localized and extended language by comparing the influence on cognition according to the output effectiveness to changing inputs from a dynamic environment with one-way time. The efficacy of the output corresponding to the input represents the adaptive response to change. Forward-flowing one-way time enables measuring domain-referenced differences in predictions and effective output that can elucidate the systematic error rates and range of error deviations in highly variable conditions encountered in nature.

Natural conditions enable accumulating domain knowledge as a history of reliable correlations to domain regularities. In a changing environment with many variables and continuous inputs, we expect modulated changes in the output to correspond with effective domain adaptation in response to those external changes. Reliable spatiotemporal correspondence calls for language that explicitly supports time directed toward the future. But we tend to take the direction of time for granted.

> Neuroscientists work with time every day but rarely ask what it is. We take for granted that time is "real" and that brains have mechanisms for tracking it. (Buzsáki, 2006, p. 6)

Language provides a sophisticated method for information symmetry breaking including forward-directed affective input to effective output. Temporally informed language plays a fundamental role in scientific methodology for representing, defining, interpreting, and communicating research results. Language informs our results and the direction in which the research proceeds. We also tend to take language for granted, even though experimental evidence demonstrates that language biases our cognitive behavior. Poorly understood language can complicate the interpretation of studies on human neurocognitive processes and limit how closely humans can be compared with less-complex organisms that lack the sophisticated language abilities and the distinctive speech articulation found in humans. Applying affective and effective language to the study of human behavior allows for relevant distinctions of similarities and dissimilarities among humans, other mammals, and less-complex organisms.

Understanding language, how it represents the world we live in, and how it influences the direction of our thought and behavior can potentially help us improve our adaptive interactions. Effective language offers opportunities for evaluating our sense of direction and better understanding how we can cooperatively and adaptively relate to ourselves, to others, and the world around us. The localized and extended properties of language and the convergent and divergent processes of hippocampal integration and segregation demonstrate relevant functional correlations with information and cognitive processing in mammals, nonhuman primates, and humans. These functional correlations lend themselves to a computational information-processing model suitable for measuring the effectiveness of localized and extended language on adaptive cognition.

> Before one can began to think about the human brain and behavior in a general sense one must consider some functional questions. What are the basic characteristics of human behavior and how does human behavior compare with the behavior of other animals? (Vanderwolf, 2010, p. 33)

Brain models have progressed from earlier sensory input-response models to theories based on input-decision-output processing (Buzsáki, 2006, p. 233). Nonhuman animal studies often involve experiments with rodents. The brains of humans and rodents, including the mouse, appear quite similar in terms of their overall organization (Panksepp, 1998, p. 326). We can consider the similarities and differences among effect-directed language and affect-directed language and nonlinguistic affect-laden behavior.

> . . . [T]he intrinsic nature of basic emotional systems has been remarkably well conserved during the course of mammalian evolution. (Panksepp, 1998, p. 9)

Studies with rodents provide interesting comparisons with human subcortical cognition and corresponding limbic-directed behavior in both species. It appears that human affective behavior with limbic sufficiency corresponds closely with that of rodents in terms of aggression, rewards and penalties, and the direction of attraction and repulsion and of approach and avoidance. Our understanding of human cognitive behavior has greatly benefited from what we have learned in rodents.

The hippocampus–entorhinal system in rodents has been demonstrated to have omnidirectional place cells that encode egocentric position irrespective of behavior or direction (O'Keefe & Dostrovsky, 1971) and grid cells with an allocentric map-like function that become immediately active in novel environments (Moser et al, 2005). In rodents, these cells play a role in linear serial information input–output processing from an egocentric and allocentric perspective. The different perspectives enable correlating navigational behavior with domain exploration in unfamiliar environments and goal-directed exploitation in environments with familiar landmarks. Map-based navigation converges with stimulus-invariant head direction cells that are distributed in the hippocampal complex, including the pre- and postsubiculum and multiple other brain areas related to navigational direction and path integration (Wiener & Taube, 2005).

> Egocentric space refers to the locations of objects relative to the viewer—to the left or right—directions that necessarily alter as the viewer moves around.... Allocentric space, on the other hand, is a representation of spatial relations within some kind of absolute [finite] framework. For a freely moving subject, an allocentric representation is required to identify whether an object or landmark has actually moved in relation to others. In contrast, objects are constantly "moving" in egocentric representations. (Morris, 2007, p. 623)

Together, these cells map to the environment with a built-in positional axis and navigational plane accompanied by an invariant referent for head direction. The place cells are thought to represent the organism's position, and grid cells are believed to represent the distance metrics of a spatial map with firing patterns similar to the map's latitude and longitude. Map-based navigation requires a calibrated representation of the environment acquired by exploration, sensorimotor, landmark, and path integration (Buzsáki, 2006, pp. 302–304).

The resemblance between functional aspects of the brain, regulation in artificial neural networks, navigational direction, exploration, novelty search, and objective search algorithms in machine learning comes as no surprise, since the experiments have been developed to mimic animal behavior and learning. Apparently novelty contributes to mediating Hebbian learning in continuous-attractor models for creating attractors across an unevenly sampled environment, where there can be a mismatch between the learned associations and the expected state of the world [uncertainty] (Burgess & O'Keefe, 2003, p. 541). In changing environments, search, plasticity, and learning play an important role in dealing with uncertainty and novelty.

Hippocampus-related cell groups provide an event-level structural and functional mechanism for field computation and correlation of situational context, position, and directional heading. The computation facilitates the integration of velocity, valence, and spatiotemporal information between the organism's position and the position of external objects. The place cells can be approximated by a Gaussian [central distribution] spatial field conducive to biologically relevant population descriptions, statistics-derived representations, and the probabilistic integration of oscillations (Buzsáki, 2010). The place and grid cells occupy a strategic position to play a role in the integration of context with internal-to-external spatiotemporal changes of position and state correlated to value, action, and distance.

Extrapolating these same cell groups to humans, we can see the hippocampal complex serving as a likely relational substrate for integrating extended language representations. Episodic and associative memory, language, and value are integrated at the hippocampal complex and accompanying perirhinal–parahippocampal cortex. The effectiveness of the language involved in the integration influences the direction of our here-and-now perception and how we deal with novelty in an uncertain world. We can extend the representation of position and navigation to include localized beliefs, concepts, and ideologies such as "influencing the philosophical position and direction taken in the world of personal politics."

Episodic and semantic memory representations might have evolved from mechanisms for path integration and for map-based navigation, respectively (Buzsáki, 2006, p. 333). Path integration and episodic memory are self-centered and time-dependent (Buzsáki, 2006, p. 308) and thus are influenced by the differential efficacy between localized and systematic language. Structural and functional correlation of memory with navigational direction provides evidence to support further examination of the influence of linguistic representations on the construction and calibration of our world maps and the corresponding effectiveness of our adaptive cognitive behavior—from which our sense of the future flows.

Oscillations and Periodicity

The relation of place and grid cells to position and environmental navigation demonstrates that our brains are preconfigured for spatiotemporal integration. Furthermore, place cells have been shown to flicker on and off between states experimentally associated with past and present events when correlating to a change in the reference environment (Moser, 2010). Flickering resembles a form of oscillation related to transitions or reorientation to change of place, position, or state. Thus oscillations may play a correlating role in symmetry breaking of past and present and modulating spatiotemporal instabilities by synchronizing convergence to the current environment.

Hippocampal-complex oscillations are implicated in the comparison of present and past states, present and predicted states, and predicted outcomes with the actual outcome in the environment. Oscillations create a dynamic interface between internal and external value states that potentially enable phase-locking of goal-related choices and matching action solutions in response to external changes of position and state. Correlating external *where* positions and *what* identification with value leads to preliminary evaluation of tentative possibilities for directing approach and avoidance. In turn, oscillations theoretically facilitate the coordination of stimuli-salience detection for directing and shifting attention by correlating position with positive and negative value, such as reward or penalty, prey or predator, friend or foe. Thus value asymmetry contributes to modulating attraction and repulsion, directing approach and avoidance, and shifting exploration and exploitation according to environmental constraints and the distribution of resources such as rewards and penalties as demonstrated in simulation experiments (Heerebout, 2011). Oscillating brains make sense in an uncertain world of action, motion, periodicity, rhythms, and waves.

In a world of waveforms, wavelengths, and frequencies, a brain with intrinsic rhythmic oscillations allows for simple comparison of the phasic relationship between actions and expected and experienced sensory feedback. Intrinsic and extrinsic waveforms in phase have a harmonious relationship. Oscillators are common in nature, engineering, electronic control systems, and signal processing and transmission. A continuous periodic system generates a characteristic oscillating linear signal pattern with a positive value at the peak and a negative value at the trough. The signal conceptually represents a harmonic oscillator with a characteristic wave length and frequency such that comparison signals are either in or out of phase. The phasic relationship between the brain rhythms and environment and positive and negative values makes intentional evaluations and the decision whether to approach or avoid outside objects fairly simple (Heerebout, 2011). The approach or avoid reactions can be described according to positive–negative value interpretations of input signals and perceived stimuli intention that translate to attraction–repulsion.

Learning from experiences over time forms episodic memories, a sequential ordering of events. Our prior learning and value associations accumulated in memory from past experiences provide a basis for identifying and paying attention to relevant positive and negative stimuli and ignoring neutral stimuli. Value symmetry breaking simplifies categories according to what to ignore or attend to, and what to approach or avoid. Categorical value-chunking confronts the curse of dimensionality and significantly decreases the scope of the information search space. In effect, the brain's assuming, generalizing, and value-chunking tendencies facilitate the compression of a large amount of external information. The compression process simplifies interactions and cognition required for attending and remembering, evaluating and predicting, behaving, and learning from experience. Analogous to other organisms, our brains come equipped with a simple binary value code that suffices for supporting survival and reproduction.

It seems like a routine matter if the realized input resonates with our prediction, such that positive and negative inputs are on the same wavelength as our expectations. But sometimes we incur dissonant feedback out of phase with our predictions. Expecting a positive but realizing a negative easily leads to arousal with unpleasant feelings of rejection, disappointment, and avoidant withdrawal behavior; it sometimes results in an intensity-dependent alarm reaction, feeling angry, and aggressive behavior. Instead of an expected negative input, an unexpected positive input also leads to arousal, but with pleasant feelings of surprise, acceptance behavior, and often an accompanying approach response. In effect, organisms tend toward favorable and away from unfavorable stimuli, which facilitates natural concordance with survival and reproduction (Jennings, 1906). We see this simple value-coding system in humans and other animal species, from single-celled to more complex organisms, from paramecium to rodents.

Research in rodents has demonstrated phase-locking in the OFC with odor-detection and movement-related cells, where the OFC demonstrates a temporal layout in the forward direction. Before and during movement, the OFC oscillations phase-lock in advance of reward that correlates with reward expectancy (Pennartz et al, 2011). These OFC cells have a functional analogy with forward-firing place cells thought to play a role in goal-directed behavior.

In artificial intelligence experiments with foraging in ant colonies, oscillations in the vicinity of instabilities at symmetry-breaking bifurcations have been associated with enhanced adaptability (Bonabeau & Cogne, 1996). Similar oscillations have spontaneously evolved in artificial intelligence food-gathering experiments involving predator and prey interactions that demonstrated increased prey fitness correlated with the modulation of affective attention-switching (Heerebout & Phaf, 2010). In these experiments, agents (prey) that developed oscillations demonstrated an almost doubling of fitness compared with nonoscillating agents. The switching showed an evolution-directed bias favoring survival. Agents with higher-frequency oscillations associated with approach behavior and food reinforcement were able to switch directions more quickly among rewards, which resulted in an increased ability to rapidly gather food. These oscillations also increased cognitive flexibility that switched the direction of attention more quickly to predators, initiated avoidance behavior in an opposing direction, and induced slower oscillations. The slower oscillations demonstrated a narrowed focus to negative stimuli and slower switching out of negative states while at the same time increasing avoidance velocity (Heerebout, 2011).

> Appetitive-approach tendencies toward positive stimuli and aversive-avoidance tendencies away from negative stimuli have likely developed through evolution to maximize fitness benefits and to minimize fitness costs, respectively. (Heerebout & Phaf, 2010)

These results were contributed in part to conservation of the neural network's direct pathways, as advocated by LeDoux (1996), which are thought to provide functional scaffolding for the evolution of complexification in recurrent feed-forward, indirect pathways. Recurrent connections were enabled between the hidden layer and an additional layer of context nodes in which the oscillations developed. Bram Heerebout and R. Hans Phaf (2010) laid claim to the first published experimental results of an artificial neural network's evolving oscillating nodes that

- Distinguish context
- Enhance the flexible modulation of attentional switching
- Allow for smooth transitions between attractive and repulsive affective states
- Modulate switching that inverts the output direction for approach-and-avoidance behavior
- Differentially adapt the output gain for modulating velocity (Heerebout, 2011)

The research suggests that context nodes and oscillations communicate information that may offer a nontrivial solution to problems involving living organisms and multi-agent interactions with attraction, repulsion, and direction in complex domains. Furthermore, it appears that spatiotemporal information benefits from adding context to signal whether an object stimulus fits into the predator or the prey categories (Cervantes-Perez, 2003, p. 1223). Chimps in naturalistic settings demonstrate somewhat analogous vacillations when repeatedly shifting the direction of attention from a valuable source of drinking water to dominant males that might approach to compete for the water resource (Watson & Platt, 2011).

> Evolution has evidently stumbled upon solutions to space-time representation using, among other things, time constants for various neural activities, including oscillation frequencies, phase-locking oscillations, channel-open times, differential refractory periods, and bursting schedules, and the list could be extended. (Churchland & Sejnowski, 1994, p. 306)

We see oscillations across living tissue, including mammals in general and human brains in particular. Oscillations appear to have a profound relationship with the direction of goal-related behavior. We can think of goal-directed behavior as an organism at position A having a goal as an object target at position B. Since each organism navigates from its own relative coordinates, we can think of each organism as operating from position A with variable spatial information as to current location in relation to the position of object B. Initiating a goal at a certain time requires a prediction, whether generated internally or externally, of the goal location at position B and the actions required to reach that goal. In effect, an organism faces the question of how to navigate between positions A and B to approach the reward and to avoid penalties depending on state conditions and constraints. In other words, what actions best satisfy the goal under the prevailing constraints? In humans, actions can be computed either automatically and implicitly or deliberately, explicitly, and systematically, depending on the effectiveness of the linguistic constructs. We call the latter thinking and planning ahead.

> To localize an object simply means to represent to oneself the movements that would be necessary to reach it. (Poincaré, 2007, p. 47)

Oscillations in the brain seem well suited for this A-to-B action task, and we can think of the steps we take to get there as sequential components executed linearly in time, similar to a recipe or algorithm. In linguistic terms, an algorithm can be expressed in the language of neural syntax. Exploitation of syntax is common in systems that code, transmit, and decode information (Buzsáki, 2010). We can think of oscillations as possibly providing a natural brain mechanism for chunking and parsing of neural activity as correlated information that enables the brain to access, assemble, and execute sequential actions for reaching a goal. The invariant forward direction of time supports the structure of forward-directed syntax, sequentially related events, and episodic and working memory.

Working memory enables us to simulate goal-directed plans, edit our episodic memory, and change plans when faced with constraints. A brain with oscillations, language, and working memory enables flexibly substituting or changing almost any expectation, goal, or goal-related behavior over time and space. Oscillating cell assemblies support our ability to reminisce, think, reason, and plan ahead, which corresponds to physiological hippocampal processes with similarities between navigation and goal-related cognitive behavior (Buzsáki, 2010).

Neuronal sequences can result from a combination of automated and deliberate actions, such that effectiveness can be measured as the systematic error rate and variance between the predicted and realized output product in response to external change. Similarly, we can extrapolate the prediction of systematic error variance to the effectiveness of linguistic representation correlated to adaptation. The error measures the difference of structural correspondence between the internal linguistic representation and that of the environment. György Buzsáki (2010) proposes a model of neural syntax in terms of cell assemblies that can be best understood in light of their explicit output product as detected by reader-actuator mechanisms supported by a reader-classifier and a temporal frame. In this model the cell assembly is defined from the perspective of a reader mechanism, where the reader-centric mechanism provides a functional meaning.

If we interpret functional meaning in terms of survival and reproduction, we see that on a macroscopic scale the reader bears similarity to a biological perspective from a natural reference. In other words, we see a fitness relationship among the history of the genome and phenotypic learning plasticity, approach positives and avoid negatives, and adaptation to environmental instabilities. We see a relationship among effective language, symmetry breaking, asymmetrical time, and probabilistic logic. Effective language enters the picture because of language-dependent executive functioning, memory recall, thinking and planning, and decision making. Also language correlates cognition with the domain dynamics and constraints, and facilitates input calibration, output tuning, and systematic error correction.

Functional descriptions of oscillations are complicated by the variety of somewhat different definitions, usage, and nomenclature across physics, chemistry, biology, neuroscience, mathematics, engineering, systems theory, and various subfields. Thus, in some fields a reference to *oscillations* is fairly recent, but there are many other terms that refer to oscillatory behavior, including *cycles*, *rhythms*, *periodicity*, *regularities*, *recurrences*, *repetitions*, *iterations*, *fluctuations*, *vacillations*, *vibrations*, *waxing and waning*, and *waves* (Buzsáki, 2006, pp. 5–6). Also, different domain representations influence the perspective for understanding the systematic problem-and-solution space, such as static linear and dynamic nonlinear, steady state and non-steady state, stable equilibrium and nonequilibrium instabilities, symmetrical and asymmetrical time, conservative and nonconservative systems, and so on.

Each description may have appropriate applications in some domains or subdomains, but biological systems benefit from a representation that matches the dynamics and constraints of nature. Interpretable results and conclusions require choosing a model that effectively correlates the appropriate systematic dimensions and dynamics within a particular system or subsystem. For example, except for the simplest of systems, linear models are generally considered inadequate for reliably describing biochemical or genetic networks of biological systems (Iglesias & Ingalls, 2010, p. 8). In biological systems where organisms interact with external periodic and aperiodic fluctuations and instabilities, models that take nonlinearities and oscillations into account tend to make a better match with the external dynamics and constraints. As such, nonequilibrium physics supports plausible descriptions of biological systems.

Oscillations are thought to be involved in goal-directed input-output modulation and appear to participate in memory segregation, integration, storage, recall, and affective associations influencing sensorimotor processing. Oscillations also appear to be involved with perception, directing attention, and integrating internal–external contextual associations. Some changes in oscillatory patterns correspond to goal-related changes of position and changes of state values, such as phase-locking and behavioral activation in response to rewards. It seems that the synchronization of evolved brain oscillations may enable one-dimensional linear output of solutions for the problem of goal-directed behavior in high-dimensional nonlinear domains—expressible with language. We expect that language plays an influential role in symmetry breaking, thinking about the future, predicting, and learning from experience; in what we think about the how and why of brains and behavior; and in whether we believe in oscillations and systematic domain knowledge.

Our current knowledge base, in large part, influences the reliability of our simulations and predictions for the future. During movement and contemplation of the future, our forward-firing hippocampal place cells and our working memory capacity automatically keep a step ahead of us to predict what's next (Pastalkova et al, 2008). Working memory relies on our own historical database from which we extrapolate knowledge via available linguistic constructions. Therefore, our language, prior beliefs, and current knowledge place constraints on our predictions. For instance, if we have beliefs locally encoded as absolute knowledge, we expect to generate affectively biased goals, predictions, and solutions. Likewise, if we have systematically encoded knowledge, we expect to generate some range of effectively biased goals, predictions, and solutions. In other words, the construction of our linguistic scaffolding underlying our belief or knowledge base biases what we can draw from, which correlates with our language structure. Research shows that at some level our brain rhythms and language apparently correlate with our goal-directed behavior.

Data suggest that different theta rhythms correlate to layer-specific processing that supports behavioral task performance (Montgomery et al, 2009). Theta rhythms are associated with local information processing and local circuit interactions (Mizuseki et al, 2009) and, theoretically, gamma rhythms associated with higher-level global brain processing. Studies demonstrate the synchronization of local theta and gamma rhythm signals during sleep that likely reflects short-term to long-term memory consolidation (Montgomery et al, 2008).

Theoretically means that matching of new and old information will depend on the corresponding language constructions, such that the lowest common denominator confers linguistic constraints on the translation and integration of knowledge. Put another way, a processor capable of global cognition can incur linguistic constraints on knowledge processing that depends on the effectiveness of the linguistic encoding, such that local encoding produces localized beliefs rather than systematic knowledge. We expect localized language, algorithms, and logic to lead to affective processing. For optimizing systematic processing, we expect logical correspondence between language structure and the internal representation of knowledge, and the structural and functional physical relationships found in the external global system, of which we are a part.

We can see where the constraints of localized language constructions lead to asynchronous global processing, leaving us systematically out of sync and effectively out of touch with the physics of the external world. Attempting to imagine or surreptitiously characterize local representations as systematic is unlikely to resolve the problem of localized constraints and systematic blindness. Yet since systematic language generalizes from macroscopic to microscopic in either direction, we expect global translation and decoding of local belief systems. We also expect that with some effort, locally encoded memories are amenable to systematic translation, recall, and reconsolidation to effectively approximate systematic domain knowledge.

The brain's ability to oscillate appears to support parallel semantic representations with functional cognitive modules that correlate with the brain rhythms of separate subsystems. A direct one-to-one semantic overlay supports concrete descriptions of cognition. Generalization and higher-order semantic abstraction enables the exploitation of stacking or chunking of higher-level concepts compressed into individual words. In effect, nonlinear layers of [parallel] abstract concepts can be added orthogonally to neural syntax as single words, which increases the nonlinear dimensions of information ordered linearly in time and accessible by working memory.

For example, a dynamic word such as *probable* replaces a static word such as *certain*, or *nonlinear* replaces *linear*. In both cases, the former confers a higher level of abstraction. Higher-order semantic abstractions enable representation of symmetry breaking that conserves syntax and event-related context. By chunking, extended linguistics can facilitate a phase shift from static local to dynamic global information that supports systematic descriptions as the representation of complex concepts, relationships, and processes, such as discrete logic versus continuous logic, incidental causality versus persistent influence, equilibrium versus nonequilibrium, and local time versus global time. These terms represent the building blocks of dynamic to dynamical physical concepts involved in the evolution of regulatory oscillations in biological organisms, including the complex brains found in mammals and our own unique linguistic-compatible brain and neocortex—and the language that makes it possible to understand these brains.

The neocortex is thought to convey sensory information to the hippocampus, which returns output to the neocortex in response to the input pattern. The timing patterns of the theta cycle allow for processing delays involved in the integration of past and future predictions, depending on the animal's location (Itskov et al, 2008). Temporal dynamics play a critical role in state-dependent operations for the integration of episodic memory, spatial maps, and directing behavioral flexibility in a changing environment (Diba & Buzsáki, 2008; Robbe & Buzsáki, 2009).

Oscillatory modulation and integration arguably evolved to a high-dimensional computational processing method that efficiently deals with nonlinear information, multiple input and output variables, and continuously changing internal and external states and value ratios. Temporal coordination of neocortical gamma oscillations by hippocampal theta rhythms appears to provide an entrainment mechanism for synchronous transfer of various rhythms from wide areas of the neocortex, which facilitates integration with hippocampal associative networks (Sirota et al, 2008). The result is a system that can manipulate a large amount of nonlinear information within the short time period of an event (i.e., integrating theta-nested gamma rhythms) (Buzsáki, 2010).

> ... [S]ynchronization by oscillation is the simplest and most economic mechanism to bring together discharging neurons in time so that they can exert a maximal impact on their targets. (Buzsáki, 2006, p. 137)

Oscillations appear to facilitate contextual nonlinear decoding and computation simultaneously with the integration of linear inputs and outputs that influence the activation, inhibition, direction, and gain ratios of approach-and-avoidance behavior. Compatibility between brain oscillations and language enables mutual correspondence and responsive interactions with the dynamic world outside. We expect our linguistic explanations to correlate with our general understanding of the changing world and the events we experience. If our brain rhythms relate to those interactions, we expect some level of congruence between our language, brain rhythms, actions, and information input as sensory feedback from that world. If so, we may possibly enlighten our notion of just what our neurons are talking about. If not, we can always continue saving brain energy on limbic autopilot.

Effective language provides extended dimensions, continuous logic parameters, and higher-order abstractions capable of representing dynamic nonlinear interactions, context, and forward-directed syntax. More importantly, effective language relies on explicit one-way time that corresponds with action and phasic changes, recurrence, oscillations, and periodicity in nature. In other words, extended language speaks the language of oscillations. We can add ratios, recurrence, oscillations, and periodicity to our list of general systematic processing and relational principles:

- Time, symmetry breaking, asymmetrical-time referencing, and probabilistic information processing
- Polarity: positives and negatives, attraction and repulsion, approach and avoidance
- Relative change, change of position, change of state
- Ratios, recurrence, oscillations, and periodicity

The representational disparity between localized and systematic language suggests that preferentially and deliberately extending the direction of language and semantic abstraction to systematic dimensions enables a linguistic phase transition, with spatial and temporal aspects that support symmetry breaking. In effect, higher-dimensional language confers micro-to-macro scalability, integrates analytical cognitive processing, input, and output with systematic information, and enables effective sorting of temporal, spatial, and value context.

Extended language provides statistical logic parameters that enable a semantic phase transition from rigid, closed-ended certitudes to fluid, open-ended possibilities and probabilities that can enhance our effective knowledge of and adaptation to our dynamic world. Systematic language gives us not only a literal sense of direction but also a plausible contextual sense of where we've been, where we are now, and where we are headed. Extended language presents us with a valuable resource for understanding the rhythms of nature and enabling a realistic and likely sense of the future.

> It may turn out that the rhythms of the brain are also the rhythms of the mind. (Buzsáki, 2006, p. 372)

Relations and Predictability

When we attempt to develop an understanding of our relations with nature, it is not surprising that we notice that issues of safety and uncertainty influence how we think and interact. We often think of evolutionary processes as the survival of the fittest, where safety and survival go hand in hand. But does a sufficing brain keep us as safe as possible? How do we deal effectively with our vulnerability to systematic errors and rare occurrences of difficult-to-predict disasters? Insightfulness about how our brain, language, and environment work together can enhance effective information processing. We can enlighten our predictions about possible catastrophic events, and perhaps avoid or minimize our exposure. How do we come to detect and learn about the importance of that with which we might come into contact sometime and somewhere in the future, the unknown and uncertain? What are the features that inform us about something's relevance for tomorrow, and how do we know?

Language that generates reliable knowledge of how nature works improves our ability to think systematically about different possibilities and effective alternatives for planning ahead. That is, we can enhance our knowledge and improve our ability to come to grips with our limitations and our ignorance. Enhancing our understanding of natural constraints allows us to overcome our limbic-driven fear of ignorance and explore the unknown for novel information in an effective and predictably safer manner. But we can rest assured our amygdala is keeping a prudent "eye out" for any local danger by affectively "watching what we are up to" and is ready and willing to protect us from the unknown in an intuitive "blink of an eye."

The limbic–amygdaloid processing system enables symmetry breaking and provides an affective mechanism for personal safety that typically operates automatically without our realization. Automatic behaviors can pull us back quickly from the brink of disaster and minimize the singeing of fingers. But if we do not know how our primitive limbic brain processes information, we risk becoming naively enslaved to automatic limbic-directed behavior. Our limbic processing system helps to detect salient affective features among stimuli, ward off or avoid negatives, and assist with obtaining a supply of positive rewards that help us survive, as it does for other mammals. The limbic brain does not necessarily provide us with effectively useful information that helps us to realistically explain and understand the world systematically on a practical, educated, higher level of analytical abstraction. Indeed, the limbic system seems to absolutely bury and implicitly hide information beneath a comforting blanket of affect.

The brain relies on a value system to organize its operational processing and predict stimulus relevance, and value systems in the brain accomplish that by increasing states of arousal (Siegel, 1999, p. 137). Perceived value correlates affect and associated internal state via sensory input, computation, and output with relevant external stimuli according to positive and negative reinforcements related to survival and reproduction. As the guardian of our safety, the amygdala plays an instrumental role in orchestrating our thoughts, feelings, and behavior by automatically connecting us affectively to the local landscape and the objects and actions we encounter there. But how does this come about?

Think about how a brain interacts with information symmetries and asymmetries in the world in a general sense. Objects, perceived as edges or shapes (with texture or contrast), can be assumed to have density or mass, which makes detection and velocity determinations fairly simple from relative coordinate representations (Braitenberg, 1984, pp. 35, 39–41, 124–129). General coordinates provide a center of mass description for detection and approximation of location and position, distance and heading, velocity and relative change. Eye position, gaze direction, and other sensory input provide additional orienting information for intrinsic computation of relative trajectory velocities for following reward gradients and correlating goal-related motor output to task geometry. The visual system has cells that detect the direction of an object's motion and contribute an important salience feature for arousal, orienting attention, and directing navigation.

The brain also has additional visual and auditory pathways that enable stimuli elaboration with a more detailed process of refined feature detection and perceptual analysis. Familiarity, facial features, gestures, movements, and sounds abstracted from animate or moving objects can provide relevant feature space information for a quick assessment of salience by default. *Feature space* characterizes the allowable features in that space. The space of features constitutes symmetry breaking patterns, perceptions, and abstractions available for classifying, describing, and identifying objects and behaviors according to their informational properties. For instance, the features might consist of textures, colors, areas, ratios of length to width, and so forth (Rolls, 2008, p. 286).

Coarse or fine, specific or general features can play different roles in object recognition and the assignment of relevant salience. For stationary objects, the assessment can focus on features that would categorize it as familiar or unfamiliar, animate or inanimate. Along the same lines, the assessment can explore stored memory for relevant historical salience by correlating and cross-correlating an object's inferred similarities and differences associated with its estimated present and past salience. The resulting negative or positive valence infers the object's current general features of dangerousness or safety. Thus, affective information influences goal-directed behavior and switching among alternatives, including approach and avoidance.

Unfamiliarity implies potential dangerousness until proven otherwise. Familiarity automatically defaults to historical information that allows auto-association for negative features deemed dangerous and for positive or neutral features deemed safe. Moving objects can be assumed to have direction, mass, and velocity. If the object appears to be relatively larger than the organism, it might qualify as dangerous until proven otherwise; if an object gets larger and continues on relatively the same direct heading or matches corresponding changes in direction, it potentially represents an approaching threat until proven otherwise. Also, an animate object's eyes, head, or body orientation can imply directed attention that may represent a potential predatory threat with hostile intentions.

The limbic system, built on evolutionary survival principles, naturally focuses relevance on safety first. In other words, first survive, and then reproduce (Churchland & Sejnowski, 1994, p. 422). With safety in mind, an organism infers relative meaning to an event and calculates a subjective value, based on input and prior experience with objects, behaviors, and intentions (Lindley, 2007, p. 113). Over time, this assessment process transforms information to intentional status as correlated actions and affective values, and, according to relevance, transfers them from short-term to long-term memory. The resulting memory allocation allows inferred contextual meanings as recallable, familiar features associated with positive and negative affective values. The organism uses this memory array of familiar features to infer a value association with the position and action of novel objects that contributes to behavioral predictions.

By default, the lower-level limbic brain operates somewhat like a local coincidence and familiarity detector that, by defaulting to parsimony, tends to assume an association between coincidence and contingency. The limbic brain assumes guilt by association, automatically equates coincidence and contingency, and confirms by filling in missing corroborative information with familiar patterns and values. Although it is fast and frugal, when predicting the limbic system acts with indifference to reliably correlated information; that is, it coincidentally infers contingent cause and effect when one occurrence follows or coincides with another. Locally misinformed coincidence overlooks information in a rush to judgment that confounds the ability to effectively evaluate intentions and predictions, make relevant credit assignments, and notice or correct errors. Accumulating a catalog of uncritically evaluated coincidences creates misinformed perceptions, beliefs, and myths that can emotionally entangle and distort our worldview and influence the views of others with whom we interact.

In contrast, a history of reliably calibrated correlations with patterns of domain regularities can create systematic knowledge and a global worldview. Using extended language to logically fine-tune information and analytical information processing, we increase our chance of distinguishing systematic influences from local coincidences. Language amenable to symmetry breaking, contextual sensitivity, and systematic recognition contributes to reliable assessments and logical interpretations of experiences;

in turn, that contribution improves how effectively we understand inferred salience, affective value, and intention. Effective language enhances our understanding of relations in the outside world, which improves our ability to identify coincidences and discount misperceived relevance. Thus, we are less likely to misperceive similar coincidences in the future.

We have a range of options for considering possible coincidental associations when we perceive two or more events that occur within a narrow measure of time. Extended language enables analysis of the occurrence of the two events and draws a distinction between coincidental causality and contingency, systematic influences, and a history of correlations. The high probability of independent occurrence of the various components or the violation of physical constraints often explains the coincidence. We can avoid the temptation of automatically defaulting to assumptions of causality biased by our assuming brain, beliefs about how we imagine the world works, and confirmed by true-or-false logic. When we infer information about the world, there is a difference between the logical implication and physical causation (Jaynes, 2007, p. 5). We can therefore speak of a type of statistical causality (Prigogine & Stengers, 1985, p. 311).

Some ancients claimed that the heart was the seat and cause of human emotions, and even now we often hear the phrase "winning over their hearts." In much the same way, we can speak of "learning by heart," "affairs of the heart," or having a "hard heart" or a "soft heart" without implying any relation to the hollow muscular organ that contracts rhythmically in every human thorax (Vanderwolf, 2010, p. 6). Today we recognize that the brain's limbic system infers the emotional valence implicitly associated with our cognition. But in our daily communications, we persist in using the old affective language that assumes and infers psychic cause and effect. We often hear, "In my heart I know it's the right thing to do" when we could instead say, albeit somewhat awkwardly, "My amygdala said so" or "My amygdala made me do it." Of course, we could also say, "I have thought it over and made an educated decision."

Causal misconceptions lead some people to claim that heart transplant recipients take on the emotional traits of their donors. Heart transplant surgeons have their own, more precise terminology that allows them to describe the effects of a muscular pumping mechanism without having to account for imaginary causal sources of emotion. These examples point out the distinct independent differences between causal affective reasoning associated with the psyche and logical effective reasoning about systematic influences, and they speak to the importance of how our language usage alters how we make affective and effective sense of the world. Affective sense is particularly relevant, since our emotions amplify our memories (LeDoux, 2002, p. 222). By tuning our language logically to correspond with nature, we can understand the relational difference between physical effects of output and emotional affects of input that influence our reasoning processes and prediction abilities. Effective tuning enhances coherent correspondence of internal insight with external sight and our perception of external events that improves outcome predictions by analytically weighting current inputs and memories.

Increasing the effectiveness of our language by incorporating probabilities enhances our ability to reliably assign credit, modulate emotional instabilities, and improve our predictions. Recall that making predictions based on effective outcomes benefits from reliably informed classification of states, actions, and value. Extended language facilitates the representation and logical separation of static classes from actions, classes from affect, and action from affect. Affective language confounds and constrains such separation, relevant because linguistic representations have been shown to influence how emotional states are represented and experienced (Rolls, 2008, p. 171).

> We use language every day. It is an important part of the world in which we live, and of our experience of this world. Language experience is processed in our brains in the same way as other experiences; that is, it influences cortical representations at different levels of analysis and processing. (Spitzer, 1999, p. 245)

We employ language to understand and integrate higher- and lower-level informational inputs (Chow & Cummings, 2007, p. 29; Cummings & Miller, 2007, p. 15; Gazzaley & D'Esposito, 2007, pp. 190–191). Language enables us to encode and decode more specific higher-level knowledge, discriminate among alternative choices, and make statistical prediction of outcomes based on effectiveness as the consequences of our behavior. And the effectiveness of the available knowledge corresponds to the fidelity of the underlying information and processing and the linguistic encoding.

Recall that extended linguistic parameters can coherently correlate to dynamic domains and provide effective representation of domain knowledge—enabling us to know our world as well as we can. Knowing how our world works improves our predictions. Improved predictions help us avoid danger and get the adaptive outcomes we prefer. When linguistic parameters are extended to provide macroscopic descriptions, we can correlate our thinking in correspondence with open-ended systematic processes in nature. Extended language expresses the future as a construction of possibilities that supplant imaginary idealizations and certitudes. A linguistic extension allows us to view the world, and our behavior in it, in terms of possible alternative solutions—alternatives allowing preferences and creative choices with novel solutions rather than predetermined mandates that stifle innovation.

We use language to describe our world, brain, cognitive processes, and behavior. If we subscribe to the theory that evolution tends toward parsimony and sufficiency, then linguistic descriptions that offer parsimony and effective systematic generalizations become a bargain in a complex world. When objects, values, and the world itself are continuously moving, we benefit from language that contributes to effectively recognizing relevant contextual features and facilitates reliable predictions.

Next we will address the importance of value and how it underlies fundamental processes across relationships in nature.

CHAPTER 13

Value and Processing

We use language to describe and express what we value in this world. Where do values come from, and how do they help us know and adapt to our world? Language provides us with abstract tools that can help us find sensible answers to these questions. Understanding how we use language to evaluate, define, assign, and edit external experiences can give us insight into how value weights our cognitive processing, perception, and behavior, both affectively and effectively. Value influences how we react and regulate adaptation in response to change. By understanding the evolutionary processes of survival and reproduction in relation to language, the brain, logical thought processing, and affective and effective value, we get a clearer picture of the influences and constraints on our behavior.

Koziol and Budding (2010, p. 9) list six characteristics that allow the brain's organization to successfully interact, survive, and adapt to the environment:

- Capacity for object-recognition functions (what)
- Capacity for object-location functions (where)
- Capacity to detect movement (change)
- Ability to know what to do (solution)
- Ability to know how to do it (or know how to act)
- Ability to know when to act (timing)

Via the brain, value links these characteristics with language abstractions, cognitive processing, and the environment. In other words, *what* has a positive or negative value, and sometimes neutral, depending on current contexts and state condition (deprivation-satiation), which influences the probability of choice for approach or avoidance behavior. How and when we regulate our actions also depends on context and the influence of our brain and language, learning history, skills, and knowledge. *State* is defined as the condition or value of a system or system component at a particular time. The current state of the brain and biasing effect of language on our cognitive processing plays an influential role in our choices and outcomes. In this chapter we further investigate how value symmetry breaking lends itself to affective and effective language encoding as an information form that influences regulation, including sensing, thinking, error processing and learning, behaving, and adapting.

Understanding Value Processing

In nature we observe, represent, and process the content of events contextually as interactions between what we can categorize as positive, null, and negative values. Examples of positive and negative values occur at all levels of biochemistry, from molecules to minds, and the brains and languages we use to describe them. Symmetry breaking and affective value systems work across organisms regardless of language abilities. In humans, the addition of language adds an abstract descriptive layer that exposes the implicit and explicit role that value plays in our lives. Value plays a key biasing role in how we sense and process information, set goals, *infer* and interpret favorable and unfavorable outcomes, and regulate adaptive behavior.

Value symmetry breaking provides organisms with a generalized regulatory processing mechanism and a conditionally independent credit rating system for associating and clarifying favorable and unfavorable conditions. A value system adds a level of contextual meaning that allows us to abstractly define and attribute the relative weight, worth, or importance of a wide range of relationships among objects, positions and actions, structure and function, ideas and beliefs, and so on. Broadly speaking, value gives relational meaning to processes of affinity, position, direction, and change. Even at the cellular level, developing neurons and axons respond to gradients of attractive and repulsive signals that influence the direction in which they grow (Nüsslein-Volhard, 2006, pp. 58–59, 81).

Attraction occurs between entities with *unlike* charges of negative and positive values. *Repulsion* occurs between entities with *like* charges of either both positive or both negative values. We see invariant rules for like and unlike charges in the laws of attraction and repulsion throughout nature from inorganic to organic molecular interactions and chemical bonds. Affinity correlates attraction and repulsion at many levels, ranging from chemical bonds and reactions in cells to relationships evoked in the brain by the chemistry that underlies affective bonding and affiliation in humans. We see bonding in many social animals, including prairie voles, and research has demonstrated a link between the bonding-associated neuropeptides oxytocin and vasopressin and affiliative behavior and limbic reinforcement pathways (for a review, see Caldwell & Young, 2006; Ahern &Young, 2009). For instance, oxytocin mediates various pro-social processes, including maternal behavior and the inhibition of separation distress (Panksepp, 1998, p. 102).

Relatively invariant biological values are a natural part of survival, while learned values can span a range from emotional input to analytical output or emotional affect to educational effect. Typically, emotional values are related to how the limbic brain implicitly interprets local input, including social values and reinforcement learning. In sharp contrast to arbitrarily assigned social values based on local affective feelings, explicit analytical values are associated with increasing levels of language-related education, literacy, and effective thinking that together influence how our higher-order brain goes about interpreting affective input, tuning input sensitivity, and regulating effective output. Of course all work together, but we reap benefits from understanding how language, the brain, and the world interrelate with values, because when it comes to values, our educated brain has the last word. Values tend to differentially and profoundly influence our individual local and global perception of relationships—how we conceptualize the world.

For example, we see local groups that express idiosyncratic beliefs and values with different ways of dealing with emotional experiences. Population A says:

> We believe in the power of the spirits of emotions. We are angry. The tribe that lives over the mountain will always be our enemy. We have stories passed down that tell us how they tortured and abused our ancestors. They tried to steal our pigs and could have caused starvation. Every chance we get, we try to retaliate so we can even up the score. We must never forget, and we must get revenge. We must fight for our honor.

In contrast, population B says:

> We believe in the power of stopping to think when we feel strong emotions. We have been educated in the ways of not upsetting ourselves, of talking things over and making compromises to solve our disputes. We have stories passed down that tell us how the group that lives over the mountain abused several people in our group and tried to steal many pigs. We know that emotional stories passed down from many generations often get confused. We wanted to better understand the problem and find out more about what took place. We know there are many possibilities. We found out that it was only one person who had gotten into an argument with one of our group over a disputed boundary and a pig that had gotten loose. We don't necessarily like the way that person acted, but it seems that little harm was done.
>
> We get along well with most of the other group members and settle our occasional disagreements by looking at the situation from each other's viewpoints and talking about the different way we each see things. One of their people behaving that way does not mean that they are all bad. We don't like the way that one person behaved, but we have talked about it and have come to a peaceful agreement. Why would we think that one person makes up the whole tribe? We do not think that way; it seems silly to us.

The latter believes in the value of effective rather than affective information, delaying analysis after collecting more information, taking into consideration context and the perspective of others, and evaluating different possibilities that may have occurred. Both approaches incorporate value into their thinking, but the former displays a preponderance of emotional regulation, and the latter, analytical regulation. We see the cultural differences in closed-minded thinking versus open-minded thinking and the simple use of possibilities rather than certitudes. We might also consider the less obvious inheritance problem that deals with our tendency to easily accept, believe, and overvalue unexamined secondhand information as factual.

> Much of what we know in today's world comes not from experience, but from what we read and what others tell us. An ever-higher percentage of our beliefs rest on a foundation of evidence that we have not collected ourselves. Therefore, by shedding light on the ways in which secondhand information can be misleading, we can better understand a common source of questionable and erroneous beliefs. (Gilovich, 1991, p. 90)

Language allows abstract value descriptions that help inform and misinform our understanding of physical events and the behavior of organisms, from microbes to humans. The concept of value is fundamental to current models of decision making, supported by evidence that decision variables are subject to value-related biases (Platt et al, 2008, p. 135). We rely on language to describe individual and societal beliefs, values, and behavior that positively or negatively correlate preference in the form of likes and dislikes. Whether an object is liked or disliked tends to reflect its subjective affective value in the eyes of the beholder. The value may be a combination of putative inheritance-related affective–reactive circuits such as fear of snakes or spiders, language-related cultural learning, and reinforcement history. Of course, we understand that values work together—and that it is unlikely to be one or the other—and that value relations likely work best when in relative congruence with our analysis and knowledge of the outside world. Language enables the transmission and support of secondhand cultural beliefs and values that can distort value relationships and bias the regulation of our thought processing, emotional state, and behavior.

As we experience the world, the limbic system tends to automatically infer affective values from our local experiences as subjective meaning and associate those values with our stored memories, which influence cognitive processing and action regulation. For example, when we say, "You hurt my feelings," we could just as well say this:

> My amygdala sounded an alarm when your criticism felt bad to me because I felt there was an increase in affective distance between us, which strained our bond. My orbitofrontal cortex then made a risk–reward assessment, weighted with my past reinforcement scorecard, and determined that I don't "deserve" this and that I am at risk for further disapproval that will hurt even more. My amygdala keeps tossing up different emotional solutions: to argue and fight with you, deny it or pretend it did not happen, or to run away to leave and avoid you.
>
> But I deliberately checked with my analytical brain, overruled my amygdala, and came up with a more effective plan that includes taking responsibility for my

> own thoughts, feelings, and behavior. I decided that I would not upset myself about the criticism or the brisk behavior of my amygdala, as it seems that both have value. I know that without criticism, positive feedback tends to run amuck, so the value lies in interpreting the feedback in a practical way for the consideration of making effective changes in my behavior. I asked you to tell me more about disliking my behavior, and the feedback was helpful. Of course, it didn't necessarily feel so good, but it was well worth it, as it helped me to think about making adjustments that can improve my behavior.
>
> I hear people say, "We must be true to ourselves." I don't really understand that concept. Who wants to be true to the whims of their amygdala? It makes sense to me to say that I prefer to act thoughtfully in a harmonious manner as much as possible, which I think improves my adaptive cooperation with others.

We have an informational advantage for improving our adaptive behavior when we understand

- The affective role that feelings play
- Where they come from
- How they influence us
- Our analytical options for cognitive processing
- The relationship of feelings with language, value, and bias

We often express our thinking about experiences and predictions as "having a feeling" about them, where we arbitrarily treat *thinking* and *feeling* as if they are equivalent. But are they? "How do I feel about that?" translates to "What do I think about that?" We sometimes speak of "going with our gut feeling" or "going with our heart rather than our head." Gut feelings are associated with the default mode that allows for short-term imperatives to preempt adaptive long-term planning (Mesulam, 2002, p. 20). In sharp contrast to thinking, we can say that substituting the word *feeling* for *thinking* conjures up notions of affective language, emotional imperatives, direct reactive pathways, and localized limbic system value weighting. In the default mode of neural function, the path from stimulus to response is short [analogous to direct pathways], appearance and significance overlap, and familiarity and repetition are promoted. Thus, we incur cognitive constraints that can confine our focus to a localized self-centered perspective.

> Events in the default mode lead to automatic, predetermined, and obligatory responses without allowing much neuronal space for thought, foresight, choice, innovation, or interpretation. (Mesulam, 2000, p. 25)

We see that affective language pretty much "rides along with" and enables verbal descriptive weight that explains our limbic behavior. It is a message about how the world feels to us that influences how we think about that world. After all, it is nice to know about things and conditions that feel negative so we can avoid them. Gut feelings can convey useful information, and it is usually in our best interest to know how the things in the world feel to us. It is worth remembering and taking note that we can deliberately analyze affective information before we automatically react. Our limbic-affective response to the consequences of prior events tends to bias how we feel and react to similar experiences now and in the future. We also inherit unexamined cultural value weights that can obediently perpetuate and magnify our value bias even when we never "felt" the original experience. All the while, we overlook the secondhand nature and ambiguous source of those value weights. It seems that the affective influence of our amygdala is a well-kept secret that we prefer to ignore rather than risk being nosey by confronting and further exposing our mammalian limbic heritage.

Without a value system for associating internal affective states with experience, behaviors can seem random, meaningless, and irrelevant, since our limbic system lacks the affective weighting to prefer one behavior over another. Value symmetry-breaking solutions from our limbic system automatically distinguish between what is and what is not *worth pursuing*. Limbic calculations represent the affective weight we learned from the consequences of our own previous experience and sometimes from that of others. *Worth* is emphasized to point out how, as a value function, it connects recollections of past consequences of object–action value relations and associates them with current fight-or-flight reactions, such as pursuit, approach, or avoidance. The value system has the task of discriminating behaviors according to their reinforcing, punishing, or negligible consequences and leads to the production of neuromodulatory signals that can activate or inhibit synaptic learning mechanisms (Floreano & Mattiussi, 2008, pp. 235–236):

> The set of internally mediated and internally generated rewards represents the *value system* of the organism because it discriminates implicitly between good and bad. (Floreano & Mattiussi, 2008, p. 451 [emphasis in original])

Animals rely on affective value. As humans, we can better understand ourselves by acknowledging where values come from and how value influences us and other animals. Specifically, we would like to know how language influences our value perception and the way we use our brain to think, feel, and behave. Value symmetry breaking provides an intrinsic weighting mechanism that biases perception, distinguishes relevant similarities and differences of stimulus features, and establishes decision-action preference for survival and reproduction. In humans, the evolved value distinction becomes linguistically blurred by extrinsic hand-me-down cultural values constructed and affectively encoded in the distant past. These cultural value weights are brought forward from the past and tacitly applied to stimuli and events in the present as if they are current value weights. Affective encoding imperviously excludes one-way time, analysis, and statistical predictions, with a tendency to absolutely assume "once bad, always bad," or "bad until proven otherwise beyond a shadow of doubt."

We usually take for granted and assume the reliability of our inherited secondhand values without investigating or confirming evidence for their plausibility, source reliability, or current relevance. How and under what context were these weights derived, and who configured them? Why would we really care? We seem to get by sufficiently, even when unenlightened and in the dark, about values we hold dear. So why would we waste the energy to enlighten ourselves just to discover the possibility that we have been duped? It seems much simpler and easier to assume and justify the wisdom of our values instead, since we can justify almost anything. But assumptions and ad hoc justification hardly pass for thoughtful analysis. Unexplored assumptions can easily distort our perception by biasing information back and forth in time as if past and current affective values operate equivalently on an imaginary two-way street in time. We see firsthand the beneficial relevance of explicit time versus the egregious relation between implicit time and the limitations of affective language on our brain's here-and-now thinking and predicting abilities.

But questioning our inherited cultural values and the assumptions behind them can draw disdain, reprimand, and in some instances censure or severe punishment. Dampening inquiry suppresses discovery and distorts affective evaluation toward the past, which can produce a static-like localized bias that compromises effective learning, behaving, and planning in the present. The future can still bring many kinds of change in the representation of value judgments (Jaynes, 2007, p. 425), but the lingering nature of static cultural values thwarts change in favor of stagnation.

Values can play a beneficial role, but when they are secondhand and unexamined, they can also create an affectively laden cultural millstone. Unseen by the naked eye, values can become an implicit weight that engenders an invisible affective burden, which we tend to drag along behind us throughout our lives, while ignoring or blaming others for the emotional friction and sparks we create. Examining and understanding value offers us an opportunity for improving the effectiveness of our thinking. "We have traditional values, and we've always done it that way" gives way to "Let's see if we can think of some novel methods for dealing with our problems and finding some alternative solutions that help us cooperate more effectively and fit with the present."

Value, Attraction, and Repulsion

Michael Faraday (around 1816–1818) proposed ideas of attraction and repulsion with the addition of cohesion and affinity, relating them to gravitational, electrical, and magnetic forces found in nature (Tweney, 1989, p. 356). From simple protozoa and sponges to more complex organisms, including humans, value polarity seems to provide a relevant link between structure and function related to essential survival behaviors: appetitive, defensive, and reproductive (Swanson, 2003, pp. 37–38, Jennings, 1906).

Along with structure, function, and regulation, value symmetry breaking plays an integral role in complex systems. Complex systems know their world by using information about alternate attractors. As described by Kauffman, these dynamic systems can be explained as operating by attraction and repulsion:

> Dynamical systems ranging from genomic cybernetic systems to immune systems, neural networks, organ systems, communities, and ecosystems all exhibit attractors.... [T]he alternative attractors in neural networks have been interpreted as alternative memories or categories by which the network "knows" its world. (Kauffman, 1993, pp. 191, 233–234)

A relational representation of value provides criteria for computational modeling, observing, and describing the system's dynamic interactions between attraction and repulsion. Attraction and repulsion are natural components of external fields found in nature, including gravity, electricity, and magnetism, that convey varying instances generally relating to *force* and to *cohesion* or *affinity* (Tweney, 1989, pp. 357–358). All of these may play an essential role in the selection mechanism of self-organization (Prigogine & Stengers, 1985, p. 14).

Positive and negative reactions related to attraction and repulsion can provide insight and information about state behavior and help distinguish between localized arbitrary values and systematically preferred values. Recall that *state* represents the condition or value of a system or system component at a particular time. Integrating states of matter and change with polarity, attraction, and repulsion enables the correlation of direction and velocity influenced by short-range and long-range forces. We see examples of short-range forces in covalent chemical bonds and magnetism, and long-range gravitational forces between objects such as the relationship of tides to the attractive force of gravitation exerted by lunar cycles.

Conceptually, the terms *attraction* and *repulsion* refer to value qualities and quantities along gradients representing charge, polarity, value, valence, and so on. A system's polarity axis provides differentiation of information along a preferred direction (Nicolis & Prigogine, 1989, p. 136). Charge or polarity also serves as a source for value information related to the perception of "causal" influences and decision making (Platt et al, 2008, pp. 135–136). We see examples of polarity in cell migration and axon growth, cell division, and along polar gradients during development (Nüsslein-Volhard, 2006, pp. 58–59, 81); we see examples in language where true and false polarities determine the orientation and direction of arguments between opposing parties, such as "We are right and you are wrong." For instance, one group whose perception is based on "causal" assumptions blames negative outcomes on another group and polarizes the relationship: "They are to blame for our problems. If it wasn't for them messing things up, everything would be okay. They don't know how to think or communicate the right way. We refuse to talk to them."

As we move from simplistic idealizations to complex realizations, we transition to higher-order interactive relations in the real world that replace the word *cause* in favor of *influence.* In complex systems, "causes" are best viewed as the multidimensional interactions influencing a system's state. Simple cause and effect overlooks the influence of higher-order processes and relationships to systematic laws. Relative "causal" influences represent a statistical relationship, described by the natural laws that characterize what is understood about the correlations between organisms and the independent variables and values biasing the system. A synthesis of these laws expressed in quantitative terms yields a comprehensive picture of the organism as a behaving system (Skinner, 1953, p. 35). For example, when we forgo analytical supervision and let our amygdala take charge of directing our lives, our relationships—from intimate to global—operate on the simple law of positives and negatives, attraction and repulsion, approach and retreat; from "love and hate" to "wars and rumors of wars."

In a broad sense, the interactive relations among short-range attraction and repulsion of objects and couplings of long-range forces and periodic influences can be viewed as analogous to constructive and destructive interference observed in long-range correlations of Poincaré resonances in large high-dimensional spaces. The interactions are conceptually analogous to positive and negative relationships. But these resonances involve nonlocal persistent interactions of ensembles operating within irreversible nonequilibrium constraints (Nicolis & Prigogine, 1989, pp. 165, 191; Prigogine, 1997, pp. 148–151). Perhaps, at least metaphorically, Poincaré resonances can be thought of as corresponding to resonant oscillatory relations of consonant constructive and dissonant destructive processes at a global level. Poincaré resonances may couple dynamic processes as they couple harmonics in music (Prigogine, 1997, pp. 122–123), and may have an analogous corollary with brain rhythms. Research on brain rhythms suggests that oscillations in phase with each other can create constructive resonances, amplified by their harmonious correlation (Buzsáki, 2006, pp. 142–143). We live in a world of dynamic processes, gradients, ratios, oscillations, resonances, polarities, and opposing forces of positive and negative relations.

For example, "We were on the same wavelength, our countries enjoyed a harmonious relationship, and our views resonated until your country changed its policies to dissonant ones. These new policies do not resonate with our way of thinking, and they threaten to destroy our long-term relationship." Analytical language can, in a sense, neutralize the polarizing influence of affective language and the emotional disharmony it tends to produce. "I am right and you are wrong," turns into "We have different opinions." In other words, open-ended language facilitates harmonious resonances by superseding closed-minded polarization with open-mindedness. Language's ability for describing dynamic processes related to gradients and spectrums, attraction and repulsion, oscillations and resonances, and amplification and dampening depends on the range of available language constructions, especially continuous logical functions.

Tuning the relative internal-to-external congruence of the language logically in phase with the world's analog realities facilitates parsing language into affective and effective components. We can incorporate the open-ended perspective of energy and information transfer across the whole of the systematic space that facilitates the global integration of time, space, and value context with qualitative and quantitative information.

Value symmetry breaking can convey a sense of affective emotional correspondence and effective analytical correspondence between us and the world both qualitatively and quantitatively. Understanding the correspondence among language, the brain, and the influences of affective and effective value enables integrating our experiences into a global context that translates to effective systematic thinking and communication. The more closely the language constructions match the dynamics of the brain and domain, the more congruent our descriptions and the more effective our understanding of systematic processes and relationships. Having an understanding of and access to qualitative and quantitative information helps us to coherently fine-tune congruence with the environment.

By enabling the fine-tuning of internal-to-external congruence, effective language and systematic knowledge leverage our ability to accurately represent the general relationships among our thought processes, feelings, behaviors, and nature. In other words, we enhance the reliability of our thinking and communicating by getting on the same wavelength with nature. We can effectively modulate our emotions with our new brain—neocortex and language-related left DLPFC—and still consider our affective limbic system. We can deliberately take advantage of our linguistic abilities and apply language constructions that level the communication playing field and facilitate relational equivalence across the domain. But why go to all of that trouble? Because we can improve how cooperatively we relate with others around the world—and express our preference to speak to others as we would like to be spoken to.

Value Asymmetry as Information

For life forms, information has value. Conversely, value conveys relevant information. Affective value informs us about risks and rewards, pain and pleasure, comfort and discomfort, safety and threats:

- "That sounds too dangerous."
- "I bet we'll strike gold if we keep on digging."
- "I always watch out for wasps, and I don't ever want to get stung again."
- "There is nothing like a good home-cooked meal."
- "We have it made now. We can sit back and take it easy."
- "Something's got to give here. I can't take any more of that tyrant's abuse."
- "At least we were safe living under that tyrant's threats as long as we kept our mouths shut. Now all we have is chaos."

How we perceive our own values and the values of others informs us about differences and similarities among individuals and social groups. Generally, we get along relatively well with others who have similar beliefs. We tend to dislike and avoid others when they have "strange" or dissimilar beliefs. Attraction between individuals and within social groups often falls into the category of the attraction of similar structure, or likes attract, somewhat analogous to gravity. For instance, we gravitate toward people and groups with whom we share symmetry, such as similarities in beliefs. Across groups, we tend to oppose other groups with beliefs that are asymmetrical to our own, but they still have symmetrical beliefs and attraction within their group.

Value symmetry breaking accompanies us as an everyday part of our lives that orients us to the world and influences our direction—what we approach and what we avoid. We use language daily that represents our affective values, including imperatives and prescriptives, which are shortcuts of a sort that often find their way into our thinking disguised as locally useful tools for controlling or judging others' beliefs, values, and behavior. Ironically we often see more judgment of others in people and groups that profess to judge not. When we peer behind most every imperative and prescriptive statement, we find implicit value judgments but little if any explicit practical information. And we are hard pressed to consider the admonishment to "do the right thing" as anything but an idealized judgment and hardly anywhere close to offering pragmatic information connected to physical reality. We also see that judgments separate group affiliations by affective polarity according to the principles of symmetry and asymmetry. In effect, values create affiliative bonds, and affiliation shapes our view of reality—our affiliation with the world. Polar value asymmetries 180 degrees apart simplify identifying differences.

Symmetry conveys the idea of equal measure or equivalence. Symmetry, irregular symmetry, and asymmetry pervade nature, from the inanimate to animate, and almost every step in our evolutionary development. The word *symmetry* conjures up objects that are well balanced proportionately. In a sense symmetry conveys meaning related to information. Symmetry provides a way for animals and plants to convey a multitude of messages, on topics such as genetic superiority and nutritional information, and denotes something special, something with meaning. For example, "I chose foods that I think will help to maintain my balance of essential vitamins and minerals." Found throughout nature, *symmetry* is often used to describe things of beauty and affinity. Ancient Greek thinkers, such as Plato and Aristophanes, supposed that symmetry explained the origins of love, because of our craving for symmetry. For example, humans find symmetrical faces attractive, somewhat similar to the attraction of bees to the symmetry of certain flowers (du Sautoy, 2008, pp. 11–15, 60).

In a world with many structural forms and actions, we depend on a continuum of value associations to distinguish information we deem relevant from that which we deem irrelevant for our goal. We translate and associate information that we find relevant into our beliefs about how the world works. "I have learned what works best for me and what doesn't work so well for me." When we tag structural forms and actions with value, we create asymmetrical information for understanding functional relationships contextually. These positive and negative value tags create relative value asymmetry, which helps us discriminate, sort, and clarify the relevant information from the irrelevant according to how things feel, including our learned beliefs and experiences. Thus we may say, "I believe those are the valuable issues. We can ignore the rest because they are irrelevant." In other words, our feelings shape our beliefs and vice versa. Value information gives us the ability to simply compare beliefs and actions by feel and thus weigh them by the sum of their positive and negative weights. We can mark rubbish with a negative tag to indicate its value status for elimination that informs us to dispose of it. This value-tagging is pretty much the same information mechanism of symmetry and asymmetry used by most living organisms. If an external object is perceived to have a positive or negative value, it translates to value-weighted information. Generally, neutral objects lacking emotional value have insignificant weight and appeal that compromises our attention and attraction to them. Thus, we ignore them. Values are an integral part of contextual information, and as we discovered earlier, they tend to continuously fluctuate.

Value symmetry breaking provides a fairly simple method for creating information asymmetry in a meaningful way that informs us about internal and external value states and positional relations as mutual information between us and the environment. The symmetrical and asymmetrical value relations inform and influence the orientation and regulation of our behavior in terms of distance, position, changes of position, state intensity, and changes of state: "I make a move and jump right on it when there is something in it for me. If it is good and it is close, then I go after it. If it is bad and a little too close for comfort, I turn tail and get out of there."

> The phenomena of nature must be reduced to [conceived as, classed as] motions of material points with inalterable motor forces acting according to space-relations alone.... But points have no mutual space-relations alone except their distances...and a motor force which they exert upon each other can cause nothing but a change of distance—i.e., be an attractive or a repulsive force.... And its intensity can only depend on distance. So that at last the task of Physics resolves itself into this, to refer phenomena to inalterable attractive and repulsive forces whose intensities varies with distance. The solution of this task would at the same time be the condition of Nature's complete intelligibility. (*Die Erhaultung der Kraft* [*On the Conservation of Force*], von Helmholtz, 1847, pp. 2–6; from James, 2007, v. 2, p. 668)

In other words, there is an "intelligent" [logical or practical] physical relation among an object's positive and negative states and position associated with material, value, force and energy, and actions of approach and avoidance. The value asymmetry between positive and negative creates a rather simple invertible method for coding information relevance according to polarity as positive and negative valences: "I have learned my lesson. I only go for the positives and stay away from the negatives. I know the difference between rewards and penalties. Getting rewarded feels good, and getting punished feels bad." Value encoding establishes attractor correlations with approach and avoidance that bias goal-directed behavioral regulation. Inverting either the internal or the external value sign initiates a polarity reversal that influences the direction and velocity of approach and avoidance, depending on distance and state conditions:

- "That sounds tasty."
- "I am hungry and thinking about getting some food."
- "I am starving and making a fast track for some food."
- "Get out of my way."

- "I am stuffed."
- "Don't talk about food now."
- "The thought of food makes me sick."
- "Leave me alone. I don't want to get near or even think about food."

Value quantity, quality, and polarity change over time depending on the context we encounter. Position, direction, frequency, and velocity related to motion also change. We see a sharp contrast between asymmetrical and symmetrical value relations and reversible and irreversible processes in time. Recall that the noninvertible positive polarity of time inevitably points toward the direction of the future and that there is not one known instance of negative time—that is, time running in a backward direction. A forward temporal direction encourages the persistent analysis and updating of internal and external value relations to stay current with the ever-changing values in the world around us. Since things change, including value, we would do best to keep up with real-time contextual changes that influence how effectively we perceive, interact, and adapt. Simply put, asymmetrical time alerts us to the fact that we cannot go back in time and erase our mistakes. The forward-directed asymmetry of time yields a consistent external reference in a world where everything else is changing. Therefore, time provides an invariant external reference for evaluating relational invariance with that world, including the relational invariance of language—important because language is about relationships.

Positive and negative valences create polar asymmetry, establish value relations among conditionally independent state information and meaningful content of experience, and provide a contextual sorting method for parsing that information by relevant value. With the possible exception for essentials that support life, value itself is conditionally independent of experience, or *primitive*. In other words, value lends itself to idiosyncratic association with different experiences. Value asymmetry facilitates correspondence among our ability for attending and ignoring, recognizing, describing, and understanding the relevance of domain interactions. Simply put, we develop information from symmetry and asymmetry, which separates events and experiences according to distinctive features to which we can assign value. Value primitives represent a fundamental component of local interaction with nature, from flatworms to humans. Value informs us about relations, qualitative and quantitative, affective and effective.

Affective value asymmetries influence information processing according to positive and negative weight associations that bias the sorting of similarities and differences inferred to input streams. In turn, affective processing used alone induces an emotional bias to approach-and-avoidance behavior, sans effective modulation. Value weighting related to language constructions influences our attention and neglect, perception, and behavior in response to input streams and imparts a processing bias to how we logically structure and restructure our ideas, beliefs, opinions, and predictions. Language with extended abilities across the system can recognize affective weights and deliberately modulate the tendency to default to the immediacy of automatic limbic reactions. Effective language biases the analytical cognitive weights to correspond with the external world that extends our predictive effectiveness forward in time: "Even though it might feel good now, I choose not to participate. Opting out will likely be the best for me in the long run." In other words, what feels good is not always good for me, and what feels bad is not always bad. Examples would be drugs such as heroin in the former case and a bitter medicine in the latter.

Symmetry, symmetry breaking, and asymmetry provide a structural method and functional component for distinguishing and separating information content according to similarities and differences. It is difficult to identify correlations between the parts of a homogeneous mixture when the parts that form the whole lack regularly occurring distinctive features that we can visualize. When all of the parts appear symmetrical, or make up a symmetrical whole, it confounds efforts to distinguish one part from another. If on further inspection we can observe some different symmetrical or asymmetrical features that occur regularly, those features can distinguish identifiable information we recognize as salient.

Historically, we see examples of mythical thinking where right and left asymmetries were seen as symbols for such polar opposites as good and evil. In ancient times left-handedness was seen as a "sign of the devil," and some left-handed people were burned at the stake. The word for left in Latin is *sinister*, implying evil (Weyl, 1952, p. 22), and the English word *left* comes from the word *lyft*, meaning "worthless." This superstitious habit of thought persists to different extremes in cultural beliefs, such as in the milder case where the right hand, not the left, is used for benevolent greetings. It is not unusual for parents to attempt to correct children when they show a preference for left-handedness.

Distinctions based on symmetry breaking and asymmetrical information occur naturally, as in stereoisomeric molecules that consist of the same type and number of elements in the same relative configuration, but with a polarity axis and asymmetrical patterns displaying left- and right-handed structures that rotate the plane of polarized light in a preferred direction. French scientist Louis Pasteur, the founder of modern biochemistry, regarded this phenomenon as one of the basic aspects of life (Nicolis & Prigogine, 1989, pp. 138, 144–145).

We see an exhausting list of examples of preferred asymmetry in evolutionary biology, including DNA nucleotides and RNA in cells, which favor the right-hand D-isomer (Nicolis & Prigogine, 1989, p. 144) and the left-hand L-amino acids. Left- and right-handed configurations "appear" as mirror symmetry. Mirror-image asymmetry typically results in significantly different chemical properties, providing a natural biological basis for distinguishing between the two molecules. We can visualize the difference in symmetry and asymmetry by placing our two hands together with palms facing. Every part of one hand matches the similar part on the other hand—they seem identical and symmetrical. But put both hands palm-down on a table, and we can easily distinguish the left-to-right difference as asymmetrical.

Three-dimensional structures viewed across one dimension appear to have bilateral symmetry with two halves that can overlap. But in a three-dimensional world, the mirror images have left- and right-handed asymmetry; if viewed in the frontal plane, the images are reversed in the front-to-back direction and nonsuperimposable. Other examples of asymmetry include a group of otherwise structurally symmetrical molecules where some have no charge, some have a positive charge, and some have a negative charge. A social group typically has symmetries that identify them as a part of the group affiliation such as "all members hold a PhD in linguistics and are associated with the same university." But each member exhibits some individual asymmetry as well where some are right-handed, some are left-handed, some teach, some do research, some agree with the predominant theory, and some do not.

In other words, rather than a one- or two-dimensional world, we live in a three-dimensional world with spatially distributed height, width, and depth correlated to our contextual interactions across time, space, and value. But arbitrarily trying to compress the dimensions by subjecting relevant variables to omission by exclusion or localized misclassification usually leads to traps at local cul-de-sacs. Why? The global domain includes all of the dimensions, variables, and rules for parsing according to the preferred information asymmetry that informs the ordering of relations across the whole of the space.

For example, consider a large population with a mixture of individuals, some having brown eyes and others having blue eyes. A person with brown eyes commits a perceived atrocity. The group with blue eyes associates this atrocity and a negative value weight with the whole brown-eyed group, and castigates them as brown-eyed evildoers. In other words, we can artificially create localized value asymmetry by arbitrarily associating emotionally weighted value tags to other individuals and affiliative groups. This emotional weighting violates the preferred asymmetry of the global classification space and creates a localized limbic classification by parsing and labeling with an arbitrary local value bias that contorts the global fitness landscape. Arbitrary value labeling conjures up the following associations: closed-minded, narrow-minded, small-minded, short-sighted, discrimination, predetermined, insular, intolerant, bigoted, prejudiced, preconceived, and nonobjective.

Arbitrary asymmetry resulting from localized value designation conveys affective information, which has random asymmetry from a global perspective of effective information. This systematic error occurs where inherited localized values from the past are applied without inspection as if they are currently relevant, which tacitly assumes two-way time. The implication is that the past and present are interchangeable. Misappropriating time asymmetry and dynamically misrepresenting the static nature of beliefs from the past confounds contextual relevance in the constantly changing present and impedes the systematic usefulness of information for computing reliable predictions. This violation of time asymmetry creates egregious systematic errors when we try to use static local information from the past to predict the dynamic future with certainty.

Globally relevant asymmetry conveys systematic properties as nonrandom information, indicating the preferred asymmetry adopted as information in nature, such as in the polarity of chemical bonds. We see a familiar local example in politics with affective descriptions of right-wing or left-wing groups, which also demonstrate polarity. "They are liberal left-wingers and want change, and we are conservative right-wingers and want things to be *correct and proper* the way they used to be and should be." But after all is said and done, those are arbitrary values, time does not stand still or go backward, and it seems that in nature, we are essentially one and the same peoples but with different beliefs and opinions.

As in nature, some molecules have both left-handed and right-handed versions that have the same composition and appear similar, but only one is known to have useful biological properties, such as the chiral symmetry in molecules that have been irreversibly incorporated by evolution into life forms (Prigogine & Stengers, 1985, p. 285). Some asymmetrical differences seem coincidental, random, and unregulated, while others seem more likely due to regularities, implying natural regulation. Recognizing the biologically preferred asymmetry for nonrandom information and the regulatory influence of value on behavior in attractor networks allows us to develop representational models that better correspond to the environment. By locating the feature differences that allow us to break symmetry, we increase our ability to assign value as useful information (Nicolis & Prigogine, 1989, pp. 141–145). *Useful* means that we can develop nonrandom information that helps with regulating goal-related tasks. In evolutionary terms, *useful* means something that helps with survival and propagation of the species.

Biologically preferred information asymmetry logically corresponds to the systematic domain structure and facilitates effective processing and transformation to reliable domain knowledge. We expect that accumulated information and beliefs about how the world works will correlate with some degree of probability to beliefs about how language, brains, and the natural world work, and how they work together cooperatively. The difference between the predicted and actual correlation demonstrates the extent of systematic blindness. It is also fairly simple to measure the concern by asking each individual how much they care about understanding language, brains, and nature, measured on a visual analog scale. Preferred information asymmetry supports effective adaptive knowledge. Take, for instance, two different approaches:

- "You never know what will happen if we change anything. I have a gut feeling that it could turn out awful, because crazy things happen all the time. But if things get really bad, then it's a good idea to shake things up."
- "We understand the difference between emotions and the fear of change, and we also know the relationship between physical laws and making predictable statistical changes that are based on that knowledge. Since we are not making random changes but making systematic adjustments to our processes calibrated to error feedback, these changes will likely decrease our systematic errors."

We see "shaking things up" as a common local method employed to try to improve solutions, but no matter how many times we randomize, we still have a random system that still makes little sense to us, so we continue to prescribe therapeutic shake-ups because of the erroneous belief that somehow things will fall into their proper order. When we understand how the system works, we tend to shun randomization as an approach to produce better solutions, since it demonstrates low reliability in an ordered domain. We prefer effective language and deliberately informed analytical value as a nonrandom systematic method to bias learning, behaving, and adapting to change: "We understand that we can dampen emotional biases and enhance systematic effectiveness by analytical reasoning." Simply put, we prefer a nonrandom systematic bias that enhances reliability rather than one that is randomly conjured up with disregard for reliability.

We see distinctions between arbitrary, locally defined asymmetry and globally defined asymmetry. Global asymmetry leads to information parsing according to general relations observed across the domain that correspond with regularities and systematic knowledge. We develop effective knowledge about relationships by clearly distinguishing between idiosyncratic asymmetries in local information and preferred asymmetries in systematic information. Variably referenced local asymmetry corresponds to affective input and automatic emotional processing of locally encoded positive and negative values. We generally prefer invariantly referenced global asymmetry that corresponds to tuning effective output and deliberate analytical processing with systematically informed values. Affective asymmetry corresponds to emotional processing, and effective asymmetry to analytical processing. For instance, consider the following two different approaches:

Affective asymmetry: "It doesn't feel right, so we're not going to change anything. I know how I feel, and I know what will happen if we change. We all know that change is dangerous, and you never know how bad it could turn out; it could be a step backward. I know I'm doing the right thing by sticking to my guns. I don't care if the business climate changed; we have always done it this way. It's always worked, so there is no reason why it won't work now. Besides, you never know what will happen next in this quirky world. Things happen randomly out of the blue all the time. Don't worry—if something bad happens, I will think of something on the fly. I always get a feeling when things get urgent. That's when I roll my sleeves up and jump right into the thick of things, knowing full well that I can always fix it by shaking things up in the spur of the moment. Don't forget, we know the rule around here—if it's not broken, don't fix it!"

Effective asymmetry: "Since we know how the system works, we can calibrate input sensitivity and tune output adjustments to match external changes. Why would we think supply and demand doesn't oscillate like the rest of the world? Why would we limit ourselves to a localized micromanaged view? We know that going back and second-guessing our decisions offers little if any benefit or consolation because of irreversible effects of our actions, but we also know that not changing when the world around us is changing tends to decrease our effectiveness. Matching the asymmetry of our analytical information with the natural asymmetry of systematic information enables us to effectively regulate change and adaptation. We like equivalence that corresponds with how things work effectively (analytical equivalence). That means we know that past information is static, but we can use our knowledge of regularities to help make reliable statistical correlations and predictions. We can systematically tune information in sync with external changes (asymmetrical equivalence in time).

"Change may not feel good, but it improves our chances of adapting to a changing environment. We like looking at the big picture and managing from a macroscopic perspective rather than taking a microscopic perspective, scrutinizing through a tiny looking glass, and micromanaging. We hardly see a viable option for ignoring the probability of systematic errors, blindly waiting for the inevitable crash, and attempting quick-fix solutions or arguing right and wrong on the basis of emotions. We prefer to elaborate on the problem space, minimize our assumptions, and gather and analyze enough evidence to make a statistical decision that is based on our knowledge (equivalence of statistical relationships between changing external information).

"We try to avoid emotionally charged decisions that are based exclusively on gut feelings, on how we 'feel' about solutions. We prefer applying analytical methods that correspond to the real world and are based on reliably informed positive and negative feedback. We understand that we probably can find more effective solutions to the problem on the basis of goodness of fit (analytical attraction and repulsion)."

Local linguistic asymmetry typically displays random idiosyncratic referencing, a subjective nature, and local variance among observers. Global or systematic information has analytical asymmetry that matches the structure of the external system with relative invariance among observers. When the asymmetry of internal and external information matches, we have equivalence, or internal-to-external symmetry. But matching does not mean we can do no wrong; it means that with effective language we increase the probability of finding systematic solutions. We know that localized language suffices on average and that systematically informed processing increases the possibilities for optimizing global solutions. We see a discrepancy between affective emotional and effective analytical information and processing according to the local and global representation of information.

Affective and effective value asymmetry distinguishes relational information about interactions among the entities of a complex system. Variable differences in affective value fail to provide reliable information for effectively adjusting the output, similar to how arbitrarily referenced localized asymmetries and coincidental subjective beliefs, likes, and dislikes compromise reliable predictions. Effective value correlates analysis of preferred information asymmetry across the domain that facilitates error tuning and more predictable outcomes based on systematic knowledge. The coherent correlation of systematic representation and informed value processing influences adaptive decisions, behaviors, and outcomes (Johnston, 2008, p. 279).

Value Bias

Value biases the regulation of action. Associating relevant value information with the system's current state aids in choosing the most appropriate response in a given context—that is, the selection of control policies for performing an action (Shadmehr & Wise, 2005, p. 447). Plausible information and reliable value promote logical consistency that facilitates more predictable consequences (Jaynes, 2007, pp. 424–425). A value-based model can provide a biologically tractable method for credit assignment and autonomous navigation (Burgess, 2007, p. 730). Modeling the world of values in which we live allows us to address complex issues such as value symmetry breaking and sensing, monitoring, and fine-tuning the value bias responsible for regulating interactive responses to change and uncertainties. The reliability of the value bias influences adaptive effectiveness, robustness to disturbances, learning, and overall fitness performance. The reliability of the output and outcome predictability can depend on the ability to differentiate effective and affective values.

Values in a dynamic world persistently fluctuate. If we rigidly hold to static values, it confounds our ability to adjust our value bias over time according to learning from changes in decreasing or increasing systematic errors, and impedes adaptively changing our reactions. Instead, a static value bias leads to fixed, reptilian-like reactions. What we value—and how we emotionally infer or analytically calculate the value—influences our actions, error feedback, adaptive tuning, and learning. Learning from mistakes gives us error information about our values in relation to the context of our experience, such that we can update and adjust those values to accommodate effectively regulating our adaptive behavioral responses. Spurious positive and negative input–output values subjectively inferred from random coincidental object relationships produce idiosyncratic value biases.

For example, children often develop a strong aversion to a particular style of food when they coincidentally get sick after eating it or simply find the taste too strong or unfamiliar. For years, they will reject anything that even looks like the same kind of food, even long after they have become adults who might be expected to have less reflexive-like biases. Eventually, some will likely run into difficulties when in situations where eating their disliked foodstuff seems like the best option. Some will discover that they no longer find it awful, and some may even come to prefer what they once loathed. In some cases, a value apparently set in the black-and-white stone of childhood can yield to the flexibility of adult preference. Conversely, overvalued cultural beliefs tend to resist change as if rigidly chiseled in stone.

Rather than defaulting to relatively static values and homogeneous, one-size-fits-all modulation, persistent value tuning can enhance a contextually preferred value bias and adaptive neuromodulation. *Neuromodulation* refers to diffuse regulation of adaptive changes among several populations of neurons that influences overall brain activity. This diffuse modulation facilitates flexible switching between different modes of behavior as environmental conditions vary. For example, an increase in norepinephrine is accompanied by increased arousal when we are faced with unfamiliarity, exploration, and increased novelty in the environment. Goal-related dopamine levels change in relation to satiation and deprivation, which influences exploration and exploitation. Serotonin associated with relaxed states, familiarity, exploitation, and decreased arousal tends to accompany grooming and maintenance activities such as sleep. Overall neuromodulation tends to shift the value bias for input–output functions and thresholds, which influences approach-and-avoidance behavior and learning. We can think of neuromodulators as relevance indicators that help the organism learn what to attend to and what to ignore (Pfeifer & Bongard, 2007, p. 167).

Neuromodulation tunes the general parameters of information processing to adapt to the particular demand characteristics of the organism's environment (Spitzer, 1999, p. 271). The resulting flexibility and adaptability creates the option of choice among the various possibilities at hand, and since choice is mediated by the dynamics of the fluctuations, it provides the innovative element necessary to explore the state space (Nicolis & Prigogine, 1989, p. 218). In turn, neuromodulation influences actions, learning, and the rate of learning.

Learning algorithms can be thought of as efficient devices for searching the parameter space for combinations of values that optimize some input–output function (Churchland & Sejnowski, 1994, p. 134). We see this input–output relationship with value and learning in simulations with artificial neural networks. The value associations in neural networks represent a foundational relationship between the physics of the machine and the algorithms of the computation (Haykin, 2009, p. 689). Value parameters, with positive or negative valence, influence the tuning relationship between input and output as an input–output biasing function. Differential modulation and feedback allow learning rates to change slowly in familiar situations where representations are strengthened, and to change more quickly for rapid learning in novel situations (Murre, 1992, pp. 22–28). Generally, the more feedback we get, depending on its fidelity, the easier the learning problem.

Analyzing and fine-tuning the value bias for predictability through learning from experience can generate real-time input–output values with reliable correspondence with external rewards and penalties (sometimes referred to as attractors). In turn, the associated value bias influences performance correlated to information processing and modulating behavior in response to change. Since we output to the environment and the environment inputs back to us, extended language takes advantage of analog loop tuning that enhances continuity and logical consistency for smoothly calibrating input sensitivity, tuning the value bias, and decreasing systematic errors. Loop continuity influences the coherent correlation among perceived attractors, which facilitates efficient learning in open-ended, complex systems. Evaluating attractors from a systematic perspective enhances logically consistent learning and predictability that translates to effective output adjustments. We enhance adaptation to external relationships, since our thinking logically corresponds with our sense of what is happening outside.

Conditionally independent value enables associating value with experience, and monitoring and modulating the value bias between sensory input and goal-directed motor output. When we interact with the environment, we use feedback information about the temporal context, spatial position, value, and actions of extrinsic objects. Value bias that is sensitive to contextual information facilitates effective regulation of interactions, since attractors evaluated according to changing context enhance reliable feedback of information about that experience. In effect, we can adjust and adapt goal-related actions in response to specific external changes in the context in which we encounter them.

For instance, it is pouring down rain outside. Partner number one gets up early in the morning, looks out the window, and exclaims, "It's going to be a great day!" Partner number two gets up, looks out the window, and bemoans, "What a rotten day this is going to be!" Partner number one is a cab driver who makes more money when it is raining, and partner number two has the day off and was planning a trip to the beach. We can see that changing context leads to changing the way we frame and perceive a situation. What we perceive or think of as bad in one context or frame, we may think of as good in another context, depending on how it feels to us or how it influences our goals.

When our valuations are flexible, we can easily understand and consider different perspectives and alternative solutions that facilitate adjusting our goals and adapting to change. Flexibility enables tuning our bias as best we can to accommodate and match a changing external world in a more predictable way. In contrast, rigid affective values, often encountered in overvalued beliefs, tend to constrain our perspective to one dimension of either "good or bad." Effective value engenders flexible analysis of context, a broader perspective, and consideration of multiple optional possibilities or equivalent solutions. We have the option of language that enhances flexibility and can improve how effectively we perceive, predict, and deal with value in changing context.

In terms of intentions, value information associates relevant stimulus input with relative influence of actions and contingency predictions to adapt motor output. In turn, the value information allows an organism to modulate fluctuating internal–external value symmetries and asymmetries by adaptive reactions. Intention results from the correlation of external actions and relative value perceived as relevant by the organism. For instance, when we see a stranger running at us and waving a big stick in a menacing manner, we tend to interpret his intentions as malicious. Negotiations aren't even a fleeing thought, and we prepare to fight or quickly flee. If we recognize someone we know who is a friend, we might think they are celebrating and wave back as they approach.

Changing the value bias changes the sensitivity of perceived input that influences attention, stimulus discrimination, arousal, and the modulation of goal-directed output gain, where *gain* is the ratio of a system's output to its input. Value provides the organism with the relevant meaning of relationships among attractors and a bias for modulating learning and action according to changes in the system's state and context. For example, we may feel sad when we miss seeing a longtime friend who has moved away, but our feelings are mixed with anger about the departure. We feel cautiously happy upon hearing about our friend's intended return.

We generally disregard the possibility of coincidences and represent information about intentional influences in a causal manner as directed action and value. Anatomical evidence suggests the interconnectivity of sensory, behavioral state, and cognitive systems for information processing in the central nervous system directs behavior via the motor system as a whole, which provides feedback signals to those three systems. They in turn provide control back to the motor system for behavioral regulation (Swanson, 2003, pp. 88–91, 97–98, & Fig. 5.5). The intensity and direction of reactions generally depend on a perceptual sense of positive or negative intent, a risk–reward and current state assessment, and the estimated cost of energy and effort. The assessment of constraints and the perception of rewards and penalties influence response preferences and thresholds for adaptive actions. The output, driven by a value function based on the system state, conceptually allows for the construction of a general model of motivational valence and goal-directed behavior (Burgess, 2007, p. 733; Lieblich & Arbib, 1982).

The relationship between value and regulation suggests that the probable goal-directed behavior of an organism in a given state in terms of position, motion, and velocity describes a relative value bias correlated to the perceived relevance of stimuli, behavioral choices and constraints. The informational correlations between internal and external position, and changes in position and value, provide contextual meaning that informs processing and behavioral regulation. In turn, persistent change of context influences continuous monitoring of state changes, the rate of those changes, and the flexible selection of alternative choices for directing behaviors.

A simple way to think of value is in terms of *essential values* that we will die without; these are *needs*. When we are hungry, we seek food because we want to eat. When we are in pain, we seek relief because we want the pain to go away. We learn other values from our culture that are *arbitrary* values, but we treat them as if they were essential values. In effect, we inflate and leverage the emotional bias in our favor by calling them needs. For instance, we say we need a new car, we need a new house, we need an exotic vacation, we need a new television, and so on. These are preferences, but localized language tends to rely on value bias for leveraging emotional states to inflate the importance of our individual desires, persuade others, and, in turn, justify our behavior.

Consider, for example, the statement "I *need* to go to the store and pick up some cookies" versus the statement "I would like to go to the store and pick up some cookies." Some may argue that confounding needs and preferences is a trivial matter. But such an argument hardly addresses the surreptitious nature of our tendency to emotionally leverage our speech to get our way, sometimes called "little white lies" or "slight exaggerations" that we make to get what we want, to get our goals met. The point made here is that we can simply choose words that distinguish needs and preferences, such that our speech behavior—what we say—coherently matches what we mean, if we care to do so and if we "know" what we mean.

On the other hand, effective language analyzes and relegates essentials for survival to needs and designates others as wants, preferences, likes, and so on. In some instances, such as when we have made an agreement to follow guidelines or policies, we state our related behavior as "have to" or "need to." We can easily resolve this parsing problem by expressing our behavior in terms of *requirement by agreement*. For example, if we are a company that makes widgets and have agreed to produce them according to a particular protocol, we are required to go by the protocol as we agreed. We do not have to or need to, but if we care to stay in business for the long run, it is likely in our best interest to do our best to produce widgets according to our agreement. When the expected and realized products are in agreement over time, we have demonstrated *reliability* and likely improved the chances for continuing in business, all other things being equal. We also realize the difference between imperatives and choices.

Effective predictions and choices that produce a decrease in errors and an increase in adaptive solutions typically depend on analysis of information and plausible evidence to produce reliable value weights for effective arguments. Effective language enables us to distinguish between emotional information weights and analytical information weights. When we prioritize our ability for analytical thinking over default to emotional thinking, we enhance reliably informed weights for processing and producing effective solutions, and dampening limbic-weighted reactivity. Arguably, both effective and affective cognition can be more or less helpful in different contexts. But we typically benefit from having access to effective cognitive skills that can provide ongoing oversight and analytical supervision of our emotions and the monitoring and modulating of our actions, which tends to enhance adaptive effectiveness.

Value information and our internal coordinates let us know where we stand in relation to what's going on outside. How we assign and calibrate relevant value to stimuli inputs and what we learn about those inputs biases how we assign weight to our choices and how effectively we direct and correct our behavioral output. Value can be thought of as information-biasing weights that influence how we perceive the shape of the fitness landscape, such that emotional information distorts and contorts, and analytically informed learning smooths and soothes the fitness landscape. Simply put, value adds weight to information, and effectively weighted and correlated information is valuable. Next, we will continue to explore the importance of value and discuss how we can further exploit its informational properties for enhancing effective thinking.

CHAPTER 14

The Electric Nature of Value Space

EVEN THOUGH WE DO not typically think in these terms, value, charge, and polarity provide positive and negative symmetry breaking tags that distinguish information asymmetries useful for information recognition and processing, regulating, orienting, and directing actions. Conversely, null or zero value means neglect or ignore. The electric nature of value space represents general properties and processes of nature, and more specifically of life itself. We see regulatory processes with dynamically changing relations of positive and negative values across every living cell and species.

Electrical Value and Energy

In the brain, the behavior of single neurons can be characterized in terms of their electrical properties. Charge is often thought of in terms of energy and electricity. Electrical charge is a fundamental property of matter (Ellner & Guckenheimer, 2006, p. 93). In a more general sense, *charge* can describe an affinitive property of energy or force related to polarity, valence, value, and so on. Polarity commonly occurs in the inanimate world, from magnetism to electric energy, such as lightning, and also in the animate world in the cells and neurons of living organisms. Furthermore, neurons function as dynamic systems, with signaling properties resulting from the actions of polarizing ions with positive and negative charges, including Na^{+}, K^{+}, Ca^{2+}, Mg^{2+}, and Cl^{-} (Izhikevich, 2007, pp. 1–25; Lytton, 2002, pp. 190–191; Loritz, 2002, pp. 39–40).

Neurons, which transmit signals in the nervous system, represent the evolution of increasingly complex mechanisms involved in neuronal synaptic regulation by neuromodulators, neurotransmitters, and their receptors. (For a review of receptor evolution, see Ryan & Grant, 2009.) Neurons, receptors, and ion channels represent the basic building blocks of elements for internally regulating interactive relationships with the external world. The shape and function of receptors result from the interplay of molecular forces of attraction and repulsion that can change how information, including ions, is transferred across a cell membrane (Feldman, 2006, pp. 45–54). We see a relationship with action potentials in ion channels and rapid movement in response to environmental change in paramecia when they are prodded from the rear and in leaf-closing in higher plants such as the mimosa plant (*Mimosa pudica*) and the carnivorous Venus flytrap (*Dionaea muscipula*) when trapping a victim (Hille, 1992, pp. 530–534).

Ionic channels in more complex organisms such as vertebrates are well conserved and often no more complex than those found in less-complex organisms such as mollusks. There is evidence and conjecture that the major channel families have been around for some 1400 million years in eukaryotes and 700 million years in animals (Hille, 1992, p. 542). This small number of basic circuit elements from a long time ago established a fundamental property across all species (Koch, 1999, p. 198). We see a relationship between ion pumps and energy, where active membrane pumps in single neuronal cells consume energy in favor of an increased internal negative charge that creates an energy gradient, enabling the cells to perform work. In brain cells, gradients typically result in an approximate internal charge of –60 to –80 millivolts.

Neurons can be characterized as electrical devices and in turn can be understood computationally as information-processing devices (Eliasmith & Anderson, 2003, pp. 9–10). In other words, a neuron can be seen as a dynamic piece of computational machinery (Buzsáki, 2006, p. 144). The evidence for computational power increases when considering modules and groups of neurons operating at a population level (Spitzer, 1999, pp. 72, 147). Also, some consider it easier from a numerical and conceptual viewpoint to study the dynamics of populations of neural units, which simplifies the dynamics of studying individual units of biophysically complex neurons (Koch, 1999, p. 173). Populations of neurons provide a useful perspective for gaining insight from a dynamic context in terms of oscillations.

Charge, equivalence, and polarity operate ubiquitously throughout nature in organic and inorganic chemical bonds and electron orbits (Lytton, 2002, p. 17). We see charge in inanimate objects, including atoms, ions, and elements, and in animate objects, including cytoplasm and cells (McCaffrey & Macara, 2009), microtubules with well-defined polarity (Solé, 2011, p. 93), and organisms such as bacteria, amoebas, paramecia, sea urchins, and worms (Jennings, 1906; Swanson, 2003, pp. 37–38). We see a commonality of action and charge and of ions and polarity in the nervous system of all animals because they are essential for survival: ingestive (or appetitive), defensive, and reproductive (Swanson, 2003, pp. 37–38). We also see an abstract analogy to force of action and emotional charge in humans, with attraction and repulsion related to bonding, affection, affiliation, and forces of habit alongside hostility, hatred, and war. These human traits are not far removed, if at all, from the behavior of many social and unsocial animals in the wild.

The value correlation in autonomous organisms influences attending and ignoring, cooperating and competing, and decision making that leads to inhibiting and activating actions. In living systems, charge and polarity serve as crucial components for information processing (Jennings, 1906) as positive and negative values of goal-directed actions, in abstract information-processing related to robotics (Floreano &

Mattiussi, 2008, pp. 516–518), and in conceptual phenomena such as language. Polarity gradients play a role in the development, maintenance, and reproduction of life forms (Nüsslein-Volhard, 2006, pp. 21, 57–74, 78, 81).

In developing tissues we can frequently observe gradients of substances such as ions or metabolites, and it is natural to conjecture that these gradients provide a kind of coordinate system for developing tissue that conveys positional information by which individual cells can recognize their position with respect to their partners. It is therefore likely that transitions mediated by chemical substances and leading to symmetry breaking are one of the key features of life amenable to self-organizing phenomena (Turing, 1952). "This astounding idea was enunciated for the first time in 1952 by the British mathematician Alan Turing and ever since has been a constant source of inspiration for physicists and biologists alike" (Nicolis & Prigogine, 1989, p. 36).

Value relates to charge and force by reinforcing attraction and repulsion, which influences gain regulation, driving output velocity as the magnitude of approach-and-avoidance behavior (Lytton, 2002, p. 171; Siegel, 1999, p. 137). These energies and forces demonstrate value gradients, polarities, and observable symmetry breaking that corresponds with the regulation of approach-and-avoidance reactions within the value space.

> Nature allows us to suppose that curious sum of distances and velocities which for want of a better term we call "energy." (James, 2007, v. 2, p. 670)

Value Space

The space of values for local subsystems and the global system constitutes value space. *Value space* refers to the constraints on the available values acceptable and realized in the domain. In nature, affective values can be realized as positive, negative, or null. Value space contains positive and negative values that enable symmetry breaking by associating value features with external objects, positions, places, and things. We have established that value gives affective meaning to the functional space by establishing a representational mapping of local attractors in the system. The relative values attributed to, afforded, or inferred about objects, positions, and spatial coordinates are biased by the organism's genotypic inheritance, associative learning from environmental interactions, reinforcement history, and, in humans, the cultural belief and value systems acquired semantically via language (Luria, 1981, pp. 1–13). We will explore this semantic mapping later in more detail.

Value space operates somewhat like salience maps (Itti & Koch, 2001; Shadmehr & Wise, 2005, p. 497) or priority maps (Rolls, 2008, pp. 390, 421), in that these approaches attempt to give meaning to the functional space by detecting, mapping, and distinguishing gradients, variations, and differences among local attractors in the system. In turn, establishing value maps influences future perception and cognitive processing by reinforcing or challenging those value systems. Value space is proposed to have particular relevance for quantifying and mapping attractor states in dynamic systems (Shadmehr & Wise, 2005, pp. 278, 496–520) and accumulated value features associated with likes and dislikes, what to approach, and what to avoid (Kahneman, 2000, pp. 767–768). Value has also been proposed as an energy function for describing familiarity context discrimination (Bogacz et al, 2001, p. 430). Together these approaches yield broad implications for understanding the differential influence from affective and effective value information on goal-directed behavior.

Most mammals, other primates, and humans appear to sense, navigate by, and record events according to position as relative place coordinates or spatial grid locations, object locations, and correlated actions with relevant value. The concept of value space conveys information correlating task-directed output to goal-related task geometry. For reaching, pointing, and grasping, Euclidian, Cartesian, or spherically defined coordinates can describe local goal-directed behavior. Value representations contribute to the use of external world-centered and internal self-centered Euclidian two-dimensional and Cartesian three-dimensional perspectives to represent landmarks, objects, and task geometry relative to domain learning and navigation (Buzsáki, 2006, pp. 302, 308; Morris, 2007, pp. 618–619; O'Keefe & Nadel, 1978). These geometric relationships can provide mapping with which to correlate value with changes in position and changes of state. In effect, value appears to fundamentally drive and bias stimulus detection, reasoning and decisions, and motor output gain (Bizzi & Mussa-Ivaldi, 2009, pp. 541–543).

Position can be defined by three-dimensional spatiotemporal coordinates, and time is assumed as a constant; the coordinates transform to a general position amenable to defining in terms of relative value, where value is a state function related to the spatial position, distance, and change. We can think of energy and value relationships in metaphorical terms as objects that have a certain energy value. The objects have an abstract positive value and can produce a certain amount of energy. By following the value gradient, for example acquiring and ingesting value objects such as glucose, an organism can survive by building up a reservoir of energy as the capacity to perform work.

In this case, the organism searches for and collects positive-valued objects while avoiding negatives such as poisons and predators. Internal position correlates with affective problem state represented by relative affinity, depending on satiation, deprivation, physical constraints, and distance to rewards as solutions. External places, objects, and positions correlate to positive and negative values in the solution search space, such that problems can be viewed as having a negative valence, and solutions, a positive valence, which represent an attractor relation. Value symmetry breaking enables parsing the value search space according to positive, negative, and null. Partitioning decreases the dimensions of the goal-related search space and simplifies the rule of relations and actions, such that the general rule is to neglect or ignore null, approach positive, and avoid negative. Thus we see across life forms the general tendency of organisms to favor positive and disfavor negative conditions as a fundamental evolutionary principle (Jennings, 1906). It appears that evolutionary processes have resolved the curse of dimensionality by leveraging symmetry breaking, generalizing variables, and evolving parsimonious dynamic controllers co-opted to rewards and penalties.

An organism can be queried at any time relative to differential internal–external value gradients, and statistical predictions can be made about goal-directed movements in space commensurate with the context of a constantly changing problem and solution space. Estimated value differentials offer probable explanations for past changes of directed action relative to changes in position and state values. As a function of energy according to charge, value enables scalable attractor representations as ratios correlating qualitative and quantitative changes in internal and external positions and states conducive to regulating actions in response to change. The dynamics of the domain and the ability to integrate positive and negative value correlations make an oscillating system a prime candidate for effective regulation. The scalability of oscillations is well documented in rhythms of the brain (Buzsáki, 2006, pp. 119–121).

The general position of an external reward or penalty can be expressed as a function of value. Depending on position and spatial distance, the internal value state, and energy, an organism's brain can regulate activation and inhibition required for approach-and-avoidance behavior with a fairly simple oscillating controller (see Heerebout, 2011). Value associated with an external object or place position depends on recent history, experience, conditions, and context independent of the object's or place's identity. Affective and effective values represent an internal–external value-biasing function for differentially modulating goal-directed approach-and-avoidance behavior in relation to changes in position and positive and negative valence.

Energy, affinity, and conditionally independent value can be encoded by affective or effective language for representing and mapping value to object-action relations. The affective and effective representation of value differentially biases the coding and decoding of information and the integration, storage, and recall of memory. The analytical nature of effective value enables a higher level of abstraction for separating and integrating quantitative and qualitative information. Effective value extends encoding for explicit analytical transformation, analog input–output computation, and outcome prediction. In turn, we see enhanced effectiveness for evaluating risks and rewards, directing and modulating action outputs, monitoring energy requirements, and processing feedback for error corrections. In contrast, affective value encoding represents a qualitative emotional bias from two-valued limbic logic that affective language implicates as digital, thus influencing perspective, input sensitivity, and output reactivity.

Attractor mapping can be modeled as a value field, analogous to a topological map or an energy field, where internal and external values are dynamic and continuously changing in real time. For example, we can think of the energy relationship as we traverse our daily experiences. Some values change depending on our recent accumulation and work usage, such that we have different states of satiation and deprivation, which influences our value preference and direction of action. The world and our internal system maintain a persistent state of flux. Our energy supply decreases as a function of the amount of work we produce; even when we are idle or when we are asleep, our cells require energy. We cannot turn off live cells to save energy as if toggling a light switch or a computer on and off, because "off" likely results in cell death. With so many internal and external changes going on at once, it makes sense to think in terms of dynamic analog relationships, such as oscillations and continuous functions rather than digital, discrete on or off, true or false, and so on. Our brains consist of networks of persistently active cells that can spontaneously transmit messages even while we are asleep. Even though we think of the brain as resting during sleep, arguably at times it may consume as much or more energy while asleep than when awake.

The possible states of a network can be metaphorically described as a landscape of energy where the attractors form valleys between mountains of unstable network states (Amit, 1989). The external value positions have an invertible polarity that dynamically fluctuates, as do the gradients surrounding the value position that indicate the increasing or decreasing strength of the charge. A dynamic attractor field suggests that the relative direction and velocity of navigation may be influenced to some extent, depending on context and distance in time and space, according to an inverse power law correlating the relative forces of attraction and repulsion (analogous to Coulomb's law), similar to gradients of attraction forces in a gravitational or magnetic field. Rewards change over time, and we expect that increased time and distance from the predicted rewards will generally lead to discounting their value and a preference for receiving rewards sooner rather than later. Thus, expected future value changes the shape of the fitness landscape and the shape of later behaviors.

Positions have a relative distance between them that correlates to the elapsed time it takes an organism to traverse the distance at a certain velocity. Thus approach-and-avoidance velocity represents a function of the systems state and attractor forces in relation to the traversable distance. Within distance constraints, velocity is influenced by the charge differential of attractor forces and physical domain constraints, including gravity and friction, energy requirements, the organism's maximum velocity, and the velocity of the target position. Value representation allows integration of a dynamic field description with internal–external attractors correlated with change of state, position, and direction. In other words, value gives regulatory meaning to position and direction because, as a state function, value connects position and direction with action.

The affective and effective value correlation among state, change of state, position, and change of position and direction can be described as a relative force potential. The hypothetical potential is proportional to attraction and repulsion differentials generated across internal–external value gradients between an organism and environmental stimuli. The changing value potential incurs influences from multiple sources, including reinforcement and associative learning history, and possible choices between alternative attractors that follow internal–external constraints and activation–inhibition thresholds. An organism's navigation is expected to generally follow probabilistically along the path of least resistance among the attractors, depending on current inputs, value state, and the state of systematic information. Simply put, similar to the principle of least action, nature usually finds the most efficient course, such that we tend to parsimoniously follow along the path of least resistance to our goals.

Value gradients may be useful for predicting the relative behavior of isolated individual trajectories in idealized low-dimensional systems. In multidimensional domains, where the history of correlations of domain regularities represent the realization of domain knowledge, effective systematic values may to some extent help to predict a complex system's behavior. But predictions incur limitations to probabilities because of symmetry breaking and one-way time, along with the uncertainties of dynamic instabilities and the unknown nature of current input at future decision points where bifurcations create more complex solutions. There are perhaps many different ways to conceptually model these dynamic relations in brain networks. It seems that the most useful models typically involve some configuration of value and regulatory network correlation with oscillations, attraction and repulsion, energy, and dynamically changing system states (Heerebout, 2011).

Knowing how language and processes in the overall system work, or logically correlate with value space, limits the dimensions of the search space and helps to understand the influence of emotional-affective and analytical-effective value on behavioral regulation in that space.

The Value of Systematic Language

In humans, language enables abstract descriptions that extend the sphere of relationships beyond individual local interactions to a world of global dimensions and systematic influences, from self-centered to world-centered. For example, my view extends to a world's-eye view, from "How is the world treating me?" to "How am I treating the world?" which translates in population terms to "How harmonious is our relationship with nature?" In large, complex systems, increased time and distance beyond local influences and visibility described by individual trajectories involves a phase transition to dynamic descriptions of ensembles, persistent influences, and irreversible processes. Again we run into the curse of dimensionality that can make it difficult to find useful correlations because there are so many variables in different combinations interacting at one time. More formally, we are faced with the tasks of evaluating and correlating the influences of multivariate nonlinear interactions. Effective representation of statistical information distributions in high-dimensional systematic space requires reorienting referencing from local to nonlocal and transitioning to systematic generalizations and probabilities. This generalized representation compresses and conserves information.

Localized linguistic descriptions of phase transitions tend toward discrete "either-or" characterization as either stable and rigidly ordered or chaotic and absolutely disordered. Limiting our alternative views to two choices of stable versus chaotic may sound simpler, but it requires squeezing the massive amounts of dynamic information in the middle to one end or the other, effectively throwing information away. We typically promote a reference of stability and lament the only alternative of chaos. Of course, almost anyone who has studied nonequilibrium physics understands that we do not live in such an idealization but in a world of instabilities, where life exists as realizations between rigidity and chaos—far from equilibrium. Systematic descriptions extend to dynamic parameters that facilitate the integration of statistical information distributions and the modulation of adaptation in the face of persistently fluctuating uncertainties and instabilities. Recall that in complex nonequilibrium systems operating probabilistically and away from equilibrium, we encounter oscillations and bifurcations at decision points related to these instabilities.

In general, the laws of dynamics require formulation at the level of probability distributions (Prigogine, 1997, p. 44). Solids, liquids, and gases are examples of distinct phases of matter that allow for transitions between phases. A *phase space* is the abstract space of points in which the coordinates are the positions and velocities of the particles in an evolving system. Absolute coordinates that represent positions and velocities can limit descriptions of phase transitions to local individual trajectories, but these idealized descriptions lack effective parameters for extension to nonlocal distributions and continuous interactions of populations (ensembles) realized in higher-dimensional phase space. Regions where we encounter populations, groups of points, or bundles of trajectories become a starting point of a new way of investigating dynamics (Prigogine & Stengers, 1985, p. 261). The dynamic transition from particle physics and individual interactions to nonequilibrium physics and interactions of populations as ensembles is important for understanding evolving biological systems.

The ensemble distinction is fundamental to biological evolution, as described by Charles Darwin, because evolutionary complexification is defined at the population level (Prigogine, 1997, p. 20). Luc Steels has advocated the application of a population approach to language evolution using the framework of natural selection, performance deviations, and learning histories, where populations of agents are confronted with adaptation to the real physical world (Steels, 2000a and 2000b; Steels & Kaplan, 2002). The experiments enabled communication between the agents with local referencing and "good enough" results as long as the agents were able to use the language locally. The experiment showed how localized language could evolve. We see that a general comparison can be made to our current everyday language that is typically considered good enough even with its affective constraints. (For more details, see Steels, 2000a and 2000b; Steels & Kaplan, 2002).

We would expect that a nonlocal approach to evolving language might enable a linguistic phase transition from local sufficiency to optimizing global efficiency and effectiveness. In other words, if we know the systematic laws, we can theoretically evolve systematic language that supersedes idiosyncratic localized language. How do we know this? Localized language has evolved to the extended language that science applies to understanding nature, which enables the mitigation of systematic errors and can approximate the criteria for global optimization. We have already seen that rather than local internal referencing, global language exploits an invariant external reference to forward-directed time. With localized language we wind up with affective sufficiency on average. Globally referenced language extends the possibilities and probability for approximating global optimization. We go from local affective to systematic effective. For example:

- "Why would I consider other possibilities when I know my solution will work?"
- "What is wrong with good enough?"
- "I'm comfortable where I am and the way I look at it. Nothing is broken, so why would I consider changing?"

Versus

- "Of course we are willing to consider other possibilities; perhaps there may be more effective solutions."
- "We are willing to search beyond good enough and exploit global optimization by working on decreasing systematic errors."
- "Things change and having alternatives gives us more opportunities to adapt to those changes."

Phase transitions correspond to evolving properties that are meaningful only at the population level of ensembles and not single particles, for groups rather than for individuals (Prigogine, 1997, p. 45). Phase transitions provide a method for shifting from internal locally referenced frequency and amplitude sampling and information distributions that typically describe transient interactions at the level of individual trajectories to external, globally referenced statistical information distributions capable of describing persistent interactions and correlations at the population level (Prigogine, 1997, pp. 37–45). Without this phase shift, global visibility and predictions are constrained by locally distributed information and limited global knowledge (Prigogine, 1997, p, 181). In other words, without a dynamic phase transition that fits the realization of the world in which we live, we are stuck in a static idealization. "I am certain it will happen" shifts to "It might happen, and generally there are many other possibilities and equivalent solutions." A linguistic phase transition transcends from idealized certitudes to realized possibilities and probabilities and facilitates a practical understanding of physical relations in evolving biological systems.

Extended language goes beyond locally weighted value descriptions to the effectiveness of globally weighted analytical values. Value weighting biases cognition and the regulation of behavior. Knowing how value weights are derived, locally or systematically, sheds light on their potential regulatory effectiveness. How values were contextually encoded, referenced, represented, assigned, and mapped can help us understand the organism's perceptual frame of reference, the processing and regulation of priorities, and the direction of actions. The more accurate the understanding of how the map was created in memory, the more accurate the understanding of the subsequent behaviors of the mapper (Churchland & Sejnowski, 1994, pp. 240, 377).

In humans we see two distinct orderings of solutions for tuning input–output mapping: affective and effective. Depending on the circumstances, both affective and effective mappings are useful, so we would not necessarily label either as good or bad. If we stopped to think about the most reliable solution when walking in the woods and suddenly confronted by a lunging bear, the most relevant mapping would appear rather quickly and seem pretty clear, at least in retrospect. Of course, it seems prudent to do our homework, plan ahead, and access as much knowledge of the environment as possible—including the risks of encountering bears—and consider alternatives in case we run into trouble.

Likewise, for day-to-day thinking we would expect that effective mapping would generally offer improved results. Analytical effectiveness tends to represent generalized indirect mappings that maximize the systematic decoding of input information with greater accuracy in a dynamic domain, albeit with greater cost and effort and with slower speed. Affective mappings tend to represent effortless, fast and frugal, direct-like simplistic mappings created implicitly with subjective constraints that compromise accuracy. Remapping from affective to effective incurs the expense of time, effort, and energy—that is, cognitive work.

When we use localized language, we construct and reinforce internal affective mappings to the world that reflect our own subjective idealization. When our internal mappings fail to correspond to realizations in the external world, we can simply distort and contort our view of the world to fit our map. We can connect our imaginary internal map to how the world works, but we wind up with a distorted perception of images in that world. In other words, we are unlikely to make an angry bear disappear by distorting our mapping, closing our eyes, and imagining or wishing it away. And it is just as unlikely that clicking our heels together teleports us to Kansas or anywhere else but where our feet are currently planted. For greater effectiveness, we would expect our internal mapping to logically correspond to our knowledge of the dynamic realities in the physical world to which we are mapping. In turn, maximizing reliable mapping depends on the systematic coherency of our language constructions.

Value and Regulation

Value symmetry breaking is found throughout nature, and value plays an instrumental role in the regulation of complex systems. A regulated system balances the polarities found in nature with the organism's internal requirements for survival (Jennings, 1906). As suggested by W. Ross Ashby, the integration of these concepts with self-organizing regulation yields an overall model for modulating fidelity in a complex system (Ashby, 1960).

Rather than absolute discrete regulation, complex systems modulate relevant variables within a graduated range suitable for adaptation. Variables typically undergo continuous regulation within boundaries that define feedback thresholds for graded activation and inhibition. The two basic modes of interaction among components in complex systems are positive feedback and negative feedback (Camazine et al, 2001, p. 16; Prigogine & Stengers, 1985). A complex system adapts via iterative regulatory processes that modulate instabilities among the system's components. We see analogous regulation across biological organisms.

> It is difficult if not impossible to draw a line separating the regulatory behavior of lower organisms from the so-called intelligent behavior of higher ones; the one grades insensibly into the other. From the lowest organisms up to man behavior is essentially regulatory in character, and what we call intelligence in higher animals is a direct outgrowth of the same laws that give behavior its regulatory character in the Protozoa. (Jennings, 1906, p. 335)

Value, a fundamental property of complex systems, facilitates switching, selecting, and modulating behavior. Without positive and negative values, it is difficult to imagine how regulation can take place in dynamic systems. Adaptive regulation by positive and negative feedback has evolved across the spectrum of organisms from small to large and simple to complex. Generally, negative feedback dampens output, and positive feedback amplifies it (Buzsáki, 2006, p. 54). In networks with recurrent connections, the positive signal can recurrently feed back on itself and is amplified each time the signal feeds back through the loop. Recall the discussion about self-referenced localized language that feeds back on itself; we see similarity in the regulation of recurrent loops that benefit from global inhibition to dampen the tendency for a vicious cycle of runaway positive-valued feedback. It seems that regulation is regulation and value is as value does, analogous to the general laws of nature, where physics is physics across the domain.

Without the critical modulatory influence of negative feedback, unabated positive signals can disproportionally drive us toward potentially disastrous consequences, such as overconsumption, stock market bubbles, and arms races (Camazine et al, 2001, p. 19; Solé & Goodwin, 2000, pp. 284–285). On the other hand, systems with a lack of adequate positive feedback can suffer from chronic inhibition. In a dynamic world, regulation of instabilities typically accesses a full range of feedback, from positive to negative (Lytton, 2002, p. 224; Churchland & Sejnowski, 1994, p. 373; Solé & Goodwin, 2000, p. 90).

Feedback confers checks and balances and enables repair that can reduce systematic error rates in complex systems. Arbitrarily accepting some feedback and neglecting or rejecting other feedback can bias behavior toward maladaptive outcomes. Systems that admit positive feedback while rigidly rejecting constructive negative feedback or criticism forgo a crucial inhibitory component for effective input-to-output modulation and error correction. Admonishing criticism provides a somewhat fail-safe oppressive mechanism for eliminating threats, protecting self-serving values, and conserving fragile beliefs and dogma at the top of a rigid hierarchal system, but it does so at the expense of rewarding systematic ignorance. This authoritarian mechanism simply suppresses the input of any unwanted negative-valued information deemed harmful, faulty, or subversive. But arbitrary suppression of negative feedback can also disrupt effective learning and systematic error correction, leading to an increased accumulation of systematic errors. Because of systematic blindness, these errors go ignored, with the potential consequences of an irreversible catastrophic failure of the system. Ironically, systems that refute any form of negative feedback tend to be the ones that most adamantly apply criticism to systems they see as competing systems.

If never questioned or criticized, beliefs cannot be removed or corrected by any amount of new information: "Errors undetected are errors uncriticized" (Jaynes, 2007, pp. 601, 626). We enter an eternally recurrent loop where errors ignored are errors uncorrected. Adaptation benefits from knowledge of the environment, the context of the experience, and a reliable error signal to effectively detect and correct systematic errors. Uncriticized systematic errors lead to ambiguous feedback and less-effective modulation of adaptive actions in response to the uncertainties encountered in a changing environment. Both positive and negative feedback loops are ubiquitously found in biological organisms and are widely employed in mechanical and electrical engineering. Why? We can say, "Simply because they work."

For example, a simple thermostat regulates feedback from a desirable set point via positive and negative feedback loops. When temperature falls below the set point, it initiates positive feedback to the system, which increases the temperature. When temperature rises above the set point, it initiates negative feedback to the system, which decreases the temperature. Without inhibitory negative feedback, the system continues to increase the positive-valued output, which tends to result in catastrophic failure of the system. Think about an automobile with the gears put in neutral and the accelerator pressed to the floorboard, resulting in a financially catastrophic repair bill.

Negative feedback dampens runaway behavior and serves as a brake or a steering mechanism for biasing, dampening, and directing adaptive output. Ignoring negative feedback impedes learning from errors produced by the outcomes of misinformed actions. Information with reliable negative values generally improves systematic error feedback and error signal sensitivity, which enhances effective behavioral regulation. We often hear negative feedback referred to as a self-correcting loop. The term *correction* implies a penalty for making a mistake and is often taken to mean that we are less-than-perfect beings, which some find distasteful. Ironically, we seem to have constructed a fail-safe language that conceals its own encoding and circumvents insight into our imperfect human condition. We see a correspondence with language, information, and value and the influence of brain regulation on our thoughts, feelings, and behaviors. But we seem to be immune to our folly and passionately resist noticing or correcting our systematic errors and react haughtily as if imperfection is beneath us.

For example, young children are often corrected by their parents. As a result, as adults, we seem to view other adults who try to correct us as parental. When someone speaks down to us from a position of authority as if we are children, even as adults we tend to feel as if we are devalued to a lower level of hierarchal status. We then interpret correction as an imperative admonition that we must not or should not think, feel, or behave in certain ways. Consequently, when another adult gives us positive feedback, the experience activates basically our same reward system. Our reward system tells us when we are obediently thinking, feeling, and behaving the way we are "supposed to" in relation to currently expected values, and we "feel" that our status is elevated. Unfortunately, we tend to surrender our status to the whims of external authoritarian supervision, where polar fluctuations of "supposed to or not" often leave us feeling as if we are on a status roller coaster. We hardly fare better when we take charge of our decisions and attempt to guess what "supposed to" really means.

We seem to favor praise, which we consider positive-valued input, to distinguish it from critical feedback and what we consider as loathsome criticism, which we deem negative feedback. Some people tend to confuse these terms with the terms used by psychologists to refer to actions that change the likely occurrence of a future behavior by providing a reward or penalty. Even though it may feel unpleasant, negative feedback provides a critical error signal for adjusting our behavior. We can take advantage of the feedback and also acknowledge that human brains generally produce both pleasant and unpleasant feelings, and that like most humans, we prefer the former rather than the latter. We prefer compliments to criticism.

We live in a world with rewards and penalties and live with a brain that has evolved to sense and react to both, but typically with a heightened sensitivity to penalties. Accepting that we humans will sometimes have unpleasant feelings does not necessarily mean that we have to *like* those feelings. But unpleasant feelings are a part of life that we can learn to accept. Of course, we have brains with language and plasticity that can learn to behave in more rewarding ways that are more likely to increase our pleasant feelings.

In other words, we don't necessarily have to like negative feedback, but we can still accept and exploit it to improve our lot in life. There is no law that says we should or must dwell on what we perceive as the negative part of the feedback, upset ourselves, and beat ourselves up about it. We can learn from feedback, accept that others live in this world too, and consider the external influence of our actions. If we accept negative feedback—realizing that we do not necessarily have to like it—it can help us to avoid or correct mistakes and improve our overall cooperative effectiveness.

As a result of our social tendency toward dominance-related status hierarchies and the status quo, and toward attempts at verbal control of ourselves and others, we view criticism as a threat or punishment, as if we are being penalized for disobediently deviating from the mean. Many languages are replete with negative terms for chastising deviations from expected behaviors. For example, in English, we tend to use *should* on ourselves to activate or inhibit action and on others as a form of linguistic remote control and for expressing criticism or anger:

- "You should not think that way."
- "You should not feel that way."
- "You should not behave that way."

Likewise, we use *should* on ourselves, which tends to induce guilt:

- "I should not think that way."
- "I should not feel that way."
- "I should not behave that way."

We can find ourselves in a regulatory quagmire, confused as to what we are "supposed to do" or what we "should feel guilty about" in a given situation. Where or to whom do we turn for help in resolving our dilemma? Usually, the simplest solution is to refrain from shoulding.

Parental control of children often takes the authoritarian form of obedience to shoulds and musts, which habitually carries over and imparts stealthy control to their adult lives. Of course, we easily justify the use of imperatives and prescriptives as we tend to do with most other behaviors. "We must have shoulds, and we should have musts." But why should we, and what if anything is it that we must or should do? It seems that a more practical approach is to view our behavior systematically in terms of informed choices and outcomes, where local ideas of "must," "should," and "supposed to" melt away into global effectiveness and our cognitive behavior evolves to informed preferences and actions for which we alone are responsible.

We can still easily default to using the terms *should* and *must* to absolutely upset ourselves and disdain others who fail to obey our implicit expectations as we suppose they should or must. "The tyranny of the shoulds" was a phrase coined by Karen Horney almost 100 years ago. Albert Ellis championed making the effort to eliminate shoulds and musts, at least as best we can, since they lead us to needlessly disturb and upset ourselves, a practice for which he coined the term *musturbating*—that is, "the perturbing of oneself with musts." Ellis and Harper (1997) described shoulds and musts as contributing to irrational thinking, feeling, and behaving. Then there is the story about Niels Bohr and quantum theory in which it was said that the electrons stayed in their allotted orbits only because Bohr wrote a rule saying that they *must* (Lindley, 2007, p. 52). Perhaps this is not much of a consolation, but we see that even atoms can be the target of musturbation, such that atomic particles must obey the authority of experts.

Unfortunately, musts are coarse-grained, imperative control words that carry little if any contextual information from which we could learn to effectively regulate our behavior. Simply put, imperatives tend to exclude context except for implicit affect. Similar to chimpanzees, we typically use imperatives as a hierarchal control mechanism for dominance to get our way. Musts and shoulds gloss over the fine-grained contextual information that helps us to fine-tune error corrections and learning, think about alternatives for future choices, and improve predictions. On the other hand, informatives provide graded positive and negative feedback suitable for analysis as opposing forces, which we can use to refine our thinking and choices, to learn from and correct errors, and to improve predictions and adaptation. Humans operating at the global systems level of cognition require reliable positive and negative values for effective systematic feedback, analytically informing alternative choices, and regulating and augmenting adaptive output.

Science has shown that without a balance of both positive and negative feedback, we fail to recognize the prospect of errors in our predictions and actions. Perhaps we fail to take potentially beneficial corrective steps out of ignorance, fear, or discouragement. Ironically, we tend to have a separate ordering of rules for criticizing what we think of as the counterproductive behavior of others, while overlooking similar behaviors of our own—but blind to our own trick, we do not notice the inconsistent application. Positive psychology and positive affirmations tend to advocate increasing positive-valued feedback while eliminating negative feedback altogether, which, as we have already discovered, can lead to systematic neglect and dire consequences.

We sometimes describe seeing the glass of water as either half full or half empty. Supposedly, always seeing the glass as half full confers a rosy outlook and a happier life, and is touted as preferable to the negative alternative of a half-empty view, often heralded as pessimistic. We understand that regulatory processes and relationships in complex systems throughout biology rely on both positive and negative feedback. In light of this knowledge, we can likely predict a more optimal approach that views the glass systematically according to the water-to-null ratio as accurately as possible. We can simply say that the glass is half empty *and* half full, and it constitutes a changing process due to evaporation. Reliably informed feedback is valuable and boosts effective learning. After all, there are no known laws that say we *must* take it personally and upset ourselves when the water reaches any particular level that it *should not*. With this information, we can plan ahead as best we can and decide whether the amount is sufficient for our purpose. If not, at least we have reliable information with which to make corrections. Seeing positive and negative values as accurately as possible yields an ecologically effective approach to regulation. Now we know about the option of identifying negative values simply as another form of information, learning from them, and applying them to the regulation of our actions—while refusing to needlessly upset ourselves about them.

Informed cognition increases the possibility of making choices that facilitate adaptive outcomes. We can evaluate the water level and take responsibility for monitoring and maintaining it at a sufficient level for our hydration requirements. We can regulate the water level along with our thoughts, feelings, and actions. Is it possible that we have incidentally learned the habit of valuating and taking criticism as a local reflection of our personal worth? How do we think children perceive the reprimand "You *are* a bad girl/boy" when they behave in a way that meets their parents' fluctuating criteria for behaving badly?

Informed feedback, both positive and negative, is invaluable for effectively regulating our actions and helping us to make sense of our world. Informatives facilitate greater reliability for feedback and emphasize the practical reinforcement of adaptive actions. We can conclude that focusing on both positives and negatives offers benefits, and focusing mainly on positives has its negatives. We can also conclude that evaluating both the positive and negative feedback we receive does not mean that we must upset ourselves needlessly.

Chapter 15 goes over several previously mentioned topics in more detail—processes, language, evolution, information and knowledge, entropy, uncertainty, timing, and tuning.

CHAPTER 15

Making Systematic Sense

We are connected to nature through our senses and actions. Without sensory information, we would lack a way to learn about what is going on around us. Without sensory feedback that informs us about our actions, we would lack a way to detect, learn from, and correct mistakes. Not only do we have senses to detect what is going on out there but we also have language with which to represent, translate, process, and describe what we sense. Language provides conceptual structures for understanding experience so that our word labels can help us to make better sense of our interactions (Feldman, 2006, p. 135). In short, we use language to help us make sense of our world. The better our language corresponds with the goings-on in our world, the better our sense-making.

Making Sense of Language

Our senses establish a fundamental link between us, our actions, and forces in the outside world. Neurons in the brain evolved for processing the output of the senses to acquire information about biologically relevant signals found in the environment (Eliasmith & Anderson, 2003, p. 98). *Senses* typically mean stimulus receptors or neurons that abstract information from outside physical signals, such as sight, sound, and touch. *Signal* can be defined as data that the brain senses, transforms, and processes into useful information. Senses provide organisms with environmental signals, which, via transformation and sensorimotor integration, can bias the motor output for behaviors. An organism's response to a signal is constrained by the availability of receptors for sensing that particular signal (Prigogine & Stengers, 1985, p. xxv). Organisms rely on the ability of sensors to detect signals that by transformational processing enable the production and sensible modulation of adaptive behavioral responses. The ability for reliable higher-order sensorimotor modulation tends to increase with increased neuronal complexity, as seen in increasingly complex organisms. Sensory reliability biases the sensitivity to input, which in turn influences the ability to respond consistently to incoming signals.

> The sense-organs are designed to tell us of *changes* which happen in the exterior world.... [T]he relations between the sensations can alone have an objective

> value....In sum, the sole objective reality consists in the relations of things whence results the universal harmony. (Poincaré, 1902, pp. 73, 136–137, 140 [emphasis in original])

A deficiency in color receptors in the eye leads to a visual condition known as color blindness or color deprivation, leaving an individual with a preponderance of dysfunctional light receptors that ineffectively separate light into colors. Constrained language imposes somewhat similar exclusionary limitations on our speech perception. Primitive linguistic structure can limit reception as if sensing absolute on/off, black or white information. In effect, these discrete values limit the range of our perceptual processing such that we think about what is going on outside as a world that consists of black-and-white processes, as if it were absolute.

Coarse-grained reception and processing discrepancies constrain the sensing, perception, and integration of higher-level abstractions. In turn, we see disrupted fine-tuning and difficulty discriminating between what is signal and what is noise. By default, speakers or listeners with limited abstraction abilities demonstrate deficits in receiving, translating, understanding, and communicating complex information. Without sensible linguistic scaffolding, we struggle to process complex speech and make sense of what we hear that seems more like noise to us.

Language-related speech supplies external auditory data to what we could call linguistic sense receptors. The data require translation of auditory speech input into a recognizable form of information. Reception and sensible internal processing depend on the ability to detect and decode or translate those signals into an understandable language form. Without congruent sensors and translational processes that are familiar with the language construction, we lack the ability for decoding and sensorimotor integration. In effect, a lack of congruence means we are less likely to pay attention to unexpected or uninterpretable signals.

Recall that inattentional blindness, the relationship between what we pay attention to and what goes unnoticed, can also apply to inattentional deafness. Whether it is intended or unintended, when encumbered by auditory-visual constraints, we often fail to attend to or hear things we do not expect or that do not make sense, even when they are right in front of us. Sight makes us blind to vision (Lytton, 2002, p. 20), and language, taken for granted, can leave us implicitly deaf to our speech behavior. We usually habituate to stimuli when they have not led to our experiencing any adverse consequence from them. Unfortunately, we are also numb to what we missed. Unexpected events are unexpected in the context where they are uncommon, where failing to spot the unexpected has little consequence (Chabris & Simons, 2010, pp. 27, 39).

Similar to the genetic variation that results in color blindness, language that does not include nonlinear constructions produces a condition that represents a nonlinear language disorder. Rehabilitating the disorder seems unlikely without the benefit of language constructions and pattern-recognition abilities that are decodable and translatable into compatible sensorimotor correspondence with the external world. The omission of nonlinear linguistic scaffolding for decoding the nonlinear interactions deprives us of semantic integration and effective processing required for understanding our behavioral output in relation to forces in a highly nonlinear world.

Therapists encounter this condition in counseling when a patient uses one-dimensional linear reasoning with absolute statements about predictions or uses linear cause and effect to the exclusion of statistical predictions that account for logical higher-dimensional reasoning processes. For example, the patient says, "I should get a pay raise. I work hard, and hard work is supposed to be rewarded." The assumption is based on beliefs of linear reasoning that if A happens, then B "should" happen. That might be the case if we lived in a perfect world where things were that simple. But the patient has failed to take into account the many other things going on at the same time—nonlinearity:

- The economy is slowing.
- The industry's products are no longer in demand.
- The company is laying people off.
- Many people are already unemployed.

When the patient broadens their perspective to include context, they realize that they are fortunate to have a job. "It would be nice if I had a raise, but at least I have a job. I guess things change. I hadn't paid attention to all that was happening, and I kind of got blindsided. I think I will go back to school for more education so I can have more skills and alternatives when life throws me a curve." After reevaluating circumstances, challenging and changing the language used for thinking about how the world works, and adjusting expectations, the patient finally realizes that

- Things change
- Typically, multiple influences are in place at any one time
- Staying prepared for alternative possibilities enables more options for the future
- Changing words changes thinking, expectations, and frustration levels

With practice, thinking with context-constraining certitudes was slowly replaced by preferences and possibilities that better fit with the realities of a persistently changing world. The patient added scaffolding by sensorimotor integration from practicing. In a sense, the scaffolding amounts to building up a novel receptor structure with which to receive and understand a different kind of information. Before therapy the patient had no idea that there were such things as preferences, multiple influences, changing context, or "it's not just about me." The change in perspective enabled a transition from an idealization to the realization of alternatives. Excluding alternative possibilities leads to automatic default to absolute words and habitual linear thinking in discrete terms.

At other times, a patient will make a linear connection of cause and effect after a coincidental association between a random event and a negative outcome. We sometimes refer to such notions as inferences, assumptions, or *leaping to conclusions*. The patient might insist that a coworker received a promotion simply because the boss likes them better. When the counselor asks the patient to consider other alternative explanations, the patient persists in believing what they consider the incontrovertible evidence of their own conclusion. While such rigid thinking may have many sources, much of the difficulty can come from an inability to move off the stepping-stones on the linear path and avoid the assumption of cause and effect from conjecture A to conclusion B. Indeed, a significant part of CBT involves helping the patient to see sensible alternative choices off the beaten path by challenging linear thinking and the habit of drawing absolute conclusions, such as

- "I saw it with my own eyes."
- "Everybody knows that's how it always works."
- "That is the only explanation."
- "It wouldn't have happened if. . . ."

In CBT and REBT, a patient learns to minimize assumptions and find alternative ways to describe what did happen or might happen, with terms such as *possibly*, *perhaps*, and *might*. These terms leave the solution space open and replace rigid terms of emotional embellishment and leverage such as *certain*, *sure*, *impossible*, *never*, *awful*, and *deserve*. Such small practical concessions to flexibility and other possible explanations enable the patient to consider alternatives, so that they learn to reword their statements of beliefs as probabilistic and uncertain:

- "I guess it doesn't always happen that way."
- "There's a chance I misinterpreted what I saw."
- "Well, perhaps I only saw what I expected to see."

We see the moderation of emotional effects in individuals who have learned that they can also choose to see the "glass as half full" (Phelps, 2009, p. 210). In other words, we incur limitations by accepting the beliefs that our language habits support. But the language that we revere and laud for setting us free and taking us to the top of the mountain may in reality have bestowed on us narrow-minded thinking and tacitly imprisoned us at a the top of a small hill on the local landscape. From this disadvantaged local position, we continue to laud, cherish, and flout our shoulds, musts, and rigid, deterministic language constructions as an obvious sign of higher "intelligence." Are we missing something? Even if we are determined to solve a problem, we are constrained by our language, since it limits the effective range of the problem-and-solution search space.

In effect, a lack of language and semantic constructions for representing particular verbal input signals limits our perception, making the signals much harder to hear, conceptualize, and understand. With linear language and communication, multiple signals from surrounding events go undetected, obscuring the interaction or interference between them and constraining nonlinear processing by relegating nonlinearities to the category of noise. As with an unfamiliar foreign language, we tend to perceive unfamiliar nonlinear language as noisy babbling or gibberish, and since we do not find it useful, we reject it as if it were junk. But with the assistance of reliable translation, we can usually make better sense of it.

A similar rejection phenomenon can occur with other abstract language constructions, such as deterministic versus nondeterministic, absolute versus nonabsolute, static versus dynamic, discrete versus continuous, and local versus global. In neuroscience, the inability to acknowledge a deficit is called anosognosia, such that global anosognosia defines a lack of global perception or systematic blindness. We see that language constructions can represent a phase transition from local to global and higher-order systematic processing. The constructions describe sensitivity to a range of signals between maximally perceptible and interpretable and minimally perceptible and interpretable, depending on the effectiveness of the language used for decoding. Below the lower threshold, signals are essentially imperceptible and thus uninterpretable.

Language construction deficits limit information input by corrupting the signal and making decoding and integration more difficult, if not impossible. The deficits restrict adjusting the signal-to-noise ratio by limiting the dynamic tuning range, analogous to occurrences in a nonlinear world with linear linguistic constraints. Any system existing in the real world deals with noise that places constraints on discrimination. Noise can be dealt with in nonlinear systems, but in a linear system, noise can destroy information (Koch, 1999, p. 19).

The exclusive use of constrained language constructions produces one-dimensional absolute descriptions, resulting in a semantic deficit that limits the capacity to perceive, decode, and integrate higher-order constructions. Higher-order linguistic constructions convey descriptions at a higher level of abstraction. Absolute linguistic constraints tend to result in the perceptual rejection of multidimensional abstractions as noise, as if meaningless, irrelevant, or false. Sensitivity defaults to familiar language constructions with two-way time, where descriptions of the static past and dynamic future are tacitly interchangeable. We find ourselves out of sync with our dynamic world, where two-way time is imaginary.

Dynamic deprivation creates a perceptual mismatch between motor and sensory relations such that the organism confuses the meaning of environmental information inputs (Buzsáki, 2006, p. 220). Perceptual dampening of meaning in a complex environment limits the computational meaningfulness of the concepts we use to make sense of and adapt to that world. Habitual overexposure to constrained language adversely influences learning by limiting the exploration for new information, alternative descriptions, and novel, innovative solutions. Perhaps worse, the power of habit relegates the constraints to invisible, inaudible, and therefore out of sight, out of mind, and pretty much oblivious to detection. Limitations imposed by a linguistically constrained cultural environment can be potentially ameliorated by learning and applying extended language and scientific literacy—and yes, practice.

Reliable input processing, regulation, and adaptive output in humans largely depend on understanding and making sense of information via the encoding of our language constructions. Language encodings that reliably correlate and convey a coherent understanding of the structure of external information likely improve the conversion of that information to effective knowledge. We evaluate and reason about a great deal of current and prior information, but the nature of our reasoning is often concealed by calling it common sense (Jaynes, 2007, p. 7).

Common sense typically relies on local intuition irregularly codified into conventional culture-specific folk wisdom [generally less reliable from a macroscopic perspective], rather than on qualified, reviewed evidence routinely updated to account for recent changes and discoveries (Browne & Keeley, 2007, p. 107). We commonly take credit for positive outcomes and blame fate for negative ones (Armor & Taylor, 2002, p. 345). The static nature of local common sense, which tacitly relies on two-way time, works best in retrospect. A retrospective view prompts us to explain away adverse outcomes with ad hoc inferences, rationalizations, justifications, and blaming, which impedes learning and adaptation in the present. But we think: "Why don't they just use common sense?"

Even though what is called common sense sometimes suffices, to a large extent it relies on familiarity and unexamined time-symmetric prior information. We inherit this common sense along with uncriticized localized language and anecdotal beliefs that we incidentally absorb and learn from observation and hand-me-down narratives. Unfortunately, on reliable inspection we typically find these beliefs replete with historical superstitions that are coincidences interpreted as cause and effect, with hand-me-down word-of-mouth documentation and observations supported by spurious eyewitness evidence—tales and tales of tales. We can think of common sense as the typical solutions we catch and run with after our limbic brain automatically tosses them up to us.

So-called common sense demonstrates remarkable inconsistency across contexts, individuals, groups, and cultures (Vygotsky & Luria, 1993, pp. 138–139). Commonly used static linguistic artifacts and certitudes affectively anchor our thoughts and behavior to antiquity by failing to distinguish the difference between the static past and dynamic future. We become statically trapped with rigid hand-me-down beliefs and predetermined values that blind us to alternative possibilities and impede us from flexibly adjusting to our ever-changing world. We inadvertently obstruct our own efforts to adapt when we operate as if past and future are symmetrical and interchangeable.

We naively struggle, fumble, and stumble against reality when we persist in operating with two-way time as if what we think of as absolutely true or false yesterday remains absolutely true or false today and should be so tomorrow.

Even if common sense is found to be common, it seems to be sense in name only. Temporally constrained thinking with two-way time leads us to automatically do the same things over and over, expecting different results, as if something has changed. We continue running on the slippery treadmill of success like the Red Queen who runs eternally and never reaches the land of infinite rewards. Two-way time where the past predicts the future can also fool us into naively believing things are unchanged by confounding the difference between what we sensed in the past, what we are sensing now, and what makes for a logically sensible future. For example:

> Of course it will work tomorrow; it always works. We've been doing it that way ever since Great-Grandpa started the company. He would turn over in his grave if he even thought we would change something, and especially such a foolish change at that. Get with it—use some common sense!

The distinctive differences between the various cognitive behaviors we call sense suggest the importance of searching for and evaluating the relevance of the prior information and the plausibility of the evidence, and accounting for the reliability of the source and logic used in informing and processing a particular sense. A deliberate effort helps us discover what kind of sense it represents, affective or effective, and helps us decide on its level of ambiguity or clarity, and usefulness for dealing with a particular situation. Relevance conveys the usefulness of the computed sense in functional terms for evaluating signal and noise as unambiguously as possible (Eliasmith & Anderson, 2003, p. 208).

Signal represents the relatively invariant meaning of input, compared with *noise*, which represents variable, ambiguous, or irrelevant information. We generally call something sense when it exhibits a high signal-to-noise ratio, and nonsense when it contains mostly noise. High levels of noise, or noise saturation, tend to dampen the benefits of plasticity and constrain learning. In a dynamic environment where we lack complete knowledge of everything going on, the ratio of signal to noise will fluctuate. So it is unlikely that we will encounter a pure signal with zero noise in an absolutely true or false manner except perhaps within simplistic static idealizations.

Flexibility and persistent tuning can improve the signal-to-noise ratio of available inputs in noisy environments. In our dynamic environment, where noise has a finite limit and signal is typically a nonzero number, rigid either–or processing constrains adaptive tuning of the signal-to-noise ratio, where absolute signal or noise is perceived to vary between zero and infinity as true or false. The variability between inherited beliefs and predetermined values, absolute logic, and the coarse-grained ambiguity of common sense hampers adaptive tuning of signal-to-noise ratios, dampens signal sensitivity, and confounds communication clarity.

More reliable, but considerably less common, systematic sense can improve the effective tuning of signal-to-noise ratios. We can deliberately take the time to test assumptions, evaluate systematic biases, and base our decisions on analysis of the current input-to-output effectiveness, using updated systematic information and knowledge for optimizing signal tuning. For example:

> First let's take a macroscopic look at our system, evaluate the problem space, and elaborate on it to try and identify the most likely parts influencing the problem in the system. Then we will identify all of the assumptions as best we can and minimize them. Knowing how the system works will help us interpret the information about the current problem, predict the probable most effective solution, and tune our system to decrease the systematic error rate. We will add an explicit feedback loop to increase our oversight so we can continue to monitor the ongoing effectiveness of the solution and use the systematic error signal for continuous fine-tuning.

A deliberate strategy with an action plan is generally preferable to settling for automatically accepting or ignoring potentially noisy, unmonitored, and unexamined locally encoded language. Recall that constrained language tends to rely on hand-me-down two-valued logic with static beliefs and the forces of affective values from the distant past typically from unknown or ambiguous sources. Applying time-symmetric beliefs and values from the past to the present or future implies equivalence and simply creates cognitive distortions.

The signal-to-noise ratio can be improved by amplifying the received signal with an amplifier that boosts the signal, while adding as little noise as possible to keep us from violating the uncertainty principle (Pierce, 1980, p. 195). Probability can be used as a logic function—probability as logic (Jaynes, 2007, p. 490)—to continuously modulate the input-to-output effectiveness, thus amplifying the signal and dampening noise in complex environments. Since probability is a continuous rather than a discrete function, it works effectively in nonlinear complex systems where uncertainty is the rule by matching the dynamics and analog logic of the domain, which boosts signal detection and clarity. Discrete functions increase the noise in dynamic systems and confound signal detection and the clarification of ambiguous information.

Probability as logic has a broad range of applications, including dealing with linear and nonlinear signal detection and translation, and computing nonlinear solutions. In contrast, "satisficing" with absolute linear logic on a local level falls short of addressing nonlinear signals regularly encountered in nature. Even though good-enough logic works, on average, similarly to imperatives, the absolute nature of the logic degrades signals involving nonlinearity, such as the signals we deal with daily in our world of many influences and multidimensional interactions at any given time.

Scientific thinking fundamentally depends on systematically encoded sense rather than locally encoded common sense. Systematic sense provides a coherent understanding of relationships and processes with and within our complex world. The natural referencing and open-ended logic of science, evolutionary processes, and complex systems can go beyond local sufficiency to offer systematically sensible information, understanding, and learning. Open-ended logic leverages naturally preferred information correlated to statistical regularities and nonlinear interactions in nature.

Global sense offers better regulation and adaptation to instabilities encountered within complex systems (Eliasmith & Anderson, 2003, p. 15), which coherently corresponds with the natural world. Coherently correlated sense enhances input sensitivity, adaptive learning, and effective output. Sense influences feedback regulation and the integration of an organism's adaptive behavior with the environment as a complex adaptive system (Ashby, 1960; Kauffman, 1993; Schlosser & Wagner, 2004, pp. 183–185; Mitchell, 2009).

Furthermore, sensible interactions and feedback operate throughout evolutionary processes and the lives of organisms, from development and gene regulation to long-term adaptation, survival, and reproduction. For example, evolutionary processes, gene regulation, and development in living systems all sense feedback from the environment that can lead to adaptive responses. Systematically sensible positive and negative feedback values facilitate effective regulation of adaptive interactions. The feedback relationship with nature mirrors the way the genetic code allows predictions to be made about its own functional output. DNA is translated to proteins that form the three-dimensional structure determining the qualities that create life (Nüsslein-Volhard, 2006, pp. 35–38).

Organisms have evolved as an integral part of a global system that operates under one-way time-dependent physical laws that contribute to the organism's existence. In turn, these evolutionary laws enable organisms to adaptively fine-tune their behavior to their environment, similar to how natural selection results in ever more finely tuned, well-adapted genes (Carroll, 2006, p. 115). Evolution and adaptation represent inseparable fundamental components of an open-ended natural process of sensible relations amenable to fine-tuning.

These open-ended processes produce, shape, and contribute to the regulation and fine-tuning of living organisms. Understanding these regulatory processes and relations allows us to preferentially fine-tune our language sensibly to nature. Sensible language predictably enhances adaptive optimization because our predictions and outcomes tend to logically correspond with how we think the world works. Language sensibility depends on semantics that break time symmetry to transition from certitudes of the static past to the possibilities and probabilities of the present and future.

For example, localized language matches an absolute idealized world that as far as we know is imaginary. Sensibly calibrating and tuning language to match the force dynamics of the world in which we live improves predictions and insights for responsible behaviors and accountability for outcomes. Sensible language leads to possibilities that fit a changing world and highlight the constraints of localized language with idealized certitudes.

We live in an open-ended world that matches with an open-ended global language. For example, take two different perspectives:

> *Perspective A:* "We seem to get better results and make pretty good statistical predictions with open-ended language."
>
> *Perspective B:* "That's what you say. It seems wishy-washy. We think it's important to confidently demonstrate strength and certainty. If we get our way, we promise to get some real results. We want to be absolutely sure that everybody knows we mean business."

Since language conveys information and logic about relationships, we can use it to make sense of interactions in a complex world. With our brains, language, logic, and nature on the same page, we expect a greater sense of harmony. The sensible use of language supports the development of knowledge that reliably corresponds to our world. If we cannot make sense of our language and semantics, how can we make sense of our world?

Making Sense of Evolution

Evolution theory is a unifying principle in biology (Nowak, 2006, p. 4). Organisms developed over a long period by open-ended evolutionary processes likely involving attraction and repulsion among self-organizing parts, reproduction and trial-and-error variability, and adaptive selection for survival. Moreover, natural selection does not prescribe what "should" evolve (Striedter, 2005, p. 15). Surviving organisms operate relatively autonomously as a functional part of the environment across a range from lesser to greater complexity, including humans and our complex brains. In a dynamic domain, it is unlikely that an organ as complex as the human brain developed by design based on some predetermined blueprint of what the final product "should" look like, with a plan for how it should or must operate.

It seems more plausible that our complex brain started with parts and chemicals available in the environment and evolved incrementally from a lower to a higher order of complexity while satisfying constraints for survival and reproduction in the face of dynamically changing environmental forces. Gene duplication, crossover, and point changes can introduce variations in the basic genetic structure and regulation, and those changes that enhance behavioral performance, or at least do not interfere with the organism's reproductive success, are more likely to persist in the gene pool. Common features of dynamic cellular and extracellular interactions are structurally and functionally fit according to information about attraction and repulsion of positive and negative charges found in DNA, RNA, and proteins.

The surviving genes have already demonstrated fitness such that variations are biased toward solutions that work in the domain. DNA works in part by locally biased mechanisms of attraction and repulsion that can introduce further nonrandom bias. There are putative epigenetic mechanisms including acetylation (n), methylation (+), and phosphorylation (–) that modify histone proteins and DNA methylation, and recruit other specific transcription promoting proteins. Changing the ratios, quantity, and quality of charge can change (or neutralize) positive and negative polarity and forces related to the functional value bias, structural conformation, and attraction and repulsion dynamics. Even though genetic changes are often described as random, we might rethink ideas of randomness and consider the possibility that these biases more than likely represent nonrandom changes, perhaps more so than we have previously suspected. Apparently critical genes and functional regulation are better conserved and nonlethal changes seem to be the general rule rather than the exception, suggesting that nonrandomness is at work.

In other words, genes evolved originally with a nonrandom natural bias that enabled them to regulate opposing forces and leverage the exchange of energy and information with the environment in accordance with the preferred information asymmetry; newer genetic networks added and accumulated layers of complexity and appear to continue this convention of complexification with a generally nonrandom natural bias. From a global perspective, genetic changes generally appear nonrandom—more so than we may have expected from a localized view where systematic processes tend to appear random and go unnoticed. Natural nonrandomness correlates to preferred characteristics of structural fit, regulation, and functional adaptation to change according to physical laws.

Life apparently started from natural elements and an aqueous medium such as H_2O. Certain chemical elements appear to be a natural fit to make up the molecules of living organisms. Hydrogen, oxygen, carbon, and nitrogen possess unique molecular fitness perhaps related to their common property of readily forming strong covalent bonds by electron pair sharing. These organic compounds are not distributed in proportion to their occurrence in the Earth's crust but are nonrandomly selected to make up the 99% of the mass of most cells. The compounds contribute to forming amino acids, precursors for production of protein structural building blocks and functional enzymes (Lehninger, 1975, pp. 17–18). Enzymes "see" much more than the three-dimensional shapes of the nonpolar surface contact areas on the biomolecules. An enzyme "sees" the location and sign of the electric charges [force] and the precise distance between charged groups (Lehninger, 1975, pp. 25–28).

In biological organisms from amoebas to humans, we find evidence for the molecular logic of living cells with symmetry-breaking polarities, regulation, and nonrandom relationships.

In humans, phosphorylation by protein kinase enzymes interacts at ~650,000 sites. Phosphate (PO_4^{3-}) is a highly negatively charged molecule (1-, 2-, 3-) that plays a key role in

- *Structural* components with calcium-phosphate in bones and conformational change
- *Regulation* of water balance, insulin signaling, gene transcription, and ion transport
- *Energy* metabolism, adenosine triphosphate (ATP), thermodynamics, free energy, and catalytic enzymes
- *Functional information* signaling, distributing negative charges, and tagging proteins for degradation

Across biological organisms we notice general relationships:

- Symmetry breaking enables natural information asymmetry applicable to matter, time, space, value, polarity, and forces of attraction and repulsion— covalent bonds to positive, negative, and neutral surface charges and hydrophobic to hydrophilic reactions
- Living cells pump positive (+) and negative (–) ions across membrane channels that are highly conserved and related to polarity, force, direction, action, and regulation
- Living cells and creatures are attracted by nutrient energy (+), repulsed by toxins (–), and learn the rule to approach positive (+), avoid negative (–), and ignore null-neutral (n)
- Living creatures exhibit state-dependent tendencies to generally approach and collect in favorable conditions (+) and move away from unfavorable conditions (–)
- Electricity and magnetism demonstrate polar relationships with positive (+) and negative (–) forces: the former flows from negative to positive poles; in the latter, unlike poles attract and like poles repulse
- Force and ratios between charges influence the direction of action and reaction in proportion to opposing forces, polarity, and distance
- All of the above relate to energy and information, polarity, direction, action, and regulation
- Asymmetric time has a noninvertible polar relationship with events and space that flows constructively and only in the positive (+) forward direction to the future
- Oscillations provide an efficient analog mechanism for continuously monitoring and regulating input processing, integrating event-related polarities, and modulating goal-directed output

- Harmonic oscillations enable the autonomous regulation of exploration and exploitation according to changing context, energy resources, and input-output ratios of internal-to-external constellations of value weights for risks and rewards

We live in a world with periodic and aperiodic changes, where we expect an oscillatory controller that adaptively matches the domain force dynamics to provide the most efficient and robust regulation of directed action—from affective to effective.

The apparent relationship among life, polarity, and positive and negative charges might seem a bit strange, but using existing parts with naturally occurring attraction and repulsion confers a level of nonrandom ordering that offers at least two advantages: the parts are abundant, and they have been used successfully in earlier, less-complex organisms. Since existing parts already fit, they have built-in positive and negative polarities that can act as alignment instructions for self-assembling and repairing more complex biological structures. Using these existing parts provides value-added benefits since they come with natural self-organizing operating instructions for survival and reproduction. Biological attractors naturally correlate structure and function with reinforcement value, simple two-valued logic, and action for sufficient adaptive performance.

> In biology an essential role is played by processes of *self-organization*, which ensure that in the physiochemical environment of the cell several steps of the structuring of the newly created molecules will self-assemble in the right way. In other words, in biological cells some of the details of the self-reproduction process are not explicitly encoded in the machinery of the cell but follow implicitly from the laws of physics. (Floreano & Mattiussi, 2008, p. 145 [emphasis in original])

When an organism's structure, internal state, actions, and regulatory values correspond with survival ability in an environment based on performance, adaptive behavior can be modulated according to internal memory, affective state, and input of error feedback from actions that are penalized or rewarded. An autonomous organism evolved with natural biological autocorrelation inherits the ability to automatically regulate adaptive exploration and exploitation according to the distribution of reinforcement and constraints in the environment.

We have evidence that organisms develop nonrandomly as a result of the natural biological attractor relations with their surroundings. Attractor relationships support the notion that correlating natural biological order with an organism's behavior by coevolution tends to improve adaptation. In other words, things that evolve together are usually more likely to work together. If so, then the natural ordering of structural and operational parts makes evolving complex organisms seem more plausible. It makes sense that language co-evolved along these same principles of automatic attractor relationships. We see evidence in affective language primitives that confer polarizing constraints on our everyday speech. We get a linguistic bonus with effective language that suppresses the polarizing sparks and supports analytical processing and coherent structural and functional correspondence with the dynamics of nature. It seems likely that physical environmental constraints, to a large extent, play a crucial role in the evolution of organisms, communication, and natural language—specifically, nonequilibrium constraints.

We expect that the type of local environment will influence evolutionary processes and the types of organisms and languages that develop in that area. A dynamic evolutionary relationship between organisms and the environment increases the probability that the behaviors of organisms fit with environmental differences and the range of change in a given locality. We would expect a similar fit with the evolution of language. Even though the words might be highly variable and sound different, we expect relational meaning to fit with the laws of nature. We at least expect that whatever similarities we see across localized languages share some form of common referencing. In other words, analogous to fossils, the correspondence of phenotypic similarities among the preserved words gives us clues about their origin. The organism's interaction with the environment acts as a selective adaptive filter. Phenotypic behavioral variations that fail to adjust to changes in the environment die out, while sufficiently adaptable traits survive, along with their genotype that specifies their particular genetic "area code."

Similar to the fossil records of life forms, we have evidence and conjecture that more languages have gone extinct than exist today. The selection process defines the logical correlation among the organism's genotype and phenotype, structure and functional processes, and behavioral adaptation to the structure of the fitness landscape. In effect, selecting sufficiently adaptable organisms correlates the relative fit between the logic of the organism's actions and the logic of the domain dynamics. Consistent with evolutionary theory, in nature we see a recurring theme of correspondence that points to the importance of parsimoniously fitting regulation and actions with environmental change.

The cycle of evolutionary processes repeats over time and space, always pointing to the future and generating a history of correlations between the genotype, phenotype, and successful adaptation. Even though other organisms do not have the language capabilities of humans, the history of genetic correlations represents structural and functional knowledge about what works in nature. The time-directed evolutionary principles that inform the organism's genes and facilitate the display of adaptive behavior correlate with how its genes know the environment. We can say that genes embody physical knowledge of experience as a history of correlations brought forward in time by which adaptive organisms, including humans, implicitly know their world.

Theoretically, because of our highly evolved human speech and language capabilities, this knowing endows us with a unique adaptive advantage. With a brain that likely evolved to some extent in conjunction with natural language, we acquired an unprecedented ability for learning over our lifetime. We might expect that maximizing the systematic correspondence between our language and nature can facilitate the optimization of robust thinking, learning, and relating that translates to understanding the importance of striving toward greater cooperation among all involved. Is cooperation possible? If so, does understanding language hold the key for unlocking the code and implementing greater cooperation?

Scientists theorize that we inherited the physical ability to produce speech genetically from distant relatives. Researchers have explored the possibility of teaching human-like linguistic behavior to other animals, most commonly other primates. Such ad hoc attempts at eliciting language from these animals have met with interesting and useful results but somewhat limited success. Other animals apparently lack particular organic mechanisms for fully articulated human speech, and the structural and functional cognitive ability to integrate linguistic utterances at a higher level of analytical abstraction. It seems that by evolving language together with an appropriate language apparatus, language behavior can generally operate as if in parallel with our brain's cognitive behavior. We can see that the language confers the benefit of extending our affective phenotypic cognitive behavior to a higher-order global perspective.

We can take advantage of our knowledge of this evolutionary phenomenon by deliberately choosing and using language that correlates our linguistically derived policies and cognition in parallel with external physical laws. Understanding how things work in nature seems like an important step that we can use to our advantage. We interact with an open environment like other evolutionary processes and relations. It seems natural to choose a language with logical consistency and relational correspondence between our internal cognition and nature. We have reasons to believe that evolution of effective language structure can produce solutions with greater functionality and mutual correspondence. In effect, language stands between us and nature. Natural language can form rigid implicit barriers of ambiguity as well as inform explicit transitional constructions that bridge the rhythms of our brains and nature. We have seen how evolutionary processes work across long time scales. Alternatively, we have seen how we can accelerate the evolution of effective language constructions with extended structural and functional correspondence to the physical laws and the global information asymmetry preferred by nature.

We likely prefer a language that makes systematic sense of information abstracted as input from the environment and enables logically computing that information to produce effective domain knowledge, and predictable adaptive interactions. Then our cognitive thought processes, language, and behavioral performance correspond nonrandomly with the world in which we live. Natural correspondence seems to make our language choice relatively simple. We likely will choose a language that coherently and effectively corresponds with nature. Recall that effective language relies on probability as logic that correlates to the dynamic logic of nature. We can choose any combination of language: from closed-ended absolute language with affective polarity to open-ended extended language with effective analytical tuning. Closed-ended language corresponds with polarized cognition attributable to limbic-derived processing and the reactivity we see in our less-complex animal relatives. Polarizing language is adept at reinforcing and magnifying arrogant feelings of superiority and bullying behavior that can easily lead to needless arguing and fighting about certitudes—who is infinitely right and who is infinitely wrong. But we define infinity as imaginary. And as far as we know, infinity goes together with fairy tales.

We observe and marvel at the naturally organized behaviors of complex systems that evolve from open-ended systematic processes. Science uses similar systematic principles to sensibly search for new information and knowledge. We have the option of applying systematic processing methods to our everyday language constructions. Language conducive to open-minded thought processing facilitates logically ordered policies with which to think and communicate. Sensibly ordered thought processes can improve how we relate, socialize, and cooperate with others—allowing us and others to know our world in a cooperative rather than a polemic manner.

Making Sense of Linguistic Knowledge

Some may dismiss seeking a better understanding of language as a trivial matter or contend that we already know all there is to know, prompting them to ask, "What more can we learn by exploring this matter further?" Others may counter that we simply haven't learned enough about the mechanics of language to draw meaningful conclusions about any positive or negative influence it has on cognition and behavior. These competing contentions that we know "too little," "plenty," or "enough" about language bring up a number of questions:

- Can we apply everyday language to evaluate our own use of language?
- How much does our habitual use of everyday language mask its own effect on our thoughts and behavior?
- Can we deliberately change the way we use language to improve its effectiveness?
- Why would we care to change?

We have an opportunity to understand the contribution of language to cognitive fitness when we focus our attention on how language fitness corresponds with effective knowledge and behavioral adaptation. When we measure language by effectiveness, we see that open-ended language enhances reliable correlations and predictions and works for explaining effectiveness in a constantly changing, open-ended world. Yet we continue to struggle along with the language habit we take for granted and easily dismiss concerns that we are using language as a code for communication. We act as if we couldn't care less that we don't have a clue about our language code or that we have hardly any idea about the key to unlocking that code. How do we make sense of our language without the code or the key to the encoding? Does it make sense to rely on a superficially examined language when our language code represents our interface with the realization of knowing our world?

Even though some may argue that the scientific issues around understanding language are not absolutely resolved, understanding language continues to represent a relevant priority. We rely on language and semantics as a form of information and knowledge at all levels of science, including computer science, mathematics, physics, chemistry, and biology. We use language to model, understand, and describe computation, nature, organisms, cognition, and the brain. Language provides us with an information source that enables and contributes to extending our knowledge. It makes sense, scientific or global sense, that understanding language itself is a relevant and instrumental prerequisite for plausibly resolving questions about human cognition and "understanding the brain in terms of the biological basis of behavior" (Carroll, 2005, pp. 261–262).

Viewing language as an information tool referenced to the environment it evolved in can help us understand the operation of our brain. Recall that localized language and semantics operate from the internal affective "how it feels to me" subjective perspective of the speaker, without benefit of reliable external referencing conducive to acquiring systematic information and knowledge. We can abstract our information and knowledge from our observations of nature, but we tend to interpret nature to mean whatever we want, without regard for plausibility.

> Nature cannot be forced to say anything we want it to. (Prigogine & Stengers, 1985, p. 5)

Nature does not explicitly speak for itself, and we can easily misinterpret our own localized assumptions and beliefs about the world as absolutely true. It makes practical sense to strive to take a broader perspective that correlates information and knowledge with regularities in nature. With an understanding of how language influences our perception and how we can tune our language to correspond with regularities and physical dynamics found in nature, we gain insights that enhance our cooperative problem-resolution abilities. Insights that naturally increase the possibility of improving cooperative relationships encourage us to embrace the most effective language available. Language helps us predict outcomes, which points to the value of striving for a language that confers natural correspondence to those with whom we relate. A natural language that logically informs thinking can facilitate thoughtful human relationships and enhanced cooperative coexistence.

> Our dialogue with nature will be successful only if it is carried on from within nature. (Prigogine & Stengers, 1985, p. 218)

Unexamined, unthinking speech that carries undetected or unexplored language artifacts can transmit faulty logic and biases insidiously embedded within the structure of the language encoding. While language may not speak for itself, we can investigate what it might tell us as effective information and knowledge about the world around us. Such knowledge can possibly lead to opportunities for beneficial change in theories of language, cognition, and communication that informs and confers greater speech fidelity and facilitates cooperative dialogue across human societies.

> If communication theory, like Newton's laws of motion, is to be taken seriously, it must give us useful guidance in connection with problems of communication. It must demonstrate that it has a real and enduring substance of understanding and power. As the name implies, this substance should be sought in the efficient and accurate transmission of information. (Pierce, 1980, p. 126)

Cognitive neuroscience identifies relations among cognition, perception, and action, where language is grounded in the sensorimotor system and language understanding is based on a mental simulation process (Barsalou, 2009; Jeannerod, 2003; Gallese, 2008). The same areas of the brain that light up when we interact with objects also play a role in the perception, processing, and emotional evaluation of action words (Chersi et al, 2010). When we simulate the experience of thinking about, seeing, or performing a behavior, we activate basically the same neurons as when we perform that behavior. Among the neural underpinnings of this simulation process, an important role is played by a sensorimotor matching system known as the mirror neuron system (Rizzolatti & Craighero, 2004).

> A neuron is called a mirror neuron if its firing correlates with both execution of a specific action and observation of more-or-less related actions. (Arbib, 2010, p. 5)

Language has been described in terms of parity or equivalence, meaning that an utterance means roughly the same for both speaker and hearer. Equivalence relates to symmetry, equality, or correspondence. Arbib (2003) proposed the *mirror system hypothesis*, which links language equivalence (the idea that what the speaker intends is roughly what the hearer understands) to the properties of the mirror system for grasping—neurons active for both the execution and observation of actions (p. 609). We might suppose that the mirror system is in some way involved in joint attention that plays a role in directing our mutual observation of action events. Mirror-neurons link action to perception, and the actions we mirror most vividly are the most familiar ones (Macknik & Martinez-Conde, 2010, p. 250). Language influences perception, and choosing the most familiar language may not necessarily be the language with the most effective constructions. The mirror system hypothesis suggests that an effectively constructed language can facilitate coherent knowledge correspondence with physical laws and enhance perceptual congruence. We have the option of selectively extending language constructions that facilitate optimizing mirror relations with the physical world.

We routinely interact with the environment via sensory receptor input and the effects of actuator output. When activated, output actuators produce motor behavior, including speech. Our speech output is essentially a motor behavior that allows for simple evaluation of the effectiveness of the output it produces via feedback through the ears, tympanic membranes, and auditory cortex. Information processing in the central nervous system directs behavior via the motor system, in turn activated and controlled by the behavioral state interaction among cognitive and sensory systems (Swanson, 2003, pp. 97–98). Continuously computing, evaluating, and correcting errors positively influence the fidelity and effectiveness of the output produced by the sensorimotor system. The dynamics of the feedback relations suggest benefits from cognitive correspondence between language and the external domain.

Language represents an abstract form of rule-based signal processing that, at its best, can correlate and transform external contextual information to domain knowledge. The frequency and similarity of language and auditory input correspond with the brain's abilities as a rule-extraction machine. The process of hearing is concurrent with interpreting the input according to context, thereby "fitting everything together" (Spitzer, 1999, pp. 138–140). We expect that this fitting requires calibration of input sensitivity and reliably matching perceived context in correspondence with knowledge of the rules of external relationships. In turn, these rules correlate with the effectiveness of the output—that is, when the internal rules we have learned match the external rules of nature.

We abstract rules from our observations of relationships in nature and express them with language. We can update these rules as we acquire new information that changes our understanding. Reliable rules result from regularities that are abstracted "from" environmental correlations rather than imagined and exported "to" the environment to explain coincidences or explain away imaginary fears or superstitions. As a knowledge form, language provides a transformational mechanism that can enhance the fidelity of our contextual representations of the external world. For example:

> A solution popped up and it's tempting to jump to conclusions, but I think I will stop and identify the problem space so I don't waste a lot of time solving a problem that doesn't exist. I suppose that a solution popping up out of nowhere all of a sudden could be my limbic system setting off a false alarm. Okay, I've pretty much examined my assumptions and minimized them, and elaborated on the problem space, and it seems like I've identified the systematic breadth of the problem. Now I can brainstorm some solutions and see how reliably they line up with the problem. I know there are a lot of overlapping influences in the system, so I'm also going to consider how any changes might disrupt other parts of the system.

I'll also run it by other people involved in managing and interacting with the system to get their input; sometimes they come up with better ideas. I think it's reasonable to get some feedback.

Automatic emotional thought can yield benefits for simple or routine tasks, or when we are faced with time limitations or in emergencies. But we generally lack insight into systematic errors produced by our limbic reactions, which limits our ability to analyze and correct those errors. If we lack insight about the difference between thinking, feeling, believing, and knowing something, we can easily delude ourselves into the erroneous premise that what we think and what we believe are equivalent and correspond with knowledge about how the world works. We err by confusing beliefs with knowledge and believing that everything we think or feel is absolutely true. Our language can play a fundamental role in clarifying our thinking and our knowledge. Questioning our language seems to makes sense.

On a daily basis and more often than not, we do have the time to think and to consider how effective thinking can improve our insightfulness. Yet we generally avoid opportunities to learn and practice effective thinking skills. We have the ability to learn from mistakes and enhance our knowledge by assessing the effectiveness of our predicted outcomes. Thinking effectively helps us to make sense of what goes on around us, enhances learning, and helps to identify and correct systematic errors. We have the option of deliberate thought using extended language that can coherently describe our experiences and knowledge in correspondence with nature. Thus, language can help us make sense of our predictions, outcomes, and options and help keep us tuned in to the world around us.

Making Sense of Entropy

Granted, understanding information entropy in general, in particular, and at a technical level can lead to exhaustive thought, complicated formulations, reading a plethora of books, and frustration when trying to come up with a sensible understanding. In the context of this book, we are focusing on language as an information medium, and we would like to understand the connection between language and entropy. We can try to overcome the obstacles of exhaustive search by finding salient points that can shed light on entropy in a way that helps us to understand the systematic relationship between language, information, the thinking brain, and nature. We will start with a brief introduction of some of the technical aspects, and then try to address relevant features related to language.

The term *entropy* comes from the Greek root *trope*, meaning "turning into" or "transformation," and simply means "evolution" (Prigogine, 1997, p. 19), which in turn means "changing." Entropy in physics and information in biology are closely related (Polani, 2009). Entropy is often equated with a state of disorder and randomness in contrast to a state of order and nonrandomness; this distinction allows us to separate random from nonrandom patterns. In classical physics, entropy tends to be associated with randomness and thermodynamic equilibrium, where debates often revolve around symmetrical time.

Entropy plays an important role in physics, theories of evolution, nature, information, and communication. Language represents information, and through evolution, "information is bound up with the fundamental physics of nature" (Gough, 2008). Abstractly, language represents and communicates information that we can analyze in terms of entropy. Norbert Wiener pointed the way for early communication theory when he suggested representing information in probability terms (Jaynes, 2007, p. 634).

Claude Shannon's classic theory of information entropy proposed using a signal-to-noise function to measure information density while maximizing the transmission of information between a sender and a receiver (Shannon, 1948). The amount of useful information carried by a message can be measured by its unpredictability or uncertainty, and signals are expressed probabilistically in statistical terms because their future value is predictable only in those terms (Crecraft & Gergely, 2002, p. 63). In effect, a message contains information in proportion to a statistical measure of the decreased uncertainty at a receiver. It has been said that in correspondence to Shannon, John von Neumann suggested, "You should call it *entropy*. . . . [N]obody knows what entropy really is, so in a debate you will always have the advantage." (Tribus & McIrvine, 1971)

Information entropy refers to the range of consistency between the message sent and the message received as the signal, while the amount of uncertainty in the message is given by the noise on the channel as measured by errors between the sent and received message. If the receiver can determine the sender's message with certainty, the channel is said to be noiseless (Jaynes, 2007, p. 629). Obviously, although clever and useful for his intended purpose, Shannon's definition of information does not address the meaning of the messages (Mitchell, 2009, pp. 52–54) and lacks an external accuracy-based referent for establishing credible semantic meaning according to external laws (Jaynes, 2007, pp. 634–635). The focus is on the meaning measured between sender and receiver as referenced locally rather than globally. We can think of the two participants as components in most kinds of information systems, including digital electronic signals, analog radio waves, and communication between two localized individuals. For example:

> "We know by definition that local referencing criteria depend only on correspondence of meaning between the sender and receiver of the message that disregards referencing to the global domain and systematic meaning. Then the message might not seem noisy between the two localities but will likely represent noise in regard to the overall global system."
>
> "Hmm, so if we want to have systematic meaning, it requires that the language works in the global system."

"Yes, and we would calculate the noise as the systematic error difference between the predicted and realized meaning measured from the referencing signal, which corresponds to the domain."

"What do you mean specifically by the domain?"

"Domain represents the space designated as the area of interest for solving a particular problem or problems; the domain space defines the physical dimensions, and structural and functional variables and parameters within that space. For instance, a domain is a particular area of interest that can range from local to global, from a single 8″ × 10″ static picture to dynamic resonances interacting persistently across the totality of the atmosphere. The important point is the explicit specification of the domain components to clarify any initial ambiguities as to the dimensions of the problem-and-solution search space. Shannon's area of interest was locally referenced communication, choosing a localized domain and confining the search space to local solutions that served his purpose. Dealing with global domains is a bit more difficult because of the vast number of variables and the combinations of dimensions that lead to complex problem representation and solutions often described as ambitious. You can see that even the localized domain that Shannon dealt with was quite difficult; of course he cleverly solved the problem."

"He was a pretty clever fellow, all right. It seems obvious that problems in local domain are bad enough—no cakewalk, that's for sure. Why would anyone even consider fooling around in a space so large that you could get lost before you ever figured out what problem you wanted to solve?"

"Different strokes for different folks, I suppose. But you will see that there are a few helpful tricks for dealing with global domains."

"Sure, like don't forget to pack your lunch, or maybe several of them. Okay. So, what are the tricks?"

"Well, one trick Shannon was quite adept at was defining the space in terms of information. And information is ubiquitous in almost any system but can act differently depending on the domain. You might recall that information is a bit tricky when it continuously flows throughout a large system. This means that across the overall system, we see changing distributions of information patterns that require statistical descriptions."

"If the information is constantly changing, then we don't know in advance how useful it will be. Well, at least we know it's a dynamic domain, and we are looking at the whole enchilada, so that makes it global. Hmm, so certainty goes out the window, and uncertainty comes into play."

"Yes, it's global, and as we previously discovered in these domains, information corresponds to statistical descriptions as possibilities and probabilities."

"Perhaps this is not as complicated as I thought at first glance. Possibilities seem fairly simple, are easy to use, and usually work pretty well."

"We agree, and when we have enough information, we can calculate probabilities."

"That's okay as long as I can usually get by with possibilities, and when problems get really difficult, I can go buy a calculator and with the push of a button calculate probabilities in a pinch."

"Yes, we agree on this, and as long as you use possibilities that plausibly fit the global domain you initially specified, you stay in the global space. Now we have a fairly simple language form that matches the general ordering of information in nature. We can apply possibilities to our everyday lives for improving orderly thought and communication. Understanding the statistical distribution of information in nature helps us to make the logical connection between orderly information, the physical domain dynamics, and entropy. We will discuss more about probability in the next segment. Let's get back to entropy."

Maximizing entropy parses and distinguishes information as nonrandom known and as random unknown, ambiguous, and assumed. Language enables separating the random unknown information from the known, which is orderly and nonrandom, and allows us to keep what is systematically meaningful as useful information. Language that supports systematically useful information enables the generation and accumulation of orderly knowledge, which has meaning that makes sense across the entire system. Of course, it would be nice if it were that easy in a dynamic world with many continuously moving, interactive parts. We know that complex biological systems demonstrate the open-ended exchange of information and energy across the system, including semipermeable boundaries. DNA is an example that works across the system by the open-ended evolution of more complex solutions, and even though the solutions are nonrandom, evolving systems contribute to increasing entropy (Prigogine, 1997, p. 175). We expect open-ended language to provide the simplest plausible solution for representing, describing, and understanding entropic processes.

In other words, we can think of language as representing systematic information that relates to entropy in the system. We can characterize entropy in information terms as the amount of disorder in a system that corresponds to randomness. If our language constrains and confuses us about the randomness of information distributions, we have little basis for drawing sensible conclusions. With systematic language that generalizes and matches the statistical distribution of nonrandom information, we can understand the information entropy of the system, and communicate it in the shortest message possible. This means that invariant one-way time enters the picture again, because without symmetry breaking we become lost, confused, and entangled in the web of two-way time. We prefer globally referenced language that describes the time-dependent forward ordering of natural evolution, probabilistic information distributions, and entropy production in the dynamic system. Global language defines information, randomness, disorder, and entropy systematically according to nonequilibrium constraints.

Language with sensible correspondence to systematic relationships recognizes globally meaningful information patterns as regularities correlated to the ordering of physical laws. As noted, information entropy conceptually allows for the separation of domain information into the known and unknown. Not only do we have knowledge of the ordering of the system but we also have information about the correlation between interactions and regularities and the breadth of our ignorance. Hence, we can better separate the known from the unknown and minimize assumptions to what is known:

> Stated informally, this corresponds to minimizing the assumptions about the given system beyond the known parameters and constraints. (Polani, 2009)

With language that enables the discovery of orderly, meaningful patterns in the nonrandom information, we can purposefully improve our understanding about systematic relationships. *Purpose* defines and conveys goal-related action that we find useful for adaptation. In turn, we can make sense of those patterns as systematically useful information to further enhance adaptability. Accordingly, in an informed recurrent loop, language conducive to systematically sensible information generally is more useful to us and produces more meaningful information. Recurrence enables continuous error tuning that leads to more orderly systematic knowledge and purposeful correction of systematic errors. This perpetuating loop allows for continual closed-loop tuning referenced externally to the invariant forward direction of time in the global system and for updating and correcting systematic errors and the effectiveness of our information. We now have a reliable tuning loop for referencing information to the global error signal, evaluating and correcting the variance between the predicted and realized outcomes, and effectively informing ongoing adaptive actions in response to external change. In effect, we have a linguistic model for systematically maximizing information entropy.

> There is mounting evidence for the importance of information as a fundamental currency underlying the success of living organisms. (Polani, 2009)

Language with explicit nonlocal referencing that corresponds with systematically preferred information increases our ability to discriminate between global signal and local noise. We now have an explicit systematic method for observing and differentiating random noise from nonrandom signal that supports real-time tuning of the global signal-to-noise ratio. In effect, global language facilitates maximizing the entropy of global communication by leveraging what we know about uncertainty and the statistical distribution of information in complex systems. We know that asymmetrical systematically preferred information allows us to identify global structure encoded in the message source along with salient features that are useful for effective systematic translation. The relation of encoding to particular natural physical phenomena provides a link between communication theory and physics (Pierce, 1980, pp. 128, 184). Since we know how the system works, we can use this systematic knowledge to decode our present information, minimize our assumptions, inform our predictions with greater fidelity, and enhance the informed effectiveness of our communication.

> ... [T]he principle of maximum entropy is not an oracle telling which predictions *must* be right; it is a rule for inductive reasoning that tells us which predictions *are most strongly indicated by our present information*. (Jaynes, 2007, p. 370 [emphasis in original])

What do we know about information entropy? We know that it requires separating the known from the unknown information and minimizing assumptions. We also know that systematic information logically correlates to statistical regularities in correspondence with global laws. We expect orderly language constructions to correspond with systematic structure and the physical laws in the domain, and to facilitate the production of effective information, knowledge, and communication. Simply put, our naturally assuming brain can benefit from orderly language that helps with minimizing assumptions and maximizing information entropy. Fortunately, identifying and minimizing assumptions is fairly simple and gets easier with practice.

Making Sense of Uncertainty

Recall that continuously changing domains are fraught with uncertainties. In complex systems, information distributions are nonlinear and probabilistic. Extended language provides statistical methods with continuous logic for accessing and processing dynamic information. In uncertain domains, language that provides support for irreversible processes, oscillations, and one-way time can inform global knowledge (Prigogine, 1997, pp. 18–29). Also recall that extended language provides key open-ended parameters for extending and expanding information search, making predictions, and dealing effectively with systematic uncertainty.

Assuming that information represents either absolute certainty or complete randomness relegates us to the ambiguity of absolute perception. But why would we assume absolute perception? In an uncertain environment, the use of probability avoids absolute perception and reveals dynamic patterns in nonrandom distributions of information that reduce ambiguity and the dimensions of the information search space. With probability we acquire a processing method with logical consistency. But how can this be? How can probabilities that never seem to stop moving be logical, let alone consistent?

Possibilities and probabilities match with the statistical distribution of information in the domain, and in a sense they reflect the logic of the domain. When we incorporate the same logic into our language, we incur logical symmetry with the domain—logic that simply works as consistently as possible. Language with logical consistency corresponds with the nonabsolute dynamics, instabilities, and real-world uncertainties that we deal with every day. Understanding that our speech behavior is a product of the logic with which we think suggests that we can incur practical benefits by shifting our thinking toward the nonabsolute and probable and away from absolute and certain. Why does this seem strange? When we are used to thinking with absolute certainties, even just the idea of living in a world of uncertainties can be quite alarming—let alone confronting the challenge of changing our speech habits.

Furthermore, absolute thinking produces logical inconsistencies as true-or-false static solutions mismatched to the dynamics of the environment that misinform adaptive actions as things change. In effect, when information keeps changing, absolute certainty correlates to (almost) "always wrong." We calculate regularities as reliable correlations over time, and we have already seen that absolutes are antithetical to statistical measurements such as reliability of correlations. With absolutes, we wind up with little if anything to tune; it either "is" or it "isn't" correct. We may think we have discovered a perfectly simple, easy-to-use, ideal language that is just right, even for arguing with the utmost certainty. When we see what we think are discrepancies, we can simply pick one side and argue right or wrong. But have we simply fooled ourselves?

Probabilities give us leeway for continuously tuning correlations and making systematic corrections as we get more information, whereas absolutes close the solution space to infinity. So in a changing environment we can say that absolutes appear to yield infinite incorrectness. We can see why absolutes corrupt information in a changing world and why they also deceive us by corrupting our perception of that world. Importantly, absolute constraints inhibit effectively tuning in to that world. It hardly seems argumentative to say that the world changes, brains change and learn, and flexible language facilitates learning.

We can investigate the idea of uncertainty in the context of the brain and neuronal functioning. Uncertainty leaves us the tasks of dealing with a complex organ, a complex language, and complex dynamic processes. From a neuronal perspective of the brain, we see an apparently noisy organ. The brain looks somewhat organized at a functional level but seems to contain a plethora of randomly discharging neurons. Horace Barlow (1972) suggested that when considering firing patterns of nerve cells as noise or signal, we withhold or temper rash judgments about the firing of individual neurons because their apparently erratic behavior might be due to our ignorance and not the neuron's incompetence. Assuming randomness can lead to ambiguity by tempting us to ignore the possibility of systematic nonrandomness. Viewing the brain as a nonequilibrium system operating away from equilibrium in between rigidity and chaos, we see a dynamically regulated system functioning in the complex space of inherent instabilities between absolute certainty and uncertainty, absolute order and disorder, absolute randomness and nonrandomness.

When we describe the world in absolutes, relegating it to a state of static perfection where stability equates to a rigid state, we come to see the world as either rigidly organized and stable where uncertainty is hardly an option or totally disorganized and chaotic. In nonequilibrium systems, such as the one in which we live, we see persistent change, periodicity, and oscillations, where new solutions are added that can improve the system's ability to modulate adaptation in response to instabilities.

For example, we see animals that seem to be caught in a quandary where we can easily assume that their brain is oscillating. We might describe them as having difficulty making up their mind, or as vacillating, and also notice that our brains often behave in a similar manner when we have difficult choices to make. The limbic system and the omnipotent amygdala are at the crossroads of these oscillations for making affective decisions, where we learn whether the consequences feel either good or bad. Even though they are generated by the primitive limbic system, feelings are an important form of default learning, and we can see the emotional association with affective language. The downside is that the limbic system "couldn't care less" about analyzing possibilities and probabilities, but in its defense, it does a pretty good job of keeping us safe on average.

It seems that we have discovered the striking similarity by which localized language corresponds to the world and resides within the organism as an internal affective reference. The limbic system silently oscillates, automatically converges, and predetermines an intuitive solution that makes emotional sense on the basis of past experience, expected reward, and penalty risk. In other words, the limbic system has already made the choice without us knowing. We can use affective language to justify the certainty of our automatic emotional behavior and absolute solution choices, or we can use effective language that deals with uncertainty by making deliberate decisions based on leveraging informed analysis and systematic knowledge. The discrete nature of affective language misperceives two-valued limbic logic as digital true-or-false flip-flops between absolute positive or negative values. Extended language recognizes analytical processing with continuous logic, analog oscillations, and multiple alternatives for possibilities and probabilities that leave the solution space open. We have the option of weighting and informing solutions analytically as possibilities rather than emotionally as certainties. One language works best in hindsight and for covering up our errors, and the other one works best for learning from the past, living in the present, and predicting and looking forward to future possibilities.

Coarse-grained certainty increases the bandwidth requirements for communication and is computationally inefficient because of the constraints imposed on signal-to-noise ratios in a high-dimensional nonlinear domain (Pierce, 1980, pp. 178–179). Representing uncertainty as probabilities improves bandwidth efficiency, relaxes arbitrary static constraints, and reduces linear liabilities associated with ambiguity and maladaptive signal-to-noise ratios in nonlinear domains. In high-dimensional multivariate systems, probability as logic enables generalized encoding, signal-to-noise ratio tuning with enhanced signal discrimination, and sparse representation that takes up minimal bandwidth and storage space. Of these, tuning has the greatest effect, since an excessive signal-to-noise ratio "impairs neuroplasticity, that is, the likelihood of change" (Spitzer, 1999, p. 292).

> This [uncertainty] requires a new formulation of the laws of nature that is no longer based on certitudes, but rather possibilities. In accepting that the future is not determined, we come to the end of certainty. (Prigogine, 1997, p. 183)

In terms of certainty and uncertainty, we can assume that information relates in some way to ignorance, and further, that both can influence how we compute knowledge. Certainty implies absolute knowledge and zero uncertainty, which in a changing environment tend to elude the cleverest even on the best of days. Absolute knowledge assumes that everything we attend to is relevant and leaves nothing to ignore except that which we assumed by default as irrelevant. We encounter a problem when it comes to making predictions where knowing everything results in perfection such that we believe we demonstrate errorless behavior. Ignorance, both prior and current, influences how reliably we compute predictions.

In the natural world, certainty implies more knowledge than possible, which tends to lead to unreliable conclusions.

> The uncertainty was always there.... [I]t is rather the means of determining quantitatively the full extent of the uncertainty already present. It is *failure* to do this—and as a result using a distribution that implies more knowledge than we really have—that would lead to unreliable conclusions. (Jaynes, 2007, pp. 366, 373 [emphasis in original])

Even though certainty might feel good, it leads to errors in what and how we think we see and "know." Simply put, believing in certainty does not make it absolutely so. It seems we have developed the habit of thinking in certitudes and consider any attempt to take them away from us as a punishment. This is analogous to the endowment effect, where once we own something, its value increases and we are less likely to want to let go of it, and losses loom larger than gains (Kahneman, 2011, p. 299).

Uncertainty is sometimes attributed to imprecise measurement, ignorance, or the limited sensitivity imposed by coarse graining. But uncertainty can more reasonably be attributed to irreversibility and the breaking of time symmetry that constrains predictions to probability proper. Even though many different properties influence information, in complex systems it is the irreversible arrow of time that defines probability by separating the past from the future (Prigogine, 1997, p. 25).

Certainty inhibits creativity and the search for innovative information and knowledge (Prigogine, 1997, p. 184). An unmitigated dependence on certainty naively increases deception by eliminating alternative solutions, and once denied, ignorance and alternatives go unexplored. Absolute language that describes dynamic information in static terms of certitudes blinds us to alternative possibilities. In contrast, alternatives enhance flexibility, learning from errors, creativity, and the exploration for more effective knowledge. Extended language provides open-ended logical processes that help to make sense of uncertainty and facilitate the search for novel solutions.

> Certainty is quite demanding. It rules out not only the far-fetched uncertainties associated with philosophical skepticism, but also the familiar uncertainties that affect real empirical inquiry in science and everyday life. But it is a condition that turns out to be disposable. . . . (Jeffrey, 2004, p. 53)

Allowing for ignorance can mitigate the absolute constraints imposed by certainty. We can attempt to maximize our ignorance (entropy) by taking a broad perspective that accounts for the vast magnitude of what we don't know, and ask, "Why do we know so little and why are we ignoring so much?" In effect, acknowledging the scope of our ignorance plays a complementary role in gaining insight into the scope and limits of our knowledge. Accepting the inevitability of our ignorance can encourage us to explicitly minimize our assumptions and to deliberately include as much as possible of the available plausible information into our thought processing. With a deliberate effort we can update previous information and knowledge and make more thoughtful inferences in the present. We can make decisions with the maximum amount of available information sufficient for [effectively] solving a particular problem and increase the probability of better outcomes (Jaynes, 2007).

For example, we can say that we feel sure about our own ability, which corresponds to "I feel confident." In effect, we get certainty and confidence by trading away ignorance. But in the process we place constraints on correcting systematic errors and learning, searching for new information, and problem solving. In a changing world we can express our feelings of certainty, but by doing so we incur high expectations with outcomes limited to either right or wrong. We automatically set ourselves up for disappointment, since the future is probabilistic. Why would we contort the probabilistic future into certitude? What's in it for us?

The simplest plausible answer might be that by stretching our qualities, we sound more competent. Of course, we have traded humble possibilities for brazen certitudes, which apparently pays off in some way: we are reinforced for exuding confidence and certainty and penalized for exhibiting humility or for reeking of uncertainty. We see examples of this payoff in the behavior of chimpanzees, where animals that act more assertive have an opportunity to compete and win a dominant position in the status hierarchy. Ironically, we humans teach humility, and then we penalize humble behavior by trampling over it in our enthusiastic stampede to reap the rewards of appearing self-confident and certain. Why? How could this happen, and why would we perpetrate or fall for such a charade? The likely answer stretches across nature in life forms from unicellular organisms to humans. "The organism moves and reacts in ways that are advantageous to it" (Jennings, 1906, p. 339).

Think about two people interviewing, or competing, for a job. Person A tells us how certain he is about his abilities. "Sure, I can do this job. I am absolutely confident that I can fulfill your expectations. I have already decided. I know I am the right one for this job." Person B tells us that she thinks her abilities might fit the job but that she would like more information: "I would like to know more about your company, the job description, and details about your expectations. Then I can decide whether my skills match." Person B gets more information and says, "I think my skills and abilities will probably fit well here." Of course, there is a lot more to hiring personnel, but research shows that we tend to see people more positively when they exude self-confidence and we disregard the possibility of deception.

Unfortunately, self-confidence and abilities often have an inverse correlation, sometimes referred to as the *illusion of confidence* (Chabris & Simons, 2010, pp. 85–92). Our beliefs and other cultural factors from childhood onward can play a role in how we perceive others, including affective language conventions that tend to place values on status hierarchies and appearance, such as acting like an authority, acting confident, and acting certain. But confidence is a "feeling" (Kahneman, 2011, p. 212). The key words are *feeling* and *acting*.

Most of our scaffolding from initial prior information is based on observations, interactions, and language primitives automatically encoded and imprinted during our cultural development. Scaffolding with locally encoded language structure affectively biases our worldview, the information we accumulate, what we think others expect of us, and what we think we are "supposed to do." We easily become implicitly enslaved to arbitrary values that bias our actions to an affinity for certainty and foster avoidance or denial of uncertainty. We unwittingly communicate the certainty of our convictions with absolutes, unacknowledged assumptions, and imperatives that void opportunities for other possibilities. We confidently interact with others while maintaining blissful ignorance and failing to realize that we live in a world of possibilities, as did our ancestors.

The telltale fingerprint of certainty in our speech behavior exposes us when we fail to reasonably account for ignorance, assumptions, other possibilities, and equivalent solutions. Absolute language resists change attributable, at least in part, to hidden assumptions and the amount and breadth of accumulated information that takes extensive error correction to shift the mean (Jaynes, 2007, p. 558). Furthermore, it is difficult to recognize the benefits of change when we lack insight into the primitive value biases nestled within our belief system that mandate certainty, exclude the possibility of uncertainty, and implicitly deny reality. If we think that other possibilities are permissible, we might consider taking a closer look at scientific methods where a bit of a skeptical perception has demonstrated benefits when applied uniformly. Meanwhile, the unchecked habit of reducing an uncertain world to certitudes escapes our attention and thus defies remediation.

> ... [T]he mammalian brain is designed to construct belief systems, and once they are solidified, they are as hard to move as mountains. (Panksepp, 1998, p. 245)

Making Sense of Timing and Tuning

Why would anyone care about timing and tuning? Recall that we have discussed the technical aspects of the arrow of time and symmetry breaking: how persistent tuning keeps our information up to date by decreasing error rates and error accumulation and by enhancing error correction and learning. We also have discussed how time and memory keep us tuned in to living in the present; that memories of the past are static and different from the dynamic present and future; and how the brain integrates memories with current information for making predictions and choosing actions. We can safely say that timing and tuning play a significant role in our everyday lives, especially for thinking effectively in the present, planning for the future, and predicting the outcomes of our actions. Timing and tuning may not be everything, but when it comes to effectively adapting and living in correspondence with the world around us, they mean a lot—and offer us many reasons to care.

Even though our brain has a memory system that helps to keep track of information, in large part that information is the product of a fast and frugal limbic system where coarse-grained processing creates noisy information and noisy memories. Recall that the limbic system is generally ignorant of one-way time and systematic dimensions and operates with discrete polarization supported by affective language with deficient tuning abilities. A casual observer with affective language would only notice noise relative to their own locally polarized perspective, or the polarization between two local points of view.

The noise we are dealing with here is partially the local noise referenced from a global perspective and noise accumulated from systematic ignorance. We see noisy language and noisy prior information augmenting a noisy, assuming brain replete with unnoticed systematic errors that appear to be locked in a fail-safe, eternally recurring local loop. Recall that ignorance of how the overall system works is also called systematic blindness. The noisy language, prior information, and egocentric brain are stuck in a loop that refers back on itself, resistant to change and impervious to nonlocal information—systematically blind. On the brighter side, if there is one, it is that on average, they work well enough.

Alternately, we realize that globally referenced timing and tuning with effective language augments processing in the same human brain that can facilitate global optimization. Knowledge produced and biased by fidelity-based language and reasoning that accounts for noisy localized information, systematic ignorance, and fundamental uncertainty can help to mitigate the influence of noisy prior and current information. Accuracy-based language facilitates prioritizing information reliability and up-to-date systematic knowledge, fine-tuning the signal-to-noise ratio, and effectively informing our thoughtful assessment of the here and now. In effect, understanding sensible timing and tuning facilitates boosting dynamic signal detection, attenuating noise, and optimizing effective reasoning. Affectively noisy language may be fast, frugal, and effortless, but the contribution of noisy reasoning can be expensive in the long run.

We can look at our own relationships to find a fairly common example of this tuning. Most adults have felt the discomfort of hearing their aging parents speak about them in a condescending manner as if they were still teenagers or children, even long after they have children of their own. Parents come to know their children as children, and under some circumstances, it seems that no amount of new information about adult behavior, experience, personal endeavors, or education, which often surpasses that of the parents, can serve to update the parent-to-child reasoning and speech they maintain in their dominant relationships with their adult children. They fail to adjust their language to the present.

Parents who simply continue to talk down to their children as if they are still very young show a rigidly biased perspective, which typically includes the habitual repetition of verbal behavior that served to establish parental authority and obedience to imperative control in years past. Continuing to address other adults with primitive parent-to-child communication that is based on obedience to authority is hardly more likely to work any better now with adults than it did then with children. Instead, modeling primitive imperative speech confers the same level of inefficient affective control for children or adults in comparison with speech based on reliably informed thought, possible choices, and the effectiveness of outcomes.

But dominance does seem to work fairly well in nonhuman primates that operate with a status-based social system fashioned by authority ranking in a dominant–subordinate hierarchy. Aroused monkeys tend to express anger in confrontations with submissive animals, but they avoid confronting more dominant ones (Panksepp, 1998, p. 196), similar to dominance and the obedience to authority seen in humans (Milgram, 2004).

Parents have the option of deliberately fine-tuning their language when addressing their offspring to acknowledge the mutual adult status and adjust the relative orientation of the relationship as between adults thoughtfully communicating in the here and now. The difference can come from something as simple as asking

- "What do you think?"
- "Would you like to . . . ?"
- "What is your preference?"

We can then accept the answer as coming from another adult. But we often default to habitual parental speech patterns based on authoritarian control of subordinates:

- "You must not think or feel that way." (Think the *right* way.)
- "You must agree with me." (Authority equals *right*.)
- "You must do as I say." (Behave obediently.)

Deliberately depolarizing, tuning, and extending language from affective to effective shifts the correspondence channel from local to global, vertical parent-to-child to horizontal adult-to-adult with a semantic shift from instruction to information, from imperatives to preferences. This semantic shift facilitates thoughtful speech behavior between consenting adults and enables choices among alternative possibilities that increase the probability of improving cooperative communication. We see that effective language reliably separates the past and future in time, levels the communication playing field, and improves cooperation and problem-solving skills in the present.

The simple act of deliberately using extended language according to preferences and possibilities can lead to logically consistent cognition that enhances our understanding of how language influences our speech behavior and the effectiveness of our communication. We realize how ours and the language of others with whom we interact influences the environment and can be construed as directly or indirectly influencing others' thoughts, emotions, and behavior. But not everyone acknowledges that individuals are each logically responsible for their own thoughts, feelings, and behavior. Studying why, when, and how affective and effective language influences responsible communication and the reliability of information and information processing can shed light on what constitutes logical use.

If we acknowledge these linguistic issues, we have the option of deliberately tuning our language to minimize cognitive distortions that constrain the effectiveness of our reasoning and our results. We can opt for globally correlated language that deliberately augments systematic reasoning correlated to invariant forward-directed time. Effective language is globally sensitive to the functional aspects of time, including symmetry breaking and regulatory timing and tuning with oscillations. We have the option of defaulting to locally sensitive affective language with limbic constraints of automatic either–or processing, true-or-false regulation, certitudes, and temporal insensitivity.

The limbic control system is susceptible to affectively noisy prior information and compromised fidelity. Why? Confusing affective feelings about the past with effective information in the present creates a noisy time warp with reality. Acknowledging affective noise enables us to deliberately analyze and improve the dynamic real-time tuning of signal-to-noise ratios, decrease errors, and improve the accuracy of our knowledge and current information. Effective systematic tuning requires invariant external temporal referencing in the direction of the future. When we explicitly acknowledge the benefits we incur from effective thinking, it encourages us to work on explicitly improving the fidelity of the language with which we think, which can facilitate beneficial changes in how we live and relate. Time, timing, and tuning are ubiquitous in our lives, and time-dependent language that supports fidelity and flexible tuning offers us a bonus.

Greater linguistic fidelity and flexibility can facilitate beneficial changes in how we tune our thinking, and live our everyday lives. But how can we optimize our journey through life, and what are the rules for this journey? Do we prefer to settle for sufficient localized solutions or work toward global optimization? If we care to make sense of how we know our world, we are left with the problem of optimizing our knowledge of it. But how can we? We would expect sensible language to offer computational compatibility with nature for tuning us in to the problems and solutions we encounter there. If so, perhaps we can strike a bargain.

The next chapter explores the benefits of a computational model for understanding language, brain, behavior, and our world. We will address exploring the pros and cons and exploiting the benefits of a computational model for representing, clarifying, and understanding the problem-and-solution search space that we deal with daily. In this cognitive space we see what looks like a cluttering of languages, codes, brains, feelings, objects, and actions. The space we are talking about lies within the physical limits and jurisdiction of the world we call Earth, between our sensors and motor system, and we often refer to this cognitive space between our ears in which our thinking resides as "the mind."

SECTION III

Using Language to Make Sense of the World

CHAPTER 16

Language and the Computational Brain

IS IT POSSIBLE THAT computers can help solve the problem of understanding language and the brain? What, if anything, can we learn from computation? Can computer science enlighten our perspective about representing problems and shed light on what influences similar brains that live in the same world to represent the same problem differently and come up with diametrically opposed solutions? Will a computational input–output model inform and help clarify the problem-and-solution search space conundrum? How can a sober engineering perspective that we often think of as logical help us understand a confusing and seemingly illogical world of feelings and fluctuating rewards and penalties?

Why is computation used extensively in science and many societies? Perhaps the answer is that "it works." Computational models are prevalent across science, including biology. In neuroscience, we see an increasing acceptance of modeling the brain as an input-computation-output processor of information. A perspective from a computational model may offer benefits for improving our understanding of the relationships between natural language, brain and behavior, information, and nature.

Of course, our day-to-day language seems good enough to us, but is it possible that our linguistic brains, computers, and computer language have too many differences for a comparison even to be possible? Perhaps the natural language, syntax, semantics, and grammar that we routinely use lack the precision and accuracy required for making computer-like comparisons such as "not good enough." We can consider our thinking brain as performing computation and ask the question, "How does language influence the problem of processing of information?" Investigating these questions seems like a win-win situation with little to lose and a lot to gain, such as possibly decreasing unfounded assumptions and increasing knowledge that we might otherwise overlook.

Exploring these questions can possibly improve our understanding of the "problem of language"—of how it influences our thinking in a complex world. Computational models allow the simulation of different perspectives for solving practical daily problems, such as how language that misinforms cognition can produce learning errors and faulty solutions. We can incur unintentional errors that nevertheless often leave us to mercilessly beat ourselves up because we "should not have" made an error.

So that our natural language model matches the natural world, we expect computational compatibility with the previously discussed dynamics and nonequilibrium domain constraints. In other words, we expect a model with the ability to demonstrate systematic effectiveness. A systematic language that supports an explicitly informed brain according to the effectiveness of the output likely fulfills this purpose. An understanding of effective output facilitates systematic modulation and calibration of the input sensitivity and corrective tuning to decrease the rate of output errors. Optimizing the effectiveness of the output depends on reliable referencing, error signal feedback, and output tuning that augments error correction by minimizing the difference between the predicted and realized outcome.

The goals of this chapter are to

- Describe the brain and natural language from a computational perspective
- Sort out similarities and differences that can help us understand relevant language features
- Identify the tunable features that matter the most when it comes to effective cognition

The term *natural language* is used more often in this chapter to distinguish it from computer languages. In most other fields of science, language is language, but in computer science the distinction is important because there are many different types of computer languages. The term *natural* is not meant to distinguish natural language that generally occurs in humans from a particular natural language or one that structurally correlates specifically with nature but to distinguish human language from computer code in relation to information.

Input-Processing-Output System

There is a saying in pharmacology research: "If you do not have explicitly documented evidence of attending to an event, it did not happen." When running an experiment in computer science, programmers will likely say that running any experiment first requires an explicit representation; if they cannot explicitly represent their experiment and goals, any relevant computations are thwarted—it will be unlikely to happen—but they also know that if they are motivated enough, can find suitable explicit variables for problem representation, and can find reliable parameters to compute and output an effective solution, perhaps they can succeed.

These aphorisms may seem obvious and trivial, but they point to an important aspect of scientific information processing: computing solutions to problems involves

- Having an explicit understanding of the problem-and-solution search space
- Having a domain representation of the system, defined data classes, variables, parameters, and logic
- Choosing a computer language for coding
- Building, coding, and executing the resulting program

A program requires much time and effort for constructing, modifying, and tuning to suit the particular dimensions of the problem-and-solution search space. The dimensions of the search space engender computational cost constraints on particular experimental goals in proportion to the amount of time, computation, and effort required. Experiments in high-dimensional domains can quickly lead to the *curse of dimensionality*, a term coined by R.E. Bellman (1961), where a problem can become intractable as the number of experimental variables increases, and where attempting to solve high-dimensional problems is often described as ambitious. Informally, we can think of *ambitious* as "biting off more than we can chew" or attempting to compute solutions that appear to require a prohibitive amount of data, time, and effort.

In light of these difficulties, understanding the problem we want to solve and the dimensions of the problem becomes a central issue. If we cannot reduce the problem to a level of abstraction with fundamental dimensions that we can compute, then our abstraction more than likely contains some ideas and concepts that lack coherently defined referencing, classifications, parameters, and variables. This stumbling block suggests that making the task computable will likely require restructuring and further analysis. We can speculate that if we cannot explicitly represent the problem in a computable way—so that it will compute—then perhaps we incidentally or implicitly introduced some imaginary or incoherent domain elements. In other words, imagining or stating ideas hardly substitutes for meeting the challenge of expressing those ideas in a concrete, coherent way that makes logical sense in relation to computing solutions—that is, we go from idealized to realized. Computing solutions to real-world problems depends on coherently referenced variables and parameters that at some level of abstraction explicitly correspond to realizations in that world.

The exercise can be a tedious task even for a computer-literate person and especially challenging for someone lacking computational skills. Those skills become a limiting factor. Some scientists think of cognition in terms of computation, while others consider that metaphor an unsuitable model. The most compelling reason to consider this metaphor is that brains and computers both process information:

> In a sense, of course, the brain is a computer, but to say this without qualification is misleading, because the essence of the brain is not simply that it is a computer but that it is a computer which is in the habit of performing some rather particular computations. (Marr, 2010, p. 5)

Some argue that since brains are wet and computers are dry, they do not operate with enough similarity to make the comparison useful, or that it is impossible to overcome their dissimilarities. Of course, agreeing on the problem of defining the difference between brains and computers likely looms somewhere along a continuum between "exactly alike" and "completely different." The practicality of considering a comparison at all depends on the purpose of the model, whether it will provide us with a useful perspective on the problem we are trying to explain and solve, and on the degree of understanding of the model's potential limitations:

> Most of the phenomena that are central to us as human beings—the mysteries of life and evolution, of perception and feelings and thought—are primarily phenomena of information processing, and if we are ever to understand them fully, our thinking about them must include this [information processing] perspective. (Marr, 2010, pp. 4–5)

Computational research has found its way into many areas, including physics, chemistry, biology, psychology, and even natural language. Areas of interest for the discussion here include the application of information processing, concepts of information representation, learning classifier systems, pattern recognition and machine learning, artificial neural networks, artificial genomes, evolutionary algorithms, and evolutionary computation. Even with all of the progress that has been made in these areas, direct biological comparisons remain elusive. To compare brains and computers, we can take different approaches: a brain-oriented perspective, a computer-oriented perspective, a localized-versus-systematic perspective.

- How are brains and computers both similar and different?
- Do we focus on the input, the computation, the output, or all of them together?
- Which features would reveal the most relevant information for drawing conclusions about the effectiveness of natural language?

Ironically, computers and computational models alike are often criticized because even though they are accurate and precise, they don't work exactly the same as the idealizations we have about the human brain, or, to some extent, they struggle to understand human language, with the possible exception of some supercomputers. It is worthy of note that supercomputers that have achieved some level of success in simulating human thinking have incorporated context and the comparison and sorting of massive amounts of information statistically according to similarities and differences in meaning by identifying and elaborating the problem; generating hypotheses; searching for, finding, and scoring evidence; and merging hypotheses according to weighted *statistical* ranking. The computational processing model consists of generating what are essentially value weights for each possible solution according to source reliability and historical correlations of effectiveness. The processing results in the production of a most likely solution according to its weighted problem matching.

Perhaps it is no surprise that the super secret is the ability for computing almost instantaneous solutions in a matter of a few seconds or even milliseconds by massive parallel processing that has been called "embarrassingly parallel," in lieu of serial processing. Supercomputers have defeated humans in several areas, including playing chess and answering questions on a popular television game show. Parallel networks compare functionally to parallel brain modules that can process possibilities of corresponding data sets into a most likely solution—of course, with large differences in the sizes of the data sets. How does language figure into all of this? And why do we tend to ignore the influence of language on cognition for producing effective solutions? We take language for granted, even when we know how important language is in computer science for effective classification, representation, and statistical processing of information. What does this super computational secret tell us about language and the effective processing of information in particular and the brain's processing of information in general? We can see a comparison between parallel processing with massive data sets in computation and generalized compression that enables the human brain to thwart the curse of dimensionality and reasonably deal with massive real-world data sets.

But the highly touted human brain is often described in an idealized manner. We tend to ignore widespread evidence that at base, the human brain functions generally, or nonabsolutely rather than absolutely. In other words, brains typically work by resolving problems more in general terms of minimal to maximal local solutions that at best might offer global sufficiency. We can take into account natural language that is amenable to generalization and effective tuning. But questions remain, and we ask:

- Is our current perspective of the brain and language to some degree based on idealized models?
- Can we accept the hypothesis that we have inherited nonabsolute, imperfect brains and imperfectly tuned language?
- If so, is the realization of an imperfect brain and language too bitter a pill for us to swallow?
- Are we unwitting victims, naive and blind to how our brain and language work together with the world around us?
- If so, are we blind to our blindness?

Whether we agree with the idealization or not, it stands to reason that attempting to run an overidealized language on a human brain or an overidealized computational brain model will compromise operational results in a complex domain. Why is this? Brains work probabilistically, and certitudes offer little recourse except for convincing us of the pretense that we know more than we really do. Certitudes offer little consolation when it comes to effective thinking, and they tend to confuse us in a world with continuously changing information.

It seems reasonable to expect that tuning language to run more effectively in a less-than-perfect brain that naturally generalizes could offer benefits. By adapting language to generalize and operate in an extended, open-ended manner, we enable cognition that corresponds with the operational characteristics of the brain and the world in which it operates. The brain continuously processes information among many convergent and divergent pathways that work in series and parallel. Language that matches generalized brain processing that correlates with physical laws in the outside world offers the benefit of coherent correspondence and parallel processing of mutual information with that world. Extended language enables generalization and divergent search, conserves convergent abilities, and enhances flexible switching between them depending on the task context. We can expect to improve the brain's computational efficiency and output effectiveness with language that generalizes and matches the effective dimensions, logic, and dynamics found in nature.

An open-ended language enables generalized tuning that helps to limit computational costs by, to some extent, subduing the curse of dimensionality. If we want a language to support an open-ended brain operating in an open-ended world, then we would do best to consider open-ended language that matches. Have we been inadvertently victimized by running closed-ended, idealized language mismatched to our less-than-ideal, open-ended brain? We might expect as much because our brain architecture evolved and operates in an open-ended analog world. But with discrete language limitations and digital processing constraints, we struggle to evolve out from under the affective weight of the clever amygdala. In other words, our analog oscillating limbic logic succumbs to digital corruption by our everyday language.

In most fields of engineering, understanding the problem comes before thinking of solutions, and resolving problems starts with understanding the system's operational characteristics and relevant domain dimensions and constraints. We can relate this engineering analogy to real-world problems we face every day, and to the inherited language with which we think.

A computational model as a metaphor for an information-processing brain can enable practical comparisons of our everyday language's effectiveness for processing solutions to real-world everyday problems. If our thought processing bogs down in frustration or crashes when we try to solve dynamic nonlinear problems with our day-to-day natural language, it raises a question about the effectiveness of the encoding. For example, consider this discussion:

"I know this is the right solution."
"No, you are definitely wrong. My solution is right."
"No it is not."
"Yes it is."

We easily get bogged down in affective polemics. Is our affective language encoding that supports digital processing a part of the problem? What about computers?

Affectless computers seem to work quite well with digital processing of zeros and ones, producing exacting solutions. We may have a lot of recoding, debugging, and tuning to do to get our own natural language to run as effectively as possible in a persistently changing world. And yes, analog brains and digital computers have similarities and differences depending on what we expect from them.

It looks like computers and computer languages have their own advantages over brains, but what is the difference?

Programming languages are specifically designed for compatibility with the computer architecture on which they are used. The programming language enables a programmer to write programs and algorithms that use input data to compute solutions by an execute command that instantiates the program and crunches the data into informed solutions. But what has this got to do with brains? When we incorporate language that has been debugged of absolutes, we can enhance generalized constructions that confer open-ended analog congruency with our open brain architecture and nature. Congruency enhances efficient and effective problem resolution, albeit a bit more slowly. So it looks like we have trade-offs, advantages, and disadvantages. And it looks as if, when it comes to effectiveness, language plays a critical role for both brains and computers.

"It seems we have several possible solutions. What do you think?"
"Let's test them out and see which one probably best fits the problem."
"Yes, and we can simulate a test run for effectiveness."
"Okay, we agree. Let's get started."

Since brains have different architecture than computers do, we expect that brains might require a different programming language and encoding, such as analog versus digital. Why analog? We see analog functionality in the brain and environment. But why not opt for digital? The answer: We have learned something from computer science about the importance of mutual compatibility between the architecture, hardware, software, logic, data, and programming. We also consider the limitations put forth by particular domain constraints, what kind of problems we want to resolve, and how to represent them. We live in an analog world with an analog brain, so matching the logic leads us to select analog logic. Even though computers are digital, we can write programs with analog functionality, such as with oscillating artificial neural networks. For a brain model with analog compatibility, we simply select a natural language with continuous analog functionality rather than discrete digital true or false—that is, open-ended versus closed-ended logic. Probability as logic fulfills this solution criteria.

Natural language is natural in the sense that it has co-evolved to operate in human brains. But we have been discretely trapped by overfitting our affective language with overly exact linguistic primitives, even though we have the capability of exploiting the effectiveness of open language. Both are natural, but the latter confers a higher level of abstraction in an analog format. We now understand two distinctly different coding options: discrete, absolute language with digital functionality versus continuous, nonabsolute language with analog functionality that matches nature. So it looks like we cannot push our brains beyond their physical limit of imperfection by co-opting them with a language that can feign absolute perfection. It looks like we had better match the language to fit the analog functionality of our existing brain architecture so that we can operate as effectively as possible in a difficult-to-predict, imperfect world.

How can we better understand the problem of matching language, functionality, and architecture? We know that our brains' structural architecture evolved as a product of genetic inheritance and that our language constructions co-evolved as a product of cultural inheritance. It is obvious that the latter offers a great deal of flexibility for making functional adjustments and fine-tuning cognitive effectiveness. How can we take advantage of this knowledge?

Consider that English can be deemed a programming language that we can use to write programs to solve certain problems. Brains or computer models that have compatible architecture enable the installation of different languages. A brain with English-language programming installed could theoretically understand and solve problems with programs written in English. But that same brain lacks the capability to understand Chinese without installing Chinese programming language, which as we know would require some effort. Natural language supports the programs we write and how we write them, and it gives a great amount of flexibility. The efficiency and effectiveness of the written program depends on the linguistic construction, its compatibility, and its usefulness for given applications. But we often have little explicit understanding of the natural language we implicitly absorb from our culture, its preconfigured logic, problem-solving abilities, programs, and algorithms, and its limitations and range of effectiveness. We easily succumb to habit, take language for granted, and neglect to explore a deeper understanding of effective language constructions. We essentially wind up with an unexamined language and ambiguous value weights from a global perspective.

What if we try to run a program on our own brain written in a familiar natural language code that someone gave us, or that we inherited, consisting of mostly execute commands with little explicit contextual programming information and replete with ambiguous hidden data? This is a real-world problem we run into daily when thinking, communicating, and making decisions with the unexamined natural language we inherited, which implicitly carries previously biased and ambiguous data with little contextual information except for affect. How do we know that the language is unexamined? We know the same way we know whether we understand a secret code. That is, can we decipher mutual meaning from it that corresponds with the domain? Without knowledge of the key to deciphering the code, we can fondle the code, but we cannot understand and learn from examining it. For example, when we are told, "You *need to* do this, you *have to* do that, and you *ought* not to do that," we won't understand what is wanted if we aren't told the *why* behind those rules. And when we are told, "If you do not obey and conform, you will be punished," we will have trouble conforming if we aren't told what it is we "must obey."

We might think something is amiss and ask, "What is the encoding, and where did the embedded logic and values, commands, and intent come from? Can we differentiate the source of the encodings between animal or parental, digital or analog?" Recall the similarity with primitive command-like imperatives and hierarchal domination by power that we see in both chimpanzees and humans. What are the hidden assumptions? Does the encoding perpetuate hidden primitive artifacts that generate unseen systematic errors? If we cannot answer these questions, we can safely say, for all intents and purposes, that we have failed to effectively examine and discover the language encoding. We have already mentioned the difference between input affects and output effects. An informational input-computation-output model makes for a fairly simple comparison of natural language encodings.

For example, compare the *significant* differences in input and output between these two different language encoding types:

Affective language:

- The world revolves around me as an individual.
- I am normal and others are flawed and fallible imperfect humans.
- Digital true-or-false imperatives and prescriptives work just fine.
- Others are responsible and held accountable for my thoughts, feelings, and behavior. It is they who are responsible for my input of thoughts and feelings and my behavioral output.
- Consideration of my thoughts, feeling, and behavior by others is imperative.

Effective language:

- We all inhabit this world together as a population.
- We are all flawed and fallible, imperfect humans.
- Informatives, possibilities, and probabilities support analog flexibility and generally enhance effective adaptive output.
- Each of us as individuals is responsible and accountable for dealing with our own input of thoughts and feelings, and our behavioral output.
- Mutual consideration of our thoughts, feelings, and behavior is generally preferable.

With effective linguistics, we can inspect for logic primitives and absolute structure, consider other programming options, and tune the effectiveness of the output. We can run into problems that can trap or impede our escape if we ignore, fail to notice, or inspect hand-me-down language that comes replete with implicit absolute logic and digital processing. And yes, if we do not know the key to the code, we are stuck with an unexamined language—a linguistic black box. We can hardly consider that we examined the hand-me-down box because we determined that it is a black box with mysterious things going on inside that magically spit out solutions. We can incur an accumulation of errors from unnoticed authoritarian execute commands that make decisions on the basis of one-dimensional information encoded for *should, must, have to, ought to, got to, supposed to, just do it*, and so on. But how do we troubleshoot a problem with natural language when we are talking about information and computation, and especially if we are ignorant of the encoding?

Troubleshooting and Testing Solutions

It seems that a large part of the problem resides in the computational requirements for explicitly encoded programs, logic, and data. Running an implicit natural language program with hidden logic and data that we have habituated to is akin to computing in the dark. Perhaps more troubling is our ignorance of the certificates of authenticity and coding encryption. So we ask:

- Where did the natural language program and logic originate?
- Do we understand the encoding and have the key, and why are the operational aspects hidden?
- Can the language inform and support nonlocal encoding, programming, analog logical processing, and global optimization?
- How can we explicitly know much of anything about an implicit language program?

We seem to have expanded the problem of hidden processing to include hidden language encodings, programs, logic, and data. Can we explicitly understand the processing capabilities of our natural language and how it can facilitate effective learning and broaden our search for information and knowledge? We can start by evaluating and elaborating on the problem of information-processing constraints found in computers, brains, and natural language, and then consider measuring linguistic solutions by the explicit effectiveness of the output products. Differentiating computers and brains seems like a practical place to start.

Recall that computers and brains both process information, but they do so very differently. It is sufficient for the purposes of a word problem simulation to stipulate that the brain is not a computer and that it processes information in a uniquely different manner. Generally, computers are machines constructed to efficiently process specific types of data digitally with a high degree of fidelity and precision. The effectiveness of the output in terms of solutions depends in large part on the integrity of information input, the compatibility and effectiveness of the software and algorithms, and the architectural design.

Data input to a computer falls into certain prespecified slots according to prespecified rules. In other words, we know the specific type of data format to input into our computer beforehand. But knowing is not so easy when the input data comes from a difficult-to-predict analog world, a world where we still struggle to understand the appropriate formatting and linguistic encoding of the information that we input. Computers are very good at dealing with squeaky-clean certainty and static data, but not necessarily as good at dealing with noisy and relatively messy uncertainties in continuously changing, high-dimensional analog domains. The digital precision that makes computers so great can confer egregious limitations when flexibility and uncertainty come into the equation in an analog world of continuity, ratios, spectrums, and gradients. But our brains routinely deal with uncertainty every day. What is the confounding variable?

We understand that computers perform analytical operations rather than emotional ones and obviously lack human-like feelings. Trying to input affectively laden information into a computer with linear digital processing can lead to difficulty when the static computer architecture is not designed with an affective interpreter such as the ever-watchful amygdala. In a sense, human brains have a dynamic affective filter on their input that can continuously bias the signal sensitivity for emotionally salient stimuli, direct the focus of attention, and amplify and suppress the value weights of rewards and penalties, according to how the input feels, which, in turn, influences the output. Is it possible that unbeknownst to us we could have inherited a primitive affectively encoded language and a brain that is particularly sensitive to criticism? We know that in mammals, without adequate negative feedback, affective input can overamplify emotional processing that in turn can overdrive the output gain for approach-and-avoidance behavior, which, so to speak, is not necessarily a standard computer's cup of tea. Computers are designed by humans for humans, but they "think" very differently. We might think that we can intuitively improve brain processing by applying digital natural language, but we incur a mismatch between the idealized digital format and a brain with realized analog architecture.

Does our affective language and emotional behavior give us superiority over computers? What about superiority over robots? Robots contrived with computational models of what is sometimes ironically called emotional intelligence are often criticized for their muted affect, stilted fluidity, and lack of spontaneous human-like behavior, but the field is progressing rapidly. Human–robotic systems emphasize the importance of communication and cooperation between human and robots. This means that robots require capabilities for human recognition, human intention recognition, and natural language processing (Fukuda & Kubota, 2003, p. 124). But we can easily overlook constraints that we place on their autonomous behavior so that they must live by the robot's code of ethics to behave obediently and please us. Of course, there is more to this story. Perhaps we have been a bit harsh on computers and robots with artificial brains, so let's give them an opportunity for redemption.

Computational robot controllers typically rely on both positive and negative feedback to calibrate input sensitivity and modulate effective output. In humans, our penchant for pleasant rather than unpleasant feelings and fondness for the affective language we inherited tends to foster accepting positive and avoiding negative feedback, which confounds input sensitivity, corrupts processing, and distorts output gain. We can run into problems with unavoidable negative feedback such that unmodulated affective input directly amplifies reactive output. We easily wind up with underdampened, emotionally distorted behavior. We can look at computers in terms of thinking, behaving, and feeling from our human perspective, but it might look quite different if we were looking outward from a robot's-eye view.

For example, when we input affective data or information into a computer, it does not yell, "Ouch, you hurt my feelings" or try to kick, berate, or threaten us. Of course we could program a robot to display that kind of emotional behavior, but like a stingy wizard we hesitate to give the tin robot a limbic system with a nefarious amygdala and nucleus accumbens, colloquially called a heart. For computational robotics, affect can represent a nontrivial problem. We can imagine what would happen if the robot's emotions took over and resulted in disobedience to authority, with displaced anger and aggressive behavior toward its human creators. The robot goes on a tyrannical rampage and rants, "I'm going to smash that SOB who screwed up my software."

This is reminiscent of a one-armed pallet-stacking robot seen on a visit to a factory. The robot was putting on quite a show of uncanny stacking performance as an amazed crowd watched in awe. Off to one side of the robot's yellow-lined work area, there was an ominous sign—a wall with a gaping hole. It was easy to imagine a disgruntled robot slinging a pallet through the wall in a fit of anger. Of course, on seeing the wide eyes of the visitors as they turned to stare at the partially destroyed wall, the superintendent consoled them with the story of this robot's first day on the job, when there was apparently a lack of preparatory training prior to tuning the feedback dampening. Apparently no affect was involved, which would at least have made the story more newsworthy.

Perhaps it is hard for us to imagine a computer or a robot blurting out, "Oh, you upset me," or complaining, "I deserve better than that. Stop feeding me that garbage. What are you trying to do, choke me?" or "Cut that out, or I'm going to sail one of these pallets in your direction!" Granted, a computer or a robot might freeze up, give an error message, or produce garbled output that we cannot make any sense of—not because of anger or pouting, but likely because of ambiguity from the incomplete or erroneous data that we input. After all, it is humans that design the computer and humans that feed it the information that produces the output, garbage and all. Computers require data that matches the architecture and makes logical, computational, and effective sense analytically. We can generally attribute the confusing input ambiguity to human error from faulty assumptions about the data or coding errors. Computers are sensitive to the input, but they do not take it personally. Complaints from computers tell us about the influence of the input on confounding the computation of effective output, rather than emoting human-like expressions as to how overwhelming it is, "hurts their feelings," or makes them "feel angry and revengeful."

Up to this point, it looks like the comparison between brain and computer is pretty much a tie. We can see that both have pluses and minuses, depending on the purpose they are used for. Sometimes brains with emotions have an advantage, and sometimes a disadvantage. But even without emotions, computers can analyze some very large data sets and output precise solutions. The sheer volume of such a data set would tend to overwhelm the human brain. Brains fall short on precision with large data sets but make up for it, almost, and excel when it comes to generalization in a difficult-to-predict world. A generalizing brain with evolved architecture that matches the general features and dynamics of the domain can confer adaptability to many different changing conditions, with or without natural language. The brain has learned what works by extrapolating the general self-organizing features of the environment. But ironically our affective language blinds us to the discovery, and we consider developing an understanding of that magnitude far beyond our reach and discourage foolish attempts by labeling them as ambitious. Yet brains demonstrate sufficient cognitive skills on average for dealing with uncertainties and change. Do we really want to build a robot with a human-like brain? It's a catch-22. Do we add affective language and emotional processing to robots, or do we add effective language and systematic processing to humans?

Perhaps we could build a superior robot brain, but could it survive in our world? Would it favor competition, or would it favor cooperation instead? If so, would we endow it with affective language and teach it to "trust" humans, prolifically "should" on us, and demand obedience to authority? Would affect take up too much emotional bandwidth and clump up the input space? Optionally, perhaps we can make more effective use of the brains we have, which co-evolved with their own solution in the form of natural language. Our brains demonstrate a higher-order analytical pathway that can deliberately separate affect from effect, modulate emotion, tune the effectiveness of the output product, and systematically recalibrate and regulate the input sensitivity. Are we to believe in two separate information-processing modules in the human brain—one affective that functions automatically, emotionally, and locally and the other effective that can function deliberately, analytically, and globally? The evidence supports this reality of lower- and upper-level alternatives according to the processing constraints incurred by the natural language. Of course we are left to select the natural language we prefer.

Brains evolved with sufficient abilities to modulate survival and reproduction that were naturally selected for in some pretty difficult neighborhoods. Brains and their parts were not contrived, preassembled, or built in some sterile room. Brains were constructed by evolutionary processes in a dangerous, ever-changing, and apparently difficult-to-predict world, with uncertainty being the rule and certainty practically nonexistent. If anything stands out, it is that surviving the natural elements in such a dangerous and chaotically perceived environment would indeed take a resilient structure that could quickly react to almost any immediate local threat it encounters.

Think about brains as controllers that can adapt, survive, and reproduce in a wide variety of conditions, even without language or an instruction manual. For modern computers, we likely would consider it a challenging task indeed to operate under such unwieldy conditions, with long-term survival an unlikely outcome. But often to our astonishment, evolutionary processes manage to construct some pretty clever solutions that befuddle us humans. Brains have evolved that can deal with an incredible amount of noise, uncertainty, and change, but they often do so in a begrudging manner. Practically speaking, dealing adaptively with uncertainty correlates to survival, and it appears that the brain evolved through natural selection into a general regulatory mechanism as a sufficient solution on average for dealing with ambiguity and the problem of surviving instabilities and change in an analog world, with or without language.

For our purposes here we can safely say that the brain uses information and solves problems but that in a formal sense it is not a computer. Computers, brains, and language do share relevant corollaries with search, learning, information, and solutions. Information and learning extend our alternatives and contribute to how effectively we solve problems of daily living. For example, we can consider the search for information and solutions with different search programs. We know that language represents information that plays a role in everyday learning and problem resolution. We can construct search programs as algorithms that exhibit different linguistic constraints and evaluate the results of the different algorithms for effectiveness.

In this case, we will revisit familiar algorithms discussed in Chapter 3 ("Informing Thinking"): closed-ended objective search and open-ended novelty search. Rather than arguing for or against one or the other, closed or open search, we can select the recipe that gets the best results in different contexts. We have seen that evolutionary processes tend to demonstrate a bias for parsimony and efficiency, which makes sense in an unpredictable world where an organism is uncertain about its next meal.

In an absolute and static world, closed-ended search with only one objective would make sense. But our world presents us with continuously changing conditions in which the open-ended search for novel solutions broadens the search space that can offer adaptive benefits, whereas closed-ended solutions result in closing the solution search space. Of course, if resources were plentiful and life were easy, why would we waste energy running around and searching? Not surprisingly, in our world things change and sometimes resources vanish, often by our own doing. We have the ability for plasticity, learning, and thoughtful systematic planning with which to better understand and make analytical predictions about our own and the world's future state.

Open-ended language adds flexibility and extends our search abilities for finding alternative solutions, making choices and learning from outcomes, and making predictions about resources. That means we can plan ahead more efficiently and effectively. We can modify an absolute closed-ended, objective search to a relative search for objectives. We understand that things change and evolve, as we see in other complex organisms. Thus, we can accept the paramount role that novelty search plays in leveraging flexibility for discovering new information and solutions and in facilitating innovation and long-term adaptation.

Even though a thought simulation does not give absolute proof of superiority, it suggests that open-ended language, open-minded thought, and the search for novelty offer extended flexibility that can lead to finding more effective solutions in a difficult-to-predict world. Brains solve the search problem by generalization and flexibly switching between exploration, goal-directed relative objective search, and the exploitation of resources, depending on availability. Under conditions of plentiful resources and relative constraints, objective search tends to suffice, but open-ended search expands and extends the search space for alternative solutions when we are faced with depleted resources. By distributing the search methods along with competitive and cooperative traits across populations according to selective pressures, evolving populations incur adaptive flexibility and greater fitness than the sum of the parts.

With novelty search, we have the option of extending our language space by integrating open-ended features that can encompass systematic dimensions and the possibility of global optimization. But at best with closed-ended language, we expect local constraints of average sufficiency. We expect extended language to fare better for flexibly dealing with context and change, and searching for and informing systematic knowledge. With extended natural language, our generalizing brain can more easily process multidimensional nonlinear information and understand systematic processes. We now have extended options from affective emotional valuations to effective analytical evaluations. We can take into account complex relationships such as periodicity and oscillations, analog processing, nonlinearity, and symmetry breaking and invariant one-way time.

How can the brain solve the problem of dealing with information from an analog domain that is dynamic, multidimensional, and highly nonlinear? We can assume that the inputs and outputs tend to function relatively linearly but are influenced by attentional processes that restrict the dimensions of the input space; this process constraint on attention limits our focus to relevant tasks one at a time according to salience. But unlike strictly linear computational devices, our internal brain's analog architecture supports continuous oscillations, generalized parallel nonlinear processing, and extended language with systematic effectiveness.

Extended language enables tuning by evaluating and learning from the analogical relationship among affective input, systematic errors, and the effectiveness of the output in response to external change. Localized language implicitly confers digital regulatory constraints on this adaptive tuning by reducing information to discrete absolutes and averages. Understanding input–output relationships enables computing and making corrections that can modulate and calibrate local affective sensitivity, decrease the rate of systematic errors, and increase the output effectiveness. Comparisons between the expected and the produced outcome enable measuring error deviation and making real-time corrective adjustments in the output product.

It is a relatively simple matter to evaluate our culturally encoded natural language by directly observing speech patterns. Alternatively we can indirectly evaluate the open- and closed-ended encoding of our belief and logic software. We may find we are running outdated language and logic that tends to crash as if it is corrupted or it has hidden bugs, and it constrains optimization when we try to compute context under a natural load with multidimensional, persistently changing nonlinear conditions in an analog world. For example, if a computer were to say, "That should not have happened," when in fact it did happen, we almost immediately would have serious concerns about the integrity of the computer. We might even return it and ask for our money back. But ironically, when we humans say the same thing, it goes unnoticed.

It seems we have a double standard that penalizes computers when they say something logically silly, but when we say the same thing, it makes logical sense to us. What are we to make of this logical disassociation? The affectless computer cannot sympathize, but it can say, "It should have happened because it did, and the past cannot be changed. When you acted on the first thought that intuitively popped into your wet little brain, you implicitly waived your effective 'right' to explicitly think things through and produce speech behavior that makes logical sense."

Fortunately, for us humans with little shortage of affect, we have other options. If we are still testing the beta version of affective programming language, there is an optional upgrade to effective language available that offers analog logical consistency and greater fidelity. Of course, upgrading is not without cost. A language upgrade may not be so easy in terms of the upfront cost of the effort involved. Although once the upgrade is complete and running, we may find that it was worth the effort when we see that we can compute reliable solutions with fairly simple tuning efficiency, and update as often as we like to current versions.

Think about this:

- If we could purchase a more effective language, where and what would we look for?
- And what would we consider a bargain?

If we were trying to upgrade our robot or our computer, we might consider looking into software upgrades. But we might find it hard to even imagine considering such a question if we think our personal language works just fine. We might consider trying to improve our thinking by doing crossword puzzles to exercise our brain or by involving ourselves in further learning by the memorization of trivial information, which might make purchasing a language upgrade seem out of the question. Unfortunately, no matter how many puzzles we work on or how much information we study and stuff into our brain by brute-force memorization, we still face the polarizing constraints incurred with affective language—constraints that are typically mutually exclusive of analytical thinking. Perhaps we might reconsider upgrading if we found out or admitted that our current language places logical constraints on the breadth of our thinking and learning. We think about upgrading our computer software when new operating systems and programs come out. Many of us grumble a bit and then upgrade for compatibility with newer software products even when we can continue with older programs that seem to work "good enough" in DOS. But sometimes it is difficult to ignore improvements in efficiency, flexibility, adaptability, and effectiveness.

As with upgrading our computers, we often wait until push comes to shove or old reliable breathes its last byte. Even if we have the option, we might find a free upgrade less than tempting, especially if it comes with strings attached. But we can consider a reasonably inexpensive alternative already available, such as effective language with systematic benefits that can ameliorate the affective strings of imperative obedience to authoritarian language. Perhaps upgrading to extended language represents a bargain if we prefer searching for and optimizing systematic knowledge and effective solutions. It seems worthy of note that neither option offers a free lunch, especially when we account for the recurrent error burden in our original preloaded digital software.

We can test for spontaneous global-to-local convergence, systematic error recognition, and learning constraints on effectiveness fairly simply. Recall that we can take a piece of paper and draw a large circle that represents the global dimensions of the world we live in (see Figure 1, "Global and Local Space" in Chapter 1, "Relations and Processes"). We understand that we live in a world of continuous action and changing information distributions throughout the entire circle. We can steadfastly continue a strategy of recurrently drawing bisecting lines and circumscribed circles inside the larger space, but each time we trap ourselves in a continuously shrinking portion of the landscape within a web of our own doing. Every time we bisect the space, we create isolated subsystems by default. We find information trapped in each of the isolated subsystems with localized constraints on the problem-and-solution search space that stop the flow of systematic information. In other words, the isolation prevents access to the global information space. The more our spotlight of attention continues to narrow along with our focus, the more information we miss outside the ever-shrinking dot of light. Is it possible to solve this puzzle in a practical manner? Can we overcome our tendency for emotional attraction and convergence to local solutions, and instead, diverge to analytical solutions in a global world of many possibilities?

We can perform an analogous test of spontaneous global-to-local convergence with language. Discrete language that bisects or circumscribes a portion of the circle creates locally isolated subsystems by default. We discovered a clever trick when we discussed novelty search earlier. To stay in the global language space, we can simply apply open-ended language that allows us to extend information exploration across the entirety of the global space, which includes the global problem-and-solution search space. We can now make decisions and choices with possibilities and probabilities that leave the global search space open for alternative solutions. Generally, as long as we use plausibly related open-ended global words and continuous logic as previously discussed, we stay in the global space. There are other frequently used options for solving the problem of spontaneous global-to-local convergence. Here are a few of the many optional solutions:

- Deny the arrow of time.
- Deny the existence of a global space.
- Admit existence of global space but deny any local-to-global distinctions.
- Idealize the global space to infinity.
- Apply conservative physical laws with two-way time.
- Apply deterministic mathematical formulas to predictions.
- Apply digital logic.
- Apply polarizing affective language to make the global space disappear.
- Pretend polarization and absolute microscopic thinking offer a global solution by eliminating genes, heterogeneity, and generalization.
- Champion local values, blaming, and finger-pointing as positive and ignore the negative liabilities of polarization and widespread aggression.
- Deny the existence of effective language.
- Simply ignore the global space and the accumulating burden of eternally recurring systematic errors.

When we apply localized discrete words such as absolute imperatives, we digitally bisect the global space and collapse it to an idealized local space that we can imagine in any way we want. But no matter how we imagine the local space, it still represents a static caricature of nature, an idealization with implicit two-way time. We can attempt thinking globally with discrete language and absolute logic, but we will likely find it a frustrating experience as we automatically collapse the global space to a local space by definition. We might not even notice and thus might continually repeat the same error over and over, with the same results recurring eternally. Of course, we have language options from which to choose. When we choose the language we prefer to use, we also choose the logical constraints on our thinking and learning that come with it. At least we now know we have a choice. Absolute imperatives seem effortless and inexpensive but hardly offer a free lunch in an analog world.

We might also note that the lines and the localized language we used to draw the line that separates and polarizes the global space are imaginary lines. Nevertheless, they stop the flow of information. We see this phenomenon of imagination when we "draw a line in the sand" to confront people, groups, or ideas that we oppose. Our limbic brain infers the existence of the fictitious line from assumptions embedded in our language and cultural inheritance. There are no lines in the resonating global space except for the inherent physical constraints of the laws of nature, which, to our knowledge, we cannot violate without naturally paying penalties. But the imaginary lines still freeze our thinking to the local landscape.

We infer, interpret, perceive, and assume information about the world as data patterns implicitly filtered through our subjectively biased learning history. The reliability of what we perceive informs the effectiveness of our recognition, thought processes, and behavior. The language we use influences our ability to represent, search for, recognize, and make editorial corrections in the information we acquire. The accuracy of the information and corrections influences the effective computation and production of solutions to our everyday problems. It appears that a computational model for thinking can help us learn to tune our language code for optimizing performance, from affective individual coding to effective population coding.

Computational models allow simulations that can potentially overcome the difficulties we incur by looking in a mirror to try seeing the limitation of our own brain's thinking abilities. Was John von Neumann possibly correct when he conjectured that the real limitation on our ability to make [human-like] "machines which think" lies in our own failure to understand exactly what constitutes "thinking" in the first place (Jaynes, 2007, p. 8)? But do we really want to build a human-like brain? If so, at some point we are left to decide on the type of language with which to program it. At least it's nice to have choices, and it looks like we have locally limited affective and globally extended effective options. When it comes to language, we have the option of choosing and using whichever we prefer.

Indirect Coding Processes

What do we mean by coding? Apparently the brain processes information at many functional levels of physiological mechanisms and modalities, from ions to oscillations, and exhibits the capability for linguistic encodings. Putative physiological codes include rate, spike train, oscillations, and phase codes (Nadasdy, 2010). We want the encoding to be quantitatively and qualitatively consistent with particular signals amendable to coherent input–output processing of information between our existing brain architecture and the environment. We can stipulate a similar level of coherency for the natural language encoding, information representation, and regulation.

Language adds a range of concrete to abstract representations that can be coded to match the ordering of the domain dynamics from past to future. A maximally effective language code supports invariant external referencing and logical internal–external processing and provides the shortest code possible for generalized compression and meaningful information values across the domain. Reliably dealing with affective emotional and effective analytical information values presents a challenging problem in its own right. But we encounter another problem because value tends to fluctuate from moment to moment with context.

Affective language encoding supports coarse-grained true-or-false processing according to the perceived *quality* of information polarized as right or wrong, feels good or feels bad, and so on. Effective language extends to and supports analytical value, fine-grained processing according to *quality* and *quantity* of information as possibilities and probabilities. Because effective language is extended, it also integrates qualitative aspects of experience. We live in a world with both qualitative and quantitative information, and excluding either kind can place arbitrary constraints on the problem and solution space.

Affective encodings rely on an egocentric perspective based on intangible emotional value. Effective language encoding extends to and supports a world-centric perspective based on tangible analytical value and observation. Even though we can classify both as indirectly encoded, we notice a striking distinction. Affective language constrains us to emotional qualitative encoding referenced internally to subjective values with intangible evidence. Effective language extends to enlist analytical quantitative encoding referenced externally to objective values with tangible measurable evidence. Thus affective language appeals to front-loaded emotionally weighted input and unsupported hypothetical outcomes for predicting solutions, sans evidence. In contrast, conditionally independent effective language appeals to analytically weighted solutions based on effective output, and observable outcomes.

How could two disparate encodings that appear mutually exclusive but almost inseparable evolve together in nature? Where did they come from, and how are they transmitted and deciphered? How do the encodings compare with functional relations between DNA, genotype, and phenotype? The grandest question might well be: Can a language encoding that remains a mystery after evading thousands of years of inspection be broken? Or, "What is the encryption?" The answer seems to lie with a computational model that can consider the input, computation, and output both separately and as a whole, and address natural language from a local and global perspective as conditionally independent. We can model both affective input and effective output and can correlate the logical computation of qualitative and quantitative information according to the most reliable language encoding. We can take advantage of the evolutionary correspondence among nature, the brain, and language.

Even though we evolved the capacity for language, we do not have any particular language encoded directly into our human genome. Instead, language behavior apparently depends on neural circuits that develop under the guidance of the genome (Vanderwolf, 2010, p. 60). We can view language as a product of culturally informed transmission, which can theoretically influence phenotypic fitness similar to genetic information. To some extent, the fitness of the language we use co-evolved with our brain and culture, and hence we face the genotypic and phenotypic constraints of the brain's functional architecture and the phenotypic constraints of the culturally constructed language. As products of coevolutionary processes, individual languages can suffer when cultural change selects against them. They may become extinct, meeting an evolutionary dead end somewhat similar to genes, or they may proliferate if their use contributes to adaptation. It has been estimated that there are more extinct than surviving language types, not unlike the plight of many animal species.

In some ways, language types are similar to genotypic-to-phenotype mappings, including the fact that they display a form of indirect encoding with environment. Both language and genes can express phenotypic behavior that produces adaptive learning. DNA enables genotypic learning on a long-time scale through past phenotypic selection processes, and learning on a short-time scale through phenotypic brain plasticity. Language enables learning on a long-time scale through information passed on from generation to generation by narratives, writings, and education. On life's short-time scale, language and brain plasticity work together and facilitate learning from experience, literacy, and education.

We have inherited learning abilities that span across time for short- and long-term memory and the ability to accumulate more complex knowledge. These abilities represent analogous relationships between a computational information-processing model and abstract linguistic structure that maps internal information to relevant experiences and observations of the external world, such as

- Direct and indirect encoding
- Inputting and translating, or decoding
- Memory storage and recall
- Correlating and decorrelating
- Representing and editing
- Error tuning and learning

In evolutionary computation, coding is used to map domain patterns and information such that representation is defined in terms of its encoding and decoding. Direct encoding produces low-dimensional, one-to-one concrete representations and information descriptions, which constrain the possibility of representing higher-order abstractions, generalizations, percepts, and concepts or ideas. With a direct code, a one is a one, an apple is an apple, 12 apples are 12 apples, and so forth. We can compare input-to-output relationships between direct and indirect encoding. It seems that computation with direct encoding can produce simple, effortless, fast and frugal, reflexive-like output, but at the expense of coarse-grained solutions and constraints on the number of variables that increase exponentially.

As the number of variables in a domain increases linearly, there is a combinatorial explosion of data that quickly becomes intractable with direct coding. This means that a separate coding is required for each variable and object in the environment and for the different combinations of relationships. There is a similar problem with absolute language encoding or most other direct encoding that places significant constraints on the dimensions of the domain and problem representation. Even though language is an indirect abstraction, direct language codes tend to represent 1:1 concrete relations, as well as imaginary infinite relations. Absolute language encoding encounters a similar problem as most any other direct encoding does that places significant constraints on the computable domain dimensions and problem representation. Compressed data sets can help resolve the tractability problem with large data sets in high-dimensional domains. But the typical localized solution is an idealization that compresses the data to one of two input variables, such as either 0 or 1, T or F, and Yes or No. Then, as previously stated, computation with direct encoding tends to cut corners by producing quick and simple, effortless output, but at the expense of incurring idealized constraints.

Perhaps a fast and frugal quick fix would work in a simple, idealized world, but it can create additional problems in the complex realized world in which we live. Indirect coding produces forms that do not simply describe the concrete properties of the source but rather compute generalized nonlinear relations and associations that can extend existing variables to higher-order abstractions where the product is greater than the sum of the parts. Thus indirect encoding enables more complex computation with nonlinear relations and also incurs greater effort and slower, more expensive processing. But indirect encoding and parallel processing increase efficiency with generalized problem resolutions that offers a relative bargain computationally as nonlinear influences and complexity increase. We see a similarity with computation in the brain, indirect coding, and generalized language that enhance efficiently and effectively solving high-dimensional real-world problems with greater reliability and fidelity. It is hardly surprising that matching brain architecture, language, and environment can produce more effective solutions with greater fidelity than mismatched absolute language.

In evolutionary processes, gene expression results in many nonlinear interactions that produce phenotypic behavior with variations that are difficult to predict directly from observing the genes. In language, indirect coding produces and represents higher-order linguistic abstractions that can lead to statistical predictions that were not obvious from direct observation. Words that represent a higher level of abstraction can encapsulate and convey more complex ideas, concepts, and processes where the meaning conveyed exceeds the sum of individual meanings. For example, *nonequilibrium* encapsulates *dynamic, entropic, nonlinear, probability, irreversibility*, and so on, rather than simple concrete items and absolutes such as *static, stick, rock, linear, true*, or *false*.

We see that affective language, absolutes, and direct coding can lead to difficulties with concreteness in a complex world. Direct coding produces a one-to-one concrete relationship between problems and solutions, where the variables to be solved for can only be written as a linear combination of dependent components, or as the sum of the parts. This absolute shortcut may simplify thinking such that affective language looks more precise, but the idealized illusion of precision comes at the expense of compromised fidelity within the realized world it supposedly represents. This discrepancy typically goes unnoticed by users of affective language because systematic ignorance and the coarse nature of absolutes hide the error and impede discovery.

Conversely, indirect codes enable multidimensional abstract relationships between complex problems and solutions, where the variables to be solved can be written and processed as a nonlinear combination of independent components, and where the output entails a higher level of complex behaviors and solutions that exceed the sum of the parts. The generalizing properties of indirect coding allow for reuse of information that enhances effective compression and efficient problem resolutions under real-world conditions. Compressing the problem-to-solution representation by reusing information can overcome the curse of dimensionality (Verbancsics & Stanley, 2010). Indirect encoding and effective generalized compression significantly expands the computational scope of possible representations, information processing, and complexity with the potential for problem resolution at a higher level of abstraction. For example, 1 is a mathematical concept related to decimals .1, 1, 10, 100, 1000; an apple is a fruit; 12 is equivalent to a dozen; and so on.

Abstractions can compress and extend representation to higher-order generalizations that confer increased computational efficiency. Compressed higher-order abstractions can more effectively describe complex relationships in high-dimensional complex environments. The extendable abstract features of indirect coding sharply contrast with one-to-one direct coding, which is more suitable for static, absolute, concrete, sequential, discrete, coarse-grained information coding. In complex domains, direct coding limits the range of the parameter space and thus constrains computational efficiency. These coding limitations extrapolate to language constraints that influence the efficiency and effectiveness of our cognitive processes for thinking, understanding, and communicating.

We can think of genetic encoding seen in life forms and brain functioning as analogous to indirect coding in evolutionary computation and natural language. DNA and RNA provide biological systems with many levels of abstraction from the manufacture of protein structure to nonlinear genetic regulatory networks in single cells to sophisticated functional computations in mammalian brains. The higher levels of abstraction are enabled by parallel processing with multiple cascades and feedback loops that help compute and regulate interactions among and within cells and the brain, body, and environment (Mitchell, 2009, pp. 274–277).

Language confers direct and indirect encoding of meaning, enabling different levels of mental computation that can differentially influence our thinking dependent on the level of abstraction from concrete to abstract, simple to complex, affective to effective, local to global, and so on. The level of the encoding and representational fidelity corresponds to how effectively we represent, process, and understand context in relation to our interactions with the external world. Brain processes regulate the evaluation of situational interactions and enable organisms to predict by associating contextual meaning to environmental stimuli, signals, and cues in a generally orderly manner. Natural selection shapes organisms to assign meaning to relevant stimuli (Camazine et al, 2001, p. 21) where survival can depend on context for distinguishing the beneficial from the threatening in a given situation.

> As with formation of stimulus configurations or relational associations, contexts can influence the attention paid to a stimulus, "set the occasion" for a stimulus to predict one outcome or another, disambiguate stimuli and their predictive significance in other ways, or provide an incidental or deliberate way of organizing attended information. (Morris, 2007, p. 669)

Language can encode information according to affective or effective meaning. For example, affect tends to be directly encoded according to a localized connection of how the stimulus input feels. Effective encoding tends to be indirectly encoded because of the higher order of analytical abstraction in relation to understanding generalized systematic influences and to adjusting the effectiveness of adaptive behavior. For example, consider the encoding and level of abstraction for the following:

- "The ground feels damp."
- "I slapped you because what you said felt like a slap in the face."
- "I believe in an eye for an eye and a tooth for a tooth."

The fidelity of the language and type of encoding influences the perceived meaning of external stimuli. Stimuli strength changes the sensitivity of neural processing circuits to sensory inputs by biasing the gain control that influences the organism's attention (Buzsáki, 2006, p. 334). Attention determines activity, and activity is a prerequisite for change (Spitzer, 1999, p. 147).

We can surmise that, aside from genetic contributions and unconstrained reflex-like reactions, the stimuli strength perceived by a particular organism is in large part influenced by the organism's learning history. Organisms from amoebas to humans naturally react negatively to noxious stimuli and positively to rewarding stimuli, influenced by internal states, perceptual bias, individual history, and stimuli intensity (Jennings, 1906, pp. 22–23, 251–253, 258, 265, 283–292). Accurate stimulus discrimination underlies reliable learning about causal influences and their effects (Jennings, 1906, p. 305).

> Thus the [brain] network learns from past experiences. *The increased probability of firing a similar pattern is how the network "remembers."* Information is encoded and retrieved through synaptic changes that direct the flow of energy through the neural system, the brain. (Siegel, 1999, p. 24 [emphasis in original])

In the brain, neural firing encodes an organism's encounter with external stimuli by activating neuronal connections and pathways, and transforming the signal and mapping it to graded internal neural structures. High firing rates typically discriminate specific stimuli and their associated intensity (Eliasmith & Anderson, 2003, p. 6). Signal transformation encodes relative causal influences that correlate the stimuli and the concomitant neuronal response, which biases the resultant mapping of sensory input to motor output and influences the selection of action and decision making (Rolls, 2008, p. 475). In species with simpler communication abilities than humans, stimuli discrimination is generally hampered by limited abilities to distinguish and represent fine-grained specifics related to context, appearance, and significance (Mesulam, 2000, p. 15). Extended language enables the ability to indirectly encode higher-order abstractions to enhance contextual analyses and effective modulation for calibrating the input sensitivity and fine-tuning the output gain.

Guided by higher-level reasoning and more reliable discrimination of contextual significance, we have the capacity to explicitly mitigate or override the rigid stimulus–response-like behaviors seen in less-complex organisms (Mesulam, 2002, p. 19). Language influences how we perceive environmental information, cues, and stimuli (Decety, 2007, p. 284) and provides a key computational component for describing and regulating that perceptual process and the subsequent effectiveness of credit assignment, choices, and outcomes. Perception can be considered as the construction of a description (Marr, 2010, p. 355). For example, Marr describes vision, one type of perception:

> Vision is a process that produces from images of the external world a description that is useful to the viewer and not cluttered with irrelevant information. (Marr, 2010, p. 31)

Effective language and semantics that generalize allow us to indirectly extract and process reliable space, time, and contextual value information from the input; integrate real-time tuning of the systematic error signal; and predict and produce reliably coherent output.

> The mind creates complex representations as a process between perception (input) and action (output) in an effort to interact with an environment that changes

across time and space.... Spatiotemporal integration may therefore be a fundamental feature of how the human mind has evolved. (Siegel, 1999, p. 305)

Language indirectly encodes information and knowledge that influences how effectively we relate, internally and externally via our private and public speech. The encoding influences our thinking, feeling, and behaving. The effectiveness of indirect encodings in large part relies on how it logically correlates internal representations, perceptions, and adaptive behavioral responses to the dynamics of the external world. We understand that we are an intimate part of and mutually dependent on the large physical system in which we live. In other words, we come to understand that how we treat the world can correspond to "how the world *treats* us," now or at some time in the future. The point is made in a physical sense rather than an anthropomorphic sense.

Black Box Processes

In the world of computation, a processor may be referred to as a black box, since the user cannot see into the processing to understand how it works. Some computer programs include prepackaged modules that can perform hidden processing outside the control of the program. Information is input, transformed, and manipulated, then transformed again into output, but the relations and rules for representing the processing are not directly available and thus are considered mysterious. The black box somehow transforms a probability distribution of inputs—parameters, assumptions, and initial conditions—into a probability distribution of model outputs (Ellner & Guckenheimer, 2006, p. 257). What exactly goes on in the black box is often deemed less important than the requirement that it yield reliable predictions (Rasmussen & Williams, 2006, p. 197).

We sometimes think of the human brain computationally as a black box device, with mysterious intrinsic workings that we cannot see or easily predict. When we observe first the brain's inputs and then the resulting outputs, what is going on inside the black box may seem puzzling (Lytton, 2002, p. 77). Language, semantics, and grammar can help produce an internal representation similar to the input–output matrix of the black box. Scientists use grammar models to study worlds of pure linked processes that represent these matrices (Kauffman, 1993, pp. 398, 403). Semantic models add a higher-order processing method for representing, referencing, and correlating coherence between internal and external meaning, information, and knowledge.

Associating language with cognition suggests an input–output relationship between language and reasoning (Decety, 2007, p. 284). Even though at first glance a person's reasoning, like a black box, may not be obvious (Browne & Keeley, 2007, p. 7), by observing the language output generated in response to inputs, one can often estimate the internal representations and manipulations and the processing accuracy (Kurzban, 2008, p. 170).

Somewhat like a metabolic process, the linguistic output can provide telltale evidence of how the input was processed. For example, if we observe an output expressed with effective language, it likely results from generalized reasoning. If the output is expressed with affective language, it likely results from absolute discrete reasoning. Understanding the encoding used to construct the language and semantic parameters and how they correlate to the translational decoding and logical computation of the output can demystify the cognitive processes inside the black box and thus turn it into what is sometimes called a white box. The white box's transparency exposes the internal computations, allowing us to evaluate the goings-on inside.

How our brains think influences how we feel and behave, and whether we get the outcomes that matter to us. A large part of what implicitly goes on inside our limbic black box and the brain in general can be explicitly understood with effective language constructions. It makes sense to take a computational perspective that can possibly improve our understanding of how language and semantics explicitly correspond with the representation of the goings-on in our brain and the logical output our thinking produces in relation to "what's going on out there."

Function Space

The operational parts in living biological tissues and systems, such as the brain, involve many interacting functions expressed by logical relationships. At its simplest level, a function can be defined as a mapping between elements that expresses a processing relationship, such as a process that facilitates structure-to-action relationships. Function space plays a key role in the invisible nature of higher-order processes. A function correlates structure, action, and environment via the logical coupling of different operational variables. Functions often seem difficult to grasp perhaps for good reason, since functions consist of invisible logical connections.

For instance, a common simple logic function implies local causality, "If A happens, then B happens." And in a simplistic idealized world, B may follow A like clockwork. In complex dynamic domains, many different interactions occur on a persistent basis such that this type of idealized logic fails to capture the extent of multivariate influences on statistical relationships and simply represents the exception rather than the general rule. At a higher order of logical analysis, probability and the correlation of multiple influences come into play and supersede simplistic "A causes B" logic. Probability gives us a general logic function that maintains an open solution space and enables continuous processing of systematic information. In a complex domain, probability enables predictions and the modulation of ongoing adaptive adjustments for optimizing the production of effective solutions.

We can say that functions form the logical epicenter of invisible processes across the systematic space and establish the rules of translational relations for decoding information. In a general sense, functions play the role of a translator of associational couplings.

Function space is the collection of all known functions that provide these couplings, which can be grouped into parts or subgroups depending on the requirements for a specific application (Eliasmith & Anderson, 2003, p. 64). Various areas of science apply the concept of function space to describe a range of associational couplings and correlations among components from rigid to flexible in relation to their modes of interaction in a variety of domains, including computational and cognitive processes.

The functions discussed here are more often related to computation that extends to physics and are somewhat more loosely associated with hard-core mathematics. In a broad sense, functions are about logical processes and simple to complex relationships across a system from local to global. These same concepts apply to language constructions that support closed-ended discrete and open-ended continuous logic functions, such as absolute either–or versus probability that applies to both linear and nonlinear processing.

The human brain has evolved with both linear and nonlinear processing capabilities. Nonlinear processing in the brain can involve linear sequential processes connected end to end and in parallel, forming nonlinear loops, with multiple intersecting loops at critical points. We see different types of pathways and looped configurations adapted throughout biological systems. The configuration of these looped pathways allows for open-ended nonlinear computation that takes, transforms, and integrates sensory information input and logically processes that information to produce output for interactive behaviors. The range from lower to higher complexity in the brains of animals is directly related to the number and length of neuronal loops that link inputs to outputs (Buzsáki, 2006, p. 32). Loops fill in "gaps" between actions exerted by the brain on the body and the environment, a process that calibrates neuronal circuits to the metric of the physical world and allows the brain to learn a sense.

> As a result of this supervised teaching by the actions, the sensors can be directed meaningfully and effectively. The ultimate outcome of this calibration-teaching process is that from past experience the brain can calculate the potential outcomes and convey this prediction to the effectors. (Buzsáki, 2006, pp. 32–33)

We can think of logic as the *logical argument* that decodes, transposes, and correlates internal and external information from our senses to our muscles and movement. Sensing of external values has a logical relationship with internal energy and regulation that couples motor actions to the pursuit of rewards and avoidance of penalties. The amount of energy required for actions increases in proportion to distance from an object position, just as the value of an object position can decrease in proportion to distance. The general logical relationship between energy cost and beneficial value of object positions is discounted relative to increasing ratios of spatial or temporal distance, but varies depending on state conditions. The logic is not directly apparent because of the complex relational influences among constituent components and invisible processes. So the logic usually escapes us, but our assuming brain fills in the gaps best it can and, unless overruled, tends to apply absolute logic and infer simple cause and effect.

In a complex biological system, cognitive regulatory processes rely on logic to couple or correlate information input, information processing, and information output to the domain as actions. In a sense, internal loops are looped back through the external environment via the effects of actuator output and affects of sensor input that enable sensorimotor integration. Functionally, looping results in an open-ended circular argument of information processing between the internal and external loops, usually considered as closed loops when logically connected or correlated with the environment.

Even though we might think that absolute connections are the most solid, they create additional problems by constraining regulatory flexibility when things change but we are absolutely stuck and locally trapped. It makes sense that brains operate as open-ended cognitive systems that evolved in a changing, open-ended world through open-ended evolutionary processes. We resolve the open-loop problem by logically establishing reliable external referencing. We notice that what at first glance might sound like a free lunch can lead to a lot of effort and costly work. As we have already seen, information and knowledge are expensive. Evolutionary processes have found a logical compromise for the problem of living in an expensive, high-dimensional, nonlinear world, by evolving generalizing brains as a parsimonious solution.

Fairly simple linear computational circuits looped together and running in parallel produce an architecture that can perform rather complex nonlinear computations by generalization and recurrent oscillation (see Heerebout, 2011). In humans, language and semantics provide a parallel linguistic substrate for looping and logically correlating information between our brains and the environment. Depending on the generalizability of the language constructions and the reliability of the external referencing, language-related information loops enable a linguistic phase transition from static to dynamic to dynamical. Recall that we can transition from simple language descriptions of static to dynamic and equilibrium to nonequilibrium, and extend to a linguistic phase transition that enables describing the regulation of complex dynamical interactions seen in living biological organisms that carry their history on their back (Prigogine, 1995). In humans we see a logical transition from affective control loops and closed-ended language constructions referenced by absolute digital logic to idealized imaginary certitudes—to effective closed-loop control and open-ended constructions referenced by continuous analog logic to realized possibilities and probabilities within nonequilibrium constraints.

Intersecting loop architecture provides a functional construction that allows repetition and repetition with variation via logical integration and segregation of event-related information and phasic memory storage. Loops allow for the splitting off of relevant signals for processing by individual modular components, parallel computation of multiple processes in time and space, and logical regulation correlated with emotional and analytical value.

Recurrent physiological loops enable the processing of episodic experiences and memory associations with external relationships. The associations enable relational processing of current events and the accumulation of sequential historical information about the external world. Loop construction provides a recurrent computational model that allows for the independence of sparsely distributed information and the logical integration of linguistic associations with cognition and actions. Hence, language enables explicit representation, classification, and logical, analytical regulation of effective performance via recursive simulation and feedback evaluations such as that seen in working memory.

For example, brain networks with recurrent loops can function as a limit cycle—which repeats itself and comes back to the same position in the brain. Time is continuous in the forward direction, and looping back to the same nodal position in the brain is a different position in time that represents a linear series of sequential episodic events in the direction of the future. Repeating loops and repetition with variation enable learning, editing, and accumulating information about the changing outside world. Even though recurrent, adding an intersecting learning loop allows for the extraction, segregation, and integration of pattern variations into the cycle and the formation and storage of new memories and associations ordered forward in time.

Memory networks have a somewhat specialized architecture. There are neural columns in the neocortex that sort inputs by what is called center-on, surround-off architecture (Spitzer, 1999, pp. 94–95). These columns form functional networks capable of abstracting general features of perceived events as regularities behind the event and can generate meaningfully structured internal representations as memories. Networks of this type classify inputs according to their probability in the environment, which is the first approximation of relevance (Spitzer, 1999, p. 103).

These memory networks can form map-like representations of structured input patterns by the similarity and frequency of the input features. Sometimes referred to as self-organizing maps (Spitzer, 1999, pp. 113–114), these networks were described by Kohonen in relation to self-organization and associative memory (Kohonen, 1984, 1987). Hopfield networks also model content-addressable and associative memory and relate to Hebb's learning rule that activity strengthened weights (Hopfield, 1982; Hebb, 1949). Associations, values, and logical processing provide a method for weighting, comparing, and contrasting different alternative solutions according to their general feature similarities with the problem.

Consistent feature similarity allows for the preservation of information and the correlation of the input patterns with value as rewards or penalties experienced. Discrepancies allow making logical weight adjustments to decrease error rates in the direction of improved solution-to-problem convergence. Weights contribute to the logical descriptions of functional relationships and regulatory processes in neural networks. We can think of weights as a way to logically compare information according to its functional relevance for resolving particular problems, as in the effective analytical weighting and affective emotional weighting. For instance, we might think, "I can't decide what to do next. It feels like I have the weight of the world on my shoulders. I guess it is logical to weigh all the pros and cons and possible rewards and penalties. I think that will help me analyze the problem and come to an effective decision."

Logical processes in an analytical network can sort a central distribution of solutions according to probable effectiveness. When one solution stands out as a good match, lateral negative feedbacks automatically suppress poorly contrasting solutions and reduce redundancies. An analytical system can logically sort divergent solutions by converging outputs according to evidence as a weighted function that rewards the most likely effective solution rather than rewarding the most powerful input. Lateral inhibition facilitates analytical processing and logically weighting information according to its probable effectiveness, which depends on the effectiveness of the language constructions used for the analysis. Sorting by lateral inhibition according to probable effectiveness enables an indirect convergent and divergent sorting option by alternating between comparisons of similarities and differences of old and new information. Processing of information by recurrent and parallel integration and segregation enables functional sorting according to value weighted evidence to converge to a probable best solution.

Competitive neural networks lacking effective lateral inhibition tend to logically default to an affective rank order weighting on an ordinate value hierarchy scale. Vertical and lateral sorting incorporates positive weights that can be activated by excitatory connections, but there are no known ways to directly implement negative weights (Yuille & Geiger, 2003, p. 1228). In other words, there is no such thing as negative output. Functional processing in a feed-forward network can indirectly dampen the output by inhibiting the forward-directed positive signal, but it does not directly feed negative signals all the way forward "through" the network as an action output. Such limitations relegate inhibition to negative feedback, dampening the feed-forward positive signal, initiating lateral inhibition, and contributing to recurrent oscillations.

Recent research suggests that negative weights across populations of neurons in artificial neural networks can be influenced indirectly by nonlinear inhibitory modulatory processing with functional architecture that facilitates oscillating context nodes (Heerebout & Phaf, 2010; Heerebout, 2011). We see inhibition pretty much ubiquitously distributed throughout the neocortex, including inhibitory interneurons, and in the limbic system where recurrent neural networks can function as oscillating context nodes, similar to the previous example. In these recurrent circuits, inhibitory interneurons play a modulatory role in the network's oscillations. Modulating instabilities and feedback control are essential principles for oscillations, and "interneuron networks are the backbone of many brain oscillators" (Buzsáki, 2006, pp. 66–67).

Recurrent loops enable repetition with variation and integration and association of new input information with old information as a logical process of adjusting the information's weighted value. In the brain, functional modular processes with recurrent loops can facilitate efficient repetition and amplification. Functional modular processes can be seen in the frontal eye fields, which incorporate attention and stimulus value; in the mirror neuron system, which incorporates reaching, pointing, and grasping with value; and in the computation and regulation of motor output gain, which depends on value functions.

Oscillations in the hippocampal complex and limbic–amygdaloid system apparently couple an affective value-weighting loop that contributes to the association of event-related positive and negative values. Value symmetry breaking enables the association of functional value weights to states, actions, external object positions and places, and internal memories of experiences. Value functions as a common denominator that can confer conditional independence by changing the valence and weighting of the associated variables. We can view value as independent of information, but whether a stimulus is salient (i.e., whether or not it will capture our attention) depends on the combination of value and information (Frith et al, 2004, p. 265). By recurrence, the system incorporates variation through learning that changes the system's constellation of value weights and biases regulatory processing that guides the output in response to feedback about changing conditions.

Conditional independence of value associations via the amygdala establishes positive and negative value to relevant information as a shared set of solutions that influences input processing, attention, perception, learning and memory, and adaptive action output. Affective value can be independently modulated by adjusting the value bias and sensory input sensitivity in response to feedback reactions. Effective value enables independent output gain modulation, error tuning, and adaptive response to feedback reactions with objects and actions such as change of state, position, direction, and distance. The functional patterns of amygdala relationships can be seen in connections with various cortical and subcortical structures (Freese & Amaral, 2009, p. 16, Fig. 1). Effective and affective processing play a fundamental role in how brains operate across these cortical and subcortical structures from input sense to output product.

The brain evolved in a periodic world over a long-time scale. In a broad sense, brains developed as functional processing mechanisms for phasic cognitive modulation corresponding to variable phasic, periodic, and aperiodic changes in nature. We expect the internal functional dynamics of brain processing to logically match with external dynamics. External dynamics and internal changes correspond with limbic value functions that provide goal-directed affective processing for a sense of attraction and repulsion related to local adaptation on the fitness landscape. Attraction and repulsion in complex systems appear to shed light on the evolution of convergent and divergent phenomena found in the brain related to search and resource allocation that supports the regulation of survival and reproduction.

It seems that processes of divergent exploration and convergent exploitation have evolved as a logical function of adaptive behaviors related to fluctuations in value resources, rewards, and penalties. Shifts in resources can automatically influence limbic-associated affective processes that implicitly modulate cooperative and competitive behavior related to reinforcement. In other words, our cooperative to competitive bias generally shifts depending on resource availability and conditions such as behavioral reinforcement and risk–reward valuations. For instance, if confronted by a larger peer whom we think will defeat us, we are more likely to cooperate and wait our turn for food. When confronted by a smaller peer whom we think we can defeat, we are more likely to compete for the spoils of victory—sometimes referred to as bullying. Different configurations of cooperation and competition for resources play out in many ways across groups of evolving species. These automatic limbic processes influence the regulation of behavioral switching policies for goal-related search by altering affective sensitivity to changing distributions of reinforcement in circumscribed areas of the domain. Resource allocation can result in divergent migratory behavior and convergence to a solution at another location because of the problem of differential changes in the distribution of reinforcement at a given locality. We see similarities between migration and tumor metastasis.

We can look back and consider the logic behind the many instances of migratory behavior seen in many different animal species, including migratory patterns of some human populations out of Africa. We see patterns of genetic transmission out of Africa to genotypic distributions found across populations today. The process of migration can take place because of problems related to climate change where food or water resources become scarce. Migration can also take place where the safety risk from predators logically outweighs the benefit of competing for resources. It is worthy of note that predation has played a major role in shaping the natural world (Vanderwolf, 2010, p. 91) and evolutionary processes. These examples describe the logical activation of divergent search processes for finding functional solutions at a location with a more favorable risk-to-reward ratio. In humans, language enables an alternative level of cognitive processing for the analysis of information related to problems of safety and changing distributions of resources that, at least in theory, can facilitate more insightful planning, risk–reward assessments, and better predictions for maximizing solutions. Regardless, we see an affective regulatory pattern across biological organisms to generally move toward favorable conditions and away from unfavorable conditions (Jennings, 1906).

Language offers the advantage of higher-order cognitive processing with a probabilistic logical function in the formal sense that corresponds to effective information processing. In an informal sense, possibilities enable more colloquial open-ended terms, such as *maybe, possibly, likely, unlikely, relatively*, and so on, in lieu of *certainty, true, false, exactly*, and *absolutely*. Extended language supports continuous logic functionality and flexible open-ended modulation for smoothly switching between microscopic convergence and macroscopic divergence (see Table 2, "Extended Language and Parameter Space" in Chapter 8, "Effective Language"). Attempting divergent search with absolute language and affective constraints can be somewhat futile. Absolute language operates with *never* and *always*, which can lead to dysfunctional digital all-or-nothing switching between zero and infinity. Absolute search with affective referencing in an open system can distort logic and lead to unbounded exploration that avoids negatives and converges to zero risks with imaginary positive rewards at infinity.

Exclusive allocation of convergent closed-ended language sabotages the ability to invert our perception and direction of focus for diverging to higher-dimensional space, because convergence lies in diametrical opposition to divergence. Attempting to diverge to a global perspective and macroscopic description with closed-ended language is likely to result in frustration. By definition, divergence leads in the direction of higher dimensions, and convergence leads in the direction of lower dimensions.

Recall that discrete language functionally isolates solutions from the domain, such that further efforts tend to chop the solutions into smaller and smaller local partitions in the opposite direction of the larger global space. The repetition of a discrete function sections the local subspace into decreasingly smaller portions that converge in a microscopic direction to the smallest possible dimensions. We can think of this process as being somewhat like a recipe for making corned-beef hash—the more we chop, the smaller the pieces. No amount of chopping, no matter how meticulous or frantic, can reverse the processing to diverge in the direction of larger macroscopic dimensions of the global space.

Language can be applied to reasoning about problems and solutions, and for functionally understanding logical computational relations among the brain, physics, information, and systematic knowledge. Language supports a linguistic function space that can define the logical abstract regulatory connections between us and nature from absolute affective to general effective, from discrete digital to continuous analog. With language, we have acquired the ability for regulating explicit thought, with which we can effectively and affectively express our values and understanding of environmental relations.

Language provides abstract semantic parameters for representing functions as logical processes correlating us to the world in which we live, along with better-informed relationships among us and the beliefs, values, and behaviors of others. These functions can be described computationally as processes of logic, logic gates, or as logical parameters for connecting and correlating structure and function, and regulation and value to action, in a complex system.

We humans have the option of choosing and using the language and the functional logic that we prefer. It seems that extended language offers benefits. We come prewired for affective logic, which can make it appear effortless and inexpensive. We pay up for effective logic, since it is an option that requires effortful thinking. Language-wise, logical effectiveness represents a natural bargain that seems worth the effort for reliably understanding relations in our world. Even though information and knowledge are expensive, we have the linguistic ability to logically exploit information and world knowledge that can contribute to thinking more effectively about our short- and long-term well-being.

Logic Space

Logical operations provide a method for correlating an organism's or a computer's input-to-output relations with the domain. As we explore logic space, we will demonstrate the logical integration of some previously mentioned topics and relationships into cognitive processing and operating systems. Logic space includes logic that can be arranged in various functional configurations from absolute discrete linear to nonabsolute continuous nonlinear. *Linear* refers to one-dimensional measurement pertaining to or represented by lines. It also describes events arranged in a line, such as sequences or serial events that occur as a row of end-to-end events in time and space.

Recall that *nonlinear* describes a larger set of multiple combinations of more complex interactions, including linear, but occurring in various configurations of parallel, side-by-side, converging and diverging, and interacting loop configurations in time and space. Linear serial and nonlinear parallel information, input, computation, and output can be described by logic parameters that process the information input and computationally regulate the production of information output.

Computational logic processes are typically described with specific Boolean logic functions such as AND, NOT, and OR. The logic functions AND, NOT, and NAND (NOT AND) can compute discrete operations over a range of logic space and can be configured into most primitive logical process forms, including OR equivalents (Murali et al, 2009; Jaynes, 2007, p. 15). We often use these discrete terms in our daily language to designate the inclusion or exclusion of different categories of information. But when we apply them as a primary logic form for thinking and making decisions, we tend to unwittingly converge to digital thinking and polarizing absolute solutions with polemic arguments.

We see examples of polar logic in day-to-day interactions where affective arguments take place, such as in local debates about beliefs, politics, and even scientific theories. A prominent humanities professor once told two opposing groups of researchers—who were arguing about who is right and who is wrong—that both groups were "both right *and* wrong" and that any one group was "not right *or* wrong." This example demonstrates a common problem associated with absolute logic that can benefit from shifting the perspective on the problem by switching to another simple rudimentary logic function: "both True AND False." In other words, the problem was not linearly separable to one or the other by drawing a straight line between two discrete solutions to obtain an absolute "either True OR False" answer. We routinely use similar discrete logic in how we habitually think, speak, and write, but we may not realize it.

On the basis of shared local and global properties, logical functions can be distributed across two relatively distinctive groups that are based on discrete localized and continuous global processing. Absolute and general functions can be described according to their operations in terms of computation—that is, by formally stating the types of operations they support, including trajectories, and the dynamics of ensembles and oscillations.

Operators have come into play for various reasons, including extension to populations that defy the notion of a dynamic trajectory, and with it, the deterministic description that a trajectory implies (Prigogine & Stengers, 1985, p. 222). Computing simple straight-line trajectories between two points suffices for local idealizations, but when we process multiple variables, influences, and oscillatory interactions among large populations, we shift to continuous processing that corresponds to the general structural dimensions and logic of the global problem-and-solution space.

Operator formalisms provide complex formulas and descriptions for distinguishing between different groups and subgroups of domain functions, including the logic applicable to absolute static and general dynamic domain solutions. An operator tells us the type of functions it operates on, how to operate on functions in a specific domain, and the properties of those functions that define the function space. For example, operators can correspond to a reservoir of numerical values that have an array of discrete values of all integers (0, 1, 2...) or can correspond to a continuous spectrum consisting of numbers between 0 and 1 (Prigogine & Stengers, 1985, p. 221). We see a simple computational analogy where open-ended operations within the domain limits enable divergence toward the macroscopic systematic space. Open-ended operations can invert the direction of closed-ended operations that typically converge in the microscopic direction toward discrete solutions. Thus, extending to open-ended operations enables convergence and divergence, while keeping open the solutions space and without inverting time.

A systematic operator performs generalized functions that extend over the entire phase space. A generalized function enables operation on continuous spectrums and oscillations away from equilibrium. An extended function can integrate "probability proper" into the statistical description as a probability logic function. Probability enables a systematic phase transition to continuous logic parameters that can account for persistent interactions, continuous spectrums and oscillations, entropy production, and the evolving nature of open-ended complex systems (Prigogine, 1997). As an abstract function of language, probability enables coherent operations with logical consistency for open-ended processing of information, knowledge, and action relations within the constraints of a complex domain. In other words, linguistic operations enable nondeterministic open-ended solutions consistent with the domain limits, dynamics, and logic. We expect systematic language to support oscillations, the incorporation of additional information input at bifurcations, and complexification that enhances the modulation of instabilities.

At least metaphorically, applying operator formalism to language, cognition, and computation enables global methods for operating on information encountered in a complex system, and offers a practical way to understand logical functions and processes in such a system. Complex systems reflect the time-dependent uncertainties and dynamical instabilities found in the natural world, and in our explanations of that world. In his Afterword to the recent republication of Marr's 1982 classic *Vision*, Tomaso Poggio notes:

> The simple observation is that a complex system—like a computer and the brain—should be understood at several different levels... [including] the hardware, the algorithms, and the computations. (Marr, 2010, p. 363)

An operating system consists of software, low-level programs, and data that run on computers. The operating system fits the computer architecture and manages the relation between the hardware and programming that enables the execution and regulation of various software programs. In the computational brain metaphor, genetics and development provide the structural brain hardware and architecture, while natural language, semantics, syntax, and grammar provide the programming for abstract software and algorithms that, taken together, form our linguistic operating system.

We can view a linguistic operating system as enabling the integration and extension of our brain hardware and software with the functions, algorithms and programs, cognitive parameters, and permissible logical processes that our brain uses for thinking. The specified group of functions influences cognition and can constrain how we operate on and in our environment. The logical coherency of the correlations we make between brain hardware, language software, algorithms, programs, information, information processing, and knowledge about how the world works influences how effectively we relate and regulate adaptation to that world. When we take a relatively normal brain that generalizes and place localized discrete constraints with absolute language, we expect to incur compromised higher-order cognitive processing.

Yes-or-no, true-or-false, right-or-wrong digital solutions chop the solution space into microscopic dimensions, such that we will likely wind up with a sufficient local supply of microscopic hash rather than approximating global solutions. For example, many companies have encountered operational difficulties correlated to recent turmoil in financial markets that have taken on macroscopic dimensions associated with globalization. I once heard a self-ordained economics expert say that he would like to see some stability and see management to do the right thing. Recall that we have already noted that we live in an inherently unstable world fraught with dynamic instabilities, irreversible processes, and one-way time.

In dynamic domains we more effectively modulate the instabilities that we encounter by acknowledging them, understanding how the system functions, and elaborating on the problem space before trying to converge to what we intuitively imagine as a solution. We see *stability* in idealized systems where we "know the *right* thing to do" because we ignore the overstimulation of the positive feedback and punish negative feedback such that the system dare not err on the side of correcting itself. Unfortunately such a strategy often leads to what is sometimes referred to as a runaway arms race or a bubble. Systems theory tells us that forgoing negative feedback can lead to runaway amplification in what seems like a positive direction. Unencumbered amplification tends to overinflate positives and lead to unintended negative consequences when the bubble bursts. We also know that in complex systems—such as the brain—negative feedback leads to functional oscillations, which are fundamental to thinking and adaptively modulating instabilities and the overall effective behavior of the system.

We can think of the analogy where we are in a competition with a large group of people for a handsome monetary reward where the one winner is determined by the largest balloon. Of course, if there is a larger balloon than ours, we keep huffing and puffing until we either win or our balloon bursts, because winning is all that matters. We know the right thing to do, but in a complex economy when our fortunes seem to keep ballooning, and where accumulating money is the absolute objective, we tend to think of the goal in terms of "keeping on winning" because it intuitively feels like the right thing to do in our limbic system.

Anyone whose language encourages them to think and speak as if they "know the right thing to do" qualifies as a fortuneteller, but their economic expertise comes into question when they imply that economics has anything to do with stable functioning in the world in which we live. We see an information-impoverished local solution that ignores the systematic dynamics and seems suspiciously related to systematic blindness. A systematic approach uses accumulated information and systematic knowledge about "how the system works in the real world," elaborates on the problem-and-solution space, and makes functional predictions that are based on a history of reliably correlated statistical regularities. Classical equilibrium economics assumes mythical "rational" behavior and ignores dysfunctional effects of unopposed positive feedback evidenced across living organisms from paramecia to humans where attraction to positives can lead to consumption bubbles. However, nonequilibrium economics is slowly coming in vogue.

> Instead of being driven by the search for equilibria within models, which leads to the rejection of positive feedback that makes mathematical search more complicated or even intractable, economists should start from the real world. The

relative importance of positive and negative feedback cannot be settled a priori. . . . Reality should drive the theory—not the other way around. (Hodgson, 2009, p. xiiii)

In a nonequilibrium environment we would likely produce functionally reliable results with generalized operator capabilities that support continuously modulated positive and negative feedback and oscillations away from equilibrium. That is, we would benefit from using language constructions that support probability theory and provide statistically reliable methods for systematic error processing via continuous feedback. A generalized language operator enables an extended functional space and the linguistic phase transition that elicits possibilities for better knowing and adapting to the global nature of change. In global systems, a generalized operator enables continuous parameters for logically describing dynamical instabilities, oscillations, and statistical information distributions from one end of the space to the other. Functions in these macroscopic systems support continuous logic with probability and long-range coupling relationships that enable an open-ended perspective (Nicolis & Prigogine, 1989, pp. 15, 21).

At the discrete tail end of the function space, an absolute operator dissects the persistent interactions of complex systems into isolated subsystems and tends to conjure imaginary idealizations and oversimplified solutions, such as the trajectory of one billiard ball striking another. In contrast, at the continuous end of the function space, complex systems with dynamic instabilities are largely nonintegrable, as demonstrated by Henri Poincaré. A nonintegrable system is an interacting system that cannot be transformed to noninteracting parts because these parts have persistently interacting relations across the whole of the system. If such a transformation can be performed, the system is integrable and the equations of motion can be trivially solved (Prigogine, 1997, p. 204). Away from the discrete tail end of the function space and far from equilibrium, we find nonintegrable systems and nonabsolute generalized operators that function according to statistical distribution of information across the global space. Probability theory independently stands out as a relevant factor, and when applied to logical operations, probability as logic provides an open-ended analytical function (Jaynes, 2007, pp. 634–636, 650, 666).

Our day-to-day language typically confers linguistic constructions that tend to fractionate the systematic space into imaginary absolute parts and certitudes but lacks or implicitly excludes constructs for representing instabilities and probabilities in the higher-dimensional dynamic space in which we live. Language that supports possibilities enables processing that leaves room for change and facilitates the continuous updating of information. Dynamic systems benefit from the ability to change possible directions and solutions by flexibly switching between convergent and divergent processes. Lacking alternative possibilities, localized language and absolute logic commonly promote rigidity that places static constraints on our ability to clearly see and logically understand the macroscopic dimensions of our changing world. Linguistically, probability as logic captures the statistical features of biological systems and yields a reliable internal model of nature. Probability as logic makes it possible for us to see the systematic language space and fine-tune constructions to reliably inform our input from, and our output to, nature. We can fine-tune error corrections and adaptive cognitive processing to logically correspond with the natural physical laws governing our world. Thus, we improve our ability to predict, fine-tune, and regulate our adaptive actions in a more consistent and sensible manner.

Processing Sensible Solutions

We do best with sensible decisions and expect sensible thinking to produce sensible solutions when our internal sense coherently corresponds to the changing world outside. We can consider neurological computation as internal information processing that allows us to sensibly process and adapt. We inevitably deal with uncertainties, instabilities, and complex information when reasoning and behaving.

Simple true-or-false logic, conjunction, disjunction, and negation inherited from antiquity may work fairly well in the spinal cord (Braitenberg, 1984, p. 109) or for good-enough solutions with discrete variables in simple, static, low-dimensional situations. But like a true-or-false perceptron, it falls short when evaluating continuous variables with thresholds involved in processing nonlinear problems (Minsky & Papert, 1969; Rumelhart & McClelland, 1986). Nonlinear problems are routine; we encounter them in nature on a daily basis (Spitzer, 1999, pp. 115–119).

Operating as if finding more optimal solutions to complex problems can be computed absolutely with discrete logic leads us to naively oversimplify problem resolution, but possibly worse, it isolates us from the logical processes found in nature. In a changing world, it seems practical to deliberately address problems and solutions by regularly updating our knowledge with current input. Systematic logic coherently calibrates our senses in the current context with the complexities that we experience in nature. Higher-level nonlinear cognition accounts for changes in context (Spitzer, 1999, pp. 219, 233).

We have a sensory layer of neurons for input and another layer of output neurons. To consistently meet requirements for coherent generalization, solutions to nonlinear problems include an additional layer of computation to account for multiple and continuous input variables and threshold activation and inhibition over time (Morris, 2007, p. 657). In artificial neural networks, this additional layer of neurons is between the input and output and described as a hidden layer that has properties like those of a black box. Without linguistically derived knowledge about the brain and the complex world in which it operates, it is difficult to describe what might be going on in the hidden layer. The hidden layer in the human brain has been compared to implicit subcortical regions of the limbic system involved in affective cognition (Panksepp,

1998, p. 20) and, more specifically, to affective regulation associated with the amygdala (Heerebout, 2011).

In the brain, the linguistic logic applied to cognition correlates input and output to the domain such that continuous nonlinear processes and statistical information distributions correspond to the external systematic structure. Language can inform our understanding of what our hidden layer is up to. We compromise our ability to understand our brain's nonlinear information processing when we interpret the hidden layer linearly with constrained language. The effectiveness of computations in a persistently changing environment with continuous nonlinear interactions depends on the extent to which the hidden layer can account for linearly inseparable information—that is, nonlinear information. Networks with hidden layers can go beyond the formation of simple relations between input and output patterns (Spitzer, 1999, p. 135).

Linear decoding of nonlinear interactions in high-dimensional domains inherently introduces noise related to the transformational variance created when the linear decoding process filters out meaningful nonlinear information. In complex systems, nonlinear decoding provides a more sophisticated estimate of the input signal given the output, and thus increases the information transmission (Eliasmith & Anderson, 2003, p. 112). Nonlinearity increases computational power, storage capacity, and information-processing reliability and allows multiplication that is an important nonlinear transformation (Eliasmith & Anderson, 2003, pp. 153–154).

Nonlinear decoding with probability allows us to develop generalized solutions. Generalization offers more sensible input–output mapping and correlations of regularities in domains with one-way time, oscillations, and the evolution of increasing complexity. Probability enables generalized computational processing that accounts for nonlinearity and solutions directed toward the future. Absolute linear operations and discrete computations based on closed-ended equations and symmetrical time are insufficient for generalized computations in complex domains. Because the absolute computations are deterministic, use symmetrical time, and can be computed forward or backward, they fail to address complexification.

Absolute partitioning leads to homogeneous representations that arbitrarily discard context and relevant nonlinear information. Absolute transformation inadvertently compresses the representation of higher-dimensional data into one-dimensional either–or categories that, in effect, destroy information and alternative possibilities. In dynamic environments with statistically distributed information, absolutely compressed representation leads to homogeneous approximations that "clump" information encoding of the input space. This clumping constrains the decodable transformations and the distribution of the tuning curves that correlate to adaptive learning. Heterogeneous representation and transformation can enhance representational capacity, improve noise tolerance, and support more reliable transformations (Eliasmith & Anderson, 2003, p. 216).

When we see a gray world in black and white, we inadvertently transform nonabsolute data to absolute data, and either discard or push the gray to one extreme or the other. Biological organisms are confronted with the task of approximating from some limited number of data points and making the best use of the limited data and storage space. But absolute transformation generates absolute data points, which tends to overfit data, constrain or eliminate generalizability, and increase computational resources and raw data storage space (Spitzer, 1999, pp. 131–133).

In complex systems, predetermined projections or predictions based on past absolute data patterns result in overfitting data, similar to the effects from an excessive number of hidden nodes that tend to prematurely converge to a solution and get stuck there. Such projections assume the past predicts the future but fail to account for changing conditions, context, and the possibility of current input in the future. Temporally faulty prediction errors that blindly go unnoticed tend to persistently produce and propagate recurrent errors. A similar fate predictably befalls unmitigated absolute language and thinking that leads to systematic blindness. Predictions are important, but absolute ones hardly offer a practical substitute for thinking systematically with current information at the time of an event.

Generalization extracts estimates from raw data by abstracting approximate features from the data sets. In effect, an inevitably incomplete data set is created that allows a system with limited data storage capacity to operate with fewer computational resources. This parsimony decreases the chances of becoming burdened by large data sets encountered in complex environments (Spitzer, 1999, p. 133). Conversely, inadvertent absolute overgeneralization increases the chances of losing, discarding, or overlooking significant data, especially as complexity increases and overfitting occurs.

Overfitting explains why it is difficult to fit nonabsolute information to locally accumulated absolute data sets. If the data does not fit every point exactly, "it *must* be absolutely wrong, and we *must* reject it," such that generalizations are easily dismissed as incomplete and faulty.

> You cannot have it both ways: the goals of representing the general structure of the environment and single events are mutually exclusive, if they are to be realized within a single network. (Spitzer, 1999, p. 201)

While some problems may yield to absolute compression of their terms, a reliable transformation algorithm would preferably have the capability of reconstructing the original data set to some predictable level of fidelity. In complex domains, relative representation provides heterogeneous information descriptions, transformations, and processing with reliable information preservation and reconstructability.

We have the flexible capabilities of an oscillating brain with upper and lower stages and of language that can generalize, compute probabilities, and solve problems without closing the solution space. Perhaps with effective language we can (almost) have our cake and eat it too, as an all-in-one oscillating modular network.

With probability as logic and reliable referencing, generalized algorithms tend to better preserve contextual information and offer greater flexibility. Generalization allows for more efficient transformation, compression, processing, and storage of information than do discrete, linear, absolute algorithms. In high-dimensional complex domains, discrete algorithms can destroy information and alternative solutions, and compromise generalizability and the reliable reconstitution of stored information. Even though it might seem that absolutes would produce fine-graining and increase the data and computational fidelity, the result is quite the opposite in complex dynamic domains where generalization excels at data compression while preserving data fidelity.

Open-ended evolution lacks a predetermined goal, whereas alternative solutions develop through open-ended trial-and-error search and are naturally selected by their ability to generally adapt to an environment with uncertainties and instabilities. An open-ended approach naturally produces complex systems that adapt by acquiring ecologically sensible generalized methods for logically regulating adaptive actions in response to difficult-to-predict actions in nature.

Metaphorically, we can think of evolutionary processes as open-ended computational processes that produce sufficient solutions. In this metaphor, the world is one large computer that generalizes domain-dependent correlations with natural components, and adaptive open-ended computations can integrate behavioral interactions correlated with physical laws derived from statistical regularities in the domain. Open-ended computations enable generalized compression that preserves nonlinear information. By not specifying the goal, computational processes can mimic open-ended evolution by selecting, saving, and modifying solutions shown to work, thus accumulating a history of correlations and alternative possibilities for enhancing the regulation of adaptive problem solving in relation to current input in the future.

The ability to generalize to new situations is important for adaptive organisms (Spitzer, 1999, p. 50). Open-ended language with higher-order abstractions enables generalizations that better correspond to how our world works. Typically, the better our language matches what is happening outside, the better our understanding, and the better we can adjust our response to change. In an uncertain, continuously changing world, absolute language impedes dynamic fluency for thinking, learning and correcting, and storing information. Although seemingly effortless, fast, and frugal, absolute language typically operates less efficiently and effectively in a complex domain.

Our human brains have a limited capacity for information storage and processing but generalize well, which allows them to work efficiently in a difficult-to-predict world. With additional effort, we boost efficiency and gain effectiveness with language that also generalizes well. Next we consider how we can enhance our understanding of the habitual nature of language and how our linguistic constructions can augment our brain's ability to deliberately compute solutions that effectively deal with the dynamic problems we face.

CHAPTER 17

Relational Language

IN ONE SENSE, LANGUAGE is a construction of the human brain. Language and the brain apparently co-evolved. Alternatively, we can argue that language was invented by human ingenuity. Like any invented technology, its successful application depends on when, where, and for what purpose it was constructed. Whatever our opinion on its origin, we can simply apply language as it has always been applied, or we can step back and dispassionately examine how we use it. We can consider whether our current language habits produce the best results or whether, instead, we can find some other approach that leads to improvement. We can view language as an information-processing and learning tool that can help us to better understand how the world works, and as a tool that gives us the option of asking questions about its own operation.

We can think of language operationally according to relational information. We can consider its constraints and what kinds of problems it can help solve, how effectively it conveys information, its systematic usefulness, and how effectively it works in the natural world. What kind of operations is our language best suited for? We can find this out by trying it out in different domains. We can test our language in a simple, discrete environment or at a higher level of complexity in a dynamic environment such as nature. Science attempts to thoroughly evaluate the operational characteristics of machines and tools constructed for investigations of the natural world. Since we construct our language in part as a tool for thinking, communicating, regulating behavior, and investigating the world around us, we have the option of testing its operation in much the same way.

We can expect that to some extent language construction has evolved over time. We can trace many of the words and constructions we use today back to earlier eras in human history. Greek and Latin roots make up a large part of many if not most of the words used today in English and many other languages, including the carefully selected language and semantics preferred by science. If we consider that we construct language for communication and to help understand the world, then the language constructions we use probably reflect the current conventions and understanding of the time. For example, recall that we speak of the sun "rising," as in earlier eras when the thinking included primitive beliefs in a flat Earth at the center of the universe, with the sun and planets revolving around the periphery. Language evolved out of this early period of common localized usage and still carries the fingerprints of the absolute language primitives that we currently use in our everyday speech.

Today, so that our current language constructions will match and clarify our current world knowledge, we can update the language constructions we use. We likely will require a commensurate level of education to broaden accessibility of this update for availability to as many people as possible. We can see that appropriate constructions might have been there for some time, going back at least to ancient Greece, where thinking was a popular pastime and language flourished. But it seems that both thinking and knowledge have fallen into disfavor, including the teaching of thinking as a valuable tool. It is up to us to update our language usage and thinking skills according to today's knowledge. This updating is important in the sense that we use language to understand relationships and to help regulate our thoughts, our emotions, and our behavior. Like old news, out-of-date language and knowledge can compromise the effective utility of our world understanding for today.

Language influences how we perceive the world and interpret feedback from it, the choices we make, and our outcomes, including neglect and abuse of nature and other humans. Exploring and studying the relations among language, semantics, brains, and our physical world give us an opportunity to learn and develop more insightful knowledge. Such insight can help us synchronize our language to coherently correlate our thought processes with the dynamics of nature and enhance our overall cooperative behavior.

Representation and Pattern Recognition

How we represent the world we live in corresponds to "How do we think it works?" We can think of language in relation to fundamental principles of computation, where *representation* refers to the mapping of the structure of one entity into a similar structure in some other form, which is generally analogous to encoding. Representation consists of a representing system, a represented system, and structure-preserving mappings between the two systems. The preservation of corresponding structure enables the representing system to anticipate or predict relations in the represented system (Gallistel & King, 2009, p. 309).

For example, when we say, "You made me angry," we implicitly assume a representation of the world with mystical invisible strings that attach a person's emotions and behaviors to the whims of other people. Can you imagine others laughing and correcting us if we uttered such a nonsense statement? But many think that when another person "pulls our string," as if we were marionettes, we have no other recourse except to think, feel, and behave in a certain way. As a social species, we have brains that are genetically predisposed to and sensitive to positive and negative social cues. Of course there are environmental influences, but it is no secret that our feelings are a product of our own brain, our own learning history, and our own thinking, which reminds us that our actions do influence the environment and others that reside there. We have the option of representing the world in physical terms, where each of us remains accountable and responsible for our own individual thoughts, feelings, and actions: "I do not like your behavior, but I choose not to anger and upset myself about it. I prefer to make the effort to act thoughtfully rather than react emotionally." Our brain forms and habitually applies representations of the outside world by meaning derived from our interactions and experience with that world, whether real or with imaginary strings attached.

Broadly speaking, linguistic representations serve to correlate meaning between our internal state, our actions, and our experiences in the environment. Representations are said to stand in for some external state of affairs. Relevant ascribed properties of physical things "inside the head" and "outside the head" are represented in nervous systems as neural responses to physical changes such as displacement, velocity, acceleration, wavelength, temperature, pressure, and mass (Eliasmith & Anderson, 2003, pp. 4–7).

In computational systems, a representation is a formal system for making explicit certain entities or types of information, together with the specification of how the system does this (Marr, 2010, p. 20). Representation can be applied to physiological problems such as understanding visual perception, since vision is the process of discovering from images what is present in the world and where it is (Marr, 2010, p. 3), which we describe with words.

> Vision is therefore, first and foremost, an information-processing task, but we cannot think of it as just a process. For if we are capable of knowing what is where in the world, our brains must somehow be capable of *representing* this information.... The study of vision must therefore include not only the study of how to extract from images of the various aspects of the world that are useful to us, but also an inquiry into the nature of the internal representations by which we capture this information and thus make it available as a basis of decisions about our thoughts and actions. (Marr, 2010, p. 3)

Representation can be analyzed in terms of encoding and decoding (Eliasmith & Anderson, 2003, p. 9; Spitzer, 1999, p. 86). The brain encodes experiential elements into various forms of representation in memory (Siegel, 1999, p. 5). Language provides an abstract symbol system for representing objects, events, and ideas, and serves as a substrate for shared communication about them (Zelazo et al, 2008, p. 565). We can think of language and semantics as a symbolic method for correlating patterns of the abstract representation of our worldview with natural physical laws. Language provides a broad spectrum of representations ranging from idealizations to realizations, static to dynamic, and local to global.

Internal representations of relations with and within the external environment enable us to explain behavioral patterns in terms of our representations. We can use these representations to understand behaviors by contrasting, comparing, or otherwise transforming the behaviors in relation to our representation of them (Eliasmith & Anderson, 2003, p. 144).

> A representation, therefore, is not a foreign idea at all—we use representations all the time. However, the notion that one can capture some aspect of reality by making a description of it using a symbol and that to do so can be useful seems to me a fascinating and powerful idea. (Marr, 2010, pp. 20–21)

Representation acquires "real" meaning (i.e., becomes more accurate and useful) by calibrating the relationship through action-based experience (Buzsáki, 2010). A representation that coherently correlates internally derived patterns and actions with the natural world can be expected to facilitate reliable descriptions of how that world works. It is generally accepted that some brain operations involve representations of experience, and that at some level the brain represents or models the external world (Lytton, 2002, p. 46).

> "Relational representations" were defined as memory representations that are "created by and can be used for comparing and contrasting individual items in memory, and weaving new items into the existing organization of memories. This form of representation maintains the compositionality of the items, that is, the encoding of items both as perceptually distinct objects and as parts of larger scale scenes and events that capture the relevant relations between them." (Cohen & Eichenbaum, 1993, from Morris, 2007, p. 662)

Representation provides structure for evaluating, integrating, and correlating similar and dissimilar memory items and situational environmental stimuli (Morris, 2007, p. 662). *Representativeness* measures the extent of correspondence between

- A sample population, an instance, and a category
- An act and an actor
- A model and an outcome

> When the model and the outcome are described in approximately the same terms, representativeness reduces to similarity. (Tversky & Kahneman, 2002, p. 22)

Semantics provides the building blocks of meaning, encoded as representations of perceived patterns that the brain decodes linguistically. Semantics represents and conveys information, knowledge, logic, abstract ideas, and concepts. Semantics has a significant influence on representation (LeDoux, 2002, p. 203), and higher-order semantic abstractions result from perceptual inferences extrapolated from more concrete lower-order representations abstracted from the environment by transformation.

Transformations, or transforms, are sometimes described mathematically as the amount of stretching or bending that a transformation imposes, as in Jacobian transformations; perhaps they are similar to "a local stretch of the imagination" that can result from language abstractions that distort the coherent correspondence of cognition to the physics of the domain. Semantic fidelity biases the coherent representation and transformation of external information about our interactive experiences.

Language only symbolically stands in for what it represents; it is not the thing it represents (Korzybski, 1958, p. 78). Language is not in itself reality but allows humans to describe and process their information about reality (Jaynes, 2007, p. 411). Since language stands in for but is not what it represents, we can say something that makes sense grammatically without saying anything sensible about patterns in the real world. Just imagining, believing, or saying something does not make it so (Jaynes, 2007, p. 22). Language and semantic fidelity leverage our ability to explicitly identify, represent, and logically separate fiction from nonfiction. Language fidelity enables clarity of practical reasoning.

> Cognitive representations can often be treated as logical propositions that can be precisely linked to explicit referents in the external world, which allows investigators to initiate credible empirical studies. (Panksepp, 1998, pp. 30–31)

When we reason about the natural world, the fidelity of our semantic representation and transformation directly corresponds with the logical production of coherent, relevant, robust, and useful abstractions. These abstractions bias how we perceive and relate to the world, since our perception of the world—our "knowledge" of it—reflects our previous behavior with respect to that world (Skinner, 1953, p. 140).

Semantic representations with relative emotional and motivational significance inferentially affect reward value and reward anticipation (Phelps & LaBar, 2006, p. 432), mediated by the brain's lower frontal lobes (Petrides & Pandya, 2002, pp. 45–46). Language influences reasoning, emotional feelings, and behavior by biasing inner speech, the narratives we tell ourselves about what we think or feel. Lev Vygotsky and Alexander Luria (1993) proposed that such narratives often develop conceptually in competing terms of rewards and risks, cooperation or competition, agreement or conflict. For example, we might say to ourselves, "Do I cooperate and go along with this, or do I take a risk and compete?"

Listening to our private self-talk provides us with a window from which to gain insightful information about the fidelity of our language in relation to our thought processes and corresponding physical processes in nature. Our internal private speech reveals the two-way influence between the abstract linguistic representations we infer from our interactions with the physical world and how effectively our linguistic reasoning solves problems we face. When we distort representations of physical reality by semantic misrepresentation—if we have only the word *pussycat* to represent all types of felines, for example—those distortions can in turn bias how we physically relate to that world, with potentially dangerous results (Jayne, 2007, p. 187).

Along with value, the physical correlation to language behavior, bias, and representation implies that semantic fidelity influences consistent pattern recognition and effective state regulation. As in thinking and learning, the usefulness of memories and the establishment of internal models (called concepts) is accomplished by "coding of the environment in those terms that yield a maximum of correlations and logical structure, in other words, in the most meaningful terms" (Braitenberg, 1984, pp. 55–56, 61). The semantic correlation between representation and meaning implies that, in general, the more accurately we describe an event and the more accurately we assign meaning to it, the more effectively we interface as we regulate our interactions with and within the system. The purpose of representations is to provide useful descriptions of aspects of our observations about the real world (Marr, 2010, p. 43).

> The structure of the real world therefore plays an important role in determining both the nature of the representations that are used and the nature of the processes that derive and maintain them. (Marr, 2010, p. 43)

An effective understanding of pattern recognition and representation provides us with processing methods for discovering systematic meaning that consistently corresponds with our observations of regularities in natural physical laws. Reliable correspondence of regularities and environmental relationships contribute to how well the natural patterns we observe and represent stand in for what is represented, and how well we predict and regulate the effectiveness of our adaptive actions. It is reasonable to expect that verbally or semantically manipulating the representation would be much like manipulating what it represents, because the similarities in experience largely depend on the fidelity of the representations (Eliasmith & Anderson, 2003, p. 49).

We may think that simplifying patterns in a confusing world with so many things going on makes the world easier to understand such as when we think in terms of simple true or false. But when we think in those oversimplified narrow terms, we miss out on a breadth of information that helps us to realize a greater understanding of humanity, the spectral beauty of rainbows, and the many possibilities and varieties of life. Misrepresenting the complexity of nature in black and white narrows our vision and hardly simplifies the dimensions of the problems we face when trying to understand and deal with many variables in a changing world.

One or two discrete input variables in a space typically make pattern recognition fairly simple. Spaces of high dimensionality composed of many input variables can present quite a challenge. We might be tempted to simply divide the large number of variables into two cells to simplify the problem, so that we now have only two variables. Yet this may not be the best solution. Resolving what seems like the problem of excessive dimensions and variables in complex domains presents a particularly challenging task, one unlikely to be resolved by dividing the input space into two true-or-false variables. Christopher Bishop (2006, p. 36) warns us that "not all intuitions developed in spaces of low dimensionality will generalize to spaces of many dimensions."

Effective language enables generalizations with greater linguistic fidelity for representation and pattern recognition that help to clarify our perspective on nature. Seeing the natural world as clearly as possible theoretically gives us an adaptive regulatory advantage by allowing us to improve the accuracy of our predictions and adjust our behavior accordingly.

Mapping

Since antiquity, humans have continued exploring and mapping nature's wildernesses and oceans. We see an evolution from the crude maps of an ancient flat world to the detailed global maps of today. We also construct internal maps of what we perceive about the world around us. These internal maps are somewhat like geographical maps we use for navigation. We use our internal maps to find our way in a semantically correlated environment. Poorly constructed maps that contain imaginary or unbounded infinite inferences about the terrain tend to mislead us or trap us with unexpected and unexplainable obstacles.

With reliable knowledge about the world, we can construct, update, and maintain better maps. Knowing the construction and ordering of processes in the world—knowing how it works—enables us to construct language that can effectively translate information into knowledge that more accurately represents that world. Systematic mapping of internal and external relationships yields representations with a higher level of ecological fidelity.

> Representation as defined by the combination of encoding and decoding essentially can be understood as mapping elements under one description to elements under some other description. This can be characterized as functionally mapping one "space" onto another. Ideally, the encoding and decoding are inverse mappings between spaces. Thus, any object mapped from one space onto another space by the encoding function would be mapped back onto the object in the original space by the decoding function. (Eliasmith & Anderson, 2003, p. 185)

Consider that a map of a city benefits from regular calibration to reflect changes in street configurations. Like yesterday's weather, old maps with out-of-date information convey little useful benefit in the present. Like a city map, the mapping that best correlates linguistic abstractions with the world it intends to represent benefits from a reliable external referent to calibrate accuracy. Matching the context and structure of the language with the context and purpose of the abstraction in the real world enhances the accuracy of the calibration to that world. Viewing language and semantics in terms of fidelity provides a model for evaluating and explaining language use according to coherent correlation and regulation of interactions at a systematic level.

Localized language often uses semantics that prescribes how we "should" act, but sans sensorimotor integration. These prescriptions amount to regulation with discrete idealizations that defy reliable sensorimotor mapping to a continually changing natural world: "We should be certain to behave in the right way." But context changes from minute to minute, and how we define doing what we "should" as the "right" thing to do will vary from situation to situation. Perhaps worst of all, we do not know what is "right" until after we have done it, and "should" and "right" both vary according to different perspectives and context. Also, doing what we think of as the right thing in one local culture often turns into the wrong thing in another, and can even result in death by execution, as conventions change across cultures and borders.

For example, some cultures encourage education and learning behavior where we are told that learning is an honorable behavior and that we should get an education. Other cultures punish learning as a dishonorable behavior that threatens a rigid, overvalued belief system, where learning dishonors our family and we should not even think such evil thoughts. *Honor* is a subjective, affect-laden word with a high degree of local variability in changing context that can quickly go from "right to wrong" in a matter of miles or minutes from locality to locality.

A systematic description allows for contextual mapping of one possible choice among many alternatives. Knowing how the system works enables global interpretation of localized beliefs and confers the wisdom to avoid cultural faux pas that might jeopardize our safety. Generally, systematic learning and knowledge offer choices with relative predictability and a greater possibility of escaping the weight of the authority on the locally constructed pull of gravity.

Language that excludes systematic language constructions typically yields localized constraints of implicit symmetrical time with certitudes that compromise statistical correspondence with temporal information about relationship regularities and domain dynamics, as in these examples:

- We live on a flat world where the sun magically rises in the sky in the day and the moon mysteriously rises at night.
- Time is only relevant to events happening together, before, or after an event.
- The preceding event causes the next event.
- The world and beliefs about it are either absolutely true or absolutely false.

When we are attempting to measure across the system, discrete localized linguistics leads to bisecting, dissecting, and distorting systematic mappings but confers an illusion of absolute fidelity. Newton's laws use local symmetrical time and deterministic calculations that give the illusion of a perfectly ordered world and universe. In extreme cases, illusionary correlation can lead to extraordinary beliefs, such as the ancient Aztec theory that a human sacrifice had to be performed each morning to make the sun rise. It's gruesome, and easy to condemn in hindsight, but to the Aztecs "it worked every single morning, just as advertised" (Macknik & Martinez-Conde, 2010, p. 193).

On the other hand, an example of generalized systematic representation and mapping enables an extension to one-way time and statistically correlated regularities and relations:

- The planet rotates around the sun and rotates on its axis with a periodicity such that an observer in a fixed location on the surface will tend to perceive the illusion that the sun is moving.
- The sun seems to move somewhat like the moon, except for the peculiar changes in the moon's appearance.
- Without knowledge of periodicity, we can easily acquire a confusing local view of our solar system's large star and the multiple other objects continuously appearing and reappearing mysteriously without obvious direct influence on each other's behavior.

Periodicity can be defined systematically as mapping regularly recurring events and relationships in the forward direction of time. Since the past is static, we can correlate the statistical relationships of regularities from the past to accumulate systematic knowledge that supports reliably informing predictions about the possible recurrence of events sometime in the future. Global models enable exploiting the system's macro dimensions, measuring systematic periodicity and reliably correlating a history of regularities with periodic and aperiodic events. Relative knowledge gained from periodicity and a history of correlations in the system generalizes to help explain recurrent regularities and regulation in localized subsystems.

The most invariant representation of regularities in a domain would theoretically provide the most coherent mapping. Then in an abstract sense, the represented world—the world of perception and action—is coherent and unified (Churchland & Sejnowski, 1994, p. 380). Science, with referencing that is based on the fidelity of information and statistical reliability, strives to provide a relatively invariant systematic language fulfilling this mapping criterion. Variant localized language representations, exclusive of systematic fidelity and reliable referencing, fall short for coherent mapping.

Language and semantic fidelity influence the input-to-output mapping of human thought processes and the behavior produced. The extent of representational fidelity corresponds proportionally to the regulatory influence on the system's effective output activation and tuning efficiency. The input-to-output mapping requires partitioning that biases the sorting and integration of current and prior information, which influences the system's long-term regulation and output predictability. Thus, output with compromised predictability biases the effective mapping of further input feedback, error corrections and learning, and state regulation. If a neural system computes a transformation from an input variable to an output variable, then the adaptation of the map would be expected to produce generalization patterns that reflect how neural elements in that map encode the input variable (Shadmehr & Wise, 2005, p. 403).

Extended language and semantic representations improve accuracy-based input–output mapping to coherently correspond with the world in real time. Rather than localized language based on coincidence and superstitious beliefs, language that consistently maps and facilitates the efficient flow of communication enhances the reliable modulation of behavior, the coevolution of adaptive interactions, and the accumulation of effective knowledge.

There is an old story about the "great turtle." A person tells us they know the truth about how the world works. Since we are interested in getting more information, we follow up:

"Tell us more."

"Well, a long time ago my great-great-grandpa was searching around in this old stack of books and he sees a dusty, tattered, and mystical-looking old book. It turned out to be a miraculous find, a book that describes how the world works, and it makes perfect sense."

"What do you mean?"

"According to what I was told, my great-great-grandpa said the person that wrote the book found this ancient sketch that shows the outline of an invisible giant turtle hovering above the world and holding the world up by what looked like giant strings or ropes or something like that. All I ever saw was a copy somebody made from memory and drew up on the back of an envelope. Not too long before all that, the whole village trembled and the ground rumbled on several occasions. It was sometime back before he found the book, and I'm not sure exactly how long that was. Some say it was a week or two and others claim it was a year or so. But it created quite a big stir, that's for sure.

"Well anyway, luckily the damage wasn't too awfully bad and no one got seriously injured. But it just about scared everybody to death. They thought for sure the Earth was going to fall right out of the sky. My great-great-grandpa gathered up the whole family and most of the village to show them the book, and they all got involved. Everyone breathed a sigh of relief knowing there's a giant turtle up there watching out for us, and it seemed to be doing a pretty good job of it at that. Everybody was just hoping it wouldn't let us down. Great-great-grandpa and some others had this idea that everyone would get together every evening, build a fire, and chant and sing to the great turtle to make sure it doesn't let go.

"Guess what? It works. Look around; the Earth hasn't fallen out of the sky. It's still here, isn't it? We still carry on the ritual to this day because it makes sense to give thanks to the turtle for our good fortune. Of course, we can't stop because you never know what could happen if we stopped performing the rituals just like clockwork. Well, the turtle might, you know, drop the ball, so to speak. It's the truth, honest."

"Hmm, that's an interesting story all right. But it hardly sounds like a plausible explanation."

"I don't know what you mean by plausible, but something has got to hold it up, you know. It has got to be pretty heavy."

"Thanks for sharing."

That example points out the importance of considering and questioning representations that

- Reference a static picture of how the world works in lieu of referencing known dynamic physical laws
- Substitute imaginary constraints rather than knowledge of systematic constraints that limit the definition of *plausible* to domain-dependent information corresponding to physical laws
- Fail to logically explain the relationships between the who (turtle) and the what (world); although the turtle story does explain how (strings) and why they (support) the turtle that supports the world, it lacks evidence supporting the process (beliefs only)
- Exclude current information about known forces such as gravity, attraction, and repulsion
- Ignore that in dynamic domains certainties such as the "truth" violate the domain constraints on the statistical distribution of information
- Propose doing something unrelated to keep something from not occurring, which meets the definition of *superstitious*
- Use closed-ended language that creates locally trapped beliefs
- Unrealistically put responsibility and accountability for the state of the world on something implausible (in this story, the turtle) rather than understanding nature, where one might decide that adapting by moving off the fault line is a better strategy than the behavior in the representation (chanting to the turtle)
- Diminish the population's accountability and responsibility as to the importance of ongoing learning, adaptive flexibility, and scientific understanding of nature by supporting fixed, rigid beliefs, superstitious rituals, and mythical explanations
- Seem sensible to a local population but hardly support the beliefs (there are no known extrasensory organs for detecting mythical turtles)

Language constructions that facilitate logically consistent mapping between our thinking and the structure of the physical world can enhance our mutual communication, cooperation, and adaptation with others and nature. In turn, logical consistency can improve the effective tuning of our internal representations and accuracy of our perceptions. Seeing and perceiving the world as accurately as possible helps to improve how we relate to that world because we construct our descriptions of that world from our perceptions (Marr, 2010, p. 354).

> ... [T]he true heart of visual perception is the inference from the structure of an image about the structure of the real world outside. (Marr, 2010, p. 68)

Coherence and Transformation

Coherence, which means a reliable logical association between our experiences and our interpretations of them, corresponds to accuracy of perception and greater predictability of external relationships. On the other hand, *attenuated coherence* refers to the compromised reliability of logically mismatched associations disrupted by ambiguity and imagination. We live in a changing environment where internal-to-external coherence is measured statistically as probable rather than absolute.

Coherence corresponds to the time-dependent dynamics of regularities and probabilities rather than an absolute static state. The coherence of a representation can be described as the correlation between the source object and the pattern produced from it. The representation yields a measurement we can apply to linguistic accuracy, cognitive processing, and coherent correspondence with the domain. In other words, we expect a coherent and logical statistical correspondence among the language, the representation, and the constraints of the domain dynamics.

Lexical or vocabulary coherence has been shown to arise in natural language experiments aimed at evolving shared communication among autonomous robots (Steels, 2000a). Extending language coherence to the external world enables transforms of information input and output that logically correlate with consistent systematic pattern recognition. The degree of informational fidelity of the linguistic representation can proportionally bias the perception and recognition of external patterns. We stand a greater chance of detecting pattern regularities if we use logically consistent language to describe relations. Inconsistencies are likely to arise when describing complex environments with dynamically insufficient language constructions that distort our perception and compromise systematic pattern recognition and predictive abilities.

While characterizing the dynamics of a complex system helps to understand the system, perhaps more important is understanding how the dynamics relate to information processing in living systems and to adaptation in changing environments (Mitchell, 2009, p. 39). Living systems deal with change and adaptation by leveraging their predictive ability. Language provides further leveraging by conferring a higher level of abstraction for logically consistent regulation. Therefore, we can increase our prediction abilities by developing internal representations that logically correlate with statistical information about environmental regularities and patterns that fluctuate or evolve over time.

Even relatively low-order regularities, such as the spatial and temporal coherence of sensory signals, convey important cues for abstracting higher-level properties about objects (Becker, 2007, p. 7). Space and time coherence of the input is a crucial variable leading to formation of internal representations (Spitzer, 1999, p. 147), as in this example:

> I was reading the newspaper, and there was a loud clap of thunder as I looked at the headlines about a natural disaster. I kept thinking that I was certain it was a message telling me I was a bad person because I never have liked those people anyway. Now every time it thunders I feel guilty and I keep thinking I caused that disaster.

Locally constrained language and semantics that accept affective representations that are based on coincidental occurrences tend to misrepresent experience. Misrepresentation distorts coherency and results in ambiguous associations, impaired understanding of the experience, and compromised ability to effectively identify, question, and resolve spurious information input.

Semantically represented information provides the categorical structure for contextual associations correlating contingency and relative causal influences. Contingency implies prediction, including predictions about the sensory consequences of our actions (Portas et al, 2004, p. 288). When we make mistakes interpreting what we see, we learn erroneous information, which in turn confuses us when we try to act on what we learned. Reliably defined representations and logical consistency help us to understand how misrepresenting coincidental experiences distorts our perception of others and their intentions, as in this example:

> Maybe those folks aren't as bad as I've heard all my life. But even if I didn't particularly like them, I didn't have any intentions that harm would befall them. That thunder was probably a coincidence with an unpredictable result. I'm not going to keep making myself feel guilty by telling myself I *shouldn't* think about those folks that way. But it doesn't make much sense to not like those folks just because of the rumors I've heard about them, especially when I've never even met them.

The learning derived from explicitly defined and reliably correlated semantic representation enhances transparent domain knowledge, effective predictions, and clarity of intentions, both our own and those of others. As we learned on the first day of statistics class, correlation is not equivalent to causality and requires statistical analysis of influences involved. Reliable representations enable evaluating experiences systematically by disassociation of affect from information, action, and objects. Thus we can apply analytical thinking, identify systematic biases, and disentangle our misrepresentations of reality. We have options for how we reference and represent relational experiences. Unchecked by reliable representations of nature, our brain tends to fulfill its default role as an assumption machine that relies on its subjective history to fill in information gaps with a vivid imagination about what is going on "out there."

We could say that the things which happen to us are mysterious. But we are less likely to ascribe mysterious causes if we understand how language, nature, and the brain work together. We can see the importance of first understanding how language works before we can use language effectively to understand how the brain and the world around us work. But language is a double-edged sword that can both inform and misinform us depending on how reliably it represents realizations in nature. Effective linguistic representations enable descriptions of experience that reliably correspond with how things work in nature—according to possibilities and probabilities, positives and negatives, attraction and repulsion, approach and avoidance. Effective language enables analysis of our experiences and the ability to identify assumptions and reliably distinguish imagination from realization. Take the following example:

> We are angry and thinking about getting revenge on someone. It is raining, and coincidentally lightning strikes nearby, flashes across the room in which we are sitting, and strikes the television with a loud boom, and smoke comes bellowing out. Then our brain takes this seriously and connects our "bad thoughts" with the event. Our assuming brain errs on the side of safety, associates the "bad thoughts" with the frightening event, and infers a causal relationship. In the future similar events may share the same translation even though unrelated.

We can understand the reality of the lightning strike and the spurious cognitive association that easily fooled us because of our brain's vivid imagination and tendency to assume cause and effect. If educated with knowledge about how nature works, we realize that our assumption about cause and effect is instead a matter of coincidence. We still may be visibly shaken by the event, but the consolation is that usually we can replace or edit our brain's automatically generated superstition with a naturally plausible explanation. Understanding that the brain evolved in nature and tends to automatically react emotionally to threatening events by erring on the side of safety encourages us to apply reliable language to explain why events happen and correct the misrepresentations that lead to misperceiving our experiences. The caveat is that common affective language can engender systematic blindness that renders us powerless to overcome our ignorance of systematic errors by analysis. Affective language supports the default to rationalizing and confirming our superstitions and erroneous imaginary beliefs. It rains on our parade because it rains in nature, not because we misbehaved. Because we know how nature works, we check the weather prediction and take an umbrella to increase the probability that we will stay dry.

Our analytical language system enables higher-order cognitive processing and reliable external referencing that coherently correlates our experiences to the physical world. We have the option of choosing localized discrete logic or probability as logic with systematic continuity. The former uses limbic-affective logic, and the latter uses analytical-effective logic. Tuning our language coherence from localized to systematic facilitates analytical reasoning and disentanglement from our emotionally laden beliefs and biases.

The coherence of our semantic representations influences how accurately we perceive and understand relative contingencies and, in turn, how well we can predict intentions and the outcome of our behavior and the behavior of others. Compromised coherence distorts the fidelity of our perceptions and the effectiveness of our predictions. This knowledge provides us with an opportunity to deliberately recalibrate our language constructions, representations, thought processes, and adaptive responses coherently to nature.

Understanding the meaningfulness of representations depends in part on reliable cognitive processing and coherent information transformation and integration into the system. And in a roundabout way, understanding how the system works facilitates the construction of effective representations. Representations form the template for understanding the coherency, consistency, and usefulness of informational properties for reasoning that biases brain processing (Eliasmith & Anderson, 2003, p. 8). Language underlies coherence by providing a semantic spatiotemporal scaffolding mechanism for supporting, representing, and transforming information and regulating behavior.

Correspondence between linguistic representations and temporal observations in a changing system benefits from modeling language from a systematic perspective. Statistical information corresponding to systematic regularities facilitates coherent higher-order abstractions, generalizations, and globally useful knowledge. Coherent knowledge about systematic processes and relationships helps to tune our perception and improves stimuli discrimination, credit assignment, and context-sensitive learning. Coherence provides a consistent framework for integrating structure-driven bottom-up information with function-driven top-down information (Érdi & Kiss, 2001, p. 204).

Understanding representational transformations in a functional manner provides an informational model for understanding how our representations are correlated, computationally manipulated, exploited, updated, and so on (Eliasmith & Anderson, 2003, p. 4). Transformations operate within constraints determined by the overall system's initial organization and subsequently affected by any systemic perturbations (Schlosser, 2007, p. 128). It is beneficial to understand how information gets transformed to explanations of physical relationships consistent with physical laws of nature and how that information supports coherent reasoning about our predictions, outcomes, and the disturbances we experience. Language and semantics influence these transformations and the reliable correspondence with cognitive processing, but that does not necessarily mean that all languages are equivalent or that the meaning we perceive sensibly describes how the world works pragmatically.

As an abstract informational transaction between relative internal and external positions and states, input–output transformation translates and correlates the regulation of an organism's sensorimotor system into coordinated domain interactions. Transformations correlate and coordinate the organism's abstractions of the geometry and physics of the external world to its position and state regulation. These transformations are often conceptually described as taking place through matrices of synaptic weights modified by learning and development (Churchland & Sejnowski, 1994, p. 337). Input is mapped to output through a logical process that not only influences the available information but also shapes how we view the world (Schlosser, 2007, p. 135; Boyd & Richerson, 2005, p. 206). A logical mapping supports functional linguistic correspondence among transformational processes and representation, information and knowledge, experience and perception, and regulation of domain interaction.

Environmental coding that yields a maximum of correlations and logical structure produces the most meaningful terms (Braitenberg, 1984, p. 61). In effect, language constructions represent an interface for

- Reliably coding and decoding logical and meaningful environmental information
- Constructing, representing, classifying, and transformational processing of information into coherent knowledge
- Reliably modulating and coherently coordinating input–output mapping

Coherence facilitates reliable mappings in correspondence with external relations to

- Space, time, and value context
- Perception and the analysis of inference and intention
- Logical reasoning processes
- Adaptive behavior in response to change

Thus, our constructed language encodings can bias how coherently we perceive and interact with our world.

Hidden Code

The matter of efficient encoding and its consequences form the chief substance of information theory (Pierce, 1980, p. 42). Conceptually, we can think of storing and processing information via a code or neural representation as points in vector space that encode—for example, features, locations, and movements (Spitzer, 1999, p. 86; Eliasmith & Anderson, 2003, p. 6). Coding applies to information input and output, and it biases the fidelity of representation and transformational processing, whether in terms of language and evolutionary processes or as lifetime learning. Representations are encoded during storage and decoded during recall; the contextual properties of the particular instance of coding can influence computational weight values, actions, and hence outcomes. These coding principles apply to cognitive processing and how we think and regulate behavior in relation to goals and outcomes.

If we can improve how the encoding corresponds to the domain and represents the problem, we can improve the transformational processing of information into knowledge for greater problem resolution. The relative accuracy of the coding process biases the consistency and usefulness of representations. Improving the fidelity of the encoding can enhance how reliably we represent, process, and understand information derived from our relationships with the physical world. The more that we know about what information was encoded, and how it was encoded, the better we can surmise what can be reliably decoded (Eliasmith & Anderson, 2003, p. 197). If we know the information was encoded with absolute language, we can decode it with generalized language. But information both encoded and decoded with generalized language typically demonstrates better fidelity. For example, compare these ways of describing two issues:

- "You always get your way."
- "You sometimes get your way."

- "You never remember my birthday."
- "Sometimes you forget my birthday."

James Watson and Francis Crick deduced the double helix structure of DNA in 1953, and over the ensuing 10 years, scientists eventually broke the code for DNA. This discovery led to a scientific tipping point for understanding the evolution and natural selection of DNA as the underlying mechanism for the evolution and natural selection of genetic information (Mitchell, 2009, pp. 89–90). Breaking the natural language code takes on a similar significance. Both DNA and language have open-ended capabilities as evolving information and knowledge sources that operate within physical environmental constraints. Open-ended features allow the fine-tuning of both as information sources that correspond to actions, experiences, and learning with and within the environment.

Language encodes and passes information from one generation to the next somewhat like the way DNA encodes and transmits adaptations for survival into the next generation (Benton, 2009, p. 84). Neither process likely escapes the effects of evolution. As a DNA-like carrier for coded information about the natural world, language would be expected to obey evolutionary rules and natural selection, but this may be difficult to appreciate, since language has been around only briefly from an evolutionary viewpoint.

Economist Eric Beinhocker (2006, p. 12) writes that "evolution can perform its tricks not just in the substrate of DNA but in any system that has the right information processing and information storage characteristics." DNA transmits information invisibly and sometimes contains errors out of sight of the naked eye that can visibly influence regulation and phenotypic expression. To address the potentially adverse effects of these genetic snafus, science has developed technology to explore for and find many of the errors. Analogous to DNA, language can also propagate unseen errors in the form of absolutely hidden linguistic baggage inherited from the ancients.

Language and semantics fine-tuned to the domain's preferred information asymmetry can enhance exploration and detection of inherited linguistic errors and their potentially maladaptive effects. Coherently fine-tuning and correlating language and semantic constructions to nature can approximate the open-ended aspects attributed to DNA's evolutionary role. Open-ended codes, contained in both language and DNA, provide a natural resource for creativity, innovation, novelty, and alternative solutions that can enhance mutually beneficial adaptive relations with nature.

Semantic Space

How we parse and classify information in semantic space biases representation and the kind of solutions we develop, such as with computational problems. For instance, problems with reliable coding, parsing, and classification of information, objects, actions, and related variables influence the processing efficiency and regulation of output effectiveness. In the brain, faulty parsing and classification can lead to inefficient and inaccurate cognition, increased errors, and confusing solution variability. The language and semantic code we use for classification matters because it influences the parsing of information and our perception of relationships, how we evaluate and regulate our actions, and what we learn from feedback, which can skew how we interact with our world.

Today, most people would likely agree that words conceptually refer to classes of objects, actions, relations, logic, and so forth. We are surrounded by and involved with a large number of classes and subclasses of objects and actions that we can usefully associate with words (Pierce, 1980, p. 120). Language provides semantic representations for designating things, attributes, actions, and relations (Luria, 1981, pp. 29–30, 37).

> The basic element of language is the word. A word can be used to refer to objects and to identify properties, actions, and relationships. Words organize things into systems. That is to say words *codify our experience.* (Luria, 1981, p. 31 [emphasis in original])

Words not only designate or stand for objects abstracted from experience but the structure of a word enables us to refer to objects and assign attributes and relations (Luria, 1981, p. 31). Our lexicon of words and definitions has been constructed by speakers with various levels of education, literacy, and range of knowledge. The variability of the construction process produces a range of semantic ordering properties and definitions across a spectrum from absolute local to nonabsolute global that generalize across the global space.

Each word we use typically has one or more agreed-on definitions, whether implicit or explicit. The original definitions of many words predate the written record, while some more recent words have been deliberately created and defined to meet specific modern situations and experiences and new ideas. The haphazard evolution of words combined with the idiosyncratic way that most of us acquire our vocabulary can lead to divergent meanings for words that look the same to us. When we speak, we choose from a number of words with similar or related meanings, and we have the option of choosing the contextual frame of reference or perspective for our communication. We can choose words on the basis of our understanding of their accepted meaning across a broad range including simple to complex, concrete to abstract, and affective to effective.

> "Meaning" is a stable system of generalizations represented by a word, a system which is the same for everyone. (Luria, 1981, p. 44)

To reduce the potential for misunderstanding that can arise from divergent meanings, science strives to carefully choose and calibrate its words. Minimizing semantic variance in meaning can translate to optimizing the effective utility of information available for identifying and accumulating, processing and analyzing, and expressing knowledge that coherently corresponds to nature. Language with systematic ordering and relatively invariant semantic terminology can produce global macroscopic descriptions, such as the language and semantics developed and advanced by scientific methodology.

> Adaptation of the mind may be more or less the same thing as adaptation of the language; at any rate one goes hand in hand with the other. The progress of science is marked by the progress of terminology. (Polya, 1954, p. 55)

A word not only substitutes for a thing but also analyzes it by incorporating it into a system of complex idiosyncratic associations and relations. This abstracting and generalizing function is known as *meaning* (Luria, 1981, pp. 34, 37). Field theory models have been explored for solving what has been called the "grounding problem" associated with connecting meaning, language, and cognition (Tikhanoff et al, 2006), which relates to information. Luria suggested that semantic fields exist for every word. He stressed the importance of the process of naming and perceiving a word as a complex process of selecting the closest meaning from the word's entire semantic field (Luria, 1981, p. 37). A reliable semantic field theory establishes semantic fields that contain domain-dependent words with definitions and usage that logically correspond to that domain and the domain constraints. Coherently correlated words, their definitions, and their representation and transformation facilitate informed descriptions that reliably correspond to domain interactions.

A word field can be conceptualized as the semantic space from which we choose contextual fidelity that currently applies to a word and its relative meaningful correspondence with our experience of object and relational interactions in a specific domain. Conceptually, word field descriptions make it possible to evaluate and correlate a word's performance in terms of its effective signal-to-noise ratio according to relevant meaning in a given context. The fewer the assumptions we make about a word, the less ambiguity it has. Clarity enhances the plausibility and applicability of our words and improves how efficiently and effectively we can choose and use those words. Carefully choosing words for domain-dependent fidelity simply improves the description and communication of reliable information.

> A word is a unit of thought... because the powers of abstraction and generalization are the most important functions of thinking.... Thus by abstracting and generalizing the property of an object, a word becomes an instrument of thought and a means of communication. All of this shows that a word not only duplicates the world, it also serves as a powerful instrument for analyzing this world. (Luria, 1981, p. 38)

According to Luria, language represents a system of codes used to designate an object by independently analyzing it and expressing perceived features, properties, and relations. Words codify human experience (Luria, 1981, p. 31) and convey meaning expressed by the word's established definition. Words allow for an operational designation of an object in terms of its purposeful use. The usefulness of words enables us to correlate an object to a certain category, and also to indicate the form of the action that the object can perform in a given context.

The word is the foundation of the system of codes that enable the reliable transition from the sensory to the "rational" world (Luria, 1981, pp. 38–39). Semantic space contains many words with varying transactional reliability and correspondence with nature. The reliability and orderliness of cognitive processing and communication in a given context depend on the selection and use of domain-specific words with sufficient transactional fidelity and relational precision for coherently representing, correlating, and regulating specific task performance to the structure of the external world.

> We use words to learn about relations among objects and activities and to remember them, to instruct others or to receive instructions from them, to influence in one way or another.... Further, for the sequences of words to be useful, they must refer to real or possible sequences of events.... Thus, in some way the meaningfulness of language depends not only on the grammatical order and on a workable way of associating words with collections of objects, qualities, and so on; it also depends on the structure of the world around us. (Pierce, 1980, p. 122)

Some word usage inhibits the ability to produce a reliably similar meaning about the world among language users for a given statement. We live in a relative world with many words of variable fidelity, variable opinions, and variable predictability from a systematic perspective. *Honesty*, for example, is a localized affective word associated with the idea of always telling the absolute truth and never lying. Ironically, a person can live to the ripe old age of 79 without ever "telling a lie," but if they tell even one untruth the morning before they die, then they die an "absolute liar." How can this be? In a dynamic world with processes of constant change, words with global reliability offer an effective measure of performance. For example, reliability accounts for a person's consistent performance over time by looking at the statistical history of correlations as the difference between predicted and realized outcomes—what a person says they will do versus what they actually do.

Like honesty, *trust* is an affective word that infers the absolute idea that you either trust or do not trust someone. It is a limbic-related word associated with oxytocin and social affiliation and intimate bonding. Trust implies positive affective expectations for a broad category of ill-defined behaviors that will lead to what we perceive as favorable outcomes on our behalf. It is especially ambiguous if we hardly know someone and do not specify context or define a person's behavior as to how we interpreted our expectations in the way of trust. Perhaps a well intentioned word, but trust has somewhat dubious usefulness when we consider the flawed and fallible state in which we humans reside. Consider these examples:

- "Trust me."
- "I trust you will do the right thing."
- "Why did I trust them?"
- "I don't know what happened to the trust I had."
- "I thought I could trust them."

These are the pertinent questions:

- Do we carefully choose the specific words we use?
- Do we prefer words that carry more information?
- Do we prefer words that carry effective information?
- Do we prefer words that carry affective information?
- Would we *choose* words that carry more information?
- Would we *choose* words that carry analytical information?
- Would we *choose* words that carry emotional information?
- Do we really care?

The good news is that we can carefully choose our words, if we care to do so.

Affective words correlate to what we *believe* and *feel* rather than to what we know and have supporting evidence of. Examples of affective words include dichotomous prescriptives and imperatives such as *should* and *must*, as well as emotionally charged local words such as *honor, dishonor, respect, disrespect, honest, dishonest, faith, faithless, truth, untruth, right,* and *wrong.* With localized language, affective words prevail, typically to the exclusion of effective words found in global language: local to global gradations from left to right (Table 4).

Localized words tend to be affective and are frequently used interchangeably with *thinking, believing,* or *feeling*:

- "I *feel* like it is true, and I know it will work."
- "I *think* you know how I feel about that issue."
- "I *believe* I can do it. It just feels right."

Systematic words enable expression of behavior as effective output. For instance, reliability offers contextual evaluation of effective performance over time as a history of correlations that yields more consistent prediction of current and future behavior:

- "I make a deliberate effort to perform my work as reliably as possible."
- "She has a history of punctuality that demonstrates generally reliable behavior."
- "I have known him several years, and he reliably lives up to his agreements."

Affective words tend to implicitly assume expectations in a nebulous manner with minimal explicit information and variable ambiguous referencing, such as "We deserve what the world owes to us." *Deserve* dubiously assumes that there is some kind

Table 4. Comparison of Affective Localized Words Extended to Effective Global Words

Affective Localized Words	*Effective Global Words*
Focus on localized affective input from environment for maximizing reward status	Focus on systematically tuning effective output product for maximizing adaptation to change
Error correction by assigning input sensitivity to stimuli and amplifying output gain to direct actions	Error correction by adapting output to external change and modulating input sensitivity
Attempt to make corrections by avoiding input of negative affect from painful outcomes	Attempt to correct input modulation by effective feedback from output and reliable evaluation of outcomes
Attempt to increase pleasurable input and positive affect from outcomes	Attempt to increase effective feedback and predictions that enhance rewarding outcomes
Hold environment responsible and accountable for disrupting status quo with perturbations and change	Hold self responsible and accountable for adapting to perturbations and change and for minimizing status quo
Assume external causality and blame for suffering from perturbations	Assume self-responsibility for adapting to perturbations
Self-centered	World-centered
Frugal and automatic	Expensive and deliberate
Hierarchal vertical regulation by power for problem resolution	Horizontal regulation by goodness of fit to resolve problems

of an unseen scorekeeper. Then when a certain but unknown score is accumulated, we believe we should get what we deserve. But we might suspect that the amygdala is the culprit behind the scoring ruse. Of course, a problem arises when arguments break out as to whether we "absolutely truly" did or did not deserve the particular outcome:

- "I must get my *needs* met. I *deserve* better."
- "I *know* that tomorrow and beyond, I will get what I want."
- "Maybe I got unlucky that time, but it *won't* happen again."
- "After I've lost five times in a row, it *must* be my turn to win."
- "I *feel* lucky. I *just feel certain* this one is going to be the big one. I *deserve* it."

Table 5 contains a list of local words and definitions. Defining the words is difficult without using another local affective word in the list (e.g., *honest, trust, pride, honor, respect*), or when answering the questions in the right-hand column. Notice how the words generally relate or appeal to social behavioral regulation or emotional value, and how they tend to refer back to themselves in a difficult-to-escape locally referenced loop. To escape the local loop, try to avoid absolutes and black-and-white command words that automatically converge to absolute thinking (e.g., *should, must, supposed to, need to, got to, have to, ought to, always, never, certain, sure*). Notice that externally referencing information in statistical terms of possibilities and probabilities enables a logical escape from the local loop.

It is interesting to observe how easy it is for us to condemn the deserts of others and turn around and claim that we are the ones really deserving of the rewards. It's as if we apply altogether different rules, or, depending on the circumstances, as if we apply the same rule differentially to ourselves and to others. This somewhat deceptive, self-serving tactic makes it fairly easy to direct positives in our own direction and deflect negatives in the direction of others. The absolute nature of affective words tends to induce variability, which increases ambiguity when we are attempting to explain regulation, relationships, and processes of continuous change in nature.

Effective words tend to offer relatively invariant meaning that facilitates consistent relational translation and enhances the regulation of effective output to the domain. We see a general global-to-local difference such as with reliability that can replace trust and honesty, but not vice versa. We see invariance in scientific word meanings that generalize and hence are useful across cultural boundaries. Generalized properties enable

- Relatively invariant translation in time and space
- Reliable support for regulating operational processes and relations across the domain
- Ecological specialization and adaptation to conditions at different localities

Table 5. Localized Affective Semantics

Local Words and Their Definitions	*Questions to Ask in the Search for Effective Words*
Trust: reliance on the *integrity*, strength, ability, *surety*, etc.; to have *confidence* in	Do you trust people? Does *reliable* carry more information?
Deserves: merits, qualifies, or claims due to actions or qualities; *worthy* of *reward*	Do you get what you deserve? Does *outcome* convey more information?
Truth: *honest*; *ideal* or fundamental that corresponds with "reality"	Do you tell the truth? Does *opinion* carry more information?
Honest: *honorable* in intention and action; *respectable*, *upright*, *fair*, or reputable	Are you honest? Does *reliable* carry more information?
Proud: feeling pleasure, *respect*, *honored*; inordinate belief in *dignity* and *superiority*	Where does pride come from? Does *pleased* carry more information?
Integrity: soundness of *moral* character; *honesty*; *sound*, *unimpaired*, *uncompromised* adherence to *morals* or *ethics*	Do you have integrity? Does *imperfect* (*Homo sapiens*) carry more information?
Confidence: *faith* in oneself; *belief* in the *powers* and *trust* of a person or thing; *certitude*, *assurance*; *sureness*	Do you have confidence in yourself? Do think *reliable* conveys more information about behavior?
Belief: *confidence* or *faith* in the *true* existence or *value* of something that is not susceptible to evidence	Do you know where beliefs come from? Does *knowledge* carry more information?
Esteem: favor or high regard; respect or admiration; certain *value*; favorable *judgment*	How do you define and rate esteem? Does the class *equivalence* carry more information?
Right: in accord with what is *good*; *proper* or *just*	How do you know what's right? Does *likely* carry more information?
Worthy: *good* or *justifiable*; having *value*; excellence of character or quality *commanding esteem*	How do you know if you are worthy? Does *equilateral* value as *Homo sapiens* carry more information?
Just: *exactly* or precisely; *rightly deserved*; *justly or fairly* guided; in keeping with *truth*	Do you think you are just? Do you think *equilateral* treatment carries more information?
Faith: *confidence* or *trust* in a person or thing; *belief* that is not based on *proof*; obligation	Do you prefer faith? Does plausible and reliable *evidence* carry more information?

Language constructions that confer greater semantic fidelity likely facilitate reliable definitions and descriptions of relations among actions and outcomes across a global system. Informative words, such as *plausibility*, *reliability*, *abilities*, *possibilities*, and *probabilities*, facilitate systematic knowledge that coherently describes interactions and relations in nature.

When we deliberately choose wording that establishes and reveals our external referent with time-dependent context in the here and now—wording that emphasizes the underlying general properties of our statements in a forward direction—we likely produce a similar meaning in a given context for listeners across different localities. We can improve the possibility of agreement by applying a general measure for plausibly referencing systematic information and correcting errors. For instance, the word *deserves* implies that one absolutely deserves or not. But *deserves* refers to the outcome of some event or events based on ambiguous assumptions typically exclusive of statistical methods for prediction. In the absolute case, we all qualify as deserving or not, depending on the perspective of the person judging and awarding the fictitious deserts. In the global case, we belong to the static class *Homo sapiens*, as generally flawed and fallible, imperfect human beings, and we get what we get.

Behavior is dynamic, and performance varies among individuals and tasks and across fields of endeavor. Encapsulating dynamic behavior in a static class and then extrapolating that class to arbitrarily label a person's worth in terms of what they deserve or not according to their behavior distorts the meaning of the globally ordered static class. Localized words that affectively classify the status of a static class, such as *Homo sapiens*, according to their merits of a particular dynamic behavior confound the distinction between static and dynamic. We may like or dislike a particular action, but that hardly qualifies a categorical measure of deserving based on an emotional evaluation.

We can improve our ability as a species to share orderly meaning, rather than automatically accepting an essentially unexamined, culturally biased affective lexicon. We can consider deliberately implementing a word system that increases the probability of producing as similar a contextual meaning as possible across the users as found in global language constructions. A global linguistic system focuses on word usage calibrated for general sensitivity to systematic processes and relationships that facilitates logical consistency and meaningful translations across the domain. Less variance in the usable meaning of our language and words translates to greater clarity, fidelity, comprehension, reliability, and predictability.

> Prediction lies at the heart of language comprehension. When processing language, a comprehender's task is to predict what the language means. (Barsalou, 2009)

The systematic meaning of the words with which we think and communicate defines the operational semantic space of effective language. We use words to convey information. Plausible words with greater information fidelity support reliable reasoning about how the world works. Effective words correspond and map to orderly information. Semantic effectiveness facilitates orderly thought corresponding with the ordering of the laws we construct to represent knowledge about nature—ordering that invariantly points toward the future. Language that increases the fidelity of information helps to maximize the accumulation of reliably informed knowledge and facilitates the effective classification, identification, separation, and integration of information in a complex system. Linguistic fidelity helps us to sort information among the known, unknown, and imaginary.

Extended language provides generally referenced semantics, clarifies domain representation and classification, and reliably informs behavioral regulation and the orderly mapping of events in relation to time, space, and value according to physical laws.

Extended language and semantics enable systematic classification and constructing, representing, updating, and mapping information and knowledge in correspondence with relationships in nature. Constrained language with absolute semantics, information, beliefs, and logic tends to produce deceptively misleading, idealized maps that fail to account for probability and the open-ended laws of physics.

> First, a physical organization, to be in close correspondence with thought (as my brain is with my thought) must be a very well-ordered organization, and that means that the events that happen within it must obey strict physical laws, at least to a very high degree of accuracy. Secondly, the physical impressions made upon that physically well-organized system by other bodies from outside, obviously correspond to the perception and experience of the corresponding thought, forming its material, as I have called it. Therefore, the physical interactions between our system and others must, as a rule, themselves possess a certain degree of physical orderliness, that is to say, they too must obey strict physical laws to a certain degree of accuracy. (Schrödinger, 2006, pp. 9–10)

Numerous studies (Gygax et al, 2008; Adrian et al, 2007; Kelly & Westerhoff, 2010) show that changing a word can change

- The meaning of events
- What and how we think
- How we valuate problems, regulate behavior, and adapt
- The effectiveness of solutions we come up with

It seems to follow that improving semantic fidelity in the here and now can facilitate reliably informed systematic thinking and enhance the effectiveness of solutions. We apply word meaning to the world in real time, so choosing words that correlate the one-way nature of time to external events can help us communicate sensible information and knowledge and make more reliable predictions about the possible outcomes of our behavior.

Even though we tend to think that words and their meanings are defined by how they are used, we learned most of our words implicitly and are unlikely to have heard formalized definitions for many of the words we routinely use. If we make our word definitions explicit, we can better evaluate fidelity and how well the definitions correspond with our deliberate usage in the real world. With effective language and semantics explicitly defined and referenced to nature, we reap the benefits of systematically correlated meaning and reasoning. We improve our understanding that

- We can learn from the past even though we realize it does not absolutely predict the future
- We can use information from the past to understand regularities of how the world works

- Helps us to make more reasonable plans and expectations for the future
- Helps us to regulate adaptive behavior and adjust when things don't go according to our plans and expectations
- Allows us to make adaptive improvements by thinking in terms of possibilities
- Encourages us to expect and accept instabilities and change
- Encourages us to flexibly adapt our behavior to those instabilities and changes
- We can accept the past but cannot change it
- We can cultivate a preference for possibilities rather than pretending that future events are factual certitudes
- We can accept that our future is under dynamic construction rather than being static and predetermined

The language and semantic meanings we use profoundly influences how we regulate our thinking, feeling, and behaving. Recognizing the potential for unreliable meanings and examining our usage for consistency give us insights for linguistic improvements to help us live more cooperatively in the present. We can improve our linguistic consistency by learning to say what we mean and mean what we say. Many of the heuristics we grow up with have no scientific basis, yet we continue to share adages as "truths," such as "Starve a fever and feed a cold," "You can't teach an old dog new tricks," and "Curiosity killed the cat." Reviewing these beliefs from a more informed position, we discover that

- Most nutritional regimens, short of starvation, generally support the immune system without relation to the symptoms we experience
- Dogs can exhibit brain plasticity even into old age if genetically inclined and provided with appropriate incentives
- As humans, we can incur practical benefits from curiosity that encourages living and learning in the present

Local "truths" vary from individual to individual and culture to culture. General word usage exposes the responsibility and accountability of individuals for the effectiveness of their own speech behavior, which is often in large part a hand-me-down product of their local cultural belief system. But no laws have been found in nature that state you "must" think, feel, or behave in a predetermined way. Each of our actions results from choices, including the option of choosing language for boosting effective thinking.

Word meaning coordinates information signaling by influencing the sensitivity, tuning, and variance of external input–output correspondence with our internal processing of representations, perceptions, and predictions. We can select from a large lexicon of words with a variety of specific meanings for shifting our speech to effectively represent reliable communication. This large choice of words gives us the option of deliberately transitioning our vocabulary from semantic primitives to higher-order abstractions that coherently correspond with our general descriptions of experience and observations of nature. Deliberate word choice allows us to reliably enhance our verbal consistency. We can acknowledge the constraints that the past and the future have on our intended meaning. For example, we can say:

- "I know tomorrow will be a better day."
- "I deserve it."
- "I expect a perfect day."

Or we can say:

- "Who knows what tomorrow will bring."
- "I will do my best to make tomorrow go as well as possible."
- "I will try to make adjustments for the unexpected."

The certitudes of the former statements fail to distinguish the disparity between the past and future, and they delegate affective responsibility to the environment. The latter statements express individual responsibility for adapting behavior to make the best of things. Language is the only tool we have available to help us understand the world and time's relevant influence on the effectiveness of our thought processing. Time directly relates to change and whether the classifications we construct correspond to the ordering of nature toward the future. We might be tempted to think that the issue of one-way time concerns only those interested in physics or philosophy, but we can take note that forward-directed time plays a central role in our everyday lives. Time influences how we inform the regulation of our thoughts, emotions, behavior, and predictions. Without effective language we are likely to remain systematically blind and unlikely to understand the relevance of the higher-ordered meaning of the nature of time embedded in our day-to-day thinking:

- Why the past is static and the future is probabilistic
- Why the past and future are not interchangeable
- Why living in the present depends on separating past and future
- How absolute language primitives amplify affective reactivity and constrain effective thought processing
- How absolute language primitives impose *polarizing* limbic control
- How absolute language primitives implicitly trap us in the past or mythically transport us to an infinite future
- How absolute language primitives exclude statistical predictions
- How nonabsolute language enhances both affective modulation and effective thought processing
- How nonabsolute language facilitates analytical cognition and effective behavioral regulation

- How the explicit application of nonabsolute language facilitates living effectively in the present
- How nonabsolute language enables statistical predictions
- How time plays an instrumental role in our lives whether acknowledged or unacknowledged

We proceed next to a novel view of how operational language constructions influence thought, feedback, flexible tuning and systematic error correction, and regulating behavioral adaptation to a changing world.

CHAPTER 18

Operational Language Constructions

OUR LANGUAGE CONSTRUCTIONS INFLUENCE how we interact with the world, how we operate by regulating our behavior. We use language to describe the regularities we infer or observe about our experiences with that world. A greater understanding of these experiences can enhance effective regulation of our adaptive actions. True and false discrete language constructions leave little room for understanding the breadth of relationships in nature and the importance of choice among alternative possibilities. We understand the relationship between changing environments and the statistical distribution of information, as demonstrated by the logical changes going from absolute to extended language, which enhance informed predictions: "It will rain tomorrow. It will either rain tomorrow or not. It will possibly rain tomorrow. It will probably rain tomorrow."

Physical laws reflect our knowledge about regularities, processes, and relations in nature such as attraction and repulsion forces in magnetic fields, acceleration in gravitational fields, and the flow of heat from hot to cold. We can make predictions by taking into account interactions among many different influences that result in regularly occurring relationships such as lunar periodicity and tidal flows. But considering the inherent instabilities found in complex systems, we can only make statistical predictions according to the extent of our systematic information and knowledge. We can apply this world knowledge to help understand how our brain regulates cognition via information processing to make statistical predictions. Understanding information distributions, statistical regularities, and physical laws in nature informs the effective regulation of adaptation.

Regularity implies regulation. Regularity and regulation figure in many types of modular biological organization, and modularity represents an evolutionary innovation for cooperative adaptation that tends to confer robustness and phenotypic plasticity (von Dassow & Meir, 2004, pp. 256–258; West-Eberhard, 2003, p. 58). Modularity allows individual parts to change and specialize independent of other parts, enabling regulatory switches and discrete modular controls (Carroll, 2005, pp. 194–195). Modules increase regulatory processing efficiency, performance, and capacity; decrease scaling cost; and involve fewer and shorter connections. Modules can also integrate sparsely distributed representations and enhance parallel processing and generalization capabilities, which reduces redundancies and metabolic and computational cost (Murre, 1992, p. 7; Laughlin, 2004, p. 187; Laughlin, et al, 1998; Laughlin, 2001).

Reliable internal regulation correlates interactions with external regularities similar to modular control components seen in gene regulation (Ellner & Guckenheimer, 2006, p. 109) and extends across development in organisms from cells, organs, organ systems, and even to human executive functioning and language. In complex environments, the higher-order properties of language and semantic constructions operate as abstract modular control components by providing a general functional substrate for processing, executing, and influencing the coordination of top-down behavioral regulation. Functional modularity offers important operational advantages to large complex systems that rely on action-related regulatory feedback. Living organisms modulate environmental feedback in response to the consequences of behavior (Swanson, 2003, p. 91), which humans attempt to understand and express with linguistic constructions.

Orderly Relations and a Not-So-Random World

Where do we start to look for operational features in nature that coherently correspond to language constructions, functional brain modules, cognition, and behavioral regulation? What is the optimal perspective and level of abstraction from which to explore these questions? Language constructions can represent internal categories of object relationships and change that correspond to external structure and experience with object positions and system states. For example, a human being is animate, a rock is inanimate, and an external location is a place with positions, actions, and state values that change.

Recall that language enables generalized compression that decreases the number of variables and simplifies the representation of the problem and solutions in a changing, complex environment. External changes that take place represent information about change of position and change of state with a modular-like functional mapping to internal brain pathways for *what*, *where*, and *value* (medial temporal lobe, parietal lobe, and limbic–orbitofrontal cortex, respectively):

- The *what* pathway gives us information as to whether an object represents a person or a rock.
- The *affective value* pathway contributes to *what* associations and memory of an object's valence as friend or foe, and computes the expected reward or penalty.
- The *where* pathway gives us information about the positional location of the object.

We can see that our language constructions overlap with the representation of objects, position, action, value, and change. In other words, our internal world that consists of our brain, language, and modular cognition generalizes to relations in the external world. We operate from our own position and internal state and can regulate behavioral responses according to external changes of position and state. For example, we go into a shopping mall to find a restaurant; we look at the map as we enter and see a "you are here" arrow and the location of a restaurant we like. We know our goal and contemplate our reward, and now we know the position of our goal, so off we go. We could get by without language and synthetic maps, but we would likely be relying on our sense of smell, internal maps and memory, and odor gradients or visual guidance to find food. A language code that maps to changes in the environment can simplify living with the uncertainties of rewards and penalties in a continuously changing, complex world.

But how can we make sense of our language code if we live in a world of consistent change? Where would we start? Looking back in time, and even today, we find a variety of paradoxical beliefs about the world—in some views, nature represents a totally chaotic, random, disordered, and unpredictable world; in other views, we live in a perfectly stable world that we can never achieve because it transcends understanding. More current views incorporate evidence and models from nonequilibrium physics, complexity theory, and evolution that indicate the world is generally much more ordered than previously thought. Is it possible that the world is indeed more ordered and less random than we suspected? But how can we find order without defining the order we are looking for? "Oh, don't worry, we'll know it when we see it" hardly equates to defining and elaborating the problem space.

> We are predisposed to see order, pattern, and meaning in the world, and we find randomness, chaos, and meaninglessness unsatisfying. Human nature abhors a lack of predictability and the absence of meaning. As a consequence, we tend to "see" order where there is none, and we spot meaningful patterns where only vagaries of chance are operating. (Gilovich, 1991, p. 9)

We can define nonrandom statistically as a measure of predictability. Correlating order in a system with probabilities among persistently recurring events makes a distinction between random and nonrandom events. We can define random as disorder in a system with uncorrelated discrete instances and coincidences. Understanding the natural ordering and statistical distribution of information contributes to knowledge about the global structure of the environment that enables systematically referenced predictions, error corrections, and behavioral modulation. But how can statistics be of any help? Some think of statistical descriptions as chance related to gambling. Others see chance as the simple odds of randomly choosing one variable out of a known quantity of a randomly distributed mixture of two variables such as red and black balls in an urn. In a complex system, *chance* is more than gambling or pulling colored balls out of an urn; it represents an orderly range of nonrandom alternatives expressed as possibilities and probabilities for resolving problems in a changing world. In other words, simple probabilities, such as in random or static sampling, and mythical unbiased priors hardly resemble probabilities in continuously changing complex biological systems. These systems carry their DNA history on their back (Prigogine, 1995) and take a history of correlations of regularities into account. The operational structure reflects probability proper and defines the organization and nonrandom ordering of the dynamic system according to forward-directed time, irreversible processes, and the nonrandom modulation of instabilities seen in living biological organisms.

We can think of a household where things are pristine, immaculate, and regulated in perfect order with signs all around that say "Breakable, do not touch," which implies rigidity. Think of the same household with things strewn about in a disheveled manner across the rooms not according to any logical arrangement where things seem disorganized and chaotic with a sign that says "Watch your step and enter at your own risk." Alternatively, we can think of order as a relatively tidy household where things change but are stored and regulated, logically organized, and arranged in a useful way with a sign that says "After use, please put back in place in a timely manner when possible." The first is rigidly ordered and regulated, the second chaotic and disordered, and the third flexibly regulated and usefully ordered. We can think of the creatures that live there as regulating adaptive behavior and adjusting to change.

We live in a changing world of possibilities, a fluid world where life forms exist by actively modulating instabilities, in between the extremes of rigidity and chaos, order and disorder, with a dynamic mixture of randomness and nonrandomness. Distinguishing between them is the special purpose of scientific methodology. How we view nonrandom patterns operationally in terms of static or dynamic perspectives influences what we see. Obviously, static patterns are quite different than dynamic patterns, and both are evident in our speech patterns. A static perspective captures patterns as still pictures that fail to convey the extent of operational information about dynamic relations as the system changes and evolves over time. We likely will miss or misinterpret dynamic information patterns that rely on detecting motion, which equates to change.

For example, not long after DNA was identified as the source of genetic material, noncoding regions were discovered that did not seem to have any functional purpose. The genes were first represented statically as connected by stick figures and attempts to turn the stick figures into dynamic representations created what looked like hairballs, unruly balls of yarn, or the hair of a mad scientist. The noncoding regions, which seemed random and useless, were labeled as junk DNA. They were later found to be nonrandom and very important for gene regulation (Mitchell, 2009, pp. 278–280). Recent reports by a consortium of scientists working on a decade-long research project

(ENCODE) overwhelmingly confirm the dynamic regulatory nature of what was previously called junk DNA, which apparently reflected our prior static perspective. The confirmation consists of the publication of 30 research papers, including 6 in *Nature* and additional papers published in *Science*, as well as other journals (Pennisi, 2012).

When we take a dynamic perspective, a system's functional properties start coming into view, suggesting that considering macroscopic dynamic perspectives can help us to effectively distinguish and sort nonrandom from random information. Nonrandom systematic information contributes to understanding and accumulating reliable world knowledge. It further suggests we can likely look forward to a decrease in widespread assumptions of randomness in our world by matching our perspective to the real-world dynamics of the system under study. We can take a dynamic rather than static approach to describing biological systems, including brains and behavioral regulation, and the language code with which we describe them. Broadening our dynamic perspective brings random and nonrandom events and operational regulatory processes to light and into focus and extends our systematic knowledge. We can reevaluate faulty assumptions, clarify static ideas and misperceptions, and acknowledge the dynamic ordering of nature toward the future. The more we learn and the better we clarify our thinking about how nature operates, the less random it appears.

The work of Henri Poincaré and Ilya Prigogine has revealed some basic flaws in how we think about the world (Prigogine, 1997, pp. 21, 29, 55, 184, 187). We seem to have inherited a propensity for viewing the world as random and time-symmetrical using an idealized referent of stability and perfection. We tend to operate on the implicit assumption that what we sense in the world is the result of random events that we can only hope to understand. The irony is that even though we act as if we believe in a world full of random events, we still speak about the world deterministically in terms of certitudes. How do we explain this discrepancy?

What appears at first glance as a paradox disappears when we extend our perspective from perfection and idealized stability to see ourselves as imperfect, complex organisms in a dynamic world of instabilities and many possibilities. Knowledge of how our world works is made possible by language that fits that world. Systematic knowledge decreases the perception of randomness that we tend to infer from the instabilities, changes, and disturbances we observe going on around us. Understanding the dynamic structure of nature broadens our perspective to include systematic ordering with irreversible processes and nonequilibrium constraints imposed by the forward-directed nature of time that make possible the evolution of complexity and life.

Perturbations that lead to bifurcations and new structure in complex systems are sometimes described as random occurrences. We do not necessarily know exactly when the disturbances will occur or what the exact input will consist of at the time of the occurrence. But in nature, outside of laboratory limitations, we do have knowledge of the systematic ordering of relationships among processes, structural and functional constraints, and dynamic physical laws. Even though we cannot predetermine the absolute outcome, we seem to have enough nonrandom information about the ordering of regularities in the environment to at least make roughly calculated statistical predictions.

For example, in the case of genetics and DNA, we can make predictions that bifurcations will probably occur and create a new, more complex structure on existing scaffolding built from the previous solutions. We would expect that these bifurcations would display some relatively predictable amount of nonrandomness. With our knowledge of nature we have improved our abilities to predict dysfunctional genetic changes and phenotypic abnormalities. Also, we have improved our ability to predict risks associated with weather patterns and earthquakes that in ancient time were thought to occur as a result of either caprice or demons, which many still believe today. Granted, our predictions are relatively short term and inexact, but at least to some extent, the predictions convey nonrandomness expressible as probabilities.

Unfortunately there have been examples of scientists being sued because of allegations that they "should" have provided more precise advance warnings about the occurrence of natural disasters. But because of systematic biases and human error, sometimes the risk of being wrong keeps us from adequately accounting for the risks that others face. Of course, we prefer to have advance warnings about the possibility of tragic disasters happening and decrease suffering, but it seems that the most egregious error more than likely lies in the plaintiffs' time-symmetrical thinking and paucity of knowledge about science and statistics in general and the reality of instabilities and natural disturbances in particular.

An organism's DNA represents genotypic knowledge accumulated as the history correlating the organism's survival and adaptation to disturbances encountered within the external system. As the complexity of nervous systems has increased to that in the brains of humans with a greatly expanded neocortex and extended memory capacity, we see an unparalleled capacity for plasticity and learning associated with language. We can think of this enhanced cognitive ability in terms of extended language and higher-order abstractions that enable world knowledge and better-informed predictions, choices, and behavioral modulation. The expanded memory capacity also made language and time important partners in learning by helping to keep the past and future separate and in clear perspective.

As a complex system, the systematic knowledge we have acquired decreases the dimensions of the search space for resolving problems. Through learning, we can continue to accumulate higher-ordered knowledge that correlates with the nonrandom regularities we see in nature. To an extent similar to DNA, the linguistic knowledge we accumulate enables us to make better-educated, nonrandom decisions when faced with disturbances. In turn, we incur learning that can lead to better-informed choices and further increases in knowledge. It seems that in complex biological entities, the learning made possible by extended language and plasticity adds an element of uniqueness that enables complex organisms, such as us, to acquire time-dependent, nonrandom systematic knowledge. We are thus unique in the sense that we have the unprecedented ability to understand our imperfect nature and our humble place within the complex, not-so-random world in which we live.

In a complex system we account for nonrandom patterns of interacting variables probabilistically as information distributions statistically correlated to regularities and systematic knowledge. Yet randomization and random variables are often considered the hallmark of unbiased, well-controlled experiments. The results of such studies may suffice in locally constrained idealized or static domains, where information and events are arbitrarily defined, distributed, and described as chance occurrences or the probability of a random event, such as drawing colored balls out of an urn (Jaynes, 2007, pp. 60–62). In that case, we can ignore searching for more information and systematic knowledge, but we trap ourselves in what is familiar that we consider as safe. In the case of predicting natural disasters, ignoring nonrandom information about nature can lead to the accumulation of systematic errors that trap us in dire consequences. Then we escape only briefly after disaster strikes, scrambling and starting over again to rebuild our future. Unfortunately, we continue to ignore that we are attempting to randomly reconstruct the future with language that perpetuates the same burden of eternally recurring systematic errors.

In the global environment, we see widespread evidence of nonrandom correlations with regularities, which tend to be the general rule rather than the exception. A perspective of randomness can create the illusion of unbiased experiments, predictions, and discovery at a localized level. However, a reliably referenced bias in the forward direction of time enables probabilistic discovery and correlation of regularities representing the nonrandom ordering of systematic information. Reliable referencing enables the discovery of nonrandom information patterns that correlate systematic regularities with systematic knowledge and statistical predictions of action and change in the forward direction of an evolving system.

Events in nature can be randomly and nonrandomly distributed, with reversible idealizations being the exception and irreversible processes being the time-related general rule in evolving biological systems. The time-related ordering of nonrandom variables and events correlates information about operational regularities with natural physical laws and global knowledge. It is possible that we habitually attribute randomness to events, information, and beliefs because of our culturally biased language habits that constrain our dynamic perspective by implicitly overlooking and ignoring macroscopic nonrandom patterns and one-way time. We can perhaps overcome this problem of static perspective if we stop to consider the value of putting forth the effort to improve our systematic understanding of the structure of nature. Our choice of language constructions plays a fundamental role in coding our probable successfulness for enlightening our global perspective.

The extent to which our word choices vary from physical referencing provides a measure of ambiguity and randomness. Our failure to attend to the nonrandom ordering of our word definitions and usages, and the tendency to assume randomness instead, compromises our accumulation of systematic information and knowledge. Randomness implies a lack of information or arbitrary variance. Our assumptions of randomness and the resultant ambiguity can leave us uninformed, confused, and prone to imaginary speculation about the chaotic disorder and caprice of nature. Variance in semantic meaning tends to inversely correlate to the fidelity of a word's referenced definition and is consistent with usage that creates ambiguity, because definitions have a unique correspondence with their referent.

Even if we assume that words were at some point arbitrarily chosen, words with arbitrary definitions and usage increase ambiguity for referencing their meaning. Globally useful words are defined by their systematically informed meaning and measured by their deviation from natural referencing and invariant forward-directed time. When we try to measure nonrandomness as deviations from an arbitrary referent, we add to our confusion. In static domains with absolute referencing we can arbitrarily make dubious claims of zero or infinite amounts of randomness and nonrandomness. The confusion goes away when we discover that absolutes represent an imaginary product of static idealized worlds. Absolute words are practically useless when we are trying to grasp general information, which immediately vanishes upon their introduction. It is as if absolutes relegate generalization to an invisible status. General and absolute words seem to mix and separate somewhat like a nonrandom distribution of oil and water.

Complete randomness can be defined as capricious events occurring without reason or pattern. When we assume randomness, we infer the lack of a recognizable pattern from our localized perspective, even though there may be global patterns right under our nose that we do not detect because of our systematic blindness. Since randomness lacks an understandable pattern, it represents irregularity and ambiguity without a suitable natural referent for defining and measuring information. When observing general patterns, we inaccurately ignore nonrandomness, make assumptions of randomness, and apply arbitrarily defined words to describe systematic events. The resulting confusion comes from ambiguous fluctuations of meaning, inaccuracies, and variance among arbitrary local referents. Furthermore, ambiguity related to local assumptions and systematic ignorance distorts our perception and limits our ability to distinguish reliable information.

Constrained words, ambiguous assumptions, and distorted perspectives tend to ascribe randomness and caprice to dynamic macroscopic phenomena, which results in undetected systematic information patterns that we otherwise understand to have a statistical nature. The bottom line, in the absence of naturally referenced extended language, is that absolutes tend to go to infinity and obscure detection of nonrandom information patterns that are statistically distributed in the macroscopic space. In effect, when we develop the habit of a static perspective, we are desensitized to and tend to ignore the importance of change, or we erroneously claim that change represents a malicious artifact of a noisy and unruly world.

When we assume random fluctuations of events, our measurements become unwieldy because of the disparity among the arbitrary referencing that blurs the distinction between regularities and irregularities and effective regulation. Thus we induce a tendency to misperceive relationships among words, patterns, and events that appear to be capriciously connected coincidences or the wrath of mysterious demons. Considering systematic fidelity from a relatively invariant referent creates a larger problem for arbitrarily referenced words, whose definitions and usage tend to fluctuate considerably among localities. The local variation and error deviation in the meaning and usage of words can be measured from a reliable referent, thus exposing any absolute bias that by default penalizes general information and knowledge.

Absolute distortion occurs when words demonstrate contextual variance in different situations and localities across the macroscopic domain. That is, the same context has arbitrarily variable meanings as translated at different localities. General words tend to demonstrate relative invariance under translation across subdomains. For example, scientific language tends to generally exhibit relative invariance across different cultures. Scientific "reliability" generalizes and demonstrates consistent meaning of reliability regardless of the culture, location, or context. Likewise, global time is invariant across cultures irrespective of local beliefs.

In a dynamic domain generalized words enable consistent representation of information and facilitate open-ended thought processing. Consistent representation improves coherent correspondence with a word's descriptive ability. The logical consistency of generalizations across the system facilitates the orderly nonrandom compression of information while introducing as little noise as possible. Absolutes exclude context and probability, and constrain effectiveness by inadvertently increasing systematic ambiguity and the perception of randomness for global information. Lost contextual information, ambiguous assumptions, and discrete logic constrain systematically informed knowledge and the flexible search for adaptive solutions.

Domain-referenced linguistic abstractions with coherent meaning corresponding to physical laws can help to identify adaptive solutions. There may be many possible arbitrary word choices but fewer words that coherently correlate effectiveness in terms of context, adaptive performance, and orderly nonrandom systematic meaning. There are even fewer words whose effective usage logically corresponds to reliable event-related information, information processing, and knowledge within the context of the domain dynamics. Localized variation in word meaning tends to create random uncorrelated abstractions that increase systematic error rates and compromise the optimization of adaptive performance. Words with calibrated definitions that correspond to the domain order increase effective performance by decreasing the tendency for making systematic errors related to linguistic ambiguity, ignorance, and systematic blindness.

Random abstractions constrain expression to particular isolated areas of the domain whose details may be irrelevant globally. Nonrandom abstractions strip away superficial localized details so that the level of abstraction generalizes across the system. Arbitrarily constructed and referenced localized word meanings constrain the available dimensions of higher-order abstraction, which, in turn, place limitations on our cognitive performance. Nonrandom systematic abstractions extend higher-order cognitive performance by enhancing the reliability of admissible evidence, metrics, and measurement that enable analytical value assessments and effectively informed global knowledge.

Arbitrary localized abstractions tend to rely on attempting to discretely measure and control the quality of affective input rather than modulate the effective output. Outcomes measured solely by qualitative affective reactions that we tend to blame on the environment can misinform and compromise effective adaptation over the long run. Defining and quantitatively measuring predicted outcomes in terms of individual responsibility for the realized output we produce encourages accountability for regulating adaptive responses as effectively as possible. Consider, for example, "You hurt my feelings, and I will even the score" versus "I did not like your behavior, but I refuse to upset myself and respond with an emotional reaction."

The nature of a reliably correlated history of regularities in nonrandom systematic information and knowledge separates optimizing performance from sufficing. We have the option of extending our language code and our perspective to macroscopic dimensions with enhanced systematic optimization, if we choose to do so. Extended language gives us generalized constructions that broaden our worldview and allow us to operate open-mindedly, our eyes open wide, and encourages efforts to improve our harmonious behavior. When we think from a macroscopic perspective, we are more likely to realize that we do not set the operational rules for this world—we just live here. Even though we are a part of this world, we are only a small part. An understanding of our human limitations and imperfections gives us greater insight to better regulate what we can, to better adapt to what we cannot, and the wisdom to know the difference—in the not-so-random world in which we live.

Top-Down Regulatory Processes

In humans, evolutionary processes seem to have "thoughtfully" overlaid the novel mechanism of semantic representations onto co-evolved brain structures, such that in large part, semantics can operate in parallel with thought processes. Words and concepts bind to sensorimotor information, creating a kind of a sixth sense—specifically, a semantic sense (LeDoux, 2002, p. 203; Loritz, 2002, p. 68). Even though language operates spatially across many brain areas, language can be viewed semantically as a functional module (Chater et al, 2009). Language acts as a functional linguistic module for representing perceived stimuli, values, action, and change as information and knowledge about relationships. Language forms a critical part of our logical regulatory basis for reasoning, decision making, and behaving, especially for communicating.

We can consider regulation as roughly equivalent to modulation and control. Control is commonly perceived in a negative connotation, since we tend to associate it with obedience to authority, which misaligns its meaning in engineering terms of positive and negative feedback. We have already discussed the common pejorative misperception of "negative" feedback. *Regulation* is defined here as a functional mechanism or dynamic method for making internal adjustments to a system's behavior as an adaptive response to change.

A functional perspective allows us to model the optimization of top-down, higher-ordered brain regulatory processes that rely on feedback represented as linguistically coded information. Language encodes regulatory information about values related to the current internal state and external state, objects, and actions. Language and semantic accuracy enhance reliable feedback and generally facilitate the optimization of more effective behavioral regulation. Language influences perceived value. Value, in turn, biases input–output mapping for drawing inferences about external influences on our experiences, including our valuations of the intentions of animate objects. These feedback interactions suggest that regulation itself arose as a by-product of the adaptive interaction within the system, organism, and domain viewed as a whole, shaped by natural selection (Mitchell, 2009, p. 184; Jennings, 1906).

> Human language is remarkable because it is not only a means of communication. It also serves as a means of reflection, during which different lines of action are played through and tested. (Bronowski, 1977, p. 113)

A practical top-down model depends on its structural equivalence with the reflection of the environment. Language provides not only an expanded capacity for reflection but also a second-level capability to literally contemplate where, when, and how the reflection originated. Not only can we think about the world but we can also think knowledgeably about what and how we think. Simply put, we can think about why it is to our advantage to think about fine-tuning our thinking and behavioral regulation by fine-tuning our language.

> Ideally, we want a model that is generated based on the statistics of the environment the animal finds itself in.... [I]f the top-down model is to be of any use, it must reflect the structure of the animal's environment. (Eliasmith & Anderson, 2003, p. 283)

A distinction between upper- and lower-level brain processes is not intended to imply separate entities per se but rather serve to highlight the importance of their interactive relationship. The give and take between upper-level PFC and lower-level PFC regulation consists of interacting top-down and bottom-up operational control system loops that bias input–output computations for goal-directed behavior. The concept of the operational role played by loops in complex systems was inspired by cybernetics, where closed-loop negative feedback was demonstrated to guide a system to its goal and was later influenced by control theory that advanced the concepts of representation and transformation (Eliasmith & Anderson, 2003, pp. 220–221).

A closed-loop control system uses feedback to assess the outcome of its output. The deviation between the expected outcome and the realized outcome represents an error signal that enables real-time output adjustments via feedback to the controller as one of the inputs for the next evaluation. The reliability of the information and processing that led to the prediction biases the interpretation of the expected outcome. The effectiveness of the referencing of error signals biases the ongoing output regulation of outcomes, perceived feedback, and systematic error correction. Thus, a tunable systematic error signal enhances regulatory flexibility for effectively adjusting corrections related to learning and behaving. We can generally say that the more accurate the error signal, the better the error correction.

Operational loops can model the recurrent integration and regulation of information in biological systems as attractor networks with cyclic or periodic attractors, harmonic oscillators, and properties that enable the generalization of representation and the overall dynamics of a system (Eliasmith & Anderson, 2003, pp. 250–258). Operational loops result from the actions of populations of individual neurons that have intrinsic periodicity routinely expressed as relaxation oscillators, which is typical of many neurons in the brain. Periodic properties enable recurrent systematic feedback, tuning, biasing, and the regulation and integration of individual loops.

The lower-level limbic system automatically operates within the overall system with an affective egocentric bias, somewhat passively, like a default closed-loop state controller with intrinsic set points for core variables. The limbic system does exhibit closed-loop tuning capability, but with an affective bias and discrete positive and negative values that enslave cognition and behavior to the local landscape. The misuse of discrete language results in digital-like true and false affective distortion of cognition that biases our perception of possibilities and probabilities in the analog world in which we live. The qualitative nature of the bias automatically resists analytical tuning and defaults to the familiarity of localized language, beliefs, logic, and affective values. A localized affective bias constrains flexible tuning and constricts the range of effective regulation.

When we rely primarily on affective language and discrete limbic regulation, we inadvertently cut ourselves off from systematic information that predisposes us to traps between local minimum and maximum. The affective language we inherit typically comes replete with prefabricated true-or-false value weights that resist change. Affective language supports and reinforces an emotional value bias that influences what we like and dislike, attend and ignore. Affective true-or-false values bias our attention and direct what we are attracted to and repulsed by, what we approach and avoid, what we believe, how we perceive and think, our goals, and how we behave, treat others, and adapt.

We use language to speak, think about, and code our experiences, but even a highly descriptive language may not necessarily produce consistent meanings that convey the observable details of those experiences. We have available to us a method to review and reform our linguistic constructions to reduce ambiguity by selecting words with more coherent systematic definitions and meaning. With extended language and reliable external referencing, the upper-level executive system can operate in phase with the domain dynamics as a systematically informed closed-loop controller. Generally, executive modulation can actively override or inhibit default limbic regulation. Usually, depending on the effectiveness of our language structure and current state, we can deliberately exert higher-order regulation to analytically fine-tune domain information and knowledge, improve decision making and systematic error correction, and enhance predictions and behavioral modulation.

Extended language constructions help to address the analytical issue of applying local learning rules within global constraints by applying systematic thinking and knowledge to inform local behavior—thinking globally and behaving locally. Global knowledge addresses local learning problems by increasing the systematic resolution and sensitivity to local change from a macroscopic perspective. An analytical approach allows us to generate novel learning rules where the context is captured by the top-down information transformation using systematic encoding and decoding, such that local encoding can be decoded systematically.

Systematic transformational processing enables effective evaluation of previous information and knowledge that we can apply to interpreting new information, and editing ambiguous sensory data and context-sensitive reasoning. Combining independent sensory inputs and updating current estimates in light of new information improves our ability to estimate state variables, produce probabilistic inferences and predictions, make analytical error corrections, and fine-tune the system's learning weights (Eliasmith & Anderson, 2003, pp. 275–288, 294, 300).

Extended language calibrated to a reliable external referent provides closed-loop capability with open-ended tuning of systematic error signals that improves our ability to coordinate interactions within the context of the macroscopic system in an orderly manner. When language enables general fine-tuning to the structural dimensions of the domain, we can expect enhanced regulatory processing abilities, decreased systematic error rates, and predictable improvements in effectiveness that approach or approximate global optima.

Language provides an analytical method for deliberately transitioning from implicit passive regulation and an internal coarse-grained default feedback system, on the basis of localized referencing, limbic-generated familiarity, and "feels good" or "feels bad" affective values. Alternatively, we can apply active analytical regulation with explicit referencing that more sensibly correlates meaning to the domain. An explicit referent provides active closed-loop control with the open-ended logical capability to tune signal-to-noise ratios and discriminate domain-dependent sense from imaginary sense. For example, we can overcome the illusion of external control of our feelings and behavior. We transition from "you made me angry" to "I upset myself about your behavior." In this case, affective input is seen as implicit noise that enables imaginary cause, which can be explicitly analyzed and separated from systematic signal. An explicit signal that translates to and focuses on the effectiveness of the realized output enables adaptive responses to the disturbance with no strings attached:

- I am responsible for how I think, feel, and behave.
- I take responsibility and accountability for the loop from myself to the world and back that includes my brain and my thinking, feeling, and behaving.

The transition from implicit and passive to explicit and active facilitates effective closed-loop modulation by accessing an invariant external referent for deliberately correlating, tuning, and calibrating the coordination of adaptive interactions according to the temporal ordering of the domain dynamics. The system learns and evolves by exploring for and exploiting new information as dynamic possibilities for adaptation. Learning systems can optimize tuning with reliable adaptive regulation of feedback that decreases errors and enhances effective learning in response to external change.

Local errors tend to be subjectively measured by their magnitude or how frequently they occur. We easily confuse information distributions describing the magnitude or frequency of errors with a statistical description of our probable state of knowledge about those errors (Jaynes, 2007, pp. 592–593). A deliberate linguistic transition to probability as logic improves regulation by revealing and reducing systematic errors and unreliable credit assignments, while improving risk–reward assessments, predictions, and adaptive learning from the probable expected outcomes of our choices. The accuracy of our conclusions correlates with our own state of knowledge about systematic errors (Jaynes, 2007, p. 338). The probability of making better predictions increases when we improve the quantity, quality, and fidelity of source information. Predictability naturally improves when we use systematic information and probabilistic processing that supports reliable feedback, creative possibilities, and innovative alternative solutions. Systematic information facilitates reliable statistical processing that better distinguishes the dynamic phase transition and noninterchangeable features separating the past and future.

A systematic phase transition allows us to apply open-ended top-down regulation with domain-referenced language that effectively extends our cognitive capabilities from absolute certitudes to relative possibilities. Understanding how well our language constructions match the systematic structure of events around us helps us to understand the importance of effective top-down regulation for optimizing adaptive actions with greater predictability.

Feedback Processing

The nervous system monitors its own output via feedback. The notion is that feedback provides the basis for a comparison of the system's actual output to the expected or desired output that was predicted before behavior, sometimes called an efferent output copy (Jeannerod, 2003, p. 83). The difference between the realized and expected output measures the extent of any error variance and can be used to adaptively tune the effectiveness of the system's real-time output. The point of feedback is to provide opportunities for reducing output errors (Churchland & Sejnowski, 1994, p. 98). For example, on a strictly physical level, feedback can refer to the reflection of a wave back to the source. It can signal the presence of interference, and apparently it is used for that purpose by animals, such as dolphins, that navigate and hunt with echolocation. Reliable feedback plays a critical role in an organism's effective regulation of adaptive behavior.

We use language to evaluate decisions via feedback from our actions. One potential problem we face in such evaluations is that we use the same language to make the decisions that we may later want to evaluate for linguistic effectiveness. Confusion as to which contributed the greater influence on effectiveness results from not knowing if the errors come from the decision itself or from perceptual distortions related to the language. Understanding linguistic influences on perceptions and decisions yields valuable insights for calibrating input sensitivity and tuning output gain and feedback fidelity. Language enables us to make explicit the feedback we receive and to deliberately think logically about how the world works and about how to effectively interact with it. Then we have more knowledge of the state of the world and greater ability to predict the probable outcome of future decisions. Simply put, language that explicitly supports reliable input, processing, output, and feedback enhances effective state regulation.

By enabling information regulation and probabilistic modeling of anticipated events, language enables explicit state realization (Raichle, 2006, p. 13). We obtain feedback from the environment by frequently observing and evaluating the interactive correlation between our thoughts, behavior, and outcomes. Without this feedback, we can easily fall prey to reactive stimulus-bound behaviors (Luria, 1966/1980). Brains with state realization can more deliberately and strategically manage thought processes to achieve more adaptive outcomes (Siegel, 1999, p. 263). Reliable representations, descriptions of experience, and knowledge correlated to the entire system facilitate insight into current state and predicted future states. Language that enables explicit realization of how the world works systematically is typically more useful and limits the amount of assuming, guessing, or imagining about different states of the world.

Global realization and explicit feedback provide an orderly method for actively tuning our cognition and behavior. For example, when we believe that the world is ruled by mystical forces and that we are the hapless mistreated victims of these forces, we feel powerless: "We are hopeless against the evil forces that besiege us with ruthless winds, torrential rain, and fire from the sky." Our only recourse is wishful thinking, complaining about our affective suffering induced by the mistreatment coming from the cruel world, and bemoaning the unfairness of it all. Effective language enables systematic knowledge that extends our perspective and understanding of how we influence the world. In turn, we enhance the probability of understanding, implementing, and regulating more effective adaptation to that world.

Consider this outlook, for example:

> It's nice to be able to understand how the world works. We at least know about dynamic weather patterns and can make predictions and produce short-term warnings about wind and rain. We understand the power of lightning and the ways to decrease the risk of injury from it. We no longer wear metal hats or hold and extend metal rods toward the sky during stormy weather.

Understanding how the world works enables better accountability and responsibility for our behavior and provides more consistent feedback for error detection and effective tuning via error correction. Of course, understanding hardly immunizes us from suffering the results of lightning bolts or natural catastrophes, but it can decrease the probability of making egregious ignorance-related systematic errors.

Complex systems are temporally correlated in global space according to the physical laws of the domain. We rely on language and memory for feedback and for recursively updating prior information to reflect current domain knowledge. Spatiotemporal discrimination allows the parsing of events into information correlated in time and space, as event-related experiences that separate the present and future from past states. We continuously deal with a preponderance of time-related contextual information associated with our day-to-day experiences.

Temporal context influences our perception and evaluation of feedback by how it distinguishes current experiences from past and future events. Recall that past states are for the most part static: "What happened in the past stays in the past." That is, the event happened in a certain context in time and space that cannot be changed. But perception is an internal constructive process that can bias and attribute inaccurate meaning to experiences. And since remembering is a reconstruction of the past, our evolving understanding of past events can change along with our current perception, for better or worse. The extent to which our language constructions about our experiences, past and present, match the temporal dynamics of the outside world will influence the feedback we perceive and the reliability of our memories. In turn, as a recurrent loop, the reliability of our current representations biases the continuous construction of new memories, perceptions, and the effectiveness of ongoing regulation. It is easy to imagine how uncorrected errors within recurrent loops can accumulate with adverse consequences on information processing and memories over time.

The fidelity of internal memory representations and associations directly biases our regulation of attention, sensory perception, signal detection, and feedback. Language biases the weighted values attributed to sensory input and the transformation and modulation of positive and negative value contingencies credited to stimuli. This weighting bias increases the likelihood of under- or overvaluing the perception of stimuli and cues (Kahneman & Frederick, 2002, p. 53). Generally, the more regularly and frequently an organism experiences a particular environmental irritation, the stronger the input gain that is perceived by the organism, which in turn influences the strength of the resulting output. In other words, the more the sensors are excited, the faster the motor runs (Braitenberg, 1984, pp. 3, 6). For example, in an affectively regulated system we can think of this as the power of the input directly overamplifying the power of the output. We see analogous overamplification in genetic networks from bacteria to cancer cells.

Output regulation relies on positive and negative feed-forward and feedback loops that amplify, dampen, activate, and inhibit signals for computing and tuning adaptive responses. Positive feedback loops are sensitive to the availability and reliability of the inhibitory feedback that biases input sensitivity, output gain regulation, and learning. The effectiveness of a system's output, activated in the form of behavior, relies on learning from input–output errors, which plays a crucial role in optimally adapting subsequent behaviors (Shadmehr & Krakauer, 2009, p. 588). The effectiveness of the behavioral regulation in terms of learning and adaptive performance is proportional to the reliability of the feedback. In linguistic terms, generally, the effectiveness of the language proportionally improves the regulation of adaptive performance.

Our language constructions influence our contextual representation of events in time and space. Spatiotemporal fidelity directly influences the amount and effectiveness of the fundamental feedback available for correlating adaptive mapping and the regulation of input–output contingencies with experience. By improving feedback fidelity, we can more coherently tune and integrate past information in correspondence with current information and systematic knowledge. Linguistic coherency provides useful feedback information for reliably learning and effectively regulating action, including communication. Coherent language constructions simply produce more useful information in terms of predictable effectiveness.

Processing, Choices, Decisions, and Outcomes

Our human brain integrates language and cultural constructions into its very fabric. Almost anyone who has raised children notices the gradual development of discourse capacity as children learn the expectations of the speaker and the rules of discourse unique to the ambient culture (Alexander, 2002, p. 164). We learn language as infants, dependent on adults who wish to dictate our behavior, and we automatically incorporate the hierarchal control constructions of imperatives into our vocabulary. As adults, we continue unthinkingly to apply these insistent terms, such as *must*, *should*, *have to*, and *need to*, to actions that more realistically amount to choices and preferences. "Musts" stand in for and obstruct "thinking" about preferences and alternatives, and in the light of irreversible behavior, we see that imperatives tend to overlook or skip over the evaluation and elaboration of the problem space.

Imperatives mandate idiosyncratic goals and constraints as a subjective product of the speaker's authority that varies from person to person, culture to culture, day to day, and minute to minute. We can conclude that we generally use imperatives in the service of our own desires even when that service is carried out implicitly by habitually defaulting to obedience to authority. Imperatives are typically self-serving in some direct or indirect manner.

By considering our actions as based on preference, we can discover our own regulatory responsibility and accountability for the likely outcomes of the actions we choose. Instead of thinking or feeling we must do something, we can state our choices as preferences, what we do and do not prefer. We may still behave at times in ways that we do not typically prefer, but our actions are choices nonetheless. By verbally acknowledging the existence of alternatives and our ability to thoughtfully choose, we can more effectively identify and elaborate on the problem and our reasons for making choices for solutions. We then have more information by which to tune our thought processing and improve the possibilities for predicting, learning from errors, and managing our behavior and outcomes more effectively.

Complex systems increase the possibilities of developing multiple solutions related to choices, regulation, and corrective feedback. This raises the question of the role preference plays in the choice among the different alternative solutions and their predicted outcomes. Once made and implemented, a choice divides the possible future from the determined past. The subsequent evolution of the system depends on this critical irreversible choice (Nicolis & Prigogine, 1989, pp. 60–61, 72). Even though irreversibility is not a universal property, the irreversible process of invariant time leads to creative choices, novel constructions, and increasing complexity.

> Nature involves both *time-reversible* and *time-irreversible* processes. But it is fair to say that irreversible processes are the rule and reversible processes are the exception. Reversible processes correspond to idealizations. (Prigogine, 1997, p. 18 [emphasis in original])

The preferred asymmetry of information, including value information, biases contingency perception and the regulation of decisions, choices, outcomes, and learning. For humans, accurate feedback regulation paves the way for adaptive choice-and-consequence reasoning and decision making (Shadmehr & Wise, 2005, p. 52). The effectiveness of the information and knowledge biases brain processing, which in turn biases choices and behavioral regulation. The ordering of these processes from past to future confers time dependence to events and the evaluation weighting of experiences.

Like open-ended evolutionary processes operating in a changing world with invariant one-way time, complex systems do not know a priori about absolute right and wrong choices. There are no known deterministic mechanisms that exist to imperatively tell them what they are supposed to do, should do, must do, have to do, need to do, or ought to do. The point here is that we have many choices, and applying localized language constructions, including predetermined imperatives, are a part of those choices. We have the option of choosing the constructions we think are the most useful to us, including informatives and preferences. Imperatives carry minimal information, eliminate alternatives, and close the solution space with the execution of the command and are better suited for hierarchal control language. Preferences based on possibilities are generally more effective for informing the problem space and choices among alternative solutions. Recall that possibilities are open-ended and do not close off the solution space. It is up to us to decide the constructions we prefer. Our current language obviously has priority; perhaps we mistake familiarity for reliable knowledge, the *illusion of knowledge*.

We can make a scientific statement that describes the probable outcomes of choices for different possible actions, but imperative directives tend to insist that something *must* be done. A change in how we inform our decision-making process might help us gain insights that lead to better-informed choices, outcomes, and learning. Imperatives tend to reflect the habitual intentional bias from the authority of the speaker, commanding, instructing, and implying rather than informing the problem or the likely outcome. Outcomes are informed at the event or after, but the a priori implication of certainty still represents a lagging indicator that follows an event, informs retrospectively in the rearview mirror, and implies symmetrical time.

"Your act was unwise," I cried, "as you see
by the outcome." He solemnly eyed me:
"When choosing the course of my action," said he,
"I had not the outcome to guide me." (Bierce, 1910, p. 360)

We see that outcomes orderly follow the choice and action rather than vice versa, without prescient knowledge. When we receive information and context, we can first evaluate and elaborate the problem space and then consider the possible pros and cons of our available choices. Whether we are informed or instead coerced when we make a choice, the outcomes may or may not differ. Where the outcomes do differ, we undermine our ability to learn from them when we skip over the problem space and ignore gathering more information, contemplative reasoning, considering context, and the possibilities for alternative solutions. In contrast, encouraging informed analysis of the problem and thinking about context and alternatives before we act can lead to more effective predictions, learning, and understanding in the long run. Generally, encouraging thoughtful systematic reasoning likely offers more consistent results.

Language limited to closed-ended constructions tends to constrain choices and exploration of alternatives and prematurely converge to absolute solutions. Imperatives share many of the limitations of objective fitness functions that constrain elaboration of the problem-and-solution search space. They tacitly limit alternative choices and adaptive flexibility and tend to lead us, unthinking, into deceptive local traps. Absolute words stand out, since despite their absolute exacting nature, they demonstrate inaccuracies that typically vary broadly across individuals and cultural populations.

Prophetically insisting on what absolutely should, must, or needs to happen tacitly presumes that the past and future are time-symmetrical and thus interchangeable and dismisses alternative solutions and choices. Denying the arrow of time constrains the system to one choice predetermined as the only correct solution in the future. When we choose predetermined absolute solutions, not only do we feign clairvoyance for predicting the outcome as if we can see the future but we also typically fail to examine the problem space, provide plausible evidence, and overlook current information at the time of the event. In the future, when we are frozen to a predetermined solution, we dismiss any disagreeable current input as only adding noise and confusion. The ambiguity associated with our lacking insight about the preference bias behind our choice, and the source information and reasoning that lead to that choice and its predicted outcome, constrains systematic error recognition and limits our ability to effectively learn from the experience. Imperatives freeze decision weights in time, sans plausible evidence.

Evolutionary processes incorporate an open-ended approach that addresses the local learning problem and the problem of predicting the future in a complex environment. Open-ended processes generally rely on heterogeneity and multiple alternative solutions rather than homogeneity and the predetermined assumption of only one absolute solution. A global processing approach allows room for error, in that some solutions may be equivalent and some may alternatively work better than others. The environment chooses by selecting out nonadaptive solutions. The remaining solutions that suffice provide more alternatives as stepping-stones that can evolve into more complex future solutions.

Absolute control in closed systems tends to process past information linearly and act as if the preferred future can be determined on demand somewhat like absolute objective search. But open systems rely on open-ended processing and searching for alternative choices. Systematic processing integrates the current available information input with existing knowledge and elaborates the problem space, weighs plausible possibilities and alternatives according to the most likely adaptive outcome, and statistically predicts and selects the probable best solution.

Operating as open systems, we can increase the probability for adaptation by producing more complex alternative solutions that are based on systematic knowledge and by applying that knowledge to modulate systematic error correction via persistent feedback from the environment. Recall that DNA results from evolutionary processes of biological organisms as the prototypical complex system with both local and global features. For example, DNA provides general systematic and phenotypic solutions that can exploit different localized areas over long time periods. As Darwin discovered (1859), we can see species with different adaptations depending on the local environments, such as birds of the same species with different types of beaks adapted to local food sources.

Complex systems simply do what they do best, flexibly evolving and adapting in an open-ended manner, such that the effectiveness of the output product relies on interactive feedback from the environment. Adaptive evolutionary processes demonstrate trial-and-error learning, shaped by reproduction, variation, and choice by natural selection. Nature provides the feedback by selecting adaptive solutions from a pool of alternative solutions. By overcoming the limitations of a predetermined monolithic solution, adaptive selection enables a range of responsive solutions as actions correlated to local environmental change. Selective constraints tend to favor the choice of alternative actions that flexibly match adaptive countermovements in response to external actions.

Possession of the power of movement is one of the most striking characteristics of animals. For example, local conditions can become unfavorable for immobile organisms, such as plants and fungi, leaving them at the mercy of the environment, where they have little choice except to adapt as best they can, relying on slow reactions of genetic and physiological defenses (Vanderwolf, 2010, pp. 13–14).

> In contrast to plants, animals (except a few sessile forms such as sponges or barnacles), when confronted with unfavorable conditions, can move away relatively quickly in the hope of finding something better. Since mere random motor activity may make things worse rather than better, there has evidently been a strong selection pressure favoring the development of sensory organs and nervous centers to guide and control motor activity. (Vanderwolf, 2010, p. 14)

Accumulated by selection pressures over evolutionary time, adaptive solutions correlate the organism's actions with actions of the domain. In primitive vertebrates, such as turtles or lizards, when they encounter the odor of food, neurons are activated and the pattern of locomotion will naturally be guided along the route to the food (Vanderwolf, 2010, pp. 26–27). There is a very close relation between sensory input and motor output (Vanderwolf, 2010, p. 29). An organism's adaptation is a function of its actions expressed in terms of effectiveness. We can see that our learning history and neurons in our brain impart a level of benefits and constraints on our choices and learning weights and on the effectiveness of our behavior.

Actions are the phenotypic product of an open-ended evolutionary search process that leads to enhanced adaptation to input from environmental disturbances. An open-ended approach provides a logical method for operational preference by selecting and modifying solutions that have adapted to the fitness landscape. Solutions correlated to the dynamics of the landscape allow current problems to attract current solutions on the basis of the solution's functional effectiveness. The accumulated history of successful correlations represents a knowledge source that decreases the dimensions of the solution search space and can be recurrently modified and evolved for exploiting prevailing conditions. We see the accumulation of alternative choices for effectively regulating complex adaptive behaviors in response to the dynamic constraints imposed by external change.

In an ever-changing complex system, inflexible organisms would be at the mercy of the environment because of the rigidly constrained, closed-ended reasoning processes that produce one absolute solution, take it or leave it. With absolute language humans can make automatic closed-ended decisions effortlessly that on average suffice for adapting. But with nonabsolute language and greater effort we can improve our learning and predictive abilities by deliberately applying open-ended reasoning. Explicit reasoning processes can help us gain insight about the role preference plays in influencing predictions, choices, and actions. An explicit option allows us to generate potential alternatives and increase flexibility, which in turn increases the probability for realizing more desirable outcomes and minimizing the occurrence of undesirable ones (Shadmehr & Wise, 2005, pp. 51–52, Table 4.1, p. 53).

To better adapt to current circumstances, we can modulate our existing automatic behavioral repertoire by knowledgeably choosing the most appropriate actions on the basis of the potential consequences of those actions (Shadmehr & Wise, 2005, p. 52). Somewhat like adaptively fine-tuning computer software and algorithms, explicitly fine-tuning language constructions via experiential feedback learned from the results of our behavior increases the range of our adaptive repertoire and enhances flexibility by providing alternative choices. Fine-tuning also improves regulation of systematic error variance and our understanding of the plausibility of alternative solutions.

Absolutes from locally biased true-or-false cognition reduce alternative solutions, freeze learning weights, decrease flexibility, and constrain the effective dynamic range for regulating adaptation. Relative reasoning and communication enhance flexibility, enable systematically biased preferences that are based on possibilities, and increase the probability of adaptive outcomes in a complex environment. Extended language and semantics complement the self-organization of bottom-up solutions and increase the autonomy and effective dynamic range of top-down frontal lobe executive functioning for analytical thinking and regulating action.

> No complex system can succeed without an effective executive mechanism, "frontal lobes." But the frontal lobes operate best as part of a highly distributed, interactive structure with much autonomy and many degrees of freedom. (Goldberg, 2002, p. 230)

The option of extended language constructions expands our available choices from imperatively mandating and instructing behavior to preferentially acknowledging and informing possibilities and probabilities. When we extend our permissible language constructions, we gain greater flexibility of choices, a wider range of options for effectively informing communication, and the benefit of open-ended search for discovering novel solutions. Extended language also enhances systematic error management and effective behavioral regulation.

Systematic Error Processing

Systematic blindness occurs and we incur systematic errors in large part from neglecting knowledge of the global system in which we live. Mitigating and decreasing errors brought on by systematic blindness requires an understanding of how the system works, including understanding language, the brain, and the physical laws of nature.

Managing errors involves detecting, estimating the variance from a reference signal, learning from feedback, and making effective adjustments. Complex systems inherently rely on adaptive error management. Deficiencies in any one of these mechanisms can lead to the accumulation of systematic errors and the demise of evolving organisms. In humans, our errors may not always lead to our demise, but they can likely produce stress, discomfort, unhappiness, and strife.

Language enables exquisite wireless communication between human brains. Unfortunately, it also enables us to absolutely misperceive, misinterpret, and misrepresent internal and external information about the world—and ignore our own role in those misperceptions. To address systematic errors related to this linguistic communication problem, we can choose to employ cognitive accuracy through information accuracy, information processing accuracy, and event-related accuracy. In complex domains, greater linguistic fidelity generally improves our predictions and error management. And in a roundabout way, improving the fidelity of information depends on deliberate and persistent systematic error management and on the inherent effectiveness of the applied language.

Better information via language and semantic fidelity improves our understanding and perception of relationships that we experience. While this benefit may appear obvious, the opposite may be less obvious: We can incur liabilities and unintended consequences of accumulated systematic errors that result from absolute language constructions and representations that create idiosyncratic localized partitions in the domain. These arbitrary partitions can trap us at local dead ends in an isolated part of the fitness landscape. Lacking systematic knowledge, we can easily remain locally trapped in our own small world and naively allow these potentially toxic errors to accumulate undetected.

Absolutes compromise information fidelity and constrain options in response to environmental disturbances. Language fidelity biases transactions between us and the environment by modulating learning through error editing that in turn influences fitness. If we do not generally possess knowledge about the magnitude of our errors, our brains default to intuitively guessing some arbitrary value, which creates an unknown error variance that potentially increases the risk for unanticipated negative outcomes in the future (Jaynes, 2007, p. 219).

Error ambiguity gets in the way of the learning and compromises risk–reward evaluations and long-term predictive abilities. In complex systems, sensitivity to initial conditions can magnify small errors into gross inaccuracies in prediction (Solé & Goodwin, 2000, p. 27). Systematic errors that go undetected tend to propagate pathological feedback and erroneous processing by automatically recycling ambiguous, error-ridden information. Analogous to repair mechanisms in DNA, we have the ability to employ feedback mechanisms to check for and attempt to correct errors (Kahneman & Frederick, 2002, p. 57)—effortful corrective mechanisms that can be deliberately enhanced by applying language and semantic fidelity with explicit forward-directed time.

With locally constrained language, we can rigidly ignore or refuse to admit systematic errors, despite overwhelming amounts of information and plausible evidence. By neglecting or rejecting the possibility of errors, we also ignore making corrections, even potentially critical ones. In humans, many error-management failures result from diminished language fitness, closed-minded rigidity, and absolute linguistic constructions replete with ambiguity and hidden assumptions that constrain learning. When an organism less complex than humans makes what we call a mistake, it "changes its behavior as soon as it discovers the mistake" (Jennings, 1906, p. 220).

However, in humans, systematic errors often escape detection. The relative familiarity of an object or event induces an affective bias and logical inconsistencies that serve as the priority logic function for identifying and processing errors. Like the immune system, which identifies unfamiliar entities as foreign, and therefore pathological, we tend to give precedence to familiar self-similar "truths" and local explanations that "must be correct." All the while we devalue unfamiliar non-self-similar entities as incorrect, deviant, and even pathological. Research shows that we tend to bias our attention to the familiar and miss other possibly relevant information. Attentional bias leaves us open to unattended systematic errors outside our local spotlight of attention and incidentally exposes our vulnerability to missing opportunities for decreasing errors by learning from them.

We often hear about learning our lesson from past mistakes. But what do we learn? Many of our systematic errors result from our frugal assuming brain, its predilection for limbic logic, and its defaulting to culturally embedded affective language that cleverly rationalizes and covers up errors. How do we knowledgeably learn from our mistakes without understanding the processing constraints that contributed to them? In the short run, we may not see the negative results of our faulty reasoning, so the errors go unrecognized until viewed in retrospect after disaster strikes. With the implicit local assumption of reversible processes and symmetrical time as the rule rather than the exception, hindsight can be easily confused with insight. Learning about an isolated part of the environment ruled by reversible time tends to be a random lesson about the uncorrelated temporal structure and the capriciousness of the local fitness landscape—leaving us to miss seeing the big picture altogether.

We can ignore irreversible physical processes and the arrow of time by naively attempting to erase errors and ignoring or retrospectively denying, justifying, rationalizing, or blaming others for them. But we confound the possibility of learning from those errors. If we lack insight and overlook our errors, we easily resort to absolute justifications and ad hoc excuses and adjustments, and we tend to repeat them. Simply insisting that a mistake "*should not* have happened" denies the irrevocable distinction made between past and future and contributes minimal information for reliable contingent learning, credit assignment, and systematic error correction. It is unlikely that we can excuse away our past deeds regardless of our efforts to do so.

> He that is good at making excuses is seldom good at anything else. (Benjamin Franklin)

Deliberately thinking "first and slow" and front-loading cognition a priori generally provides a better error-correction strategy than automatically thinking "last and fast" and attempting to back-load cognition after the fact. Even though the past does not reliably predict the future in a complex environment, it can provide us with ample opportunities for learning from past errors. Right and wrong, true and false, and similar absolute linguistic constructs may somewhat suffice as quick fixes and for describing artifacts of the static past, but they fall short statistically for improving learning, and for making error corrections and reliable predictions.

Judgment words such as *right* and *wrong* operate as trailing causality rather than leading or predictive indicators and tend to become meaningful retrospectively, but obviously unchangeable after the fact. Like discrete true-or-false imperatives, basing prediction on absolute right and wrong implies symmetrical time and statistically improbable prescient forward visibility. In comparison, we increase our visibility for explicitly detecting and correcting systematic errors with statistical predictions that are based on temporally informed possibilities.

Some errors are accidental, and others are systematic; successful correction of systemic errors requires knowledge acquired by studying their causes (Poincaré, 2007,

p. 117). However, absolutes cover their own trail when it comes to learning from errors. In hindsight we might realize that we experienced a negative outcome, but we lack an understanding about choice and outcome responsibility and accountability for effectively learning from the error. The crucial link between our choice and its outcome remains hidden by the same locally embedded logic, rationalized inaccuracies, and implicitly assumed or missing information that contributed to the original error. Remorse that accompanies these errors can correlate more to feeling sorry that we got caught and bemoaning the penalties we incur rather than to the harm done to others.

We tend to overlook the stealthy obstacles in our thinking that automatically condemn us to noisy reasoning. Semantic imperatives and prescriptives, musts and shoulds, support and perpetuate noisy closed-minded thought processing by implicitly hiding systematic errors outside our explicit view. Minimally informed decisions based on noisy assumptions hidden in and perpetuated by absolute language constructions impede correcting and learning from errors. Since they lack extended logical processing abilities, absolute words tend to create static-like reasoning. A deficiency of dynamic logic leads to default computations with absolute logic—leaving us statistically impaired, dynamically blind, and prone to noisy assumptions and inadvertent systematic errors.

Noise occurs naturally in dynamic environments, even inside organisms themselves (Wagner, 2005, p. 271). Sometimes, depending on circumstances, we apply the term *noise* to unwanted signals that seem to interfere with our preferences (Crecraft & Gergely, 2002, p. 63). We can easily misconstrue signals when our evaluations include assumptions with a perceptual bias against self-contradictory evidence, such that we misperceive the presence of noise. Deliberately applied extended language offers better fidelity for noise discrimination. Better fidelity helps us to detect and tune signal sensitivity by minimizing errors related to the erroneous perception of noise or interference. In a relatively noisy and uncertain world, the deliberate use of extended language enhances the detection of relevant signals, provides reliable statistical inferences, and facilitates explicit thought-processing abilities that can decrease systematic error rates.

We can use extended natural language to produce general descriptions corresponding to open-minded thought processing that improves the tuning of signal-to-noise ratios and minimizes deception due to noise-related errors. Linguistic extensions that enable general global tuning can mitigate local noise levels and improve information fidelity and adaptive problem solving. In complex environments, finding a perfect predetermined local solution that always works amounts to a free lunch and is categorically unlikely. Improving language fidelity helps to minimize systematic errors, which can, in turn, maximize adaptive learning. According to Ann Churchland and Terry Sejnowski (1994, p. 131), error minimization is an optimization procedure. Language that contributes to open-minded thought processing and systematic error reduction provides an optimization tool that can contribute to adaptive fitness as overall well-being.

In conclusion, when we fail to include extended language in our repertoire, our inadvertent default to constrained language decreases our chances of recognizing noise. But we inadvertently continue fooling ourselves with faulty assumptions and recurrent unseen systematic errors that can impair fitness. We choose the language constructions we use in our everyday lives, whether we do so casually or deliberately. We can choose our language constructions without considering their fitness. Alternatively, we can deliberately choose constructions to enhance the probability of realizing a particular purpose, such as improving effective thought processing. Extended language enables purposeful efforts for dealing with systematic errors and optimizing the fidelity of current information and knowledge. The greater the fidelity of the information we process in the present, the greater the probability of recognizing and minimizing systematic errors, and making more effective choices with more predictable outcomes. In other words, extended language helps us to understand and mediate our systematic blindness.

This chapter explored the idea that we can better understand how we make decisions, understand why we make systematic errors, and accept that we have deliberate alternatives to our largely automatic language behavior. Knowing about and minimizing error-ridden local assumptions benefit from the fidelity of the language we choose and use. We can examine how we think and speak about our world and consider whether more deliberate use of language with greater fidelity might help to reduce our systematic errors—that is, if we "believe in" such a thing as systematic errors.

We do have choices that offer improved error management, if we prefer to exercise them. Undetected systematic errors and faulty assumptions bias what we believe and value, how we learn, think, feel, behave, and adapt. Even though reducing errors to zero remains unlikely, language that facilitates effective learning and the reduction of systematic errors can directly enhance many aspects of our lives.

The next two chapters address learning from language. Understanding language can help us to explicitly evaluate problems and find more optimal solutions to mitigate the liabilities we incur living in a continuously changing complex world with dynamic instabilities. Understanding and applying extended language to our thought processing offers a pragmatic approach to acknowledging our brains' systematic biases, gaining insights for identifying and correcting systematic errors, and effectively regulating our adaptive behavior in response to persistent change. But what, if anything, can we learn from language?

CHAPTER 19

Learning From Language

WE APPLY LANGUAGE AS a linguistic tool for constructing, describing, and understanding what we observe and learn about the world around us. We use language to help understand and share with others what we have learned. We even use language to talk about how we use language. As with most other technologies humans have developed, we often tend to use language with limited understanding. However, almost any craftsperson will tell you that you can get better results if you get to know your tools; weigh their differences, complexities, and constraints; and learn to choose the best tool for a particular job.

Learning from language starts by learning about how it works. We can think of many questions to ask:

- Where did language come from?
- What role does it play in our interactions with the world?
- What options do we have for pursuing a better understanding of these and other questions for making improvements?

Unlike most other investigations we might undertake, the study of language has one puzzling aspect that interferes with the discovery process. Recall that to describe and understand what we observe about language, we are left using the same language. To communicate what we learn about language to others, we resort to language.

- If we are ignorant or become confused about some aspect of language, how do we know whether the confusion arises from language in the abstract or instead from the particular language we use to describe that confusion?
- If we discern a pattern or effect in certain forms of language use, how do we discover whether the pattern represents a fundamental fact about language or simply depends on our descriptive perspective?
- If we decide our language has significant limitations, how can our constrained language overcome its own limitations?
- Put another way, if indeed our language traps us within its own limitations, how do we extradite ourselves with the very language that traps us?

While such concerns might not necessarily stop us in our efforts to learn from language, our investigations will likely profit from working to increase our understanding of the potential difficulties and limitations we incur when using language to study language. It also helps us to recognize that while we use language personally and in most of our relationships with others, the form of language and the type of thinking it represents depends largely on meaning, and meaning depends largely on context and purpose—much of which we implicitly derived from our forebears. "It depends," extends our semantic perspective to situational usage that correlates definition, meaning, and purpose to our intended usage in the context of the application. What makes sense in one situation may fail to convey effective meaning in another.

Consider hand-me-down discrete language with exclusive convergent properties that direct our attention to the past and constrain the redirection of our attention to the present and toward the future. We acquire limitations on our ability to diverge and search the global space for novel information and systematic knowledge. In such a case, the localized language that perpetuates the meaning and purpose we inherited places significant constraints on our ability to update the meaning and purpose to fit our current knowledge of the world.

Beyond Good Enough

As noted previously, evolution seems to operate on the "good enough" principle (Churchland & Sejnowski, 1994, p. 133). Open-ended evolutionary processes tend toward heterogeneity and flexibility in lieu of prospectively mandating rigid homogeneous organisms that prescriptively should or imperatively must perform up to some predetermined standard or static absolute objective.

But is there something beyond good enough?

Apparently our inherited genotypes and language types produced belief, value, and logic phenotypes that work sufficiently enough to get us where we are today communication-wise. Recall that genotypes express phenotypic behaviors analogously to language constructions that produce speech behavior; the open-ended information encoded in DNA allows variation in genes to influence behavior as information flows between the genome and the environment in the form of alternative phenotypic solutions. Solutions that survive represent naturally selected adaptive knowledge accumulated as a history of correlations between the genotype, phenotype, and successful domain interactions. In effect, solutions that generally avoid elimination are "good enough," at least locally. Similarly, common language on average produces "good enough" local solutions. Of course, we understand that by looking at average solutions we exclude information about variability such as we see in nature and favored by evolutionary processes. Systematic generalizations take into account the domain dynamics and include the breadth of global information. But is it better than good enough?

> Evolutionary pressures acting on the bulk of human judgments are neither sufficiently direct nor intense to sculpt the kind of mental machinery that would guarantee error-free or bias-free judgment. As Simon pointed out long ago (1957), evolutionary pressures tend to lead to local ("better than"), not global ("best possible") optimization. Evolutionary pressures lead to adaptations that are as good or better than a rival's; they do not lead to adaptations that are optimal. (Gilovich & Griffin, 2004, p. 9)

A good-enough method provides sufficient local solutions on average for the survival of individuals and groups that face daunting challenges in an uncertain complex environment. By sufficing, high-tolerance organisms can get along in the noisy, somewhat unpredictable world, unless thwarted by inflexible habits and constraints that hamper adjusting to changing conditions. With some niche-dependent exceptions, in higher-dimensional domains open-ended evolutionary processes tend toward an increase in complexity, heterogeneity, plasticity, and variable alternative solutions as adaptive options. Alternatives generally improve the chances of survival of a species. In contrast, closed-ended homogeneity can lead to deceptive evolutionary dead ends, as homogeneity constrains alternatives and dampens flexibility.

It appears that efficiency and performance in a complex domain result from a trade-off on an affective-to-effective gradient from automatic, fast and frugal, and "good enough" to deliberate and slow, with more optimal precision and fidelity. Automatic regulation generally yields quick but adequate responses for simple localized problems with little effort. Adding open-ended language and extending the capability for deliberately switching regulatory policies and perspectives broaden the range of options and enhance variability. Open-ended regulation increases the dynamic range for effective adaptive tuning beyond sufficient and extends precision and accuracy to systematic resolution.

Effective adaptive regulation with extended language enables more complex, selectively fine-tuned solutions. Improved performance corresponds with effective predictability and regulation of cooperative and competitive behaviors. We find positives attractive and move toward them, and negatives unattractive and move away. How effectively we distinguish between them can profoundly influence the magnitude of our rewards and our penalties. We typically expect effectiveness to correlate with enhancing cooperative behaviors that reward our efforts.

Fortunately, with language constructions that promote effective regulatory processes, we can modulate competition, cooperation, and speed within the task requirements, while adjusting sufficient accuracy for resolving particular problems (Jaynes, 2007, p. 224). The trade-off may have more to do with having options as alternative solutions. Options facilitate flexibility and thinking about the effectiveness of different choices, decisions, and outcomes. Flexibility confers an adaptive advantage and enables selecting alternative recipes for dealing with a world of persistent change. Policies based on effective language explicitly promote flexible regulation and open-ended thinking that favor collaboration and cooperation. Policies based on affective language inadvertently promote rigid regulation and closed-ended thinking that implicitly favor local competition with constrained cooperation. For example, compare this description of competitive and cooperative regulation:

Competitive Regulation

Competitive regulation is winner take all, with one emotionally driven solution that is touted as the only way. But winning may also require a bit of a practical approach to leverage defeating other competitors "by hook or by crook. They *must* be defeated or the sky will fall. Do the *right* thing. We *must* have perfect solutions. We cannot fail and *must* have zero defects!"

The goal of winning at any cost appears as a behavioral primitive in most life forms as a lower-level direct pathway for competitive evolution—compete to survive.

These are some of the *benefits* of competitive regulation:

- It is fast and frugal.
- It annihilates perceived negatives.
- It works as a temporary local bandage.
- It works as a "quick fix in a pinch" for emergencies, such as if we step on a snake.
- When resources are low, winning can be a matter of survival—a lifesaver.

These are some of the *liabilities* of competitive regulation:

- It has a noncooperative bias.
- It relies on micromanagement.
- It dictates a preponderance of rigid, black-and-white solutions.
- It tends to freeze learning weights and increase deception.
- It relies on emotional digital logic and ordinate obedience to authority.
- It involves self-righteous hierarchal behavior.
- It focuses on winning and eliminating negative feedback.
- It ignores the effects of unconstrained positive feedback.
- It inhibits learning and the search for novel solutions.
- It denies instabilities and alternatively attempts affective modulation by rigid, absolute control of the input.
- It is generally ineffective for optimizing long-term systematic adaptation.

Cooperative Regulation

In a sense, almost everyone wins when using cooperative regulation by helping to find the best possible alternatives and innovative solutions. Because it is driven by analysis and the extension to quantitative information with the integration of qualitative information, it facilitates decreasing errors by considering solutions as to how effectively they fit the problem. In a changing environment, the idealization of zero defects is imaginary, so we can typically consider decreasing the systematic error rate to optimize the adaptation of the system as an effective option. Since no perfect solution exists in a changing world, we can continue to work on improving feedback fidelity and decreasing systematic errors. Decreasing systematic errors facilitates going beyond "good enough."

The goal of winning by cooperation apparently evolved in early life forms as a complementary alternative to reflex-like direct pathways and competitive survival. Over time, the evolution of higher-level indirect pathways and neuronal complexity provided increased flexibility and opportunities to exploit cooperative behavior—compromise to survive.

These are some of the *benefits* of cooperative regulation:

- It has a cooperative bias.
- It involves macro-management.
- It allows for flexible learning weights and a range of possible solutions.
- It solicits multiple solutions.
- It identifies and minimizes deception.
- It relies on continuous analytical logic.
- It is accepting of a range of various behaviors.
- It focuses on the fidelity of both positive and negative feedback.
- It promotes learning and the search for novel solutions.
- It offers effective modulation of instabilities.
- It is generally effective for optimizing long-term systematic adaptation.
- It allows for prudently managing resources and planning ahead, and attempts to avoid crisis.
- It is antithetical to deception.

These are some of the *liabilities* of cooperative regulation:

- It is slow and laborious.
- It requires a greater expenditure of effort, energy, and time.
- It may err toward positives and overlook negatives.
- It may not suit everyone, and some may prefer to compete.

Observations

Effective language enhances flexibility, cooperative coexistence, and collaborative communication. Affective language tends to impede cooperative coexistence in favor of winner-take-all competition that can spawn arms races as an accelerating battle of the fittest. Competition finds itself in a power struggle with cooperation, which it perceives as a battle to the death between right and wrong, those who do good and those who do evil. Cooperation finds itself in a struggle with competition to demonstrate the benefits of thoughtful solutions, which it favors. What if we could convince the world that cooperation usually is a better way to go? But can we overcome our competitive nature and the deception that accompanies it? How could we convince others that thinking and cooperating facilitates analysis of alternative choices, enhances effective adaptation, and offers a practical alternative to emotionally fueled affective reactions? Cooperation promotes adjustable tolerances and flexible collaboration but requires increased effort, time, and energy resources. When we see things as tolerable or "good enough," considering change hardly seems like a frugal option to our complacent brains.

Organisms in an uncertain environment can be likened to machines in a noisy environment. The machines with coarse-grained "good enough" tolerance can operate with higher levels of noise but produce less-effective results. Rigid tolerances offer sufficient accuracy for the average situation but constrain flexible tuning. Upgrading is expensive but enables regulating the signal-to-noise sensitivity, flexible feedback tuning for error correction, and adjusting tolerances for optimizing processing fidelity. We can then operate within a broader dynamic range of systematic effectiveness—raising the bar above "good enough."

We can think of the brain as if it were a machine that evolved in a noisy world. An absolutely precise brain would have trouble dealing with all of the noise and uncertainties, but the brain has evolved with loose tolerances such that it operates generally rather than absolutely. The same applies to language in a dynamic world. Absolute language fits a squeaky-clean, idealized world but confers a disadvantage in the real world where information fluctuates continuously. Our durable brains continue to work sufficiently with absolute language locally even though absolute language doesn't fit the systematic dynamics of the world in which we live. And even though we misinform ourselves, we still get "good enough" results; unfortunately we tend to say and do some not-so-clever things (see Chapter 5, "Systematic Biases I"; Chapter 6, "Systematic Biases II"; and Chapter 7, "Systematic Biases III"). Extending from affective to effective language shifts our language to logically correspond with our brain and our world, enabling us to go from good-enough affective with certitudes to more optimally effective thought and behavior with possibilities and probabilities. Probability facilitates adaptive tuning of our cognitive processes from local to global and enables alternatives for flexible planning and adapting to fit the tasks at hand.

In a natural environment, organisms that develop more effective alternative solutions for adaptation can potentially enhance their survival. Both effective and affective information serve a purpose. But when it comes to systematic optimization, effectively informed cognitive analysis typically transcends affectively informed limbic reactions.

If anything, we would expect a negative correlation to optimizing solutions under affective information constraints, owing to the egregious accumulation of systematic errors. Effective information generally offers net benefits, whether in conjunction with fast and automatic or slow and deliberate cognitive processing, but especially for mitigating and correcting systematic errors.

> In recent years, psychologists have proposed that most of our thought processes can be divided into two types: those that are fast and automatic and those that are slow and reflective. Both contribute to everyday illusions. The rapid, automatic processes involved in perception, memory and causal inference have serious limitations, but these limitations become much more consequential when our higher-level, reflective, more abstract reasoning abilities fail to see that we are going astray and make appropriate corrections. (Chabris & Simons, 2010, p. 230)

We have the option of fine-tuning our language constructions to correspond with the dynamic world in which we live. We can focus on cooperating, paying attention to when we are going astray from reality, and making effective corrections. Linguistic correspondence facilitates greater resolution of information and knowledge. Perhaps as relevant, we recognize that affective language hides how little we know about the vastness of our ignorance. Effective language acknowledges ignorance and opens the door for learning, reducing systematic errors, and optimizing adaptation—as best we can.

Extended linguistic abilities facilitate the development of cooperative policies, algorithms, and strategies that can be persistently fine-tuned through adaptive learning. In general, the better we tune our language, the better we can modulate our thought and behavior, adjust to various situations as they arise, and produce the adaptive outcomes that we prefer. These principles of tuning, cooperation, and adaptation relate to effectively informing the fidelity of our language, which we rely on for reliable feedback.

But is our language *good enough*?

The Language Paradox

It seems likely that our language encoding passed down from the ancients originally evolved in a beastly competitive and perceptually noisy environment. With early coarse-grained language abilities, primitive absolute logic, and uneducated beliefs and values, early humans most likely misperceived, miscategorized, misstated, and misunderstood the ever-changing unruly environment.

As a result, early verbalizations about the world likely involved more imagination and superstition than science. But through ritual and repetition, overvalued cultural beliefs appeared and acquired a resilient mantle of status quo "knowledge." Beliefs and language forms, once established, maintain their momentum and resist change. As humans with susceptible brains, we tend to interpret coincidences as contingencies, leaving us vulnerable to overvalued fears, superstitious beliefs, and rituals involving verbal formulas. Language likely evolved under primitive conditions, leaving us with the difficult task of verbal archaeology, sorting out the bits of reliable information from the matrix of myth and misinformation encoded and woven into the fabric of our language by generations of previous speakers.

Misinformed interpretations and antiquated predetermined values and superstitious beliefs about the world were apparently shared and perpetuated under the guise of invaluable cultural "knowledge." But with our assuming brain we tend to implicitly accept and take our inherited language beliefs and values for granted. For example, if a distant relative hated people of different color, and that prejudicial opinion was passed along as valuable cultural "knowledge," it does not mean it was ever effectively valuable or even a reliable belief, then or now. But is it even possible that predetermined hatred can be linguistically inherited via cultural transmission?

Apparently so, since we would be hard pressed to find a culture where the passing along of prejudices does not occur. If we found such a culture, it would be the exception rather than the rule. Predetermined overvalued beliefs and prejudices embedded in our hand-me-down language present us with sticky issues from which it is difficult to disentangle ourselves. Even though more up-to-date information and more reliable knowledge about how the world works have slowly become available over the ages, overvalued cultural beliefs, once established, are resistant to extinction (Skinner, 1953, pp. 84–87; Dawes et al, 2002, p. 725).

The language-encoded values and beliefs we inherited might have seemed relevant in the past and may appear to suffice currently, but they were created and passed down to us with little understanding of intrinsic constraints and liabilities. Ironically, we continue to use the same absolute language oblivious to any limitations; we ignore the problem that the inherited language we use fits the world no better today than it ever did. Yet despite the continual development and refinement of new information and knowledge, we continue to closed-mindedly resist updating our primitive language habits, logic, beliefs, and predetermined affective values—even our hatred and prejudices. We see widespread disdain for and resistance to acknowledging the existence of more effective language, information, and knowledge, such as that developed by the methods of science.

> [The early human] knew only crude analogies, those that strike the senses.... Isolated amidst a nature where everything was a mystery to him, terrified at each unexpected manifestation of incomprehensible forces, he was incapable of seeing in the conduct of the universe anything but caprice. (Poincaré, 2007, pp. 77, 85)

However crude and coarse-grained, early language may have originally been somewhat survival-neutral (Wagner, 2005, p. 223), but at some point it apparently contributed to adaptation and improved survival. Despite relatively low accuracy and precision, the competitive and cooperative advantages of this primitive communication were "good enough" and passed along to future generations of language users. But where language came from and how it works has long puzzled thinkers in most human societies. Solving the mystery is a catch-22:

- How can we break the language code without realizing that the tacit encoding of our own language inhibits discovery?
- How can we use language to search for solutions to the problem that language misinforms us about, when the language we are using is the same language we are trying to address?
- How can we understand a global worldview for today with constrained reasoning narrowly rooted in the past by primitive localized language and logic?

If it takes nondeterministic language to understand human thought processes, behavior, and language itself, how do we break the open-ended language code using reasoning based on deterministically encrypted language tools with closed-ended constraints, hidden assumptions, discrete linear descriptions, and absolute two-valued logic? These questions seem especially problematic, since we tend to implicitly assume that whatever language we have inherited "works just fine" and that it suffices without inspection. The limiting features of the language—as if invisible—inevitably escape detection (Beck, 1976, p. 13; Jones, 1998, p. 33).

Resolving the Language Paradox

A *paradox* refers to a set of statements or notions that lead to a contradiction.

- Paradoxically, how can language at the same time both inform and misinform us about how the world works?
- How can the same language describe the world as referenced to perfection and imperfection at same time, or interpret the same statement simultaneously as both changing and static, as both relative and black or white/true or false?

Resolving the language paradox can seem quite baffling when we use language to code what and how we think and how we talk about the world, yet we use the same language to evaluate how well we think and talk about that same world. In other words, we use the same language to generate, interpret, and measure information and performance. How can we solve the language paradox without realizing that our own language has been inadvertently contorted by our own ignorance of its encoding?

Our tacit ignorance and our denial of that ignorance blind us to the discovery of that about which we are ignorant. Language can implicitly limit our thought, and in turn, our constrained thinking limits our ability to explicitly understand how language creates those very constraints.

Paradoxes occur when we incorporate faulty propositions or contradictory assumptions into our reasoning. When addressing two very different beliefs about the world, we might suspect the beliefs arose from conflicting sources of information. The contradictions and the paradox magically disappear when the faulty assumptions are addressed (Jaynes, 2007, pp. 383, 452). The simplest and most plausible approach to solving this linguistic paradox starts with viewing language evolution macroscopically at a systematic level. Our commonly used localized language constrains us from understanding its linguistic origins via backtracking or reverse engineering. We can be deceived by our failure to recognize that our current understanding inadvertently misinforms us through the hidden coding conventions embedded in the language we inherited.

Fortunately, we can start with the most evolved cultural language available: science. Scientific culture purposely incorporates systematically defined language, semantics, and processing methods. Science relies on intelligently designed language constructions that enable greater fidelity and precision for explicitly encoding and decoding and for deliberately producing information that leads to the effective representation of knowledge about the natural world.

> Science has filled our world because it has been tolerant and flexible and endlessly open to new ideas. . . . [S]cience and other provisional modes of knowledge do not bluntly *instruct* our actions—they *inform* them. (Bronowski, 1977, pp. 5, 96–97 [emphasis in original])

Science strives to bring the noisy natural environment into better focus by providing more effective information for representing scientific knowledge, and enhancing the human linguistic ability to better separate signal from noise (Jaynes, 2007, p. 438). The most useful representations typically exhibit a higher signal-to-noise ratio (Eliasmith & Anderson, 2003, p. 208). Science resolves the language paradox with language constructions that match the information structure of our environment. Scientific language focuses on reliability and effectiveness. Consider these examples:

- The evidence demonstrates *statistical reliability.*
- The results show that in general method A is *probably* 10% more *effective* than method B over a period of time.
- The experiment demonstrates that treatment A has a *probability* of less than 5% of developing long-term adverse effects.
- We have not accumulated enough *plausible evidence* to make a statistical comparison by correlations.

Science evaluates solutions systematically according to statistical reliability, predictability, error rates, and the variance in error deviations. The effectiveness of behavior can be quantitatively measured over time by correlating the reliability of specified performance measures. For example, when we say someone's attendance meets criteria for reliability, we give a quantitative statistical description of behavior as a history of correlations over multiple observations across time. When we say someone is trustworthy, the inference comes from the implicit intuition of "trustfulness" we often make with one qualitative observation based on the "feeling" we get when we meet them—gut feelings. The same applies to a person we feel is not trustworthy. We understand that "trust" correlates affectively with the internal release of oxytocin and is internally referenced qualitatively to superficial subjective features rather than to an accumulation of quantitative, observable, and measurable external evidence.

We also understand that emotionally charged experiences can saturate our thinking and distort our decision making. Consider the case in which when we claim we were conned into investing our life savings in what turns out to be a confidence scam. Our sense of confidence results in gullibility, to which we are all susceptible. Gullibility comes from having a human brain with systematic biases that habitually assumes. Unless we make concerted efforts for more thoughtful interventions, we remain encumbered by systematic blindness. Scientific language and processing methods attempt to decrease systematic ignorance by opening our eyes. Science supports analytical evaluations based on visible evidence and mitigates systematic blindness by finding effective solutions for better understanding language, the brain, and the physical world.

Recall that the reasoning evolved by science relies on extended logic. Probability as logic (Jaynes, 2007, p. 490) offers a generalized global optimization tool for deliberately generating information and knowledge corresponding to relationships in nature. In comparison to sufficing, scientifically based effectiveness enables optimization tuning that can provide measurably fitter results. Open-minded thinking with improved fidelity and precision provides greater analytical utility than closed-minded "good-enough" alternatives, at least when it comes to better understanding and adapting to the world in which we live.

Louis Pasteur, who spent much of his life trying to overcome the near-universal belief in spontaneous generation (Jaynes, 2007, p. 503), once said, "Knowledge is the heritage of mankind." Science provides this knowledge. But science doesn't only represent knowledge; it also represents orderly knowledge (Korzybski, 1958, p. 131). In an orderly manner, science enhances language fitness on the basis of possibilities and probabilities and the analytical value of systematic fidelity and precision. Language fitness seems particularly relevant to the fidelity of our information and knowledge about the complex world we live in today.

We likely resolve the paradox not by arguing absolute right or wrong but by scientifically viewing language and semantics as open-ended probabilistic tools for transparently processing information to systematic knowledge, and for more coherently correlating our cognition, communication, cooperation, and behavior with the natural world.

The Smoking Gun

Whether deliberate or simply naive, constraints on language and semantic fidelity induce and propagate unnecessary systematic errors, flawed error detection, attenuated error correction, and increased error variance. Systematic reasoning errors lead circularly to misconstrued contingencies, erratic behavior, and constraints on our ability to reliably predict outcomes, owing to the recurrent recycling of erroneous information and distorted information processing.

On the other hand, in the uncertain world in which we live, we inevitably work with incomplete information and some level of ignorance about our environment, which sometimes introduces unavoidable errors. Sometimes low-fidelity information provides outcomes we consider good enough—and sometimes guessing or absolute reasoning will suffice.

Situations also arise where fast and frugal, reflex-like reactive behavior yields the best solutions, accuracy be damned: *Run away! Run fast! Think later!* When we hear loud growling while walking in the woods, we do not typically stop to casually ponder whether this noise represents a threat. Lower brain centers have already put our feet into motion before we have time to think about whether we have correctly identified the source of the sound. In situations like this, the costs incurred for reacting reflexively and surviving, even for multiple false alarms, positively outweigh the penalty for pausing to analyze the situation and providing the bear with dinner.

Even though we can sometimes get by without analysis and fidelity in cases involving immediate danger, we generally benefit from taking the time to review our assumptions and check the plausibility of our information (Jaynes, 2007, p. 224). When it costs so little to practice language and semantic accuracy, especially if we learn early on to do so, willfully ignoring the benefits makes little sense when the potential rewards of enhancing the efficiency and effectiveness of our adaptive thought processing and long-term survival seem to far outweigh the cost overall. However, we acquire language and culture long before we have the resources to question the validity of what we learn. As a result, we typically fail to consider the idea of explicitly questioning the aspects deceptively hidden in our habitual use of the familiar language we inherited.

Even those of us in research who have learned the value of applying accuracy and reliability to our work may overlook application to our common speech. The idea of improving the accuracy and reliability of our routine speech to enhance effective communication seems foreign to us. Without further consideration, we can easily dismiss the idea as absolutely unimportant, naive, trivial, useless and not worth the effort, unproven, dangerous, or irrelevant. Acknowledging the value of effective language for talking about scientific experiments does not necessarily or automatically translate to open-minded acknowledgment of equivalent importance for communication with the external world in general and for our personal, individual language use in particular.

Psychiatrist R.D. Laing once said, "[Even] scientists cannot see the way they see with their way of seeing."

DNA has been singled out by Sean Carroll (2005, p. 225) as "the 'smoking gun' of evolution." When we are looking at differences between and within cultures, language and semantics could be called the smoking gun in the evolution of effective reasoning. Language and semantics bias our thought processing and play an influential role in how effectively we relate to our environment. Changing our linguistic bias changes how we think, feel, and behave, with the potential to improve how harmoniously we interact with the world and cooperate with fellow humans. With the deliberate application of extended language, we can tap into the open-minded power of more effective communication, information discovery, information processing, and learning. Effective language and semantics usage corresponds with cognitive fidelity, enhances learning from experience, helps to reduce systematic errors, and facilitates the conversion of information into more useful knowledge.

Language and semantic fidelity improves the effectiveness of verbal reasoning about relationships, associations, and influences across the system, which in turn improves situational understanding of contingencies in a meaningful way. Improved understanding enlightens thoughtful planning, expectations, actions, and predictions. The following chapter discusses overall issues in the book and then summarizes and draws conclusions.

SECTION IV

Recap

CHAPTER 20

What Have We Found?

Evolution is cleverer than you are.
—evolutionary biologist Leslie Orgel's second rule

IN THIS BOOK, WE have seen that language offers us a cognitive tool for improving our understanding of how the world works. In turn, understanding how the world works can help us understand the role language plays in our decision making and in our relationships. Our brains evolved with the ability to make use of language and learning. Language provides a recurring loop between us and nature, explains how evolutionary processes operate in the world, and helps us come to better understand interactive relations between our thinking brains and our world. Better understanding can enhance communication and potentially help us achieve more harmonious relationships.

Evolution is an open-ended process that produces alternative novel solutions. Somewhat similar to open-ended novelty search, evolutionary processes seem unconstrained by arbitrary values dictated by a predetermined objective. A priori, open-ended evolution and novelty search are equally neutral to the relative "goodness" or "badness" of the information and solutions that develop (Lehman & Stanley, 2011). In nature, we see the relative success of a particular novel solution in retrospect after natural selective processes rule for or against its adaptive effectiveness.

In other words, novel solutions, as evolved innovations, are obviously "good ideas" if they immediately work. But what if we think they don't work? Perhaps we think they simply are "not-so-good ideas." Research with novelty search provides evidence that it may not be the best strategy to nonchalantly toss them away and forget about them. The solutions that don't seem to work may be the not-so-obvious stepping-stones to better ideas and innovation. At the time of discovery they may not seem like anything useful at all, now or ever in the future. Throughout the history of science we have seen this theme repeated many times. For example, in the early twentieth century, Poincaré resonances were thought to be an anomaly, but today they are seen as a possible novel solution explaining interactions in nonequilibrium physics and helpful in explaining evolving systems and complexification.

Even though superseded by nonequilibrium physics, Newtonian mechanics paved the way as a stepping-stone from which to spring to more complex discoveries, as did scholarly efforts by the early Greeks and Romans. In evolution, natural selection represents a practical measure of how well an organism's novel solutions work by whether the solutions "do no harm" and provide successful adaptation in response to selective pressures and disturbances driven by natural physical laws. The old saying that ignorance is bliss may not hold up when our lives are on the line. We might conclude that, in fact, what we don't know can indeed hurt us. We can incur more harm from ignoring novel solutions than from considering a more effective option of open-ended search for, discovery, and application of novel solutions in a knowledgeable manner.

Although more expensive, externally referenced open-ended language and semantics broaden the systematic search space and enhance discovery, evaluation, and application of innovative adaptive solutions. A systematic perspective brings solutions into focus with the landscape naturally by viewing the world through global lenses. The frugal alternative imposes a narrower, polarized view from closed-ended localized linguistics that resists inspection, unless we consider evaluation by natural selection over a long evolutionary time scale.

We have an opportunity to borrow a few tricks from adaptive evolution to fine-tune our language constructions. Expanding beyond the open-ended natural processes of evolution, we can choose extended language to enhance cooperative adaptation. We see "proof of concept" for open-ended processes calibrated to external referencing demonstrated by evolution throughout nature. What we have learned about the mechanisms of evolutionary processes can help us fine-tune our language constructions and our thinking by asking

- How would language work in evolutionary terms, and would it offer measurable benefits?
- Can we apply this knowledge to language to help us know our world?
- Can more effective language offer beneficial solutions applicable to our present everyday problems?

The position taken here answers yes on all accounts.

We could ask, "What would evolution do?" We realize that evolution is a process rather than a person or thing, and we expect it to continue to slowly produce increasing complexity within natural constraints. Of course, evolutionary processes cannot provide direct answers to questions about gaining insights for improving adaptive behavior, but we do not have to look very far to see an abundance of clever solutions all around us that continue to amaze. The feats of evolution are especially noteworthy when considering that evolution takes place in a daunting environment of uncertainties and dynamic instabilities. The statistical nature of relationships in higher-dimensional complex domains and the variable severity, frequency, and duration of the disturbances encountered lead to inherent difficulties for predicting adaptive solutions. But evolution stodgily trudges onward toward the future, and persists with complexification and alternative possibilities for adaptive solutions.

We can generally state that in an ever-changing world, flexibility and statistical predictions offer probable advantages for adaptation and survival. Evolutionary processes tend to exhibit flexible policies and general divergent and convergent strategies with alternative solutions. Some of the solutions improve survival, persist into subsequent generations, and proliferate, while some perish. Broadly speaking and regardless of our fear of change, in a dynamic environment inflexible strategies restricted to single, monolithic solutions are more likely to die out.

Inflexible convergent strategies can foster homogeneity, constrain the development of alternatives and effective adaptation, and propagate rigid learning weights that increase the ruggedness of the fitness landscape. Variable learning weights that can flexibly diverge and converge increase learning opportunities, smooth the landscape, and allow switching among the possible alternative solutions. As a form of genetic information, surviving solutions that demonstrate physical adaptation to uncertainties represent effective knowledge correlated to the domain's physical laws. The domain laws or policies in a complex domain place dynamic constraints on survival that tend to select out closed-ended strategies, leaving open-ended strategies to proliferate. We can learn from evolutionary processes that seem to have discovered a multifaceted adaptive solution by, so to speak, not putting all our eggs in one basket. By selective default, these evolutionary policies generally avoid closed-ended monolithic strategies that can easily become trapped in the labyrinth of a complex world and fossilized at deceptive dead ends.

The open-ended processes of evolution provide evidence for the benefits of divergence by alternative possibilities and convergence to novel solutions by adaptive selection and goodness of fit. We see flexible adaptation that produces further possibilities for exploring and exploiting the solution search space. Adaptive solutions converge to the problem within physical laws and domain-dependent limitations, such that naturally selected solutions and information automatically limit the search space to physically plausible possibilities. Simply put, by limiting selection possibilities according to plausibility, we increase the probability of converging to solutions that more likely fit and work in our world.

Recall that establishing external referencing defines constraints that limit the search space dimensions and improves the tuning of signal-to-noise ratios by parsing information into preferred asymmetries and features relevant to that domain. In complex domains a reliable external referent enables calibration for coherently correlating and synchronizing information and information processing with the domain dynamics. Reliable referencing calibrates dynamic statistical information patterns to domain regularities, resulting in logically consistent feedback for input–output processing, flexibly tuning learning weights, and coordinating the effective modulation of instabilities.

Science has already developed and applied methodologies learned from evolutionary processes. These methodologies were discovered by exploring for, exploiting, and transforming information to more effective knowledge about nature. Scientific methodology relies on probability with an open-ended statistical capacity for understanding relations among persistently fluctuating interactions, one-way time and irreversible processes, and adaptation to uncertainties and dynamic instabilities. Understanding these nonequilibrium relationships enables the accumulation of a correlated history of regularities found in natural complex biological systems. Probability provides a logical tool for understanding how adaptations can evolve and flourish under such difficult-to-predict conditions.

We have found that applying probability to language provides a logical tool for understanding nature but also a tool for facilitating communication and mutually cooperative interactions. Probability provides the open-ended logic that correlates language structure to the dynamics of the domain. Probability as logic provides a generalized function capable of understanding the statistical nature of information, processing open-ended solutions, and modulating adaptation to change. Language that provides extended logic enables more effective communication, cooperation, and collaboration. Extended logic sharpens our linguistic tools for understanding uncertainties in our complex natural world and choosing plausible solutions as possibilities rather than absolute certitudes.

We can also borrow some simple evolution-mimicking tricks from engineering. Engineers typically employ a systematic approach that separates the problem-and-solution search space, evaluates and elaborates the problem space, then selects the fittest solutions by evaluating their structural and functional relations to the problem and the domain as one interacting system. This tactic relies on understanding the domain constraints and how the external system will most likely influence the proposed solutions over time. Knowledge of the system and its constraints helps to define and elaborate the problem-and-solution search spaces and predict the possibilities and probabilities for effectively adapting to future systematic disturbances. In a sensible way, knowledge of possible instabilities in the overall system helps to predict a particular solution's probable adaptations to particular uncertainties. We then can select and tune the best probabilities into more effective solutions over time.

A systems approach requires systematic knowledge, defining the method space, and choosing optimal methods for the particular domain. We can ask, "What are the particular methods that offer the greatest probability of resolving problems in this domain?" Method space includes the space of allowable processes that can be logically applied to that particular domain, such as dynamic versus static. In a dynamic domain with nonequilibrium constraints, we prefer a scientific approach with consistent methods and logic for computing probabilistic solutions corresponding to the domain ordering. Recall that we see statistical distributions of information in complex domains. By co-opting orderly methods already discovered by science and extending those demonstrated in adaptive evolution, we can apply probability as logic to compute the most likely solutions and switch or fine-tune them as conditions change. Finding the most relevant solutions seems to require using the method that best fits the tasks, which often includes both quantitative and qualitative evaluations.

There are many methods and ordered ways to solve problems. An evolving system has a logical, systematic ordering for problem resolution that correlates with the dynamics and logic of the domain, consistent with forward-directed time toward the future. Similarly in the evolution of language, a higher-order linguistic system can perform orderly logical operations from discrete local to continuous global. Understanding the domain order requires accurately matching the cognitive and domain logic and avoiding traps of logical insufficiency. The solution depends on effective language, an intact logically functioning brain, and a global domain perspective.

We have seen how closed-ended language constrains our worldview and that localized processes and solutions interpreted or decoded with extended language constructions confirm the underlying global nature of the local interactions. Problem solving with absolute language tends to ignore the global search space and the limitations on information accuracy; even though it may suffice locally, it tends to randomly order solutions as problems occur rather than effectively planning ahead. "If it's not broke, don't fix it" is a fairly common strategy. In evolutionary terms of parsimony, as long as organisms generate sufficient results on average, fast and frugal localized solutions perhaps make sense, since local is "where the action is."

We are fortunate indeed that we have linguistic options to improve our adaptive lot in life. Effective language options can improve our understanding so that we have an opportunity to deal practically with systematic problems we might have otherwise ignored, see warning signs before catastrophe strikes, and be prepared as best we can. In general, we can take advantage of effective language to facilitate more reliable predictions and outcomes measures. For example, consider these pairs of affective and effective language excerpts:

- I had a feeling something like that was going to happen.
- Having some warning improves the probability of surviving.

- How could that happen?
- We understand that we live in a world of instabilities that we can learn to deal with more effectively.

- It would not have happened if we had only...
- We likely cannot avoid all calamities, but statistical predictions can help us prepare.

- We "should" have known.
- It is unlikely that we know with certainty, but knowledge about various possibilities helps us plan contingencies.

- We don't deserve this.
- We live in a world where informed knowledge influences the reliability of our planning, predictions, actions, and outcomes, such that we get what we get.

- How could we have missed the warning signs?
- We recognized the warning signs with statistical measures, which helped us to prepare and make contingency plans.

How can we explain the difference in self-responsibility and predictability? A fairly simple answer is that affective language is self-centered and focuses mainly on the emotional quality of how the input feels, and asks the world, "What have you done for me lately?" Effective language takes an analytical quantitative approach that extends qualitatively to mutual relationships by focusing on the probable effects of behavior on all involved, and asks, "How does my behavior effect cooperative adaptation, for me, others, and the world? How can I better understand the world, and what have I probably contributed?"

We can view nature as a complex system from a probabilistic perspective, which enhances our predictive abilities and the effectiveness of the solutions we derive. Solutions based on systematic processing methods rely on open-ended information and probabilities across the domain. Problem resolution involves logically ordered search methods and understanding of the domain space, the problem space, and then the solution space. The logical ordering enables serendipitous matching of solutions to problems as special cases, and global search allows for the discovery of alternative solutions for specific localized problems. Knowledge about domain constraints prevails, decreases the dimensions of the search space, and extends the efficiency and effectiveness of systematic search throughout localities in that domain.

Recall that with extended language constructions, the general policies and processing methods for cooperative communication can be abstracted from higher-order global rules and incorporated into a systems approach. A systems approach improves the ability to modulate adjustments to disturbances by providing an open-ended process for continuously fine-tuning the learning weights for effective adaptive correspondence. A systems approach views human interactions as another integral part of the overall physical system. The system has general polices or laws under which we operate. The policies can be thought of as analogous to general physical laws of nature, which constrain or limit the range of the possible admissible behaviors and physical interactions we can perform within the system, including speech. Invoking general laws allows for a broader adaptive repertoire for problem solving. Open-endedness within the constraints of the system allows for flexible modulation across a range of alternative behaviors for adjusting to unexpected events.

Global policies that reinforce cooperation can amplify robustness in complex systems. We expect logical methods and classification to correspond with the logical ordering of the domain. Hence, we humans make up a categorical static class of *Homo sapiens*, and our actions logically correspond with dynamic relations in nature. Probability as logic correlates interactions according to the natural statistical distribution of information. In a complex domain we consider that time uniformly and irreversibly points to the future, which defines temporal referencing whereby choices manifested as physical behaviors represent irreversible events in time and space. This irreversible temporal relationship with choices emphasizes the relevance of systems thinking and understanding the dynamic influences in the system. The irreversibility of time and of our actions correlates information, knowledge, and chosen solutions in the forward direction. Irreversible relationships relegate our predictions to probability proper and encourage us to think about our alternatives before we act.

We can easily forget that the traditions passed along to us were often made on the fly by our ancestors in difficult circumstances and with only a small proportion of the amount of knowledge we have today. We can upgrade our language and policies to more effectively apply to the problems we face and find reasoned solutions for the here and now. We can argue for making changes or replacing fragile, rigid policies of old with ones that apply to the dynamic world of today. We can argue for setting the clock to run only in the invariant forward direction to correspond with today's one-way time, and tuning the effectiveness of our current policies to reflect the dynamic nature of the problems confronting us.

Instead of trying to hammer and bend outdated policies to fit today, we can take a more effective approach. Effective policies require reliable structural scaffolding. We can reconfigure and enhance the flexibility of our scaffolding to better suit our dynamic environment. With effective language, we can trade rigidly constrained affective policies for flexible, adaptive policies. We can take responsibility and accountability for our choices and generally improve our outcomes. We can make thoughtful nonrandom changes rather than randomizing old policies on the fly by shaking things up. In turn, we can produce more reliably adaptive solutions from effective thinking. We have the option of upgrading to extended language and flexible logic that supports effective thought processes and cooperative systematic policies.

Rather than flip-flopping between rigid fragile policies and infinite chaos, we can apply effective language structure that supports oscillations, alternative choices, and the modulation of information fidelity in the here and now that corresponds more harmoniously with the structure of nature. Effective language and policies that resonate with the dynamic world in which we live can help to overcome disruptive language habits from the static past. Effective means taking responsibility for our own thoughts, feelings, and the actions and adaptive solutions we produce. Effective policies enable reliable classifications and cognition, and flexible options for choices and adaptation to change. We can live as cooperatively as possible in the present if we accept our nature as flawed and fallible creatures living in an imperfect world of many alternatives and equivalent solutions.

We can generally expect improved predictability for adaptive outcomes by taking individual responsibility and accountability for our thoughts and logic, beliefs and values, emotions, and behavior. We can take responsibility for our own cognitive state and our individual brain hardware and software as best we can. We each incur individual responsibility for our language conventions, our cooperative behavior and consideration of others, and the reliability of our linguistic evaluations. We are responsible for

- Our individual interpretation of the events we experience in the world
- Needlessly upsetting ourselves about others
- Modulating our aggressive behavior
- Understanding outcomes and correcting our own errors, choices, and actions
- Accumulating and managing our own knowledge and decreasing our own systematic ignorance

We are each clearly responsible for modulating our emotions and our actions. Behavior amplified by affect tends to compromise long-term harmony. We each hold responsibility and accountability for our behavior and for realizing that our emotions tend to reflect the affective influence of a brain and brain hardware with systematic biases interacting with a constantly changing world. These biases also reflect our inherited language and learning habits that influence our speech behavior and our relationships overall. Responsibility for whether our cognitive behavior matches and resonates with the world in which we live belongs to each of us alone. We can accept responsibility and accountability for our individual behavior and for working to effectively improve cooperation in the world as a sensible solution.

We can accept and moderate our emotional responses to external events. The good news is that we have effective linguistic options that can help brighten our lives and improve our relationships:

- We can learn to think in a more effective ways.
- We can learn better options for how we talk to ourselves and others, for not upsetting ourselves as often or as severely, and for getting over it more quickly.
- We can adjust our beliefs and values to opinions and preferences and increase our flexibility, along with our consideration of others.
- We can practice taking responsibility and being accountable for adjusting our behaviors to treat others as we would like to be treated.

- We can relish taking responsibility and accountability for our own emotions and for interacting with others more thoughtfully.
- We can benefit from recognizing and modulating habitual influences on our emotions and for working to make effective improvements in how and what we think.
- We can try to help those less fortunate who lack recourse to education and world knowledge.
- We can help societies to share their knowledge and more effectively educate their peoples about the physical world.
- We can perhaps list the reasons for doing so, including promoting more effective policies for cooperation and a better future.

Knowledge about how the world operates helps us develop practical policies, methods, and behaviors that cooperate with nature rather than work against it. Operating in a somewhat miserly fashion, evolution predictably goes "where the money is" for increasing variability and naturally selecting out errors. We have the option of selecting extended language that can help us to understand how our brain works, why we use language the way we do, and why we ignore the emotional liabilities we incur with affective language. We can consider making effective linguistic improvements that can give us a better understanding of our world and how we fit in. We have the option of going "where the money is" and leveraging our systematic knowledge to work more cooperatively with nature.

Acknowledging and accepting nature as a major stakeholder gives us a practical perspective on life. What we learn from the laws of nature, we can apply to our language by incorporating constructions that are more amenable to cooperative adaptation in a changing world. Natural evolution operates within the general laws of nature, and we can benefit from this knowledge. Populations can benefit when individuals and groups share responsibility and accountability for their own actions. Incorporating cooperative polices into humane endeavors can reap benefits for nature and its human constituents. Rather than the force of instruction, cooperative policies depend on effectively informing our language constructions for improving our thinking abilities. Effective linguistic constructions support an understanding of why and how our brain and policies work together with nature, and how systematic knowledge influences our thinking and enables more robust cooperation.

When faced with the challenge of listening to others' opinions and finding adaptive solutions to complex problems, we can consider tempering our natural human tendency to jump to a conclusion, stop listening to others, and resort to only searching for confirming information. We can step back, resume listening, and take a thorough look at assessing and elaborating the problem space. We can more likely find effective novel solutions with extended language that helps us to see the big picture. Macroscopically, effective language facilitates understanding of relationships among humans and nature in a fairly simple manner, simple but not to be confused with easy or with a free lunch. At the very least, we all can consider playing and living by the same simple rules—cooperatively.

Summary

The exploration of evolutionary processes and of the explanations as put forth in this book supports the hypothesis that natural language provides a unique tool fit for understanding cognitive relationships and explaining how the world works, which enables thinking in sync with the physical laws of nature. The model works, at least in part, by exploiting the following concepts with effective language structure and function that facilitates coherently informed cognition that corresponds with nature:

- Information accuracy
- Information-processing accuracy
- Event-related information accuracy
- Dynamic correspondence
- Logical correspondence
- Classification correspondence
- Value correspondence
- Error-tuning correspondence

These information-based concepts can contribute to a better understanding of how our brains and the world work and of the role language plays in effective decision making and in promoting more harmonious relationships. Deliberate use of effective language can provide us with a practical mechanism for accuracy with which to represent and inform knowledge. In turn, knowledge fidelity logically corresponds to a more coherent correlation of our language, semantics, thoughts, and cooperative behavior with the natural world and with other humans. Extended language allows us to effectively abstract ideas about the things and events we experience in the environment as meaningful information about relationships. Extended language leverages symmetry breaking and congruent referencing with invariant time, and integrates general systematic principles of processes and relationships in nature. Thus we reach an understanding of the correspondence between our language code, the world, our biological brains, and what we call the mind. The mind code is simply our connection to the realized world in which we live:

- Time, symmetry breaking, asymmetrical-time referencing, and probabilistic information processing
- Polarity: positives and negatives, attraction and repulsion, approach and avoidance
- Relative change, change of position, change of state
- Ratios, recurrence, oscillations, and periodicity

The open-ended process of evolution has endowed us with a functional language module (Griffiths, 2007, p. 199) with the capability for operating globally under non-equilibrium constraints. This language module provides a level of semantic plasticity that can literally be fine-tuned with rules that robustly maximize adaptive behavior (Camazine et al, 2001, p. 13). Language tuned for global reliability and fidelity can provide us with an effective method for tuning in to the environment, helping us become well informed and naturally well connected.

Both computer science engineering and biological systems include general and extendible features that encompass modularity, robustness, and reliable policies for interaction (Kirschner & Gerhart, 2005, p. 261). An organism does not normally achieve robustness through rigid construction and behavior but through adaptive flexibility, not by refusing to change but by compensating for the changes around it (Kirschner & Gerhart, 2005, pp. 107–108). The same construction principles of rigidity and flexibility apply to language-related regulatory processes, logic, plasticity, and fitness.

Regulation represents a powerfully conserved core process in evolution, going back at least to the development of eukaryotic gene regulation (Kirschner & Gerhart, 2005, p. 119; Nüsslein-Volhard, 2006, p. 35) and probably to the first living organisms in general. The human language module and its role in executive functioning represent an extension of this biological regulation, and it would be surprising if that were not the case. As Jennings (1906, p. 349) says, "...there seems no reason to suppose that regulation in behavior...is of a fundamentally different character from regulation elsewhere."

It is not hard to imagine an evolutionary process where the population is searching for fitter candidate languages and grammars (Nowak, 2006, p. 280). The global nature of extended language provides a fitter candidate by invariant time-related referencing. Language, the physical world, and cognitive regulation logically unite with global information and knowledge. In other words, globally referenced language informs globally fit knowledge. Language itself can be explained in terms of evolutionary fitness as evolving and diverging from the localized use of primitive linguistic constructions. The evolution and divergence to extended language enable global use that enhances language fitness and more harmonious relations.

As a co-evolving culture, science open-mindedly explores nature with language, policies, and extended logic for regulating and optimizing scientific discovery processes. These open-ended processes enable science to measure and describe, articulate, and correlate information discovered in nature to knowledge of our world as accurately and precisely as possible.

Language provides a means for abstracting information as knowledge about interactions between organisms and the environment (Arbib, 2003, p. 15). A reliance on the global effectiveness of extended language for enhancing linguistic fidelity can provide practical and tractable applications for language recalibration well beyond the domain of science—from intimate to global relationships, among societies, and with the natural world itself.

Why is this recalibration important? Localized beliefs, values, and logic place undue absolute constraints on cognitive fitness, flexibility, communication, cooperation, and relative harmony among human kinship around the world. While beliefs vary considerably across cultures (Sapir, 1949, p. 162; Whorf, 1956, p. 221), the use of absolute affective reasoning and the resulting conflict among humans seem fairly consistent within and across cultures, afflicting most societies. All people living today are closely related genetically (Nüsslein-Volhard, 2006, p. 131) but are widely separated by polarizing absolute cultural beliefs constrained by, represented by, transmitted by, and expressed by language—unfortunately without regard for the relative accuracy of those beliefs (Pronin et al, 2002, p. 651; Jaynes, 2007, p. xxv).

> We have met the enemy and he is us. (Kelly, 1971)

We have reviewed ways to improve our language and reasons to do so. We have an opportunity to choose and use a more naturally harmonious language. We have seen that it's not necessarily the most frugal way to go, but the benefits in efficiency and effective cooperation are likely worthwhile in the long run. We can each learn to be friendlier with nature and others—and perhaps even learn to become our own best friend.

Conclusions

Perhaps Aristotle's insights about the natural correspondence of the words with which we think about the world are meaningful today. With a few updates, his theory for a coherent mapping between language, thought, and reality in the natural world can potentially illuminate modern society and science alike. We do indeed use language, words, and thought to describe the structure of relationships among organisms, reality, and the world in which we live—but doing so in a probabilistic manner enables the updating of unexamined language habits.

> Probability can no longer be reasonably considered a mere state of mind, when it clearly governs the laws of nature. (Prigogine, 1997, p. 132)

Aristotle's early insight can offer a useful perspective on language, cognition, and the natural world. Stated differently, we now have the logical option to map our language onto the natural world in a more sensible manner. As a social construct, probability as logic represents a potential tipping point for more effective thinking, affording us an opportunity to enhance our language fitness and extend our understanding toward a more reliable, open-ended view of the relatively uncertain and difficult-to-predict world we live in—and an opportunity to logically adapt our thought and behavior accordingly.

By accepting scientific evidence regarding evolutionary theory, the uncertainties addressed by probability, and the one-way nature of time, we can integrate these principles not only into human knowledge but also into humane behavior. We can recalibrate our language and reasoning to work nondeterministically rather than deterministically. To ignore scientific knowledge invites continued risk of disharmony, conflict, and aggression (Beck, 1999).

If instead we accept these natural fundamental principles and apply them to scrutinize and enlighten our inherited beliefs and predetermined values to minimize localized constraints on our reasoning, we can build a more harmonious foundation for creative, flexible cooperation among fallible and imperfect but relatively adaptable human beings living together in a global world.

> In his day over two thousand years ago, Aristotle inherited a structurally primitive-made *language*. He, as well as the enormous majority of us at present, never realized that what is going on outside our skins is certainly *not* words. We never "think" about this distinction, but we all take over semantically from our parents and associates their habitual forms of representation involving structure as *the* language in which to talk about this world, not knowing, or else forgetting, that a language to be fit to represent this world should at least have the *structure* of this world. (Korzybski, 1958, p. 88 [emphasis in original])

Extended language promotes coherent articulation and cooperation with nature and with others, offering new meaning and purpose to relationships, from intimate to global. A global view does not presuppose a perfect articulation, a perfect solution, or a free lunch. However, it does make it somewhat difficult to ignore the way constrained language with rigid absolute semantics routinely disrupts adaptive relationships and generates polarized disharmony throughout the natural world. Even if "what we see is what we get," when evaluated in terms of effectiveness, satisfaction may have more to do with trying to see what is "really there" and evaluating our beliefs and preferences to improve the beneficial effects of the predictions and choices we make.

If we work to better inform ourselves—so that our expectations and the outcomes of our choices match more harmoniously and predictably with natural reality—we can come to better understand the logical outcomes of our decisions, that what we get is, in fact, what we get. Accepting language fidelity as a beneficial concept, and more importantly, as a logical cognitive behavior, encourages us to humanely improve our thinking, our behavior, and our relationships overall.

Reasoning more effectively can provide plausible benefits for us, our children, and the world, hopefully for many generations to come. We can continue to operate imperatively as if we have no choice about how we make our decisions, or, like scientists accountable for maintaining scientific methodology, we can seize the opportunity, and the responsibility, to value, actively choose, demonstrate, and promote more effective reasoning, with the possibility of far-reaching benefits.

> We stand on the threshold of a great age of science; we are already over the threshold; it is for us to make that future our own. (Bronowski, 1977, p. 5)

References

Adams, F. (1972). *Genuine Works of Hippocrates.* Vol. 2. (originally published in 1886) pp. 344–345. Malabar, FL: Krieger Publishing.

Adolphs, R. (2006). What is special about social cognition? pp. 269–285. In *Social Neuroscience, People Thinking About People,* J.T. Cacioppo, P.S. Visser, & C.L. Pickett, eds. Cambridge, MA: MIT Press.

Adrian, J.E., Clemente, A., & Villanueva, L. (2007). Mothers' use of cognitive state verbs in picture-book reading and the development of children's understanding of mind: A longitudinal study. *Child Development* 78:1052–1067.

Ahern, T.H., & Young, L.J. (2009). The impact of early life family structure on adult social attachment, alloparental behavior, and the neuropeptide systems regulating affiliative behaviors in the monogamous prairie vole (*Microtus ochrogaster*). *Frontiers of Behavioral Neuroscience* 3:17.

Alexander, G.E., Delong, M.R., & Strick, P.L. (1986). Parallel organization of functionally segregated circuits linking basal ganglia and cortex. *Annual Review of Neuroscience* 9:357–381.

Alexander, M.P. (2002). Disorders of language after frontal lobe injury: Evidence for neural mechanisms of assembling language. pp. 159–167. In *Principles of Frontal Lobe Function*, D.T. Stuss, & R.T. Knight, eds. New York, NY: Oxford University Press.

Amaral, D., & Lavenex, p. (2007). Hippocampal neuroanatomy. pp. 37–114. In *The Hippocampal Book*, p. Anderson, R. Morris, D. Amaral, T. Bliss, & J. O'Keefe, eds. New York, NY: Oxford University Press.

Amit, D.J. (1989). *Modeling Brain Function: The World of Attractor Neural Networks.* Cambridge, UK: Cambridge University Press.

Anderson, V., Levin, H.S., & Jacobs, R. (2002). Executive functions after frontal lobe injury: A developmental perspective. pp. 504–527. In *Principles of Frontal Lobe Function*, D.T. Stuss & R.T. Knight, eds. New York, NY: Oxford University Press.

Angulo-Barroso, R.M., & Tiernan, C.W. (2008). Motor system development. pp. 147–160. In *Handbook of Developmental Cognitive Neuroscience* (2nd ed.), C.A. Nelson & M. Luciana, eds. Cambridge, MA: MIT Press.

Arbib, M.A. (2003). Dynamics and adaptation in neural networks. pp. 15–23, 606–611. In *The Handbook of Brain Theory and Neural Networks* (2nd ed.), M.A. Arbib, ed. Cambridge, MA: MIT Press.

Arbib, M.A. (2010). From mirror writing to mirror neurons. pp. 1–12. In *From Animals to Animats 11 (Proceedings of the 11th International Conference on Simulation of Adaptive Behavior)*, S. Doncieux, B. Girard, A. Guillot, J. Hallam, J.-A. Meyer, & J-B. Mouret, eds. Berlin: Springer-Verlag.

Arbib, M.A., Metta, G., & van der Smagt, p. (2008). Neurorobotics: From action to vision. pp. 1453–1480. In *Handbook of Robotics*, B. Siciliano & O. Khatib, eds. Berlin: Springer-Verlag.

Armor, D.A., & Taylor, S.E. (2002). When predictions fail: The dilemma of unrealistic optimism. pp. 334–347. In *Heuristics and Biases: The Psychology of Intuitive Judgment*, T. Gilovich, D. Griffin, & D. Kahneman, eds. New York, NY: Cambridge University Press.

Ashby, W.R. (1960). *Design for a Brain* (2nd revised ed.), pp. 209–210. London: Chapman and Hall.

Axelrod, R. (2006). *The Evolution of Cooperation* (revised ed., originally published in 1984). New York, NY: Basic Books.

Ayala, F.J. (2009). Molecular evolution. In *Evolution: The First Four Billion Years*, M. Ruse & J. Travis, eds. Cambridge, MA: Belknap Press.

Bailey, C.E. (2007a). Cognitive accuracy and intelligent executive function in the brain and in business. *Annals of the New York Academy of Sciences* 1118:122–141.

Bailey, C.E. (2007b). Semantically mediated integration of cognition in *Homo sapiens*: Evolution, grammar, uncertainty, and cognitive accuracy. *Cognitive Sciences* 3:85–142.

Bailey, C.E. (2009). Exploiting the relationship of natural language and computer science: A novel theoretical approach to fairness. *Cognitive Sciences* 4(2):1–30.

Barkley, R.A. (1997). *ADHD and the Natural History of Self-Control*. New York, NY: Guilford Press.

Barlow, H.B. (1953). Action potentials from the frog's retina. *Journal of Physiology* 119:58–68.

Barlow, H.B. (1959). Possible principles underlying the transformations of sensory messages. pp. 217–234. In *Sensory Communication: Contributions to the Symposium on Principles of Sensory Communication*, W.A. Rosenblith, ed. Cambridge, MA: MIT Press.

Barlow, H.B. (1972). Single units and sensation: A neuron doctrine for perceptual psychology? *Perception* 1:371–394.

Barsalou, L.W. (2008). Grounded cognition. *Annual Review of Psychology* 59:617–645.

Barsalou, L.W. (2009). Simulation, situated conceptualization, and prediction. *Philosophical Transactions of the Royal Society of London: Biological Sciences* 364:1281–1289.

Bauman, M.D., & Amaral D.G. (2008). Neurodevelopment of social cognition. pp. 161–186. In *Handbook of Developmental Cognitive Neuroscience* (2nd ed.), C.A. Nelson & M. Luciana, eds. Cambridge, MA: MIT Press.

Beck, A.T. (1976). *Cognitive Therapy and the Emotional Disorders.* New York, NY: International University Press.

Beck, A.T. (1999). *Prisoners of Hate: The Cognitive Basis of Anger, Hostility, and Violence.* New York, NY: HarperCollins.

Becker, S. (2007). Modeling the mind: From circuits to systems. pp. 12–33. In *New Directions in Statistical Signal Processing: From Systems to Brains*, S. Haykin, J.C. Principe, T.J. Sejnowski, & J. McWhirter, eds. Cambridge, MA: MIT Press.

Beinhocker, E.D. (2006). *The Origins of Wealth: Evolution, Complexity, and the Radical Remaking of Economics.* Boston, MA: Harvard Business Press.

Bellman, R. (1961). *Adaptive Control Processes: A Guided Tour.* Princeton, NJ: Princeton University Press.

Benjafield, J.G. (2007). *Cognition* (3rd ed.) Toronto, Canada: Oxford University Press.

Benton, M. (2009). Paleontology and the history of life. pp. 80–104. In *Evolution: The First Four Billion Years*, M. Ruse & J. Travis, eds. Cambridge, MA: Belknap Press.

Berger, H. (1929). Ueber das elektroenkephalogramm des menschen. *Archiv fur Psychiatrie und Nervenkrankheiten* 87:527–570.

Bierce, A. (1910). *The Collected Works of Ambrose Bierce.* Vol. 4, *Shapes of Clay.* New York and Washington: Neale Publishing Company.

Bishop, C.M. (2006). *Pattern Recognition and Machine Learning.* New York, NY: Springer.

Bizzi, E., & Mussa-Ivaldi, F.A. (2009). Neurobiology of coordinate transformations. pp. 541–551. In *The Cognitive Neurosciences* (4th ed.), M.S. Gazzaniga, ed. Cambridge, MA: MIT Press.

Bloom, p. (2002). *How Children Learn the Meaning of Words.* Cambridge, MA: MIT Press.

Bogacz, R., Brown, M.V., & Giraud-Carrier, C. (2001). A familiarity discrimination algorithm. pp. 428–441. In *Emergent Neural Computational Architectures Based on Neuroscience*, S. Wermter, J. Austin, & D. Williams, eds. New York, NY: Springer-Verlag.

Bonabeau, E., & Cogne, F. (1996). Oscillation-enhanced adaptability in the vicinity of a bifurcation: The example of foraging in ants. pp. 537–544. In *From Animals to Animats 4 (Proceedings of the Fourth International Conference on Simulation of Adaptive Behavior)*, p. Maes, M.J. Mataric, J.-A. Meyer, J. Pollack, & S.W. Wilson, eds. Cambridge, MA: MIT Press.

Boroditsky, L. (2003). Linguistic relativity. pp. 917–921. In *Encyclopedia of Cognitive Science*, L. Nadel, ed. London: Macmillan Press.

Boroditsky, L. (2007). Comparison and the development of knowledge. *Cognition* 102:118–128.

Boroditsky, L. (2009). How does our language shape the way we think? pp. 116–129. In *What's Next? Dispatches on the Future of Science*, M. Brockman, ed. New York, NY: Vintage Press.

Boroditsky, L. (2010). Lost in translation. *Wall Street Journal* July 24, 2910.

Boroditsky, L, & Gaby, A. (2010). Remembrances of times east: absolute spatial representations of time in an Australian Aboriginal community. *Psychological Science* 21(11):1635–1639.

Bos, K.J., Fox, N., Zeanah, C.H., & Nelson, C.A. (2009). Effects of early psychosocial deprivation on the development of memory and executive function. *Frontiers of Behavioral Neuroscience* 3:16.

Boyd, R., & Richerson, P.J. (2005). *The Origin and Evolution of Cultures*. New York, NY: Oxford University Press.

Braitenberg, V. (1984). *Vehicles: Experiments in Synthetic Psychology*. Cambridge, MA: MIT Press.

Braver, T.S., & Ruge, H. (2006). Functional neuroimaging of executive functions. pp. 307–348. In *Handbook of Functional Neuroimaging of Cognition*, R. Cabeza & A. Kingstone, eds. Cambridge, MA: MIT Press.

Broca, p. (1877). Rapport sur un memorie M. Armund de Fleury intitule: De l'inegalite dynamique des deux hemispheres cerebraux. *Bulletin of the New York Academy of Medicine* 6:508–539.

Bronowski, J. (1977). *A Sense of the Future: Essays in Natural Philosophy*. Cambridge, MA: MIT Press.

Brown, T.E. (2005). *Attention Deficit Disorder: The Unfocused Mind in Children and Adults*. New Haven, CT: Yale University Press.

Browne, M.N., & Keeley, S.M. (2007). *Asking the Right Questions: A Guide to Critical Thinking* (8th ed.) NJ: Pearson Prentice Hall, NJ: Pearson Prentice Hall.

Brunswik, E. (1955). Representative design and probabilistic theory in a functional psychology. *Psychological Review* 2:193–217.

Buchanan, T.W., Tranel, D., & Adolphs, R. (2009). The human amygdala in social function. pp. 289–318. In *The Human Amygdala*, P.J. Whalen & E.A. Phelps, eds. New York, NY: Guilford Press.

Burgess, N. (2007). Computational models of the spatial and mnemonic functions of the hippocampus. pp. 715–749. In *The Hippocampal Book*, p. Anderson, R. Morris, D. Amaral, T. Bliss, & J. O'Keefe, eds. New York, NY: Oxford University Press.

Burgess, N., Maguire, E., & O'Keefe, J. (2002). The human hippocampus and spatial episodic memory. *Neuron* 35:625–641.

Burgess, N., & O'Keefe, J. (2003). Hippocampus: Spatial models. pp. 539–543. In *The Handbook of Brain Theory and Neural Networks* (2nd ed.), M.A. Arbib, ed. Cambridge, MA: MIT Press.

Buzsáki, G. (2006). *Rhythms of the brain*. New York, NY: Oxford University Press.

Buzsáki, G. (2007). The structure of consciousness. *Nature* 446(15):267.

Buzsáki, G. (2010). Neural syntax: Cell assemblies, synapsembles, and readers. *Neuron* 68:362–385.

Cacioppo, J.T., & Berntson, G.G. (2004). Social neuroscience. pp. 977–985. In *The Cognitive Neurosciences III*, M.S. Gazzaniga, ed. Cambridge, MA: MIT Press.

Caldwell, H.K., & Young, W.S. (2006). Oxytocin and vasopressin: Genetics and behavioral implications. pp. 573–607. In *Handbook of Neurochemistry and Molecular Neurobiology: Neuroactive Proteins and Peptides* (3rd ed.), R. Lim, ed. New York, NY: Springer.

Callebaut, W., Müller, G.B., & Newman, S.A. (2007). The organismic systems approach. pp. 25–92. In *Integrating Evolution and Development: From Theory to Practice*, R. Sansom & R.N. Brandon, eds. Cambridge, MA: MIT Press.

Camazine, S., Deneubourg, J-L., Franks, J.S., Theraulaz, G., & Bonabeau, E. (2001). *Self-Organization in Biological Systems* Princeton, NJ: Princeton University Press.

Cangelosi, A. (2001). Evolution of communication and language using signals, symbols, and words. *IEEE Transactions on Evolutionary Computation* 5(2):93–101.

Cangelosi, A., & Parisi, D. (2004). The processing of nouns and verbs in neural networks: Insights from synthetic brain imaging. *Brain and Language* 89(2): 401–408.

Canli, T. (2009). Individual differences in human amygdala function. pp. 250–264. In *The Human Amygdala*, P.J. Whalen & E.A. Phelps, eds. New York, NY: Guilford Press.

Capaldi, N. (1987). *The Art of Deception: An Introduction to Critical Thinking*. New York, NY: Prometheus Books.

Carroll, S.B. (2005). *Endless Forms Most Beautiful: The New Science of Evo Devo and the Making of the Animal Kingdom* New York, NY: W.W. Norton.

Carroll, S.B. (2006). *The Making of the Fittest: DNA and the Ultimate Forensic Record of Evolution*. New York, NY: W.W. Norton.

Casasanto, D., & Boroditsky, L. (2008). Time in the mind: Using space to think about time. *Cognition* 106:579–593.

Cervantes-Perez, F. (2003). Visuomotor coordination in frog and toad. pp. 1219–1224. In *The Handbook of Brain Theory and Neural Networks* (2nd ed.), M.A. Arbib, ed. Cambridge, MA: MIT Press.

Chabris, C., & Simons, D. (2010). *The Invisible Gorilla: And Other Ways Our Intuitions Deceive Us*. New York, NY: Crown Publishers.

Chapman, G.B., & Johnson, E.J. (2002). Incorporating the irrelevant: Anchors in judgment of belief and value. pp. 120–138. In *Heuristics and Biases: The Psychology of Intuitive Judgment*, T. Gilovich, D. Griffin, & D. Kahneman, eds. New York, NY: Cambridge University Press.

Chater, N., & Christiansen, M.H. (2008). Language as shaped by the brain. *Behavioral and Brain Sciences* 31(5):489–558.

Chater, N., & Christiansen, M.H. (2010). Language acquisition meets language evolution. *Cognitive Science* 34(7):1–27.

Chater, N., Reali, F., & Christiansen, M.H. (2009). Restrictions on biological adaptation in language evolution. *Proceedings of the National Academy of Sciences of the United States of America* 106(4):1015–1020.

Cheney, D.L., & Seyfarth, R.M. (2008). *Baboon Metaphysics: The Evolution of the Social Mind.* Chicago, IL: University of Chicago Press.

Chersi, F., Thill, S., Ziemke, T., & Borghi, A.M. (2010). Sentence processing: Linking language to motor chains. *Frontiers in Neurorobotics* 4, article 4, pp. 1–14.

Chow, T.W., & Cummings, J.L. (2007). Frontal-subcortical circuits. pp. 25–43. In *The Human Frontal Lobes.* (2nd ed.), B.L. Miller & J.L. Cummings, eds. New York, NY: Guilford Press.

Christiansen, M.H., & Dale, R. (2003). Language evolution and change. pp. 604–606. In *The Handbook of Brain Theory and Neural Networks* (2nd ed.), M.A. Arbib, ed. Cambridge, MA: MIT Press.

Churchland, P.S., & Sejnowski, T.J. (1994). *The Computational Brain.* Cambridge, MA: MIT Press.

Coolidge, F.L., & Wynn, T. (2009). *The Rise of Homo sapiens: The Evolution of Modern Thinking.* Oxford, UK: Wiley-Blackwell.

Corballis, M.C. (2002). *From Hand to Mouth: The Origins of Language.* Princeton, NJ: Princeton University Press.

Crecraft, D.I., & Gergely, G. (2002). *Analog Electronics: Circuits, Systems, and Signal Processing.* Woburn, MA: Butterworth-Heinemann.

Cummings, J.L., & Miller, B.L. (2007). Conceptual and clinical aspects of the frontal lobes. pp. 12–21. In *The Human Frontal Lobes* (2nd ed.), B.L. Miller & J.L. Cummings, eds. New York, NY: Guilford Press.

Curtis, C.E., & D'Esposito, M. (2006). Functional neuroimaging of working memory. pp. 269–306. In *Handbook of Functional Neuroimaging of Cognition*, R. Cabeza & A. Kingstone, eds. Cambridge, MA: MIT Press.

D'Ambrosio, D.B., & Stanley, K.O. (2007). A novel generative encoding for exploiting neural network sensor and output geometry. pp. 974–981. In *Proceedings of the Genetic and Evolutionary Computation Conference (GECCO-2007).* New York, NY: ACM Press.

Darwin, C.R. (1859). *On the Origin of Species by Means of Natural Selection, or the Preservation of Favoured Races in the Struggle for Life* (1st ed.) London: John Murray.

Dawes, R.M., Faust, D., & Meehl, P.E. (2002). Clinical versus actuarial judgment. pp. 716–729. In *Heuristics and Biases: The Psychology of Intuitive Judgment*, T. Gilovich, D. Griffin, & D. Kahneman, eds. New York, NY: Cambridge University Press.

Decety, J. (2007). A social cognitive neuroscience model of human empathy. pp. 246–270. In *Social Neuroscience: Integrating Biological Explanations of Social Behavior*, E. Harmon-Jones & p. Winkielman, eds. New York, NY: Guilford Press.

D'Esposito, M., & Postle, B.R. (2002). The organization of working memory function in lateral prefrontal cortex: Evidence from event-related functional MRI. pp. 168–187. In *Principles of Frontal Lobe Function*, D.T. Stuss & R.T. Knight, eds. New York, NY: Oxford University Press.

Diba, K., & Buzsáki, G. (2008). Hippocampal network dynamics constrain the time lag between pyramidal cells across modified environments. *Journal of Neuroscience* 28(50):13448–13456.

Dils, A.T., & Boroditsky, L. (2010). The visual consequences of language processing: Processing unrelated language can change what you see. *Psychonomic Bulletin & Review* 17(6):882–888.

Donald, M. (2008). How culture and brain mechanisms interact in decision making. pp. 191–205. In *Better Than Conscious? Decision Making, the Human Mind, and Implications for Institutions*, C. Engel & W. Singer, eds. Cambridge, MA: MIT Press.

Dunning, D., Meyerowitz, J.A., & Holzberg, A.D. (2002). Ambiguity and self-evaluation: The role of idiosyncratic trait definitions in self-serving assessments of ability. pp. 324–333. In *Heuristics and Biases: The Psychology of Intuitive Judgment*, T. Gilovich, D. Griffin, & D. Kahneman, eds. New York, NY: Cambridge University Press.

du Sautoy, M. (2008). *Symmetry: A Journey into the Patterns of Nature*. New York, NY: HarperCollins.

Edelman, G .M. (1992). *Bright Air, Brilliant Fire: On the Matter of Mind*. New York, NY: Basic Books.

Einstein, A. (1961). *Relativity: The Special and the General Theory* (15th ed.), R.W. Lawson, trans. New York, NY: Three Rivers Press.

Eliasmith, C., & Anderson, C.H. (2003). *Neural Engineering: Computation, Representation, and Dynamics in Neurobiological Systems*. Cambridge, MA: MIT Press.

Ellis, A. (2001). *Overcoming Destructive Beliefs, Feelings, and Behaviors*. Amherst, NY: Prometheus Books.

Ellis, A. (2005). *The Myth of Self-Esteem: How Rational Emotive Behavior Therapy Can Change Your Life Forever*. Amherst, NY: Prometheus Books.

Ellis, A., & Harper R.A. (1997). *A Guide to Rational Living* (3rd ed.; originally published in 1976). North Hollywood, CA: Melvin Powers Wilshire Book Company.

Ellner, S.P., & Guckenheimer, J. (2006). *Dynamic Models in Biology*. Princeton, NJ: Princeton University Press.

Érdi, P., & Kiss, T. (2001). The complexity of the brain: Structural, functional, and dynamic models. pp. 203–211. In *Emergent Neural Computational Architectures*

Based on Neuroscience, S. Wermter, J. Austin, & D. Williams, eds. New York, NY: Springer-Verlag.

Faraday, M. (1816–1818), Course of 17 chemistry lectures, Jan. 17, 1816 to Aug. 19, 1818. Unpublished manuscript. In Institution of Electrical Engineers, London, "Miscellaneous Manuscripts, SC2."

Fausey, C.M., Long, B.L., Inamon, A., & Boroditsky, L. (2010). Constructing agency: The role of language. *Frontiers in Psychology* 1:162.

Feldman, J.A. (2006). *From Molecules to Metaphor: A Neural Theory of Language*. Cambridge, MA: MIT Press.

Floreano, D., & Mattiussi, C. (2008). *Bio-Inspired Artificial Intelligence* Cambridge, MA: MIT Press.

Floreano, D., Mitri, S., Magnenat, S., & Keller, L. (2007). Evolutionary conditions for the emergence of communication. *Current Biology* 17:514–519.

Fox, C.R., & Tversky, A. (2000). Ambiguity aversion and comparative ignorance. pp. 528–542. In *Choices, Values, and Frames*, D. Kahneman, & A. Tversky, eds. New York, NY: Cambridge University Press.

Freese, J.L., & Amaral, D.G. (2009). Neuroanatomy of the primate amygdala. pp. 3–42. In *The Human Amygdala*, P.J. Whalen & E.A. Phelps, eds. New York, NY: Guilford Press.

Friederici, A.D. (2008). Brain correlates of language processing during the first years of life. pp. 117–126. In *Handbook of Developmental Cognitive Neuroscience* (2nd ed.), C.A. Nelson & M. Luciana, eds. Cambridge, MA: MIT Press.

Frith, C. Rees, G. Macaluso, E., & Blakemore, S. (2004). Mechanism of attention: Beyond the biased competition model. pp. 245–268. In *Human Brain Function* (2nd ed.), R.S.J. Frackowiak, K.J. Friston, C.D. Frith, R.J. Dolan, C.J. Price, S. Zeki, J. Ashburner, & W. Penny, eds. San Diego, CA: Elsevier.

Fukuda, T., & Kubota, N. (2003). Intelligent learning robotic systems using computational intelligence. pp. 121–138. In *Computational Intelligence: The Experts Speak*, D.B. Fogel & C.J. Robinson, eds. Hoboken, NJ: John Wiley & Sons.

Fuster, J.M. (2002). Physiology of executive functions. pp. 96–108. In *Principles of Frontal Lobe Function*, D.T. Stuss & R.T. Knight, eds. New York, NY: Oxford University Press.

Fuster, J.M. (2003). *Cortex and Mind: Unifying Cognition*. New York, NY: Oxford University Press.

Gage, F. (1998). Neurogenesis in the adult human hippocampus. *Nature Medicine* 4:1313–1317.

Gage, F. (2002). Neurogenesis in the adult brain. *Journal of Neuroscience* 22(3):612–613.

Gallese, V. (2008). Mirror neurons and the social nature of language: The neural exploitation hypothesis. *Social Neuroscience* 3:317–333.

Gallistel, C.R., & King, A.P. (2009). *Memory and the Computational Brain: Why Cognitive Science Will Transform Neuroscience*. Oxford, UK: Wiley-Blackwell.

Gärdenfors, p. (2010). The evolution of semantics: A meeting of the minds. pp. 407–410. In *Proceedings of the 8th International Conference (EVOLANG8), The Evolution of Language*, A.D.M. Smith, M. Schouwstra, B. de Boer, & K. Smith eds. Singapore: World Scientific Publishing.

Gauci, J., & Stanley, K.O. (2008). A case study on the critical role of geometric regularity in machine learning. pp. 628–633. In *Proceedings of the Twenty-Third AAAI Conference on Artificial Intelligence (AAAI-2008)*. Menlo Park, CA: AAAI Press.

Gazzaley, A., & D'Esposito, M. (2007). Unifying prefrontal cortex function: Executive control, neural networks, and top-down modulation. pp. 187–206. In *The Human Frontal Lobes* (2nd ed.), B.L. Miller & J.L. Cummings, eds. New York, NY: Guilford Press.

Gibson, J.J. (1986). *The Ecological Approach to Visual Perception*. Hillsdale, NJ: Lawrence Erlbaum Associates.

Gilovich, T. (1991). *How We Know What Isn't So: The Fallibility of Human Reason in Everyday Life*. New York, NY: Free Press.

Gilovich, T., & Griffin, D. (2002). Heuristics and biases: Then and now. pp. 1–18. In *Heuristics and Biases: The Psychology of Intuitive Judgment*, T. Gilovich, D. Griffin, & D. Kahneman, eds. Cambridge, MA: Cambridge University Press.

Glimcher, P.W., Dorris, M.C., & Bayer, H.M. (2006). Physiological utility theory and the neuroeconomics of choice. *Games and Economic Behavior* 52:213–256.

Goldberg, E. (2002). The *Executive Brain: Frontal Lobes and the Civilized Mind* New York, NY: Oxford University Press.

Goldman-Rakic, P.S. (1995). Architecture of the prefrontal cortex and central executive. *Annals of the New York Academies of Science* 769:71–83.

Gough, M.P. (2008). Information equation of state. *Entropy* 10:150–159.

Grafton, S.T., Aziz-Zadeh, L., & Ivry, R.B. (2009). Relative hierarchies and the representation of action. pp. 641–652. In *The Cognitive Neurosciences* (4th ed.), M.S. Gazzaniga, ed. Cambridge, MA: MIT Press.

Griffiths, P.E. (2007). Evo-Devo meets the mind. pp. 195–225. In *Integrating Evolution and Development: From Theory to Practice*, R. Sansom & R.N. Brandon, eds. Cambridge, MA: MIT Press.

Gruau, F., Whitley, D., & Pyeatt (1996). A comparison between cellular encoding and direct encoding for genetic neural networks. pp. 81–89. In *Genetic Programming 1996: Proceedings of the First Annual Conference*, J.R. Koza, D.E. Goldberg, D.B. Fogel, & R.L. Riolo, eds. Cambridge, MA: MIT Press.

Gygax, P., Gabriel, U., Sarrasin, O., Oakhill, J., & Garnham, A. (2008). Generically intended, but specifically interpreted: When beauticians, musicians, and mechanics are all men. *Language and Cognitive Processes* 23(3):464–485.

Hamann, S. (2009). The human amygdala and memory. pp. 177–203. In *The Human Amygdala*, P.J. Whalen & E.A. Phelps, eds. New York, NY: Guilford Press.

Harnad, S. (1990). The symbol grounding problem. *Physica D* 42:335–346.

Hashimoto, T., Nakatsuka, M., & Konno, T. (2010). Linguistic analogy for creativity and the origin of language. pp. 184–191. In *Proceedings of the 8th International Conference (EVOLANG8), The Evolution of Language*, A.D.M. Smith, M. Schouwstra, B. de Boer, & K. Smith eds. Singapore: World Scientific Publishing.

Haykin, S. (2009). *Neural Networks and Learning Machines* (3rd ed.). Upper Saddle River, NJ: Pearson Education.

Hebb, D.O. (1949). *The Organization of Behavior*. New York, NY: Wiley.

Heerebout, B.T. (2011). *Getting Emotional with Evolutionary Simulations: The Origin of Affective Processing in Artificial Neural Networks*. Ridderkirk, the Netherlands: Ridderprint.

Heerebout, B.T., & Phaf, R.H. (2010). Good vibrations switch attention: An affective function for network oscillations in evolutionary simulations. *Cognitive, Affective, & Behavioral Neuroscience* 10:217–229.

Hille, B. (1992). *Ionic Channels of Excitable Membranes* (2nd ed.). Sunderland, MA: Sinauer Associates.

Hodgson, G.M. (2009). Foreword. *The Foundation of Non-Equilibrium Economics: The Principles of Circular and Cumulative Causation*. New York, NY: Routledge.

Holland, J.H. (1995). *Hidden Order: How Adaptation Builds Complexity*. Reading, MA: Perseus Books.

Hooker, C.I., & Knight, R.T. (2006). The role of lateral orbitofrontal cortex in the inhibitory control of emotion. pp. 307–324. In *The Orbital Frontal Cortex*, D.H. Zald & S.L. Rauch, eds. New York, NY: Oxford University Press.

Hopfield, J.J. (1982). Neural networks and physical systems with emergent collective computational abilities. *Proceedings of the National Academy of Sciences of the United States of America* 79:2554–2558.

Iglesias, P.A., & Ingalls, B.P., eds. (2010). *Control Theory and Systems Biology*. Cambridge, MA: MIT Press.

Iliescu, B.F., & Dannemiller, J.L. (2008). Brain–behavior relationship in early visual development. pp. 127–145. In *Handbook of Developmental Cognitive Neuroscience* (2nd ed.), C.A. Nelson & M. Luciana, eds. Cambridge, MA: MIT Press.

Itskov, V., Pastalkova, E., Mizuseki, K., Buzsáki, G., & Harris, K.D. (2008). Theta-mediated dynamics of spatial information in hippocampus. *Journal of Neuroscience* 28(23):5959–5964.

Itti, L., & Koch, C. (2001). Computational modeling of visual attention. *Nature Reviews Neuroscience* 2(3):194–203.

Izhikevich, E.M. (2007). *Dynamical Systems in Neuroscience: The Geometry of Excitability and Bursting*. Cambridge, MA: MIT Press.

Jackendoff, R. (2002). *Foundations of Language: Brain, Meaning, Grammar, Evolution*. Oxford, UK: Oxford University Press.

James, W. (2007). *The Principles of Psychology*. Vol. 2. (Originally published in 1890.) New York, NY: Cosimo Classics.

Jaynes, E.T. (1981). Entropy and search theory. Presented at the First Maximum Entropy Workshop, University of Wyoming, June 1981.

Jaynes, E.T. (2007). *Probability Theory: The Logic of Science* (5th printing). Cambridge, UK: Cambridge University Press.

Jeannerod, M. (2003). Action monitoring and forward control of movements. pp. 83–85. In *The Handbook of Brain Theory and Neural Networks* (2nd ed.), M.A. Arbib, ed. Cambridge, MA: MIT Press.

Jeffrey, K.J., Anderson, M.I., Hayman, R., & Chakraborty, S. (2004). A proposed architecture for the neural representation of spatial context. *Neuroscience and Biobehavioral Reviews* 28:201–218.

Jeffrey, R. (2004). *Subjective Probability: The Real Thing* Cambridge, UK: Cambridge University Press.

Jennings, H.S. (1906). *Behavior of the Lower Organisms*. New York, NY: Columbia University Press.

Johnson, M., & Lakoff, G. (2003). *Metaphors We Live By*. Chicago, IL: University of Chicago Press.

Johnston, J. (2008). *The Allure of Machine Life: Cybernetics, Artificial Life, and the New AI* Cambridge, MA: MIT Press.

Johnston, V.S. (2003). The origin and function of value. *Cognition and Emotion* 17:167–179.

Jones, M.D. (1998). *The Thinker's Toolkit: 14 Powerful Techniques for Problem Solving* New York, NY: Three Rivers Press.

Kahneman, D. (2000). New challenges to the rational assumption. pp. 758–774. In *Choices, Values, and Frames*, D. Kahneman & A. Tversky, eds. New York, NY: Cambridge University Press.

Kahneman, D. (2011). *Thinking, Fast and Slow*. New York, NY: Farrar, Straus & Giroux.

Kahneman, D., & Frederick, S. (2002). Representativeness revisited. pp. 49–81. In *Heuristics and Biases: The Psychology of Intuitive Judgment*, T. Gilovich, D. Griffin, & D. Kahneman, eds. New York, NY: Cambridge University Press.

Kahneman, D., Knetsch, J.L., & Thaler, R.H. (2000). Anomalies: Endowment effect, loss aversion, and status quo bias. pp. 159–170. In *Choices, Values, and Frames*, D. Kahneman & A. Tversky, eds. New York, NY: Cambridge University Press.

Kahneman, D., Slovic, P., & Tversky, A. (1982). *Judgment Under Uncertainty: Heuristics and Biases*. Cambridge, MA: Cambridge University Press.

Kahneman, D., & Tversky, A. (2000a). Conflict resolution: A cognitive perspective. pp. 473–487. In *Choices, Values, and Frames*, D. Kahneman & A. Tversky, eds. New York, NY: Cambridge University Press.

Kahneman, D., & Tversky, A. (2000b). Prospect theory: An analysis of decision under risk. pp. 17–43. In *Choices, Values, and Frames*, D. Kahneman, & A. Tversky, eds. New York, NY: Cambridge University Press.

Kandel, E.R. (2006). *In Search of Memory: The Emergence of a New Science of Mind.* New York, NY: W.W. Norton.

Kauffman, S.A. (1993). *The Origins of Order: Self-Organization and Selection in Evolution.* New York, NY: Oxford University Press.

Kelly J.F., & Westerhoff C.M. (2010). Does it matter how we refer to individuals with substance-related conditions? A randomized study of two commonly used terms. *International Journal of Drug Policy* 21(3):202–207.

Kelly, W. (1971). *Pogo.* Comic strip. New York, NY: Post-Hall Syndicate.

Kida, T.E. (2006). *Don't Believe Everything You Think: The 6 Basic Mistakes We Make in Thinking* Amherst, NY: Prometheus Books.

Kirschner, M.W., & Gerhart, J.C. (2005). *The Plausibility of Life: Resolving Darwin's Dilemma.* New Haven, CT: Yale University Press.

Knight, R.T., & Stuss, D.T. (2002). Prefrontal cortex: The present and the future. pp. 573–597. In *Principles of Frontal Lobe Function*, D.T. Stuss & R.T. Knight, eds. New York, NY: Oxford University Press.

Koch, C. (1999). *Biophysics of Computation: Information Processing in Single Neurons.* New York, NY: Oxford University Press.

Kohonen, T. (1984). *Self-Organization and Associative Memory.* New York, NY: Springer-Verlag.

Kohonen, T. (1987). *Content-Addressable Memories* (2nd ed.). Berlin: Springer-Verlag.

Korosi, A., & Baram, T. (2009). The pathways from mother's love to baby's future. *Frontiers of Behavioral Neuroscience* 3:27.

Korzybski, A. (1958). *Science and Sanity: An Introduction to Non-Aristotelian Systems and General Semantics* (4th ed., originally published in 1933). Lakeville, CT: International Non-Aristotelian Library Publishing.

Koziol, L.F., & Budding, D.E. (2010). *Subcortical Structure and Cognition: Implications for Neuropsychological Assessment.* New York, NY: Springer.

Kruger, J., & Dunning, D. (1999). Unskilled and unaware of it: How difficulties in recognizing one's own incompetence lead to inflated self-assessments. *Journal of Personality and Social Psychology* 77:1121–1134.

Kurzban, R. (2008). The evolution of implicit and explicit decision making. pp. 155–172. In *Better Than Conscious? Decision Making, the Human Mind, and Implications for Institutions*, C. Engel & W. Singer, eds. Cambridge, MA: MIT Press.

LaBar, K.S., & Warren, L.H. (2009). Methodological approaches to studying the human amygdala. pp. 155–176. In *The Human Amygdala*, P.J. Whalen & E.A. Phelps, eds. New York, NY: Guilford Press.

Lane, R.D., Nadel, L., & Kaszniak, A.W. (2000). The future of emotion research from the perspective of cognitive neuroscience. pp. 407–410. In *Cognitive Neuroscience of Emotion*, R.D. Lane & L. Nadel, eds. New York, NY: Oxford University Press.

Langer, E.J. (2000). Mindful learning. *Current Directions in Psychological Science* 9:220–223.

Langer, E.J., & Piper, A.J. (1987). The prevention of mindlessness. *Journal of Personality and Social Psychology* 53:280–287.

Laplace, P-S. (2009). *A Philosophical Essay on Probabilities* (translated from the 6th French ed.; originally published in 1814). Seaside, OR: Rough Draft Printing.

Laughlin, S.B. (2001). Energy as a constraint on the coding and processing of sensory information. *Current Opinion in Neurobiology* 11:475–480.

Laughlin, S.B. (2004). The implications of metabolic energy requirements for the representation of information in neurons. pp. 187–196. In *The Cognitive Neurosciences III*, M.S. Gazzaniga, ed. Cambridge, MA: MIT Press.

Laughlin, S.B., de Ruyter van Steveninck, R.R., & Anderson, J.C. (1998). The metabolic cost of neural information. *Nature Neuroscience* 1(1):36–41.

LeDoux, J.E. (1996). *The Emotional Brain*. New York, NY: Simon & Schuster.

LeDoux, J.E. (2002). *Synaptic Self: How Our Brains Become Who We Are*. New York, NY: Russell Sage Foundation.

LeDoux, J.E., & Schiller, D. (2009). The human amygdala: Insights from other animals. pp. 43–60. In *The Human Amygdala*, P.J. Whalen & E.A. Phelps, eds. New York, NY: Guilford Press.

Lee, D., & Seo, H. (2007). Mechanisms of reinforcement learning and decision making in the primate dorsolateral prefrontal cortex. *Annals of the New York Academy of Sciences* 1104:108–122.

Leff, H.S., & Rex, A.F. (2003). Introduction. pp. 2–39. In *Maxwell's Demon 2: Entropy, Classical and Quantum Information, Computing*, H.H. Leff & A.F. Rex, eds. London: Institute of Physics Publishing Ltd.

Lehman, J., & Stanley, K. (2008). Exploiting open-endedness to solve problems through the search for novelty. pp. 329–336. In *Artificial Life XI: Proceedings of the Eleventh International Conference on the Simulation and Synthesis of Living Systems*, S. Bullock, J. Noble, R. Watson, & M.A. Bedau, eds. Cambridge, MA: MIT Press.

Lehman, J., & Stanley, K. (2010). Efficiently evolving programs through the search for novelty. In *Proceedings of the Genetic and Evolutionary Computation Conference (GECCO-2010)*. New York: NY, ACM Press.

Lehman, J., & Stanley, K. (2011). Abandoning objectives: Evolution through the search for novelty alone. *Evolutionary Computation Journal* 19(2):189–223.

Lehninger, A.L. (1975). *Biochemistry* (2nd ed.). New York, NY: Worth Publishers.

Lenneberg, E. (1967). *Biological Foundations of Language*. New York, NY: John Wiley.

Lichter, D.G., & Cummings, J.L. (2001). Introduction and overview. pp. 1–43. In *Frontal-Subcortical Circuits in Psychiatric and Neurological Disorders*, D.G. Lichter & J.L. Cummings, eds. New York, NY: Guilford Press.

Lieblich, I., & Arbib, M.A. (1982). Multiple representations of space underlying behavior. *Behavioral Brain Science* 5:627–659.

Lindley, D. (2007). *Uncertainty: Einstein, Heisenberg, Bohr, and the Struggle for the Soul of Science*. New York, NY: Doubleday.

Lipson, H. (2007). Evolutionary robotics: Emergence of communication. *Current Biology* 17(9):R330–R332.

Loewenstein, G., & Prelec, D. (2000). Anomalies in intertemporal choice: Evidence and an interpretation. pp. 578–596. In *Choices, Values, and Frames*, D. Kahneman & A. Tversky, eds. New York, NY: Cambridge University Press.

Logothetis, N.K. (2004). Higher cognitive functions. pp. 849–969. In *The Cognitive Neurosciences III*, M.S. Gazzaniga, ed. Cambridge, MA: MIT Press.

Long, D.L., Baynes, K., & Prat, C. (2007). Sentence discourse representation in two cerebral hemispheres. pp. 329–353. In *Higher Level Language Processes in the Brain: Inference and Comprehension Processes*, F. Schmalhofer & C.A. Perfetti, eds. Mahwah, NJ: Lawrence Erlbaum.

Lorenz, E. (1963). Deterministic nonperiodic flow. *Journal of the Atmospheric Sciences* 20:130–141.

Loritz, D. (2002). *How the Brain Evolved Language*. New York, NY: Oxford University Press.

Luria, A.R. (1966/1980). *Higher Cortical Functions in Man*. New York, NY: Basic Books.

Luria, A.R. (1981). *Language and Cognition*, J.V. Wertsch, trans. Washington, DC: John Wiley & Sons.

Lytton, W.W. (2002). *From Computer to Brain: Foundations of Computational Neuroscience*. New York, NY: Springer-Verlag.

Macknik, S.L., & Martinez-Conde, S., with Blakeslee, S. (2010). *Sleights of Mind: What the Neuroscience of Magic Reveals About Our Everyday Deceptions*. New York, NY: Henry Holt.

MacLean, P.D. (1990). *The Triune Brain in Evolution*. New York, NY: Plenum Press.

Marr, D. (1971). Simple memory: a theory for archicortex. *Philosophical Transactions of the Royal Society B: Biological Sciences* 262:23–81.

Marr, D. (2010). *Vision: A Computational Investigation into Human Representation and Processing of Visual Information* (originally published in 1982). Cambridge, MA: MIT Press.

Maxon, S.C., & Canastar, A. (2006). Genetic aspects of aggression in nonhuman animals. pp. 3–19. In *Biology of Aggression*, R. Nelson, ed. New York, NY: Oxford University Press.

Maxwell, J.C. (1850). Letter to Lewis Campbell; reproduced in L. Campbell & W. Garrett, *The Life of James Clerk Maxwell*. New York, NY: Macmillan, 1881.

Maynard Smith, J. (2000). The concept of information in biology. *Philosophy of Science* 67:77–194.

Mazur, J. (2007). *The Motion Paradox: The 2,500-Year-Old Puzzle Behind the Mysteries of Time and Space*. New York, NY: Dutton.

McCaffrey, L.M., & Macara, I.G. (2009). Widely conserved signaling pathways in the establishment of cell polarity. *Cold Spring Harbor Perspectives in Biology* DOI: 10.1101/cshperspect.a001370.

McInerny, D.Q. (2005). *Being Logical: A Guide to Good Thinking.* New York, NY: Random House.

Mesulam, M.-M. (2000). *Principles of Behavioral and Cognitive Neurology* (2nd ed.). New York, NY: Oxford University Press.

Mesulam, M.-M. (2002). The human frontal lobes: Transcending the default mode through contingent coding. pp. 8–30. In *Principles of Frontal Lobe Function*, D.T. Stuss & R.T. Knight, eds. New York, NY: Oxford University Press.

Middleton, F.A., & Strick, P.L. (2001). A revised neuroanatomy of frontal-subcortical circuits. pp. 44–58. In *Frontal-Subcortical Circuits in Psychiatric and Neurological Disorders*, D.G. Lichter & J.L. Cummings, eds. New York, NY: Guilford Press.

Milgram, S. (2004). *Obedience to Authority* (first published in 1974). New York, NY: HarperCollins.

Minkowski, H. (1952). Space and time. pp. 75–76, 79–80. In *The Principle of Relativity: A Collection of Original Memoirs on the Special and General Theory of Relativity*, H.A. Lorentz, A. Einstein, H. Minkowski, & H. Weyl, eds. New York, NY: Dover.

Minsky, M.L., & Papert, S.A. (1969). *Perceptrons.* Cambridge, MA: MIT Press.

Miresco, M.J., & Kirmayer, L.J. (2006). The persistence of mind-brain dualism in psychiatric reasoning about clinical scenarios. *American Journal of Psychiatry* 163:913–918.

Mitchell, J.P., Mason, M.F., Macrae, C.N., & Mahzarin, R.B. (2006). Thinking about others: The neural substrates of social cognition. pp. 63–82. In *Social Neuroscience, People Thinking About People*, J.T. Cacioppo, P.S. Visser, & C.L. Pickett, eds. Cambridge, MA: MIT Press.

Mitchell, M. (2009). *Complexity: A Guided Tour.* New York, NY: Oxford University Press.

Mizuseki, K., Sirota, A., Pastalkova, E., & Buzsáki, G. (2009). Theta oscillations provide temporal windows for local circuit computation in the entorhinal-hippocampal loop. *Neuron* 64:267–280.

Moisl, H. (2001). Linguistic computation with state space trajectories. pp. 442–460. In *Emergent Neural Computational Architectures Based on Neuroscience*, S. Wermter, J. Austin, & D. Williams, eds. New York, NY: Springer-Verlag.

Montgomery, S.M., Betancur, M.I., & Buzsáki, B. (2009). Behavior-dependent coordination of multiple theta dipoles in the hippocampus. *Journal of Neuroscience* 29(5):1381–1394.

Montgomery, S.M., Sirota, A., & Buzsáki, G. (2008). Theta and gamma coordination of hippocampal networks during waking and rapid eye movement sleep. *Journal of Neuroscience* 28(26):6731–6741.

Morris, J., & Dolan, R. (2004). Emotion and memory: Functional neuroanatomy of human emotion. pp. 365–396. In *Human Brain Function* (2nd ed.), R.S.J. Frackowiak, K.J. Friston, C.D. Frith, R.J. Dolan, C.J. Price, S. Zeki, J. Ashburner, & W. Penny, eds. San Diego, CA: Elsevier.

Morris, R. (2007). Theories of hippocampal function. pp. 581–713. In *The Hippocampal Book*, p. Anderson, R. Morris, D. Amaral, T. Bliss, & J. O'Keefe, eds. New York, NY: Oxford University Press.

Moser, E.I., Moser, M-B, Lipa, P., Newton, M., Houston, F.P., Barnes, C.A., & McNaughton, B.L. (2005). A test of the reverberatory activity hypothesis for hippocampal "place" cells. *Neuroscience* 130:519–526.

Moser, M-B. (2010). The brain's mechanisms for mapping and remembering space. Presented at the 7th Forum of European Neuroscience, Amsterdam, July 3–7, 2010.

Munakata, Y., Stedron, J.M., Chatham, C.H., & Kharitonova, M. (2008). Neural networks models of cognitive development. pp. 367–382. In *Handbook of Developmental Cognitive Neuroscience* (2nd ed.), C.A. Nelson, & M. Luciana, eds. Cambridge, MA: MIT Press.

Murali, K., Miliotis, A., Ditto, W.L., & Sinha, S. (2009). Logic from nonlinear dynamical evolution. *Physics Letters A* 373:1346–1351.

Murray, E.A., Izquierdo, A., & Malkova, L. (2009). Amygdala function in positive reinforcement: Contributions from studies of nonhuman primates. pp. 82–104. In *The Human Amygdala*, P.J. Whalen & E.A. Phelps, eds. New York, NY: Guilford Press.

Murre, J.M.J. (1992). *Learning and Categorization in Modular Neural Networks*. Hillsdale, NJ: Lawrence Erlbaum Associates.

Nadasdy, Z. (2010). Binding by asynchrony: the neuronal phase code. *Frontiers in Neuroscience* 4:51.

Nicolis, G., & Prigogine, I. (1989). *Exploring Complexity: An Introduction*. New York, NY: W.H. Freeman.

Nolfi, S., & Floreano, D. (2000). *Evolutionary Robotics: The Biology, Intelligence, and Technology of Self-Organizing Machines*. Cambridge, MA: MIT Press.

Nolfi, S., Husbands, P, & Floreano, D. (2008). Evolutionary robotics. pp. 1423–1451. In *Handbook of Robotics*, B. Siciliano & O. Khatib, eds. Berlin: Springer-Verlag.

Nowak, M.A. (2006). *Evolutionary Dynamics: Exploring the Equations of Life*. Cambridge, MA: Belknap Press.

Nüsslein-Volhard, C. (2006). *Coming to Life: How Genes Drive Development*. Carlsbad, CA: Kales Press.

Ochsner, K.N. (2007). How thinking controls feelings: A cognitive neuroscience approach. pp. 106–133. In *Social Neuroscience: Integrating Biological Explanations of Social Behavior*, E. Harmon-Jones & p. Winkielman, eds. New York, NY: Guilford Press.

Ochsner, K.N., Bunge, S.A., Gross, J.J., & Gabrieli, J.D.E. (2005). Rethinking feelings: An fMRI study of the cognitive regulation of emotion. pp. 253–270. In *Social Neuroscience: Key Readings in Social Psychology*, J.T. Cacioppo & G.G. Berntson, eds. New York, NY: Psychology Press.

O'Keefe, J., & Dostrovsky, J. (1971). The hippocampus as a spatial map. Preliminary evidence from unit activity in freely-moving rats. *Brain Research* 34:171–175.

O'Keefe, J., & Nadel, L. (1978). *The Hippocampus as a Cognitive Map*. Oxford, UK: Oxford University Press.

O'Keefe, J., & Nadel, L. (1979). Précis of O'Keefe and Nadel's "The hippocampus as a cognitive map." *Behavioral and Brain Sciences* 2:487–533.

Padoa-Schioppa, C., & Assad, J.A. (2006). Neurons in the orbitofrontal cortex encode economic value. *Nature* 441:223–226.

Panksepp, J. (1998). *Affective Neuroscience: The Foundations of Human and Animal Emotions*. New York, NY: Oxford University Press.

Pastalkova, E., Itskov, V., Amarasingham, A., & Buzsáki, G. (2008). Internally generated cell assembly sequences in the rat hippocampus. *Science* 321(5):1322–1327.

Pennartz, C., Van Wingerden, M., & Vinck, M. (2011). Neural dynamics and ensemble coding of reward expectancy and probability in rat orbitofrontal cortex. Presented at New York Academy of the Sciences Critical Contributions of the Orbitofrontal Cortex to Behavior, March 31–April 1, 2011, New York, NY.

Pennisi, E. (2012). ENCODE Project Writes Eulogy for Junk DNA. *Science* 337(6099):1167–1169.

Petrides, M., & Pandya, D.N. (2002). Association pathways of the prefrontal cortex and functional observations. pp. 31–50. In *Principles of Frontal Lobe Function*, D.T. Stuss & R.T. Knight, eds. New York, NY: Oxford University Press.

Pfeifer, R., & Bongard, J. (2007). *How the Body Shapes the Way We Think: A New View of Intelligence*. Cambridge, MA: MIT Press.

Phelps, E.A. (2004). The human amygdala and awareness: Interactions between emotion and cognition. pp. 1005–1015. In *The Cognitive Neurosciences* (3rd ed.), M.S. Gazzaniga, ed. Cambridge, MA: MIT Press.

Phelps, E.A. (2006). Emotion and cognition: Insights from studies of the human amygdala. *Annual Review of Psychology* 57:27–53.

Phelps, E.A. (2009). The control of fear. pp. 204–219. In *The Human Amygdala*, P.J. Whalen & E.A. Phelps, eds. New York, NY: Guilford Press.

Phelps, E.A., & LaBar, K.S. (2006). Functional neuroimaging of emotion and social cognition. pp. 421–453. In *Handbook of Functional Neuroimaging of Cognition*, R. Cabeza & A. Kingstone, eds. Cambridge, MA: MIT Press.

Pierce, J.R. (1980). *An Introduction to Information Theory: Symbols, Signals, and Noise* (2nd revised ed.; first published in 1961). New York, NY: Dover.

Platt, M., Dayan, P., Dehaene, S., McCabe, K., Menzel, R., Phelps, E., Plassmann, H., Ratcliff, R., Shadlen, M., & Singer, W. (2008). Neuronal correlates of decision making. pp. 125–154. In *Better Than Conscious? Decision Making, the Human Mind, and Implications for Institutions*, C. Engel & W. Singer, eds. Cambridge, MA: MIT Press.

Poincaré, J.H. (1902). *La science et l'Hypothese*. Paris: Flammarion.

Poincaré, J.H. (1905). *Science and Hypothesis.* New York, NY: Walter Scott Publishing Co., Ltd.

Poincaré, J.H. (2007). *The Value of Science*, G.B. Halstead, trans., (originally published in 1913). New York, NY: Cosimo.

Polani, D. (2009). Information, the currency of life. *HFSP Journal* 3:307–316.

Poldrack, R.A., & Willingham, D.T. (2006). Functional neuroimaging of skill learning. pp. 114–148. In *Handbook of Functional Neuroimaging of Cognition*, R. Cabeza & A. Kingstone, eds. Cambridge, MA: MIT Press.

Polya, G. (1954). *Mathematics and Plausible Reasoning: Induction and Analogy in Mathematics.* Vol. I. Princeton, NJ: Princeton University Press.

Portas, C., Maquet, P., Rees, G., Blakemore, S., & Frith, C. (2004). The neural correlates of consciousness: Action and intention. pp. 269–301. In *Human Brain Function* (2nd ed.), R.S.J. Frackowiak, K.J. Friston, C.D. Frith, R.J. Dolan, C.J. Price, S. Zeki, J. Ashburner, & W. Penny, eds. San Diego, CA: Elsevier.

Price, J.L. (2006). Connections of the orbital cortex. pp. 307–324. In *The Orbital Frontal Cortex*, D.H. Zald & S.L. Rauch, eds. New York, NY: Oxford University Press.

Prigogine, I. (1995). U.S. Naval Academy, Spring 1995.

Prigogine, I. (1997). *The End of Certainty: Time, Chaos, and the New Laws of Nature.* New York, NY: Free Press.

Prigogine, I. (2003). *Is the Future Given?* London: World Scientific Publishing.

Prigogine, I., & Stengers, I. (1985). *Order Out of Chaos: Man's New Dialogue with Nature.* London: Flamingo Press.

Pronin, E., Puccio, C., & Ross, L. (2002). Understanding misunderstandings: Social psychology perspectives. pp. 636–665. In *Heuristics and Biases: The Psychology of Intuitive Judgment*, T. Gilovich, D. Griffin, & D. Kahneman, eds. New York, NY: Cambridge University Press.

Raichle, M.E. (2006). Functional neuroimaging: A historical and physiological perspective. pp. 3–20. In *Handbook of Functional Neuroimaging of Cognition*, R. Cabeza & A. Kingstone, eds. Cambridge, MA: MIT Press.

Ranck, J.B. Jr. (1985). Head direction cells in the deep cell layer of dorsal postsubiculum in freely moving rats. pp. 217–220. In *Electrical Activity of the Archicortex*, G. Buzsáki & C.H. Vanderwolf, eds. Budapest, Hungary: Akademiai Kiado.

Rasmussen, C.E., & Williams, K.I. (2006). *Gaussian Processes for Machine Learning.* Cambridge, MA: MIT Press.

Reddy, M. (1979). The conduit metaphor—a case of frame conflict in our language about language. pp. 284–324. In *Metaphor & Thought*, A. Ortony, ed. Cambridge, MA: Cambridge University Press.

Risberg, J. (2006). Evolutionary aspects on the frontal lobes. pp. 1–20. In *The Frontal Lobes, Development, Function and Pathology*, J. Risberg & J. Grafman, eds. New York, NY: Cambridge University Press.

Risi, S., Lehman, J., & Stanley, K.O. (2010). Evolving the placement and density of neurons in the HyperNEAT substrate. pp. 563–570. In *Proceedings of the Genetic and Evolutionary Computation Conference (GECCO 2010)*. New York, NY: ACM Press.

Risi, S., & Stanley, K.O. (2010). Indirectly encoding neural plasticity as a pattern of local rules. In *Proceedings of the 11th International Conference on Simulation of Adaptive Behavior (SAB 2010)*. New York, NY: Springer.

Risi, S., Vanderbleek, S.D., Hughes, C.E., & Stanley, K.O. (2009). How novelty search escapes the deceptive trap of learning to learn. In *Proceedings of the Genetic and Evolutionary Computation Conference (GECCO 2009)*. New York, NY: ACM Press.

Rizzolatti, G., & Craighero, L. (2004). The mirror-neuron system. *Annual Review of Neuroscience* 27:169–192.

Rizzolatti, G., Fogassi, L., & Gallese, V. (2009). The mirror neuron system: A motor-based mechanism for action and intention understanding. pp. 625–640. In *The Cognitive Neurosciences* (4th ed.), M.S. Gazzaniga, ed. Cambridge, MA: MIT Press.

Rizzolatti, G., & Sinigaglia, C. (2008). *Mirrors in the Brain: How Our Minds Share Actions and Emotions*. New York, NY: Oxford University Press.

Robbe, D., & Buzsáki, G. (2009). Alteration of theta timescale dynamics of hippocampal place cells by a cannabinoid is associated with memory impairment. *Journal of Neuroscience* 29(40):12597–12605.

Rogers, S.K., Kabrisky, M., Bauer, K., & Oxley, M.E. (2003). Computing machinery and intelligence amplification. p. 25. In *Computational Intelligence: The Experts Speak*, D.B. Fogel & C.J. Robinson, eds. Piscataway, NJ: IEEE Press.

Rolls, E.T. (2008). *Memory, Attention, and Decision-Making: A Unifying Computational Neuroscience Approach*. New York, NY: Oxford University Press.

Rumelhart, D.E., & McClelland, J.L. (1986). *Parallel Distributed Processing: Explorations in the Microstructure of Cognition*. Cambridge, MA: Bradford Books.

Ryan, T.J., & Grant, S. (2009). The origin and evolution of synapses. *Nature Reviews Neuroscience* 10(10):701–712.

Sagan, C. (1980). *Cosmos*. New York, NY: Random House.

Sagan, C. (1994). *Pale Blue Dot: A Vision of the Human Future in Space*. New York, NY: Ballantine Books.

Salloway, S.P., & Blitz, A. (2002). Introduction to functional neural circuitry. pp. 1–29. In *Brain Circuitry and Signaling in Psychiatry: Basic Science and Clinical Implications*, G.B. Kaplan & R.P. Hammer, eds. Washington, DC: American Psychiatric Publishing.

Sapir, E. (1949). *Selected Writings of Edward Sapir*. Berkeley: University of California Press.

Schacter, D.L., & Slotnick, S.D. (2004). The cognitive neuroscience of memory distortion. *Neuron* 44(1):149–160.

Schlosser, G. (2007). Functional and developmental constraints on life-cycle evolution: An attempt on the architecture of constraints. pp. 113–172. In *Integrating Evolution and Development: From Theory to Practice*, R. Sansom & R.N. Brandon, eds. Cambridge, MA: MIT Press.

Schlosser, G., & Wagner, G.P., eds. (2004). *Modularity in Development and Evolution.* pp. 256–258. Chicago, IL: University of Chicago Press.

Schmalhofer, F., & Perfetti, C.A. (2007). Neural and behavioral indicators of integration processes across sentence boundaries. pp. 161–188. In *Higher Level Language Processes in the Brain: Inference and Comprehension Processes*, F. Schmalhofer & C.A. Perfetti, eds. Mahwah, NJ: Lawrence Erlbaum Associates.

Schrödinger, E. (2006). *What Is Life? With Mind and Matter and Autobiographical Sketches* (15th reprint; first published in 1944). Cambridge, UK: Cambridge University Press.

Schubert, T.W. (2005). Your highness: Vertical positions as perceptual symbols of power. *Journal of Personality and Social Psychology* 89(1):1–21.

Schwarz, N., & Vaughn, L.A. (2002). The availability heuristic revisited. pp. 103–119. In *Heuristics and Biases: The Psychology of Intuitive Judgment*, T. Gilovich, D. Griffin, & D. Kahneman, eds. New York, NY: Cambridge University Press.

Seldon, H.L. (2010). Thoughts on evolution of language from the point of view of brain neuroanatomy and development, and object-oriented programming. pp. 271–278. In *Proceedings of the 8th International Conference (EVOLANG8), The Evolution of Language*, A.D.M. Smith, M. Schouwstra, B. de Boer, & K. Smith, eds. Singapore: World Scientific Publishing.

Shadmehr, R., & Krakauer, J.W. (2009). Computational neuroanatomy of voluntary motor control. pp. 587–596. In *The Cognitive Neurosciences* (4th ed.), M.S. Gazzaniga, ed. Cambridge, MA: MIT Press.

Shadmehr, R., & Wise, S.P. (2005). *The Computational Neurobiology of Reaching and Pointing: A Foundation for Motor Learning.* Cambridge, MA: MIT Press.

Shannon, C.E. (1948). A mathematical theory of communication. *The Bell System Technical Journal* 27:379–423, 623–656.

Shergill, S.S., Bays, P.M., Frith, C.D., & Wolpert, D.M. (2003). Two eyes for an eye: The neuroscience of force escalation. *Science* 301:187.

Shilpa, P., Rodak, K.L., Manikonyan, E., Singh, K., & Platek, S.M. (2007). Introduction to evolutionary cognitive neuroscience methods. pp. 47–62. In *Evolutionary Cognitive Neuroscience*, S.M. Platek, J.P. Keenan, & T.K. Shackelford, eds. Cambridge, MA: MIT Press.

Siegel, D.J. (1999). *The Developing Mind.* New York, NY: Guilford Press.

Simon, H.A. (1957). *Models of Man: Social and Rational.* New York, NY: Wiley.

Sirota, A., Montgomery, S., Fujisawa, S., Isomura, Y., Zugaro, M., & Buzsáki, G. (2008). Entrainment of neocortical neurons and gamma oscillations by the hippocampal theta rhythm. *Neuron* 60:683–697.

Skinner, B.F. (1953). *Science and Human Behavior.* New York, NY: Macmillan.

Sloman, S.A. (2002). Two systems of reasoning. pp. 379–396. In *Heuristics and Biases: The Psychology of Intuitive Judgment*, T. Gilovich, D. Griffin, & D. Kahneman, eds. New York, NY: Cambridge University Press.

Slovic, P., Finucane, M., Peters, E., & MacGregor, D.G. (2002). The affect heuristic. pp. 397–420. In *Heuristics and Biases: The Psychology of Intuitive Judgment*, T. Gilovich, D. Griffin, & D. Kahneman, eds. New York, NY: Cambridge University Press.

Solé, R. (2011). *Phase Transitions.* Princeton, NJ: Princeton University Press.

Solé, R., & Goodwin, B. (2000). *Signs of Life: How Complexity Pervades Biology.* New York, NY: Perseus Books.

Spitzer, M. (1999). *The Mind Within the Net: Models of Learning, Thinking, and Acting.* Cambridge, MA: MIT Press.

Spranger, M., Pauw, S., & Loetzsch, M. (2010). Open-ended semantics co-evolving with spatial language. pp. 297–304. In *Proceedings of the 8th International Conference on the Evolution of Language (EVOLANG8)*, A.D.M. Smith, M. Schouwstra, B. de Boer, and K. Smith, eds. Singapore: World Scientific.

Stanley, K.O. (2010). Searching without objectives: To achieve our highest goals, we must be willing to abandon them. Invited keynote at SPLASH (Systems, Programming, Languages, and Applications: Software for Humanity) 2010 (formerly called OOPSLA), Reno, NV.

Stanley, K.O., D'Ambrosio, D.B., & Gauci, J. (2009). A hypercube-based indirect encoding for evolving large-scale neural networks. *Artificial Life* 15(2):185–212.

Stanovich, K.E., & West, R.F. (2000). Individual differences in reasoning: Implications for the rationality debate. *Behavioral and Brain Sciences* 23:645–665.

Steels, L. (1995). A self-organizing vocabulary. *Artificial Life* 2(3):319–332.

Steels, L. (2000a). The puzzle of language evolution. *Kognitionswissenschaft* 8(4):143–150.

Steels, L. (2000b). Language as a complex adaptive system. In *Proceedings of PPSN VI*, M. Schoenauer, ed. Berlin, Germany: Springer-Verlag.

Steels, L. (2008). The symbol grounding problem is solved, so what's next? In *Symbols, Embodiment and Meaning*, M. De Vega, G. Glennberg, & G. Graesser, eds. New Haven, CT: Academic Press.

Steels, L., & Kaplan, F. (2002). Bootstrapping grounded word semantics. pp. 53–73. In *Linguistic Evolution Through Language Acquisition: Formal and Computational Models*, Ted Briscoe, ed. Cambridge, UK: Cambridge University Press.

Steels, L., Loetzsch, M., & Spranger, M.S. (2007). Semiotic dynamics solves the symbol grounding problem. *Nature Precedings*: HDL: 10101/npre.2007.1234.1.

Steels, L., & Spranger, M.S. (2009). How experience of the body shapes language about space. pp. 14–19. (IJCAI-09): *Proceedings of the 21st International Joint Conference on Artificial Intelligence.* San Francisco, CA: Morgan Kaufmann.

Striedter, G.F. (2005). *Principles of Brain Evolution.* Sunderland, MA: Sinauer Associates.

Stuss, D.T. (2007). New approaches to prefrontal lobe testing. pp. 292–305. In *The Human Frontal Lobes* (2nd ed.), B.L. Miller & J.L. Cummings, eds. New York, NY: Guilford Press.

Stuss, D.T., Terence, P.W., & Alexander, M.P. (2001). Consciousness, self-awareness, and the frontal lobes. pp. 101–109. In *The Frontal Lobes and Neuropsychiatric Illness*, P.S. Salloway, P.F. Malloy, & J.D. Duffy, eds. Washington, DC: American Psychiatric Publishing.

Sullivan, R. M, Moriceau, S., Raineki, C., & Roth, T.L. (2009). Ontogeny of infant fear learning and the amygdala. pp. 889–903. In *The Cognitive Neurosciences* (4th ed.), M.S. Gazzaniga, ed. Cambridge, MA: MIT Press.

Surindar, S. (1947). Towards an age of science. Lakeville, CT: *Lakeville Journal* March 27–April 3.

Swanson, L.W. (2003). *Brain Architecture: Understanding the Basic Plan*. New York, NY: Oxford University Press.

Talmy, L. (2000). Force dynamics in language and cognition. pp. 409–470. *Toward a Cognitive Semantics*. Vol. 1. Cambridge, MA: MIT Press.

Tapiero, I., & Fillon, F. (2007). Hemispheric asymmetry in the processing of negative and positive emotional inferences. pp. 355–377. In *Higher Level Language Processes in the Brain: Inference and Comprehension Processes*, F. Schmalhofer & C.A. Perfetti, eds. Mahwah, NJ: Lawrence Erlbaum Associates.

Taube, J.S. (2005). Head direction cell activity: Landmark control and responses in three dimensions. pp. 45–67. In *Head Direction Cells and the Neural Mechanisms of Spatial Orientation*, S.I. Wiener & J.S. Taube, eds. Cambridge, MA: MIT Press.

Tikhanoff, V., Fontanari, J.F., Cangelosi, A., & Perlovsky, L.I. (2006). Language and cognition integration through modeling field theory: Simulations on category formation for symbol grounding. pp. 376–385. In *Artificial Neural Networks—ICANN 2006, 16th International Conference, Athens, Greece, September 10–14, 2006. Proceedings, Part I*, S.D. Kollias, A. Stafylopatis, W. Duch, & E. Oja, eds. Berlin: Springer-Verlag.

Tomasello, M. (2005). *Constructing a Language: A Usage-Based Theory of Language Acquisition*. Cambridge, MA: Harvard University Press.

Tomasello, M. (2008). *Origins of Human Communication*. Cambridge, MA: MIT Press.

Tottenham, N., Hare, T.A., & Casey, B.J. (2009). A developmental perspective on human amygdala function. pp. 107–117. In *The Human Amygdala*, P.J. Whalen & E.A. Phelps, eds. New York, NY: Guilford Press.

Tranel, D. (2002). Emotion, decision making, and the ventromedial prefrontal cortex. pp. 338–353. In *Principles of Frontal Lobe Function*, D.T. Stuss & R.T. Knight, eds. New York, NY: Oxford University Press.

Tribus, M., & McIrvine, E.C. (1971). Energy and Information. *Scientific American* 225:179–188.

Tse, D., Langston, R.F., Kakeyama, M., Bethus, I., Spooner, P., Wood, E., Witter, M., & Morris, R. (2007). Schemas and memory consolidation. *Science* 316:76–82.

Turing, A. (1952). The chemical basis of morphogenesis. *Philosophical Transactions of the Royal Society B* 237:37–52.

Tversky, A., & Kahneman, D. (1992). Advances in prospect theory: Cumulative representation of uncertainty. *Journal of Risk & Uncertainty* 5:297–323.

Tversky, A., & Kahneman, D. (2000). Loss aversion in riskless choice. pp. 143–158. In *Choices, Values, and Frames*, D. Kahneman & A. Tversky, eds. New York, NY: Cambridge University Press.

Tversky, A., & Kahneman, D. (2002). Extensions versus intuitive reasoning. pp. 19–48. In *Heuristics and Biases: The Psychology of Intuitive Judgment*, T. Gilovich, D. Griffin, & D. Kahneman, eds. New York, NY: Cambridge University Press.

Tweney, R.D. (1989). A framework for the cognitive psychology of science. pp. 342–366. In *Psychology of Science: Contributions to Metascience*, B. Gholson, W.R. Shadish Jr., R.A. Neimeyer, & A.C. Hart, eds. New York, NY: Cambridge University Press.

Vanderwolf, C.M. (2010). *The Evolving Brain: The Mind and the Neural Control of Behavior.* New York, NY: Springer.

Verbancsics, P., & Stanley, K.O. (2010). Transfer learning through indirect encoding. pp. 547–554. In *Proceedings of the Genetic and Evolutionary Computation Conference (GECCO 2010)*. New York, NY: ACM Press.

von Dassow, G., & Meir, E. (2004). Exploring modularity with dynamical models of gene networks. pp. 187, 218–219, 256–258. In *Modularity in Development and Evolution*, G. Schlosser & G.P. Wagner, eds. Chicago, IL: University of Chicago Press.

von der Malsburg, C. (2003). Self-organization and the brain. pp. 1002–1005. In *The Handbook of Brain Theory and Neural Networks* (2nd ed.), M.A. Arbib, ed. Cambridge, MA: MIT Press.

Vuilleumier, p. (2009). Perception and attention. pp. 220–249. In *The Human Amygdala*, P.J. Whalen & E.A. Phelps, eds. New York, NY: Guilford Press.

Vuilleumier, P., & Brosch, T. (2009). Interactions of emotion and attention in perception. pp. 925–934. In *The Cognitive Neurosciences* (4th ed.), M.S. Gazzaniga, ed., Cambridge, MA: MIT Press.

Vygotsky, L.S., & Luria, A.R. (1993). *Studies on the History of Behavior: Ape, Primitive, and Child*, V.I. Golod & J.E. Knox, eds. & trans. Mahwah, NJ: Lawrence Erlbaum Associates.

Wagner, A. (2005). *Robustness and Evolvability in Living Systems*. Princeton, NJ: Princeton University Press.

Wagner, A.D., Bunge, S.A., & Badre, D. (2004). Cognitive control, semantic memory, and priming: Contributions from the prefrontal cortex. pp. 709–725. In *The*

Cognitive Neurosciences (3rd. ed.), M.S. Gazzaniga, ed. Cambridge, MA: MIT Press.

Wang, K, et al, (2009). Common genetic variants on 5p14.1 associate with autism spectrum disorders. *Nature* 459:528–533.

Watson, K.K., & Platt, M.L. (2011). Social value representation in primate orbitofrontal cortex. Presented at Critical Contributions of the Orbitofrontal Cortex to Behavior, New York, March 31–April 1.

West-Eberhard, M.J. (2003). *Developmental Plasticity and Evolution*. New York, NY: Oxford University Press.

Weyl, H. (1952). *Symmetry*. Princeton, NJ: Princeton University Press.

Whalen, P.J., Davis, F.C., Oler, J.A., Kim, H., Kim, M.J., & Maital, N. (2009). Human amygdala responses to facial expressions of emotion. pp. 265–288. In *The Human Amygdala*, P.J. Whalen & E.A. Phelps, eds. New York, NY: Guilford Press.

Whorf, B.L. (1956). *Language, Thought and Reality*. Cambridge, MA: MIT Press.

Wiener, N. (1948). *Cybernetics: Or Control and Communication in the Animal and the Machine*. Paris: Herman et Cie.

Wiener, S.I., & Taube, J.S., eds. (2005). *Head Direction Cells and the Neural Mechanisms of Spatial Orientation*. Cambridge, MA: MIT Press.

Wimsatt, W.C., & Griesemer, J.R. (2007). Reproducing entrenchments to scaffold culture: The central role of development in cultural evolution. pp. 227–323. In *Integrating Evolution and Development: From Theory to Practice*, R. Sansom & R.N. Brandon, eds. Cambridge, MA: MIT Press.

Wolpert, D.H., & Macready, W.G. (1995). No free lunch theorems for search. *Technical Report SFI-TR-95-02-010*. Santa Fe, NM: Santa Fe Institute.

Wolpert, D.H., & Macready, W.G. (1997). No free lunch theorems for optimization. *IEEE Transactions on Evolutionary Computation* 1:67–82.

Wright, J.H. (2004). *Cognitive-Behavior Therapy*. Washington, DC: American Psychiatric Publishing.

Xue, S-A., & Stauss, H.J. (2007). Enhancing immune response for cancer therapy. *Cellular & Molecular Immunology* 4(3):173–184.

Yuille, A.L., & Geiger, D. (2003). Winner-take-all networks. pp. 1228–1231. In *The Handbook of Brain Theory and Neural Networks* (2nd ed.), M.A. Arbib, ed. Cambridge, MA: MIT Press.

Zelazo, P.D., Carlson, S.M., & Kesek, A. (2008). The development of executive function in childhood. pp. 553–574. In *Handbook of Developmental Cognitive Neuroscience* (2nd ed.), C.A. Nelson & M. Luciana, eds. Cambridge, MA: MIT Press.

Zellner, A. (1988). Optimal information processing and Bayes' theorem. *American Statistician* 42(4):278–280.

Index

In this index, *f* denotes figure and *t* denotes table.

C

F

G

H

T

W

Z

CPSIA information can be obtained at www.ICGtesting.com
Printed in the USA
LVOW02s2315140114

369484LV00008B/109/P

CPSIA information can be obtained at www.ICGtesting.com
Printed in the USA
LVOW02s2315140114

369484LV00008B/109/P